DR . I . WALKER .

WITHDRAWN

D1810823

GUMS AND STABILISERS
FOR THE FOOD INDUSTRY 4

Proceedings of the 4th International Conference held at Wrexham, Clwyd, Wales, July 1987

GUMS AND STABILISERS FOR THE FOOD INDUSTRY 4

Edited by

GLYN O.PHILLIPS, P.A.WILLIAMS

The North East Wales Institute of Higher Education, Deeside, Clwyd, Wales

and

DAVID J.WEDLOCK

Shell Research, Sittingbourne, Kent

Please return to:

**PHARMA
R & D LIBRARY**

 **IRL PRESS**
OXFORD · WASHINGTON DC

IRL Press Ltd
PO Box 1
Eynsham
Oxford OX8 1JJ
England

© Copyright 1988 IRL Press Ltd

All rights reserved by the publisher. No part of this book may
be reproduced or transmitted in any form by any means,
electronic or mechanical, including photocopying, recording or
any information storage and retrieval system, without
permission in writing from the publisher.

British Library Cataloguing in Publication Data

Gums and stabilisers for the food industry 4.
  1. Food. Additives : Gums
  I. Phillips, Glyn O. (Glyn Owain), *1927-*
  II. Wedlock, David J.   III. Williams, Peter A. (Peter
  Anthony)
  664'.06

  ISBN 1-85221-087-7

Printed by Information Printing Ltd., Oxford, England.

PREFACE

This series of books on Gums and Stabilisers are now proving to be the best source of new information to meet the needs of the users and developers of this group of materials. It is an inter-disciplinary field, embracing academic and industrial researchers, industrial producers and users. Of necessity, the exact needs of these often disparate groups do not always coincide, with the industrial user often being intimidated by the specialist eloquence of the scientist. At the Wrexham Conference, where the papers were presented, there was considerable discussion about how the most effective interface between the various interests could be achieved. The main sections are relevant to all, albeit to different degrees of specialism:

- Analysis, Structure and Properties
- Gelation and Rheological Properties
- Emulsion Stabilisation
- Current Developments

In the last grouping, the masterly review by Dr Katsuyoshi Nishinari (Natural Food Research Institute of Japan) of the current progress in Japan will testify to all their success in eliminating the barriers between academic research and industrial application. To succeed elsewhere, the same degree of integration must be achieved, despite the logistical difficulties which this imposes. This is, and will remain, the objective of the Wrexham Conferences.

In the forthcoming conference in July 1989, greater emphasis will be given to the requirements of the industrial producers and those seeking new and novel uses for gums and stabilisers. Academic researchers will be required to establish the relevance of their work in practice.

The dynamic growth of this field is also illustrated by the steady progress of the new journal FOOD HYDROCOLLOIDS. This fourth volume provides a broader treatment of subjects than is possible in original publications, and is, therefore, complementary to the journal. The market response and excellent reviews of previous volumes encourages us to believe that this book also will serve to fuel the continued expansion of this subject.

Finally, may I thank all participants for their contributions and also the continued steadfast support of the Organising Committee.

PROFESSOR GLYN O.PHILLIPS,
CHAIRMAN

ACKNOWLEDGEMENTS

This fourth Meeting owed its success to the invaluable assistance of the Organising Committee.

Members of the Organising Committee

Dr J.C.Allen	The North East Wales Institute
Dr R.Ashton	Dari-Tech Company
Mr G.A.Barber (*Honorary Treasurer*)	Technical Consultant
Mr P.Cowburn	National Starch and Chemical Corp.
Mr S.J.Dickson (*Vice-Chairman*)	Hercules Ltd
Mr D.Gregory	Grindsted Products Ltd
Dr P.Harris	Unilever Research Ltd
Dr R.Harrop	The North East Wales Institute
Dr I.Hodgson	Kelco International Ltd
Mr R.M.W.Hopkins	Meyhall Chemical (UK) Ltd
Mr H.Hughes (*Secretariat*)	The North East Wales Institute
Dr J.Mlotkiewicz	Spillers Foods Ltd
Dr R.G.Morley	Delphi Consultant Services Inc.
Professor E.R.Morris	Cranfield Institute of Technology
Dr V.J.Morris	AFRC Food Research Institute
Dr A.Onions	Honeywill and Stein
Professor G.O.Phillips (*Chairman*)	The North East Wales Institute
Mr A.Procter	CPC (UK) Ltd
Mr B.Shrimpton	General Foods Ltd
Dr D.J.Wedlock	Shell Research Ltd
Dr P.A.Williams (*Secretariat*)	The North East Wales Institute

The Editors would like to express their appreciation to Mr Haydn Hughes and Miss Linda Sneddon for assisting in manuscript preparation.

CONTENTS

Part 2: Gelation and rheological properties

Part 3: Applications

Part 4: Emulsion stabilisation

Part 5: Current developments

Poster presentations

Part 1

ANALYSIS, STRUCTURE AND PROPERTIES

Some challenges for gums and gum research

James N. BeMiller

Whistler Center for Carbohydrate Research, Purdue University, West Lafayette, IN 47907, USA

ABSTRACT

The challenge for the application of gums in food systems is much the same as it has been, i.e., to improve processing performance and the stability, texture, and appearance of the final product. The challenge for gum researchers is to use modern techniques to study structure/property relations in multicomponent systems and to use this information to develop new products and applications. Specifically, what is needed is a better understanding of the conformations of polysaccharides, both in low-moisture systems and in systems containing an excess of water, more precise descriptions of the intermolecular interactions of polysaccharides with each other and with other food-system components, better rheological characterization of polysaccharide solutions and polysaccharide-containing systems using dynamic measurements, a better understanding of the effect of environment (e.g., moisture, temperature, pH, soluble salts and solutes) on the conformations and molecular dynamics of polysaccharides, a better understanding of the relation of molecular structure and dynamics to rheological and other physical properties and to organoleptic properties, and better insight into the relation of structures to surface phenomona such as emulsion stabilization and boundary layer flow. A collateral challenge is the development of new gums from land plant, marine, and microbial sources; new gums produced by application of contemporary biotechnological techniques to plants and microorganisms; and additional chemical and enzymic modifications of polysaccharides to control functional properties. This activity, in turn, will present challenges to prove safety of the new gums and the physiological benefits of both new and existing gums.

It is impossible for me to present my thoughts on challenges for gums and gum research as related to the food industry without presenting part of the agenda for the new Whistler Center for Carbohydrate Research; for one of our goals is to increase utilization of carbohydrates, especially polysaccharides, by far the most abundant of the carbohydrates.

We in the Whistler Center take a broad view of the gum field. Gums are used in a variety of businesses, including paper manufacture, petroleum production, mining, printing, textiles, etc., and what we learn about one application is likely to pertain to others, because the properties of the gums and structure/property relations do not change, only the uses vary.

The outlook for the future of gum utilization is healthy. An analysis of the growth of the use of gums in foods, both in terms of volume of use and of kinds of available products, reveals a steady overall growth. However, my charge is not to analyze the past and to make extrapolations. Rather, my charge is to present my view of the future of the use of gums in foods in terms of challenges.

Current reasons for the addition of gums to food products should remain. They are improvement in quality or convenience and allowance of the use of new processes and processing equipment. It is easy to give examples of how gums have been used to make these improvements; it is difficult to predict what new improvements might result from new gums or new applications. The usefulness of gums is based upon their physical properties, in particular their capacity to thicken and/or gel aqueous solutions and otherwise to control water. Because all gums modify (control) the flow of aqueous solutions, dispersions, and suspensions, the choice of which gum to use for a particular application often depends upon its secondary characteristics. These secondary characteristics are responsible for their utilization as adhesives, as binders, as bodying agents, as bulking agents, as crystallization inhibitors, as clarifying agents, as cloud agents, in coatings, as emulsifying agents, as emulsion stabilizers, as encapsulating agents, as film formers, as flocculating agents, as foam stabilizers, as gelling materials, as mould release agents, as protective colloids, as suspending agents, as suspension stabilizers, as swelling agents, as syneresis inhibitors, as texturing agents, in water absorption and binding, and as whipping agents. What may become more and more important considerations are improved healthfulness through the addition of soluble fiber; the reduction of energy costs through the use of gums especially effective at drag reduction during mixing, pumping, etc. and through the use of materials that lower the temperatures of phase transitions; and gums that improve the quality of microwavable foods.

The challenge for the gum producers is to continue to provide new gum products with superior functionality in at least one area. Opportunities in the area of new gum products would seem to be (a) gums with improved secondary characteristics (not new viscosities), (b) gums that can be used as bulking and/or bodying agents in artificially sweetened food products, (c) gums with enhanced cholesterol-reducing activity, i.e., gums with improved soluble-fiber characteristics, (d) gums that can replace part of the lipid of food products, and (e) gums that produce gels with improved performance and organoleptic properties.

One obvious area of challenge with regard to improved secondary characteristics is development of gum products that are emulsifiers and emulsion and suspension stabilizers, that are readily water soluble, and which can be used in spray-drying processes, in other words, a gum product(s) that has the unique properties and protective colloid action associated with gum arabic. Gum arabic is an exudate gum, and exudate gums are characterized by variable and uncertain supply (1,2), making development of a substitute with at least similar properties of the same quality something sought after by processed-food companies. Only gum arabic of the four commercial exudate gums is used in significant amounts at this time. This is the result of two factors. The first and foremost is that gum arabic is as yet unequaled in its ability to form stable citrus oil emulsions over a wide pH range and in the presence of electrolytes, without the need of a secondary stabilizing agent, and in the preparation of spray-dried flavors from them. Gum arabic is the gum of choice for making stable flavor powders. It has other uses, but other gums can compete more effectively with it in other applications. The second factor is that efforts are underway in the Sudan to improve the quality and increase the quantity of gum arabic. Plantations of Acacia trees have been planted; new geographic areas have been opened to gum arabic production, modern biological techniques have been applied to improvement of Acacia species and strains, and efforts are underway to make the collection-for-export process more efficient and effective. Further aspects about the approaches being taken to ensure a stable, dependable gum arabic supply are discussed elsewhere in this volume (3).

One product, larch arabinogalactan, which is a "wood extractive," i.e., not a cell wall constituent, has already been shown to have most of the desirable properties of gum arabic (4,5); but efforts at its commercialization were discontinued several years ago. However, such polysaccharides do not contain the amino acids (peptide portion) now strongly suggested to be part of the gum arabic structure (6,7) and, perhaps, related to its superior ability to stabilize citrus oil emulsions. In addition, with one notable exception, viz., xanthan, new gums have not replaced existing gums; rather they have found new applications based upon their own unique properties, thus expanding gum consumption.

Other "new" properties in gums that will be sought by food-processing companies will relate to such things as specific responses of their sols and gels to changes in temperature and compatibility with foods high in fats or oils for the purpose of replacing some of the lipid.

From where will these new industrial polysaccharides come? It is possible, but not likely, that they will come from new seaweeds or agricultural crops. Rather, the sources are likely to be new microorganisms, genetically engineered microorganisms, genetically engineered agricultural plants, and new or current polysaccharides modified by treatment with chemical reagents or enzymes.

An example of a new polysaccharide from a microorganism is gellan from Auromonas elodea, a gum about which is discussed later in this volume (8-11). The high level of activity surrounding this potential food gum, as evidenced by papers in these Conference Proceedings, and elsewhere in the literature, is indicative of the great interest in new gums in part because of their demonstrated value and in part because of the realization that each polysaccharide has unique properties and, therefore, offers unique opportunities. It is also indicative of the willingness of industry to evaluate the potential of new gums for improving existing products or processes, to explore their potential applications, and to accept them when approved for food use. Gellan also provides an example of chemical modification to alter properties. The native polysaccharide, which forms weak gels, could be termed an "intermediate-viscosity" gum. Upon de-O-acetylation, it is transformed into a "high-viscosity" gum which forms stiff, brittle gels. I will have more to say about gellan later.

There is also considerable interest in several other new gums produced by fermentation processes. Rhizobia species produce capsular, extracellular, and lipopolysaccharides (12-14). In some strains, the capsular and extracellular polysaccharides have the same structure; in others, they have different structures. Several are of interest as potential industrial gums and have been described. The anionic extracellular polysaccharide from Rhizobium trifolii strain TA-1 is a poly(octasaccharide) that has a tetrasaccharide repeating unit in the backbone; each tetrasaccharide repeating unit carries a tetrasaccharide side chain that contains two pyruvic acid units as cyclic acetals (15). The capsular polysaccharide is a neutral poly(hexasaccharide), composed of a trisaccharide repeating unit in the main chain that carries mono- and disaccharide branches on the same, doubly-branched, main-chain residue (16,17). It is insoluble in cold water and soluble in hot water. Its solutions gel when cooled to 40-45°, even at low concentration, without metal ions. Its gels melt at about 50° (17,18).

These examples of new and potential commercial gums, both of which happen to be fermentation gums, are pointed out as indications of the considerable activity underway in industrial, private, academic, and government research laboratories to meet the need to provide, and the challenge of providing, new industrial gums with superior properties for specific applications.

Advances in prokaryotic gene modification and gene transfer techniques have given us tools, in addition to those of mutant generation and physiological control and new methods of conjugation and transduction, that present us with the opportunity and challenge of altering microorganisms to improve the quality and/or quantity of produced gums. One effort in the area of mutant generation has been reported. A xanthan-like molecule with truncated side chains is produced by a mutant strain of Xanthomonas campestris created by the introduction of transposons (19).

Likewise, advances in biotechnology, viz., eukaryotic gene modification, recombinant DNA, chromosome transfer, and plant regeneration techniques, in addition to those of the classic techniques of selective breeding and mutant generation, have presented us with opportunities and challenges of engineering higher plants to improve the quality and/or quantity of produced gums, including starches.

Ideas and efforts in both of these challenge areas are yet in the formulative stage. The challenges here are greater than they are with changing the nature or quantity of a protein, because in the case of a polysaccharide, an entire pathway(s) is involved. As a result, the biosynthetic pathway of the polysaccharide whose production or structure is to be modified must be known. That is yet another challenge. To understand what structural features can be modified, and which cannot, or whether the amount of the gum produced can be increased, it is necessary to know what are the carbohydrate and non-carbohydrate precursors, if any lipid-linked intermediates are involved, the site of synthesis, any transporting mechanisms (for intermediates or polymeric materials across membranes), and whether any post-polymerization modifications occur. For the most part, for most polysaccharides, our knowledge in this area, while improving, is incomplete, especially with regards to plant polysaccharides (20-24). The one experience with alteration of the bactoprenol-linked repeating unit (19), which gave a much reduced yield of polymer, might indicate that, in the case of bacterial polysaccharides, only post-polymerization modifications are possible. As a result, the question in the case of any plant or microorganism might be "Can we modify the organism so that it does the chemistry that we might wish to do, such as make acetate or phosphate ester derivatives?"

Molecular biology has also provided us with the ability to produce specific enzymes in quantity. Commercial enzymes are used to produce syrups and other modifications of starch. An organism that produces a thermal-stable a-amylase has been developed by gene transfer. The a-amylase gene from

Bacillus stearothermophilus (ATCC 39,709) has been introduced into a Bacillus subtilis (ATCC 39,701) host. The resulting Bacillus subtilis (ATCC 39,705) secretes the thermal-stable a-amylase (25). The use of enzymes, whether as mixtures in crude preparations or as specific one-activity preparations, whether from native or genetically engineered organisms, presents a greatly underexplored opportunity. There have been attempts in this area with respect to galactomannan modification i.e., the attempted conversion of guaran into locust bean (carob) gum, but none are yet commercial successes. Further aspects in this area of investigation are discussed elsewhere in this volume (26,27).

Acceptable chemical modification presents yet another challenge. As you know, chemical modification is widely used to make starch and cellulose gums. However, handling is a problem that needs to be overcome when dealing with other water-soluble polysaccharides and little work in this area has been done, particularly with regards to food-grade gums. Several derivatives of guar gum, prepared primarily for use in paper and textile manufacture and mining operations, are available for industrial use.

Approval for the use of native and modified gums in foods will probably be a growing challenge. Regulatory agencies are likely to look most favorably upon polysaccharides that are constituents of the normal diet, i.e., hemicelluloses from foodstuffs such as the cereal grains. They will probably be most concerned with genetically altered materials, including the use of enzymes from genetically engineered organisms, unfortunately for political reasons rather than scientific reasons; but new gums, such as tara gum and gellan, will be approved for food use. Another challenge is to characterize the medical aspects of soluble dietary fiber, the class to which gums belong. This subject is also covered in these Proceedings (28).

What is holding us back from post-biosynthetic modification of polysaccharides is a lack of specific enzymes; an inability to target chemical reactions to specific locations in many cases; the cost of handling high-viscosity aqueous solutions before, during, and after reaction; and an as yet incomplete understanding of structure/property relations. In order to design a molecule to be produced through the application of modern biotechnological techniques to plants or microorganisms or through post-biosynthetic chemical or enzymic modification, it is necessary to have a good idea of how the properties of a gum are determined by its chemical and molecular (three-dimensional) structure and how modifications of the chemical structure will effect changes in its molecular structure and properties.

The role of chemical structure as the determinant of physicochemical properties as applied to polysaccharides was recognized more than 30 years ago by Professor Whistler (1) and pioneered by him as a way to think about the function of polysaccharides that make them useful in their practical applications. Subsequently, through use of instrumental methods, Rees and his coworkers (29,30) gave us insight into the shapes of polysaccharides in solutions and gels. This development was followed by the application of x-ray diffraction analysis to oriented, and sometimes polycrystalline, fibers prepared from concentrated solutions of polysaccharides, primarily by Arnott and coworkers (31,32). Recent advances by the Arnott group, now part of the Whistler Center, have enabled them to determine the gel conformations of polysaccharides with a low degree of three-dimensional order. These advances include development and application of the Linked-Atom Least-Squares technique to augment sparse diffraction data (33-36), development of a system for accurate digital analysis of Bragg reflections and/or continuous diffraction data on fiber diffraction patterns (37,38), and use of continuous intensity data in structure refinement (39,40). Laboratories in the Whistler Center under the direction of Drs. Chandrasekaran and Millane are applying the improved techniques to determine the molecular architectures of the capsular polysaccharide of Rhizobium trifolii TA-1 (41,42), gellan (36,42), and kappa-carrageenan (39,42).

The crystalline regions of gellan were found by them to consist of an intertwined duplex of two parallel, left-handed, three-fold helical chains (42). The duplex is stabilized by interchain hydrogen bonds involving each carboxylate group.

The improved techniques have also been applied to analysis of the molecular structure of kappa-carrageenan (42), a molecule that had defied such analysis until now. The model that best fits the continuous x-ray diffraction data obtained from oriented fibers is a coaxial duplex comprised of parallel, right-handed, three-fold helical chains offset from the half-staggered arrangement, a shape significantly different from that of iota-carrageenan, the other gel-forming member of the carrageenan family. The size and stability of the junction zones in gels of iota- and kappa-carrageenan (43) and their molecular structures (42) are the subjects of papers elsewhere in this book.

This information is important in determining the relations between chemical structure, molecular architecture, and functional properties (see 44). The challenge now is to extend this analytical technique to systems containing other components, including other gums. Recently, x-ray fiber diffraction analysis was applied to studies of synergistic, binary polysaccharide gels formed from galactomannans plus red algal polysaccharides and galactomannans plus xanthan (45,46). No evidence for intermolecular binding was found for tara/kappa-carrageenan, locust bean gum/kappa-carrageenan, tara/furcellaran, or locust bean gum/furcellaran gels; but evidence of mixed polysaccharide junction zones was found for tara/xanthan and locust bean gum/xanthan gels. More of this kind of work is needed to meet the challenge of understanding the mechanisms of gelation and structure formation in solutions, most important properties in food and other systems. In these Proceedings we will learn more about the mechanism of gelation in binary mixtures of gums (47).

Another challenge is to determine more accurately solution conformations of polysaccharides. Recent years have seen the advent of, and considerable improvements in, our ability to model solution conformations (see, for example, 48-52). To date, calculations have been restricted to linear polysaccharides (48-50); recently they have been applied to linear polysaccharides with regular short side chains (51). The challenge is to extend these calculations to other molecules and to substantiate the models with experimental data, such as high-resolution nmr data, as has been used to explore the spatial arrangements of the oligosaccharide side chains of glycoproteins in solution (53,54).

These developments of the past 30 years pertain to solutions, hydrated molecular dispersions, and gels of polysaccharides, all systems containing >95% water. However, many foods have low or intermediate levels of moisture. Accordingly, the challenge is to extend these investigations to determination of conformations, the influence of soluble and insoluble components on intra- and intermolecular interactions, and structure/functional property relations in low- and intermediate-moisture foods.

Following the introduction of sensitive instruments, thermal analysis techniques were applied rapidly to foods and food components (55-57). Differential scanning calorimetry (DSC) is especially well-suited to demonstrate and study phase transitions associated with conformational changes of polymeric substances and has been applied extensively to investigations of granular starches (58-63) because of its ability to detect heat flow changes associated with both first-order and second-order transitions of polymeric materials; that is, both the melting of crystallites and the transition from a glassy state to a rubbery state in partially crystalline polymers are revealed by DSC. In an excellent review of starch gelatinization and retrogradation based upon application of this technique and the concepts of polymer science, Slade and Levine (64) were able to state that "native granular starches, normal and waxy, exhibit the non-equilibrium melting, annealing, and recrystallization behavior characteristic of a kinetically-metastable, water-plasticized, partially-crystalline polymer system with a small extent of crystallinity" and that "aqueous starch gels crystallized from an undercooled rubbery melt, as well as native granules, can be described by the "fringed micelle" morphological model for a 3-dimensional, metastable polymer network composed of hydrated microcrystalline junction zones crosslinking plasticized amorphous regions of randomly-coiled, possibly-entangled chain segments". Because all polysaccharides used as thickening or gelling agents have one or more structural features or physical properties in common with one or both of the starch polymers, it is easy to predict that DSC will be applied to them and to products containing them (see 8,65,66). I point out what can be learned from the application of thermal analysis techniques because it does represent a challenge; the challenge is to apply the concepts, principles, and techniques of other disciplines, such as polymer science and fluid dynamics, to investigations of polysaccharides and their solutions.

Common to gums and the starch polymers are such things as helical structures, a preference for polymer-polymer contacts over polymer-water contacts (67), the formation of junction zones by a crystallization process, and connection of the microcrystallites (junction zones) by random-coil, possibly entangled chain segments, plasticized with water. This, of course, is a description of gelation via partial crystallization. In gels formed in this way, microcrystalline hydrates form physical cross-links connecting hydrated amorphous chain segments, resulting in a fringed-micelle structure (68). As pointed out by Slade and Levine (64), the functional aspects of such systems, which exhibit non-equilibrium behavior, can be described in terms of water and glass dynamics. In this morphological model, the amorphous regions and the crystalline regions are interdependent phases (60).

Recently, it has been suggested that the three-dimensional structures of partially crystalline polymers contain two types of amorphous domains, viz., mobile amorphous domains and rigid amorphous domains, with each kind capable of exhibiting its own glass-to-rubber transition (69-72). This "three-microphase" model will probably apply to linear, ordered, gel-forming polysaccharides.

Slade and Levine (64) have further reminded us that crystallization (retrogradation) of starch pastes (60,62,73-84) occurs by the classic three-step mechanism used to describe the crystallization of partially crystalline synthetic polymers (85,86). The three sequential steps in the non-equilibrium process are as follows: (a) nucleation (formation of critical nuclei, i.e., formation of ordered [helical] chain segments, intramolecularly), (b) propagation (conversion of nuclei into microcrystallites by intermolecular aggregation), (c) maturation (growth and perfection of crystalline regions, i.e., annealing of metastable microcrystallites). Nucleation is the rate-limiting step in this process (75,76,87). An additional important factor in the gelation-via-crystallization process for linear polymers seems to be prior formation of a network of entangled chains (88). However, gelation of amylopectin solutions is dominated by formation of microcrystalline junction zones rather than chain entanglements (76). Aggregate formation and gelation of other polysaccharides in systems of various moisture levels needs to be examined in the light of these concepts. These investigations will necessarily include determinations of the effect of solutes and particulates on the crystallization process and its thermal reversible and thermal irreversible components, if any. In food systems, the presence of lipids (61), sugars, and oligosaccharides (89) will undoubtedly be important, just as they are with crystallization of amylose and amylopectin, where they affect nucleation and plasticization.

Another contribution of Slade and Levine (64) is to use information they and others have generated, again drawing upon the concepts and principles of polymer science, to point out to us that the glass transition represents a material-specific, structural transition from a glassy to a viscoelastic, rubbery state. Additionally, they have pointed out that polymer crystallization can occur only in the rubbery state, in which there is molecular mobility, and not in the glassy state. They have also discussed the kinetics of these dynamic, non-equilibrium processes. Again, these concepts must be extended to food gums in general.

While we have always taken for granted the ubiquitous nature of water and the hydration of polysaccharides, we have failed, until recently, to recognize completely the functional aspects of water. More and more, the critical role of water as a plasticizer for gums and other hydrophilic food polymers (60-62,78,79), the importance of plasticization by water in determining glass transition temperatures, the interdependence of crystalline and plasticized amorphous regions with respect to their responses to heat, and the importance of glass-to-rubber transitions to food properties (64) is being recognized. Extension of these ideas and this work will lead to a better understanding of the temperature dependence of relations between composition, structure, thermomechanical properties, and functional behavior that can, and will, be applied directly to improve product quality and stability and to utilize and control changes in properties during processing. Again, there is a special need to better understand the role of water as a plasticizer and its function in diffusion-controlled processes in low-moisture foods. In fact, this information has been used to design a process for accelerating the staling of bread for the preparation of a stuffing product (90), an example of how thermal analysis can be put to practical use.

Another developing technique that can be applied to low-moisture [and frozen (17)] systems is solid-state nuclear magnetic resonance. High-resolution, ^{13}C-nmr spectra, obtained using the cross-polarization, magic-angle spinning technique on solid samples, have been used to examine the crystalline and noncrystalline regions of cellulose (91) and amylose (92) and their transformations, and crystalline cyclomalto-oligosaccharide complexes (93). Undoubtedly, this technique can be applied more widely to semi-solid and solid food systems. As I have already stated several times, the challenge for us is to apply all applicable contemporary and developing techniques to studies of polysaccharides and multicomponent systems.

Other instrumental techniques that are being applied to structural analysis are two-dimensional nmr (17) and FTIR (84). Areas of investigation by various techniques and disciplines should include emulsion stabilization and flocculation, interfacial film properties, phase behavior, distribution of particle sizes and surface area in emulsions and suspensions, surface loading and layer thickness in stabilized emulsions and suspensions, conformations and interactions of molecules at interfaces, determination of emulsion stability, boundary layer phenomona, gelling mechanisms, and interactions of polysaccharides and proteins.

Yet another likely development is the application of kinetic analysis to polysaccharide systems. Fast-reaction techniques have been applied to studies of salt-induced transitions of solution conformations of aggregating, anionic polysaccharides and were effective in separating the primary process of conformational ordering (nucleation) from secondary aggregation and gelation (94). This approach as applied to gelation of gelatin will be described in this Book.   (95). The kinetics of starch retrogradation has been examined using rapid-scanning, Raman spectroscopy (82). While this analysis involves a much longer time frame, it is another example of a physicochemical technique that could be applied to food systems. Certainly, both more kinetic and more thermodynamic (96) studies are needed.

Finally, the recent development of instruments that can make dynamic, oscillatory, rheological measurements over a wide range of sheer rates and temperatures and under different geometries provides a long-needed tool for the characterization of gum solutions. The challenge is to apply these new instruments to refine theories, concepts, and principles of the viscoelastic behaviors of gum solutions and gels under conditions of use and to develop new products and processes based upon application of this knowledge. Studies of viscoelastic behaviors involving polysaccharide mixtures (97), and the oscillatory rheology of xanthan gum solutions (98) are described in these Proceedings.   Also, to be described is a more thorough analysis of force/deformation profiles for the characterization of gel texture (11) to meet the challenge of converting inprecise word descriptions into numbers that can be correlated with organoleptic properties and compared.

The discerning food scientist needs the ability to select the proper gum product and to apply it in the proper way to meet the specific requirements of physical and organoleptic properties of the product under development. It is the understanding of the relations between chemical structure, molecular structure, and functional physicochemical properties that will provide this ability. The challenge for us is to provide the knowledge and develop the principles and concepts upon which that understanding can be built.

REFERENCES

1. Whistler, R.L. (1959) in Industrial Gums, (eds. Whistler, R.L., and BeMiller, J.N.) pp 1–13. Academic Press, New York.
2. Whistler, R.L. (1973) in Industrial Gums, (eds. Whistler, R.L., and BeMiller, J.N.) 2nd edn., pp 5–18. Academic Press, New York.

3.   Awouda, E.H.M., and Magar, W.Y. in This Volume.
4.   Adams, M.F., and Ettling, B.V. (1973) in Industrial Gums, (eds. Whistler, R.L., and BeMiller, J.N.) Academic Press, New York, pp 415-427.
5.   Lawrence, A.A. (1976) Natural Gums for Edible Purposes, Noyes Data Corp., Park Ridge, New Jersey, pp 3-6.
6.   Anderson, D.M.W., in This Volume.
7.   Snowden, M., Phillips, G.O., and Williams, P.A., in This Volume.
8.   Crescenzi, V., and Dentini, M., in This Volume.
9.   Chapman, H.D., Chilvers, G.R., Miles, M.J., and Morris, V.J., in This Volume.
10. Sanderson, G.R., Bell, V.L., Clark, R.C., and Ortega, D., in This Volume.
11. Clark, R.C., Bell, V.L., Ortega, D., and Sanderson, G.R., in This Volume.
12. Bauer, W.D. (1981) Ann. Rev. Plant Physiol., $\underline{32}$, 407-449.
13. Carlson, R.W. (1982) in Nitrogen Fixation, Vol. 2, $\underline{Rhizobium}$, (ed. Broughten, W.J.) pp 199-234. Clarendon Press, Oxford.
14. Dazzo, F.B., and Hubbell, D.H. (1982) in Nitrogen Fixation, Vol. 2, $\underline{Rhizobium}$, (ed. Broughten, W.J.) pp 274-310. Clarendon Press, Oxford.
15. Robertsen, B., Aman, P., Darvill, A., McNeil, M., and Albersheim, P. (1981) Plant. Physiol., $\underline{67}$, 389-400.
16. Zevenhuizen, L.P.T.M. (1984) Appl. Microbiol. Biotechnol., $\underline{20}$, 393-399.
17. Gidley, M.J., Dea, I.C.M., Eggleston, G., and Morris, E.R. (1987) Carbohydr. Res., $\underline{160}$, 381-396.
18. Zevenhuizen, L.P.T.M., and van Neerven, A.R.W. (1983) Carbohydr. Res., $\underline{124}$, 166-171.
19. Betlach, M.R., Capage, M.A., Doherty, D.H., Hassler, R.A., Henderson, N.M., Vanderslice, R.W., Marrelli, J.D., and Ward, M.B. (1987) Abst. Papers, Am. Chem. Soc., $\underline{193}$, CARB 65.
20. Sutherland, I.W. (1982) Adv. Microb. Physiol., $\underline{23}$, 79-150.
21. Sutherland, I.W. (1985) Ann. Rev. Microbiol., $\underline{39}$, 243-270.
22. James, D., Jr., Preiss, J., and Elbein, A.D. (1985) in The Polysaccharides, (ed. Aspinall, G.O.) Vol. 3, pp 107-207. Academic Press, New York.
23. Shibaev, V.N. (1986) Adv. Carbohydr. Chem. Biochem., $\underline{44}$, 277-339.
24. Lezica, R.P., Daleo, G.R., and Dey, P.M. (1986) Adv. Carbohydr. Chem. Biochem., $\underline{44}$, 341-385.
25. Zeman, N.W., and McCrea, J.M. (1985) Cereal Foods World, $\underline{30}$, 777-780.
26. McCleary, B.V., in This Volume.
27. Reid, J.S.G., Edwards, M., and Dea, I.C.M., in This Volume.
28. Anderson, D.M.W., in This Volume.
29. Rees, D.A. (1977) Polysaccharide Shapes, Outline Studies in Biology, Chapman and Hall, London.
30. Rees, D.A., Morris, E.R., Thorn, D., and Madden, J.K. (1982) in The Polysaccharides, (ed. Aspinall, G.O.) Vol. 1, pp 195-290. Academic Press, New York.
31. Arnott, S. (1980) Am. Chem. Soc., Symp. Ser., $\underline{141}$, 1-30.
32. Arnott, S., and Millane, R.P. (1984) in Proc. Ital. Assn. Sci. Technol. Macromol. Summer School, 1984, Biopolymeri, $\underline{XXIV}$, 1-15.
33. Arnott, S., Scott, W.E., Rees, D.A., and McNab, C.G.A. (1974) J. Mol. Biol., $\underline{90}$, 253-268.
34. Arnott, S., Fulmer, A., Scott, W.E., Dea, I.C.M., Moorhouse, R., and Rees, D.A. (1974) J. Mol. Biol., $\underline{90}$, 269-284.
35. Arnott, S., and Mitra, A.K. (1984) in Molecular Biophysics of the Extracellular Matrix (eds. Arnott, S., Rees, D.A., and Morris, E.R.) pp 41-67. Humana Press, Clifton, New Jersey.
36. Chandrasekaran, R., Millane, R.P., and Arnott, S. (1986) Abstr. Papers, 13th Int. Carbohydr. Symp., B140.

37. Millane, R.P., and Arnott, S. (1985) J. Macromol. Sci.-Phys., B24, 193-227.
38. Millane, R.P., and Arnott, S. (1985) J. Appl. Cryst., 18, 419-423.
39. Millane, R.P., Chandrasekaran, R., and Arnott, S. (1986) Abstr. Papers, 13th Int. Carbohydr. Symp., B139.
40. Arnott, S., Chandrasekaran, R., Millane, R.P., and Park, H.S. (1986) J. Mol. Biol., 188, 631-640.
41. Chandrasekaran, R., Millane, R.P., Walker, J.K., Arnott, S., and Dea, I.C.M. (1987) in Industrial Polysaccharides, (eds. Stivala, S.S., Crescenzi, V., and Dea, I.C.M.) Gordon and Breach Science Publishers, New York.
42. Chandrasekaran, R., Millane, R.P., and Arnott, S., in This Volume.
43. Oakenfull, D., and Scott, A.G., in This Volume.
44. Robinson, G., Eggleston, G.E., Morris, E.R., and Dea, I.C.M., in This Volume.
45. Cairns, P., Morris, V.J., Miles, M.J., and Brownsey, G.J. (1986) Food Hydrocolloids, 1, 89-93.
46. Cairns, P., Miles, M.J., Morris, V.J., and Brownsey, G.J. (1987) Carbohydr. Res., 160, 411-423.
47. Brownsey, G.J., Cairns, P., Morris, V.J., and Miles, M.J., in This Volume.
48. Brant, D.A. (1980) in The Biochemistry of Plants, Vol. 3, (ed. Preiss, J.) pp 425-472. Academic Press, New York.
49. Brant, D.A. (1982) Carbohydr. Polym., 2, 232-237.
50. Burton, B.A., and Brant, D.A. (1983) Biopolymers, 22, 1769-1792.
51. Buliga, G.S., Brant, D.A., and Fincher, G.B., (1986) Carbohydr. Res. 157, 139-156.
52. Talashek, T.A., and Brant, D.A. (1987) Carbohydr. Res., 160, 303-316.
53. Montreuil, J. (1984) Pure Appl. Chem., 56, 859-877.
54. Carver, J.P., and Brisson, J.-R. (1984) in Biology of Carbohydrates, (eds. Ginsberg, V., and Robbins, P.) Vol. 2, pp 289-331. John Wiley, New York.
55. Biliaderis, C.G. (1983) Food Chem., 10, 239-265.
56. Lund, D.B. (1983) in Physical Properties of Foods, (ed. Peleg, M., and Bagley, E.B.) pp 125-143. AVI Publishing, Westport, CT.
57. Wright, D.J. (1984) Crit. Rep. Appl. Chem., 5, 1-36.
58. Donovan, J.W. (1979) Biopolymers, 18, 263-275.
59. Biliaderis, C.G., Maurice, T.J., and Vose, J.R. (1980) J. Food Sci., 45, 1669-1680.
60. Biliaderis, C.G., Page, C.M., Maurice, T.J., and Juliano, B.O. (1986) J. Ag. Food Chem., 34, 6-14.
61. Biliaderis, C.G., Page, C.M., and Maurice, T.J. (1986) Food Chem., 22, 279-295.
62. Biliaderis, C.G., Page, C.M., and Maurice, T.J. (1986) Carbohydr. Polymers, 6, 269-288.
63. Biliaderis, G.G., Page, C.M., Slade, L., and Sirett, R.R. (1985) Carbohydr. Polym., 5, 367-389.
64. Slade, L., and Levine, H. (1987) in Recent Developments in Industrial Polysaccharides, (eds. Stivala, S.S., Crescenzi, V., and Dea, I.C.M.) Gordon and Breach Science Publishers, New York, and references therein.
65. Clegg, S.M., and Morris, E.R., in This Volume.
66. Barford, N.M., in This Volume.
67. Starkweather, H.W. (1980) Am. Chem. Soc. Symp. Ser., 127, 433-440.
68. Billmeyer, F.W. (1984) Textbook of Polymer Science, 3rd edn., Wiley-Interscience, New York.
69. Wissler, G.E., and Crist, B. (1980) J. Polym. Sci., Polym. Phys. Ed., 18,

1257-1270.
70. Jin, X., Ellis, T.S., and Karasz, F.E. (1984) J. Polym. Sci., Polym. Phys. Ed., 22, 1701-1717.
71. Cheng, S.Z.D., Cao, M.Y., and Wunderlich, B. (1986) Macromolecules, 19, 1868-1876.
72. Menczel, J., and Wunderlich, B. (1986) Poly. Prepr. 27, 255-256.
73. Ring, S.G. (1985) Int. J. Biol. Macromol., 7, 253-254.
74. Ring, S.G. (1985) Starch/Stärke, 37, 80-83.
75. Miles, M.J., Morris, V.J., Orford, P.D., and Ring, S.G. (1985) Carbohydr. Res., 135, 271-281.
76. Ring, S.G., and Orford, P.D. (1986) in Gums and Stabilizers for the Food Industry, (eds. Phillips, G.O., Wedlock, D.J. and Williams, P.A.) Vol. 3, pp 159-165. Elsevier Applied Science, London.
77. Jankowski, T., and Rha, C.K. (1986) Starch/Stärke, 38, 6-9.
78. Blanshard, J.M.V. (1986) in Chemistry and Physics of Baking, (eds. Blanshard, J.M.V., Frazier, P.J., and Galliard, T.) pp 1-13. Royal Society of Chemistry, London.
79. Ablett, S., Attenburrow, G.E., and Lillford, P.J. (1986) in Chemistry and Physics of Baking, (eds. Blanshard, M.V., Frazier, P. J., and Galliard, T.) pp 30-41. Royal Society of Chemistry, London 1986.
80. Gidley, M.J., and Bulpin, P.V. (1987) Carbohydr. Res. 161, 291-300 and references therein.
81. Ring, S.G., Colonna, P., I'Anson, K.J., Kalichevsky, M.T., Miles, M.J., Morris, V.J., and Orford, P.D. (1987) Carbohydr. Res., 162, 277-293 and references therein.
82. Bulkin, B.J., Kwak, Y., and Dea, I.C.M. (1987) Carbohydr. Res., 160, 95-112.
83. Winter, W.T., and Kwak, Y. Food Hydrocolloids, in press.
84. Belton, P.S., and Wilson, R.H., in This Volume.
85. Boyer, R.F., Baer, E., and Hiltner, A. (1985) Macromolecules, 18, 427-434.
86. Domszy, R.C., Alamo, R., Edwards, C.O., and Mandelkern, L. (1986) Macromolecules, 19, 310-325.
87. Guilbot, A., and Godon, B. (1984) Cah. Nutr. Diet., 19, 171-181.
88. Miles, M.J., Morris, V.J., and Ring, S.G. (1985) Carbohydr. Res., 135, 257-269.
89. Krusi, H., and Neukom, H. (1984) Starch/Stärke, 36, 300-305.
90. Slade, L., Altomare, R., Oltzik, R., and Medcalf, D.G. (14 April 1987) U.S. Patent 4,657,770.
91. Chanzy, H., Henrissat, B., Vincendon, M., Tanner, S.F., and Belton, P.S. (1987) Carbohydr. Res. 160, 1-11 and references therein.
92. Horii, F., Yamamoto, H., Hirai, A., and Kitamaru, R. (1987) Carbohydr. Res., 160, 29-40 and references therein.
93. Veregin, R.P., Fyfe, C.A., Marchessault, R.H., and Taylor, M.G., (1987) Carbohydr. Res., 160, 41-56 and references therein.
94. Goodall, D.M., and Norton, I.T. (1987) Acc. Chem. Res., 20, 59-66.
95. Goodall, D., McBurney, S.J., Tancock, T.A., and Bailey, P.D. in This Volume.
96. Dickinson, E., in This Volume.
97. Clark, R.C., in This Volume.
98. Callet, F. and Milas, M., in this Volume.

Polysaccharide molecular weight determination: which technique?

Stephen E.Harding

Department of Applied Biochemistry & Food Science, University of Nottingham, Sutton Bonington LE12 5RD, UK

ABSTRACT

Average molecular weights and molecular weight distributions are important parameters affecting the performance of polysaccharides as thickening and gelling agents in foods and in other commercial applications. Difficulties encountered with presently widely used techniques for the determination of these parameters are highlighted. The potential of low speed sedimentation equilibrium in the analytical ultracentrifuge as an alternative is demonstrated, particularly in the light of some recent theoretical & experimental developments.

INTRODUCTION

Polysaccharides such as alginates, galactomannans, xanthan & carageenan gums are widely used as thickening and gelling agents in the food industry (1). Their potential in the pharmaceutical industry (drug delivery systems) and in oil mining (well bore technology) has also been identified (2,3). One of the most important factors governing the performance of such ´commercial´ polysaccharides is their molecular weights, and, since for a given preparation they are polydisperse, the distribution of molecular weights. For example, the performance of pectins in drinking yoghurts has been related to such distributions of M_r; some concern has also been expressed over the possible toxicity of low molecular weight species in carageenan. As a result, considerable attention has been paid to polysaccharide molecular weight determination. Unfortunately, the technique which has revolutionised protein biochemistry - SDS polyacrylamide gel electrophoresis (SDS PAGE) - cannot be applied to polysaccharides as a quantitative tool. Although certain charged polysaccharides satisfy similar charge per unit length criteria, gel electrophoresis of polysaccharides has thus far had only a fraction of the impact of its protein counterpart, through, for example, difficulties of calibration.

Other ´relative´ techniques (i.e. requiring calibration standards of known M_r and similar conformation, have however been

applied widely, as have ´absolute´ (viz. not requiring
standards) light scattering procedures. We now highlight some
of the difficulties associated with the latter, which has
contributed to considerable data variability (see, e.g., 4-6)
and demonstrate the potential of some recent advances in the
relatively under-used technique of low speed sedimentation
equilibrium in the analytical ultracentrifuge.

LIGHT SCATTERING

   For over four decades light scattering procedures have
provided powerful tools for elucidating the size and
conformation of biomolecular systems in solution. They are
particularly well suited for the analysis of relatively large,
fairly monodisperse systems, and indeed, we have extensively
applied these techniques to systems of viruses and bacterial
spores (see, e.g. ref. 7).
   Two principle techniques have been applied to polysaccharide
size determinations: (i) ´classical´ or ´differential´ light
scattering, in which the intensity scattered by a dispersion is
measured as a function of angle and (ii) ´quasi-elastic´ light
scattering (QLS) in which the short-time (ns - μs) fluctuations
in intensity at a given angle are measured. For molecular
weight measurement in (i) a double extrapolation to zero angle
and zero concentration (´Zimm´ plot) is normally employed, or
measurements are made at very low angle and a single
extrapolation to zero concentration is sufficient (´Low angle
Light Scattering, LLS); in (ii) ´autocorrelation´ measurements
of the intensity fluctuations can yield the (z-average)
translational diffusion coefficient, D_z, after an extrapolation
to zero concentration. This, when combined with the
sedimentation coefficient, can also yield the molecular weight.
   Despite their wide application, we believe these procedures
have inherent difficulties when applied to heterogeneous systems
such as polysaccharides, largely because of problems of dust and
even trace amounts of large aggregates, particularly for
measurements at low angles. QLS diffusion measurements are
often performed at 90°, to minimise such contamination effects.
Although this does not lead to any appreciable error for rigid
spherical particles, for polysaccharides extrapolation to
zero-angle is normally necessary because of the finite
contribution to the observed autocorrelation profiles from
rotational diffusion - unfortunately at low angles, the effects
of any dust/ aggregates are aggravated. Clarification
procedures run the risk of not removing aggregates or removing
part of the distribution of sizes that is actually being
analysed. A good demonstration of the effects of aggregates has
been given for glycoconjugates by Preston & coworkers (8).
Further, apparent agreement between Zimm plots and the Svedberg
equation (using D_z values measured by QLS) can be misleading in
that the same effects producing high M_r values (and high Rg
values) from the Zimm method would also produce lower D_z values
(and hence higher M_r values) from QLS.
   If QLS is being used for size distribution analysis, it is

not normally possible to distinguish genuine polydispersity (viz. non-interacting species of different molecular size) from self-association phenomena; such analyses also generally fail to take into consideration the effects of thermodynamic non-ideality. Finally, an accurate measurement of concentration is normally necessary (for both the Zimm plot and diffusion coefficient extrapolations) and the measurement of refractive increments: concentrations can rarely be measured to better than 5%.

Because of these difficulties, light scattering techniques would not, where possible, be our method of choice: if light scattering has to be used, we feel that some form of confirmation of results using an independent procedure (such as low speed sedimentation equilibrium) would be desirable.

RELATIVE TECHNIQUES: CALIBRATED GEL CHROMATOGRAPHY (CGC) AND INTRINSIC VISCOSITY

Calibrated Gel Chromatography, like SDS PAGE, is another technique that has revolutionised protein biochemistry and polymer science as a whole. The major difficulty, once an appropriate gel material and detection system has been chosen for a given separation, is in the calibration, using standards of similar conformation and M_r to the sample being analysed. For globular proteins such standards are readily available. For polysaccharides this is not so easy, because of uncertainties concerning conformation: the popular use of dextrans as standards is not generally reliable. The only reliable way would appear to be to measure the molecular weight of isolated narrow fractions using an absolute technique such as light scattering, (given the limitations referred to above). The assumption has to be made that the distribution of molecular weights is due to genuine polydispersity (i.e. the presence of non-interacting components of different molecular weight) as opposed to a self-association.

The other widely used relative procedure is in the use of intrinsic viscosity measurements, calibrated by light scattering or some other ´absolute technique´ using the Mark-Houwink equation. This procedure again requires the use of standards of known M_r and similar conformation/solvation to the polysaccharide whose M_r is required.

LOW SPEED SEDIMENTATION EQUILIBRIUM

This is a technique that has been available in various forms for over four decades since its inception by Svedberg in the 1920´s (see, e.g., 9). Its routine use in biochemical laboratories has declined, primarily with the advent of SDS PAGE and CGC for protein work. For difficult heterogeneous thermodynamically non-ideal systems - the hallmark of polysaccharides - its usefulness is however retained, particularly in the light of recent advances. Nonetheless,

unlike light scattering, there are very few centres of expertise
left with this technique, both in Europe & in the USA. This is
somewhat surprising considering alone its potential in the
polysaccharide field. The length of time required to reach
equilibrium (up to three days) and the shear size of the
machinery required when compared with light scattering are no
doubt contributory factors to its lack of popularity. In
addition, the ´high speed´ or meniscus depletion method, which
facilitates much easier data handling, is not generally suitable
for polysaccharides because of the difficulty of depleting the
meniscus of low molecular weight material, without losing
optical registration at the cell base: a pitfall is to assume
depletion conditions when this is not valid (10). I want to now
describe some of the recent analytical developments involving
the capture and analysis of the data that we have been involved
with which I feel now make the technique particularly
attractive. This includes an improved method for the extraction
of the weight average molecular weight for the whole sample
distribution and how we cope with thermodynamic non-ideality.
Other developments include both an ´indirect´ and a ´direct´
approach to the characterisation of distributions of M_r and also
a combined approach with gel chromatography. I will demonstrate
how a new method for off-line fringe analysis has opened up the
possibilty of using a series of most interesting methods derived
over the last 20 years but previously very difficult to
implement because of the severe requirements on the precision of
the fringe data.

Determination of weight average molecular weights: the ´star
average´.
     One of the most fundamental pieces of information to be
obtained from a sedimentation equilibrium experiment is the
weight average molecular weight over the whole solute
distribution, M_w^o. The distribution of solute at sedimentation
equilibrium is most accurately recorded using Rayleigh
Interference Optics: M_w^o can be obtained from the mean slope of
a plot of Ln J (where J is the absolute fringe number
displacement) versus the square of the radial displacement from
the centre of the rotor, r^2. At equilibrium in a ´low speed´
experiment the concentration at the air/solution mensicus
remains finite, but can be found without too much difficulty by
mathematical manipulation of the fringe data (11).
     For polysaccharides a plot of LnJ versus r^2 is not generally
linear, because of polydispersity (presence of non-interacting
components of different molecular weight) and, sometimes,
self-associative phemomena, both which produce upward curvature;
and also thermodynamic non-ideality (through exclusion volume &
charge effects) which produces downward curvature.
[Occasionally these effects can cancel each other out and linear
´pseudo-ideal´ plots are obtained which can be misleading]. The
principle difficulty in obtaining a value for the mean slope
lies in establishing an accurate value for J at the cell base.
This can be avoided by using a new type of point average
molecular weight, the star average, M* (11). The M* function
has many interesting properties, one of which is that its value
extrapolated to the cell base = M_w^o, and provides a more

accurate way of determining this latter parameter (to within ±
5% if a conventional light source and manual data capture
procedures are used): we have applied this function for
determining for example, the molecular weights of pectins (12)
and galactomannans (13). The M* function also facilitates the
determination of the point number average at the meniscus (11),
where the precision in the data justifies this. Point weight
average molecular weights, M_w can be readily obtained (without
the use of M*) using sliding strip fits to the Ln J versus r^2
plots. Another potential pitfall if sedimentation equilibrium
is used is a failure to allow properly for the effects of
thermodynamic non-ideality, which tends to diminish measured
molecular weights and mask heterogeneity. The effects on M_w^o
are normally minimized by using the lowest possible loading
concentration (∿0.2 mg/ml in a 30mm path length cell), and this
normally suffices. Alternatively, point weight averages can be
extrapolated to zero concentration (J=0) to yield a value
essentially independent of thermodynamic non-ideality or
associative phenomena. The ´ideal´ value obtained in such a way
may however be biased towards the lower end of the molecular
weight distribution, but this bias can be minimised by using
short columns and (in extreme cases) by extrapolating the value
so obtained to zero gravitational field.

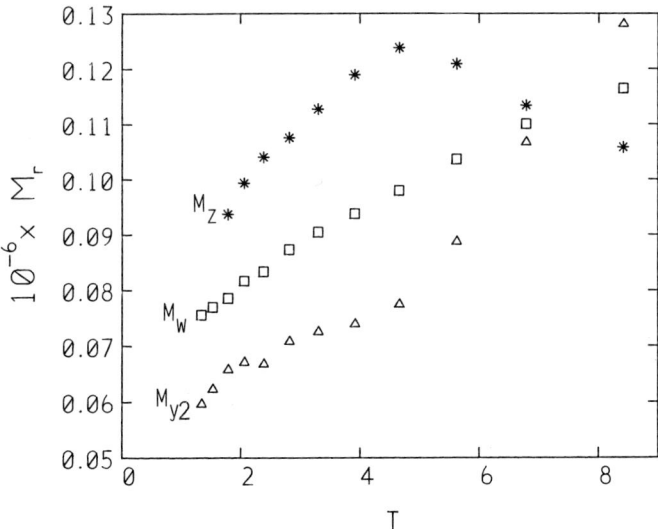

Figure 1. Point average molecular weights plotted versus
fringe concentration for a sedimentation equilibrium
experiment on a tomato fruit polyuronide (17,12). Rayleigh
fringe data had been captured on an LKB Ultroscan & fringe
increments determined using the PASCAL routine ´ANALYSER´.

Another way of dealing with non-ideality is to combine weight average values with number , z- and higher order averages (14,15) if the precision in the data justifies this: it is possible to obtain number and z whole cell average molecular weights (M_n^O and M_z^O respectively) and also point M_n, M_z and 'compound' point average molecular weights:

$$M_{y1} = 1/(2/M_n - 1/M_w), \text{ and}$$

$$M_{y2} = M_w^2/M_z.$$

M_{y1} and M_{y2} point averages are free of first order non-ideality effects: experimental values are generally reliable, however, only if a laser light source can be employed to generate the interference fringes, or if accurate on- or off-line data capture procedures are available. We have recently adapted a commercially available laser gel scanner (LKB ´Ultroscan´) for data capture and written a Fourier series algorithm ´ANALYSER´ (UCSD PASCAL) for data analysis (16,17): the improvement in precision over manual microcomparators is remarkable, and enables the realistic determination of M_z and M_{y2} values. Fig. 1 illustrates an example of this for tomato fruit polyuronides.

Molecular Weight Distributions
    There are four approaches using sedimentation equilibrium I want to highlight

1. Molecular weight ratios
    The whole cell number and z averages (M_n^O & M_z^O) have been hitherto obtainable with considerably less precision than M_w^O, but again, with the availabilty of improved off-line processing referred to above, this should no longer be a limiting factor. The M_z^O/M_w^O or M_w^O/M_n^O ratios can be related to standard deviations of distributions (whatever form they may take) via the ´Herdan relations´ (see, e.g. 18)

2. Modelling the concentration distributions for a non-ideal polydisperse system.
    Predicting the concentration distribution for a non-ideal polydisperse system at sedimentation equilibrium has until recently not been possible because of the nature of the non-linear equations characterising such distributions. An interdependent minimization procedure has now however been developed (19) and successfully applied to a particular glycoconjugate system (a chronic bronchitic glycoprotein) for a discrete distribution of molecular weights. The procedure at the present time takes a considerable toll on computer resources (19), and as a result has not yet been applied to quasi-continuous distributions of M_r which are characteristic of polysaccharides, although we are exploring its possibilities.

3. Modelling the concentration distribution: ´effective´ association constants.
    This makes use of the fact that it is normally impossible to distinguish in one experiment the effects of a polydisperse

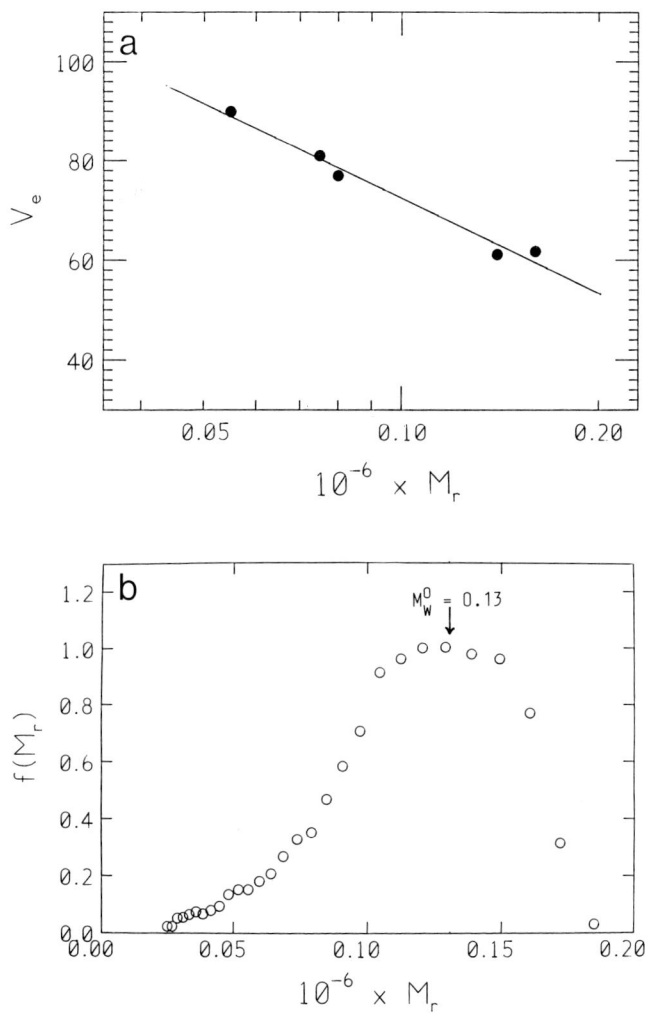

Figure 2. Molecular weight distribution of manucol DM in standard phosphate chloride buffer, pH 6.8, I=0.3

a: Calibration plot for a Sephacryl S-400 column using low speed sedimentation equilibrium on isolated fractions of manucol DM;   b: Molecular weight distribution for the whole elution profile. Alginate assayed using the Phenol/sulphuric acid method. M_w^o for the whole distribution (from sedimentation equilibrium) as shown.

system (viz. non-interacting components of different M_r and/or density) and a self-associating system which contains the same distribution of molecular weight. It is therefore possible to apply the equations of e.g., a (first order) non-ideal isodesmic association to calculate a constant (an ´effective´ association constant) which, when applied to a static system, will define a distribution of molecular weight for the polydisperse case. Again, this method has been succesfully applied to glycoconjugates (pig gastric & chronic bronchitic mucins) (20) and we are currently exploring its potential application to polysaccharides.

4. A combined approach with CGC.

This is a much simpler method to implement, but is nonetheless of considerable use in visualising a distribution. Gel chromatographic methods provide a very easy to use way for giving an estimate of the extent of size variation in a sample, once a suitable column pore size has been found. We have used it uncalibrated to compare the change in size distribution of tomato fruit polyuronides on ripening (12): the change agrees well with the change in M_r measured from sedimentation equilibrium, used here independently.

If calibrated gel chromatography is to be used, the particular column being used requires calibration using standards of similar conformation to the unknown M_r sample. We have used an approach (Ball, Harding & Mitchell, unpublished) using fractions of the same polysaccharide as standards: we isolate 5-6 narrow fractions from the eluant, measure the M_r of each fraction using short column sedimentation equilibrium, work out the calibration constants for the gel from a V_e versus log M_r plot (within the range of the gel), and thus able to convert the elution profile into a molecular weight distribution. An example of such a distribution obtained in this way is given in Fig 2 for a high mannuronate sodium alginate (manucol DM): the distribution agrees well with the M_w^0 for the whole distribution, measured using sedimentation equilibrium alone.

ACKNOWLEDGEMENTS

I would like to thank Dr. John Mitchell for many learned discussions and Miss A. Ball for expert assistance with the experimental work.

REFERENCES

1. Glicksman, M. (ed.) (1982) Food Hydrocolloids, Vol. I, CRC Press, Florida.
2. Cartlidge, S.A., Duncan, R., Lloyd, J.B., Kopeckova-Rejmanova, P. & Kopecek, J. (1987) J. Controlled Rel. 4, 253
3. Gabriel, A. (1979) in Microbial polysaccharides and polysaccharases (Berkeley, R.C.W., Gooday, G.W. and

Ellwood D.C. eds.), Chap. 8, Academic Press, London
4. Launay, B., Cuvelier, G. & Martinez-Reyes, S. (1984) in Gums & stabilisers in the food industry II, (Phillips, G.O. ed.), p79, Pergamon Press, Oxford.
5. Lecacheux, D., Mustiere, Y., Panaras, R. (1986) Carbohyd. Polym. 6, 477
6. Robinson, G., Ross-Murphy, S.B. & Morris, E.R. (1982) Carbohyd. Res. 107, 17
7. Harding, S.E. (1986) Biotechnology & Applied Biochemistry 8, 489
8. Harper, G.S., Comper, W.D. & Preston, B.N. (1985) Biopolymers 24, 2172
9. Ranby, B. (ed.) (1987) Physical Chemistry of Colloids & Macromolecules: the Svedberg symposium, IUPAC, Oxford
10. Hinnie, J. & Serafini-Fracassini, A. (1986) Biopolymers 25, 1095
11. Creeth, J.M. & Harding, S.E. (1982) J. Biochem. Biophys. Meth. 8, 25-34
12. Seymour, G.B. & Harding, S.E. (1987) Biochem. J. 245, 463
13. Gaisford, S.E., Harding, S.E., Mitchell, J.R. & Bradley, T.D. (1986) Carbohyd. Polym. 6, 423
14. Roark, D. & Yphantis, D.A. (1969) Ann. N.Y. Acd. Sci. 164, 245
15. Teller, D.C. (1973) Meth. Enz. 27, 346
16. Harding, S.E. & Rowe, A.J. (1987) Optics & Lasers in Eng. (in press)
17. Harding, S.E. & Rowe, A.J. (1987) Biochem. Soc. Trans. 15
18. Creeth, J.M. & Pain, R.H. (1967) Prog. Biophys. Mol. Biol. 17, 217
19. Harding, S.E. (1985) Biophys. J., 47, 247
20. Creeth, J.M. & Cooper, B. (1984) Biochem. Soc. Trans. 12, 618

Repeating units in the structure of pectin

Joop de Vries

The Copenhagen Pectin Factory Ltd, DK-4623 Lille Skensved, Denmark

ABSTRACT

Pectin can be characterized as constructed of repeating units. A polygalacturonan-chain and a rhamnogalacturonan-chain (with arabinogalactan and xylose side chains) together form such a repeating unit. However, this regularity is not complete: pectins from different parts of the cell wall may have different ratios of "smooth region" to "hairy region". These huge repeating units themselves consist of repeating units. Also within this second type of repeating unit, the regularity is not complete. The distribution of the methoxyl groups is taken as an example. Degradation with purified pectin lyase and subsequent fractionation of the pectin fragments suggests that some regularity exists in the distribution of the methoxyl groups: a repeating unit of 5 esterified and 1 non-esterified galacturonate residues seems likely. Pectin in statu nascendi may have a degree of esterification of 83%. This regularity, however, is masked by the occurrence of spontaneous and enzymatic deesterification. Degradation by purified polygalacturonase can be used to show the effect of pectin esterase action. It can be concluded that a proportion of the pectin consists of repeating units which themselves partially consist of repeat units.

INTRODUCTION

A complete description of the structure of a polysaccharide implies that data are available on: sugar residues present, type of glycosidic linkage, anomeric and absolute configuration, conformation of the residues, presence of substituents and molecular weight. It is not enough to know what sugar residues are present, but also the sequence of the residues must be elucidated. In many cases, the extraction from the polysaccharides' natural environment represents a major problem. Cell wall polysaccharides like pectin cannot be extracted in high yield without degradation. Moreover, the structure of pectin and other polysaccharides may change with ageing, ripening, storage etc. For these reasons, it is perhaps misleading to talk about "the structure of pectin".

This paper, however, will discuss some of the factors mentioned above. It will focus on the distribution of the neutral sugar residues and the methoxyl groups along the galacturonan main chain of pectins. The main tools in the study presented are purified pectolytic enzymes. Pectin lyase prefers highly esterified pectins as substrate, whereas polygalacturonase's optimal substrate is pectate. Unfortunately, the exact require-ments are unknown, but it is likely that pectin lyase needs a conformation of 3 or 4 adjacent esterified galacturonic acid residues (1), whereas polygalacturonase needs a similar confor-mation of non-esterified residues.

Pectins were degraded by these enzymes, and the resulting pectin fragments were fractionated by different chromatographic systems. The chromatograms obtained contain information about the distribution of the neutral sugars and the methoxyl groups. Evidence for the presence of repeating units is presented.

MATERIALS AND METHODS

Apple pectic substances were extracted as described in (2). Citrus pectins were commercial pectins from The Copenhagen Pectin Factory (GENU Pectins). Chemicals used were of analyti-cal grade. Analytical methods used, gel filtration of degraded pectins, HPLC of degraded pectins and purification of enzymes were done as described in (1, 2, 3, 4).

RESULTS AND DISCUSSION

Distribution of neutral sugars
Pectic substances always contain varying amounts of glucose, xylose, arabinose, galactose and rhamnose. Information on the distribution of these neutral sugars was obtained by partial enzymatic degradation of fractionated pectins (3, 4). Exten-sive fractionation of apple pectic substances showed that the intermolecular distribution is not a continuous one, and that the composition of the neutral sugars is constant for highly purified fractions. After enzymatic degradation of these purified fractions, the neutral sugars are present as high molecular weight material together with some galacturonic acid (1). These results, together with the results of methylation analysis (5) and partial acid hydrolysis (e.g. 6) can only be explained by assuming that the neutral sugars are present in so-called "hairy regions" on the molecule. Thus, pectin molecu-les are likely to consist of hairy regions and smooth regions. The hairy regions contain only a few percent of the total amount of galacturonic acid. From a neutral sugar distribution curve, it could be deduced that in carefully extracted apple pectin, three main types of molecules exist with one, two resp. three hairy regions (3). This suggests that the protopectin is con-structed of alternating smooth and hairy regions, i.e. repeating units. The main chain of a hairy region is a rhamnogalacturonan with alternating rhamnose and galacturonic acid residues. It is likely that this rhamnogalacturonan also has repeating units (e.g. rhamnose-galacturonic acid), but the distribution has not been established yet. Powell et al (10) suggested that the rhamnose units are distributed evenly at a distance of about 25

residues. Table 1 presents the results of acid hydrolysis in 0.5 M H_2SO_4. It can be seen that a sharp decrease in molecular weight (to about 30,000) occurs in the first 10 minutes after which the pectin seems to be more resistant.

It is to be expected that the acid-labile pseudo-aldobiuronic acid linkages (rha-galA) are hydrolyzed first and, therefore, Table 1 suggests that the distance between rhamnose units is more than 25 units.

Reaction time	Mw	% bonds broken/minute
0	120000	–
10	30000	0.43
20	25000	0.07
40	24000	0.02
50	22000	0.06
70	10000	0.09
90	16500	0.04

Table 1. Molecular weight of pectin after hydrolysis in 0.5 M sulphuric acid.

The xylose units are present in the hairy regions as short side chains bound to galacturonic acid, probably as xylogalacturonan regions (5). Thus, pectin can be thought as constructed of homogalacturonan, rhamnogalacturonan (with side chains of arabinogalactan) and xylogalacturonan.

All polysaccharides containing galacturonic acid may be constructed of the 5 building blocks listed in Table 2.

    a. homogalacturonan
    b. apiogalacturonan
    c. xylogalacturonan
    d. rhamnogalacturonan
       (with side chains of arabinogalactans)
    e. galactogalacturonans

Table 2. Building blocks of galacturonic acid containing polysaccharides.

In commercial pectins, the hairy regions are present in degraded form, because acid-labile pentose- and rhamnose-linkages are split during extraction. Fruit pectins consist of a and d with minor amounts of c. In apple pectic substances, a separate fraction e is present (5). The galactose residues in e are 3,6-linked, whereas those in d are 4-linked. The pectin from duckweed contains b (7). Sterculia gum has d and e, khaya gum d and a. Zosterine, the pectic polysaccharide from eel grass, is constructed of a, b and d (8). The scheme in Table 2 is quite consistent with the one proposed by Aspinall (9).

Distribution of the methoxyl groups

The degree of esterification (DE) of commercial pectins depends on the extraction conditions. This paper concentrates on those samples that did not lose many methoxyl groups during extraction, the so-called rapid set pectins with a typical DE =

72%. Gel filtration of pectins degraded by pure pectin lyase (PL) or pectate lyase (PAL) shows that the distribution of methoxyl groups is not extreme: all the molecules can be degraded to some extent, which would not be the case, if e.g. 70% of the molecules had DE = 85% and 30% had DE = 35%. Ion exchange chromatography indicated that only a few percent of fruit pectin molecules have a DE of less than 50% (13). In carefully extracted apple pectic substances, more than 80% of the molecules have a DE between 65% and 80% (11). The distribution of the methoxyl groups can be altered by transesterification (i.e. esterification up to 95% and subsequent cold alkaline deesterification). The action of PL on transesterified and native pectins has been studied (4), and it was concluded that the distribution of methoxyl groups is not the same in both cases. The degradation limit was not the same for the two cases: native pectin is a better substrate for PL than transesterified pectins. Fractionation of PL-degraded pectins by HPLC also showed a difference (4): in the case of native pectins, a high amount of pentamers is present. This can be explained by assuming that in native pectins the distribution of the methoxyl groups is a regular one: 5 esterified residues, 1 non-esterified etc. In other words, a repeating unit is present. However, this ideal structure is only partially present in commercial pectins. During fruit storage, processing and during pectin extraction, spontaneous and enzymatic deesterification occurs. Therefore, the DE is always lower than the theoretical 83% (resulting from the "ideal structure").

It is known that action of pectin esterase (PE) from plants and from fungi results in different distributions of the methoxyl groups (12). The action of PE can be studied by degradation with purified polygalacturonase (PG). The PG used in this study can only degrade regions with very low DE and has digalacturonic acid as end product. Figure 1. is a Biogel P_2 gel filtration pattern of a pectin treated with orange PE and (subsequently) PG.

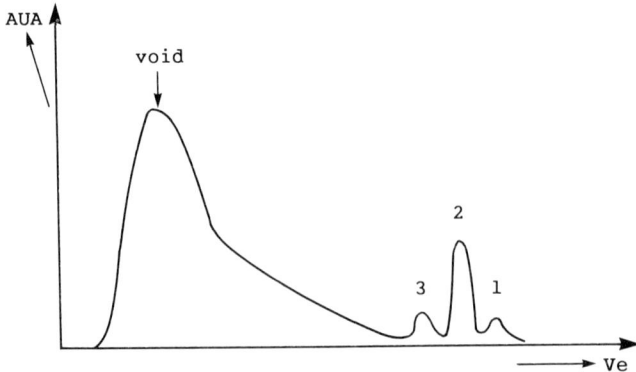

Biogel P_2 chromatograms of pectin treated with PE and PG

PG can release digalacturonic acids from orange PE treated pectins, confirming that orange PE, indeed, forms "blocks" of non-esterified residues.
Table 3. shows the amount of digalacturonic acid released pectin desterified by cold alkali, orange PE and fungal PE.

orange PE, pH = 7.0:	9.5%	digalacturonic acid
orange PE, pH = 4.5:	4.5%	"
fungal PE, pH = 4.0:	4.0%	"
alkaline deesterification:	2.5%	"

Table 3. Digalacturonic acid (as % of total uronic acid material) released by PG from pectins deesterified from 71% to 61% by different methods.

Fungal PE can be characterized as something in between block-wise and random deesterification. It is likely that short blocks - not long enough to give rise to intermolecular Ca-binding - are present after fungal PE action. At pH = 4.5, orange PE looks very much like fungal PE, obviously the degree of multiple attack for orange PE is much lower at this pH, which is far from the enzymes' optimum pH. It cannot be excluded that the enzyme in vivo brings about a different distribution of methoxyl groups than in the laboratory at pH = 7. The ratio enzyme/substrate may also influence the resulting distribution.
In conclusion, the results discussed in this paper indicate that pectins can be characterized as partially consisting of repeating units, partially consisting of repeating units.

REFERENCES

1. De Vries, J.A., Rombouts, F.M., Voragen, A.G.J. and Pilnik, W. (1983) Carbohydr. Polymers, 3, 193-205.
2. De Vries, J.A., Voragen, A.G.J., Rombouts, F.M. and Pilnik, W. (1981) Carbohydr. Polymers, 1, 117-127.
3. De Vries, J.A., Rombouts, F.M., Voragen, A.G.J. and Pilnik, W. (1982) Carbohydr. Polymers, 2, 25-33.
4. De Vries, J.A., Hansen, M., Søderberg, J. and Pedersen, J.K. (1986) Carbohydr. Polymers, 6, 165-176.
5. De Vries, J.A., Den Uyl, C.H., Voragen, A.G.J., Rombouts, F.M. and Pilnik, W. (1983) Carbohydr. Polymers, 4, 3-13.
6. Aspinall, G.O., Craig, J.W. and Whyte, J.L. (1968) Carbohydr. Res., 7, 442-457.
7. Mascaro, L.J. and Kindell, P.K. (1977) Archiv. Biochem. Biophys., 183, 139-150.
8. Ovodov, Yu. S. (1975) Pure Appl. Chem., 42, 351-369.
9. Aspinall, G.O. (1970) Polysaccharides, Pergamon Press, New York, USA.
10. Powell, D.A., Morris, E.R., Gidley, M.J., Rees, D.A. (1982) J. Mol. Biol., 155, 517-531.
11. De Vries, J.A., Rombouts, F.M., Voragen, A.G.J. and Pilnik, W. (1983) Carbohydr. Polymers, 3, 245-258.
12. Kohn, R., Dongovsky, G. and Bock, W. (1985) Nahrung, 29, 75-85.
13. Anger, H. and Dongovsky, G. (1985) Nahrung, 29, 397-404.

The structural significance of amino acids in some plant gums

D.M.W.Anderson

Chemistry Department, The University, Edinburgh EH9 3JJ, UK

ABSTRACT

This paper reviews the data available for the amino acid composition of the commercially important exudate gums and other hydrocolloids permitted in foodstuffs as emulsifiers, stabilisers, and thickeners. Controlled degradations of gum arabic (Acacia senegal (L.) Willd.) and other Acacia exudates have revealed that some amino acids are peripheral. They can therefore contribute to the functionality of the macromolecules, which, at least in part, is lost when the peripheral amino acids in the natural gum molecules are degraded or denatured by heating or other mechanisms. Greater proportions of the amino acids appear to be located within the branched galactan core region of the gum macromolecules than at their periphery. The relative proportions of amino acids located at the periphery and within the core appear to vary considerably for the gums from different Acacia species. The properties and functionality of such gums must be regarded as being those of proteoglycans rather than of nitrogen-free polysaccharides.

INTRODUCTION

Although the earliest studies of Acacia exudates ignored the possibility of the presence of nitrogenous components (1,2), with this omission being continued by some groups of investigators until comparatively recently (e.g. 3,4), all of the Acacia and other exudates studied within this Department since 1954 (e.g. 5,6) were reported to contain nitrogen, with its content ranging from ca. 0.1% (7) to 7-9% (8,9). The fact that these nitrogen contents arise largely, if not exclusively, from amino acids, peptides or proteins has long been established (10). It was also established that (a) the nitrogen content could not be eliminated easily (11), (b) the physical properties of gum solutions were dependent on both the methoxyl and nitrogen contents (12), (c) fractionation experiments involving molecular-sieve (gel) chromatography showed close correspondence between the protein and carbohydrate content (13) and (d) the proteinaceous content could be reduced, but not eliminated, by a combination of degradative solvent extractions and enzymatic digestions (13).

Early structural studies recorded that Acacia exudates were inexplicably sensitive to autohydrolysis, during which flocculent precipitates were formed (14,15). These precipitates were shown to be proteinaceous, but the identity of the amino acids involved

31

was not investigated at that time. Fractionation of gum arabic (Acacia senegal) by competitive salt dehydration gave (14) a high molecular weight fraction of high nitrogen content together with low molecular weight material with a depleted nitrogen content; this experiment was confirmed subsequently by French workers by gel chromatography (16). The fractionation of A. seyal gum (gum talha) into fractions with different uronic acid contents was achieved (11) by gradient elution on a column of diethylamino-ethylcellulose, but the ability to carry out amino acid analyses was not available at that time. Early workers recognised (11) that gum arabic was a mixture of polysaccharides of similar composition, that no single over-all formula had significance, and that only general structural features could be indicated.

At the previous Symposium on gums and stabilisers in 1985, data were presented (17) for the amino acid components of gums arabic, karaya, tragacanth, guar, carob, and xanthan. During the ensuing discussion, a questioner asked where the amino acids in gum arabic were located; this writer undertook to answer at the next Symposium. That undertaking is fulfilled within this paper.

During the interval that has elapsed, however, a serious world-wide shortage of true gum arabic (Acacia senegal (L.) Willd.) unfortunately occurred as a result of the Sahelian droughts of 1984/85. Although commercial supplies of gum arabic are again available as a result of a reasonably productive 1986-87 gum season, the gum arabic market has sustained what many traders regard as an irreversible change as a result of several factors. These include (a) the fact that many industries which were obliged to use modified starches and other hydrocolloids as gum arabic replacements do not wish to return to the use of gum arabic for economic or other reasons, (b) the present high price of gum arabic (4950 U.S. dollars per tonne ex Port Sudan, with a reduction to 4000 U.S. dollars proposed from October 1987) has led to a situation wherein gum arabic is only cost-effective in very special formulations for which cheaper alternatives to the unique functionality of gum arabic have yet to be devised, although R. and D. programmes towards this end are known to be in progress.

Some of these programmes are devoted to investigating the reasons for the unique functionality of gum arabic as an emulsifier; others to devising ways of differentiating between good and inferior qualities of gum arabic, or of gum arabic admixed (inadvertently) or adulterated (deliberately in blending operations) with non-permitted food additives or with permitted additives whose presence nevertheless leads to loss of, or reduction in, functionality. The current criteria of identity and purity are not adequate for the purpose of such discrimina-tion: some consignments of gum arabic conforming to the existing regulatory criteria simply do not function efficiently in some specific manufacturing processes. As long as the present high market price levels of gum arabic continue, and legislation concerning food additives and ingredients continues to tighten, there will remain an interest in acquiring the ability to select gum arabic consignments that will function well, and to reject those that do not. It can be predicted that, henceforth, demand at high prices for gum arabic will be restricted to gum of a very special quality: gum arabic that does not conform to this requirement may well be marketable at lower prices for other purposes in other products. If trade in the natural gum exudates is to survive in future in competition with the derivatised celluloses and starches and the biosynthetic gums, the Companies involved must become much more technologically advanced.

AMINO ACID DATA FOR HYDROCOLLOIDS

Since the previous review (17), further data for the amino acid compositions of many other hydrocolloids have become available, both for permitted food additives and for gums that have no toxicological clearance and are consequently not permitted in foods world-wide. The non-permitted gums for which data are now available include Combretum spp. (18,19), Terminalia spp. (19), Chloroxylon, Sesbania gum (21), and Albizia, Aralia, Entada, Grevillea and Lannea spp. (22). It is very important to have analytical data to assist the identification of such gums which are toxicologically unproven and which are therefore illegal in consignments intended for food use. The permitted gums for which additional amino acid data are now available include tara gum (21), gum ghatti (Anogeissus spp.) (19) and some pectins (23). It should be made clear that, although tara gum was granted the status "ADI not specified" by JECFA in 1986, it has yet to be permitted within the EEC or the U.S.A., and that gum ghatti is GRAS in the U.S.A. but not permitted within the E.E.C. Such short-term inconsistencies whilst the various regulatory committees harmonize are understandable but the situation with regard to gum ghatti is difficult to understand. The American delegation from the F.D.A. to Codex Alimentarius and JECFA is insistent on high standards of safety; yet gum ghatti, which is not on any other permitted list, remains G.R.A.S. within the U.S.A. without any toxicological evidence of any kind being available.

As a consequence of the acquisition of all these amino acid data, it has been possible to calculate (24) nitrogen conversion factors for the principal permitted food hydrocolloids. It is also becoming established (22) that the gums from some genera, particularly those within the Family Leguminosae, are rich in hydroxyproline, whereas others have either aspartic acid or proline (22) as their major amino acids.

EVIDENCE FOR THE LOCATION OF AMINO ACIDS WITHIN GUM ARABIC (Acacia senegal (L.) Willd.)

Detailed degradative studies, based on sequential (25) Smith-degradations (periodate oxidation followed by borohydride reduction then acidic hydrolysis) have been reported for the gums from Acacia senegal (26) and its close relative, Acacia polyacantha (27). These show that, as the macromolecules are consecutively degraded in four stages, the degradation products become progressively more proteinaceous. Thus for A. senegal gum, the nitrogen content increases from 0.34% (whole gum) to 0.85% (final degraded branched galactan which constitutes the core of the macromolecule). The corresponding molar ratios of

sugars/amino acids are 31/1 (whole gum) and 11/1 (galactan core).
For A. polyacantha gum, the corresponding values are %N = 0.37%
(whole gum), %N = 1.0% (galactan core) with sugars/amino acid
molar ratios of 28/1 (whole gum) and 9/1 (galactan core).   For
both these gums, the proteinaceous enrichment of the galactan
core involved serine, threonine and proline.   Hydroxyproline
accounted for ca. 30% of the amino acid content of the whole gum,
third and fourth degradation products, and ca. 45% of the first
and   second   degradation   products   from   Acacia   senegal;
hydroxyproline and aspartic acid accounted for ca. 30% of the
amino acid content of Acacia polyacantha gum and each of its
degradation products.
     The amino acids eliminated in the first degradation of Acacia
senegal gum, and hence located in peripheral locations, were the
minor amounts of methionine, arginine, iso-leucine, lysine and
tyrosine originally present, plus considerable proportions of the
original content of alanine, aspartic acid, glutamic acid,
phenylalanine and valine (27). These therefore may comprise the
amino acids or peptides available at the periphery of the gum
macromolecules, together with the chain-terminal sugars, to
establish the surface activity/functionality of the gum.   For
Acacia polyacantha gum, the depletion of amino acids in the
formation of the first degradation product involved aspartic
acid, hydroxyproline, alanine, valine, leucine, tyrosine and
lysine (26).   There are therefore very similar trends in the
degradation sequences of these two closely related Acacia gums;
the model of an amino acid-rich branched galactan core with a
sparse but functionally important distribution of amino acids at
the periphery is suggested.   Studies are in progress to test the
validity of such a model for other Acacia gums known to have
different functionalities and lower nitrogen contents than A.
senegal and A. polyacantha. The indications to date are that the
small amino acid content in such gums is concentrated at the
molecular core, with very few amino acids located at the
periphery of the gum macromolecules (28).

DEGRADATIVE STUDIES OF GUM ARABIC AND HIGHLY PROTEINACEOUS ACACIA GUMS BY OTHER METHDOS

     During early structural studies of Acacia gums (14,15) it was
frequently observed that autohydrolysis led to the formation of
flocculent brown precipitates, shown to be proteinaceous but not
investigated further at the time. Degradations resulting from
autohydrolysis have therefore been studied, together with the
degradations caused by mild acid hydrolysis and ultraviolet
irradiation.   These three degradative processes have been applied
to the gums from Acacia senegal (gum arabic), Acacia gerrardii,
A. goetzii, A. eriopoda and A. tumida; these gums have nitrogen
contents of 0.34%, 1.86%, 0.89%, 6.70% and 7.66% respectively.
Three fractions were isolated from each degradation: (a) the
proteinaceous precipitate, (b) the degraded, protein-depleted gum
polysaccharides, (c) the low molecular weight fragments (labile
sugars and amino acids) present in the degradative solution and
isolated by freeze-drying dialysates. Detailed descriptions of
the experiments, and the data obtained, have been given (29).

For gum arabic (%N = 0.34), autohydrolysis gave 0.2% of insoluble material (%N = 11.1), 5% of diffusate (%N = 0.2%) and 95% of degraded gum (%N = 0.22%). The degraded gum was enriched in hydroxyproline, serine and threonine; the insoluble material was rich in aspartic acid, valine and leucine; the diffusate was rich in aspartic acid, serine, and glycine. The qualitative amino acid results were very similar for the mild acid hydrolysis, but 40% of diffusate was obtained. Ultraviolet irradiation gave only 1.5% of insoluble material (%N = 2.87) and 1.7% of diffusate (%N = 1.97) but the distributions of the amino acids in all three fractions was similar to those in the two acidic degradations. These results complement well the data obtained from the Smith-degradation study: the same amino acids are susceptible to degradation from peripheral positions by each of the degradative processes used. Space does not permit a detailed discussion of the data obtained (29) for the other Acacia gums studied. To summarize, the data extend those for Acacia senegal; the acidic degradations and ultraviolet irradiations gave differing yields of the insoluble and dialysate fractions, and these had different sugar and amino acid contents. Acid hydrolysis of A. tumida gum gave a large insoluble proteinaceous fraction but no dialysate. After mild acidic hydrolysis, rhamnose appeared predominantly in the insoluble proteinaceous fractions from A. gerrardii, A. tumida and A. eriopoda and, in contrast, in the essentially non-proteinaceous dialysate from A. goetzii gum. As established for A. senegal gum, alanine, aspartic acid, glutamic acid and glycine are more labile than the other amino acids to degradative processes. The main effect is for the high proportions of serine and hydroxyproline in the natural gums to become even higher in the degraded gums. There are therefore points in common but also indications of fine structural differences between the gums studied, and differences in the relative stabilities of their amino acids and sugars (29).

Although much remains to be done to elucidate the structural inter-relationships between sugars and amino acids in the plant gums, it is already clear (7) that gums from different Acacia spp. contain widely differing proportions of amino acids, and this has a parallel in the diversity of sugar compositions shown by different Acacia gums. There are also very large variations in the nitrogen contents, and hence amino acid contents, of Acacia gums (7-9); hence it is reasonable to regard some Acacia gums as proteoglycans (23) and others with high nitrogen contents as glycoproteins (30). It was recognised ten years ago that the Acacia genus provides a wide range of natural hydrocolloids, some being more acidic than others, some more proteinaceous than others, some more viscous than others etc, and that these variations lead to differences in the technological performance and properties of these gums (31). It now seems clear that in such technological matters, the functionality of each gum may be dependent more on a minor proportion of amino acids than on major proportions of sugars.

ACKNOWLEDGEMENTS

It is a pleasure to take this opportunity to express thanks to
all the sponsors who have contributed to this Gum Research
Programme since 1959, and to the successive research students
who, as a result, were enabled to exercise their analytical
skills and assist in our steady increase in understanding of the
compositions, structures, and functionality of the plant gums.

REFERENCES

1. Smith, F. (1939) J. Chem. Soc., 744-753.
2. Stephen, A.M. (1951) J. Chem. Soc., 646-649.
3. Churms, S.C., Merrifield, E.H. and Stephen, A.M. (1983)
    S. African J. Chem., 36, 149-152.
4. Churms, S.C. and Stephen, A. M. (1984) Carbohyd. Res.,
    133, 105-123.
5. Hirst, E.L. and Perlin, A.S. (1954) J. Chem. Soc. 2622-
    2627.
6. Anderson, D.M.W. and King, N.J. (1959) Talanta, 3, 118-126.
7. Anderson, D.M.W., Gill, M.C.L., Jeffrey, A.M. and McDougal,
    F.J. (1985) Phytochemistry, 24, 71-75.
8. Anderson, D.M.W. and Farquhar, J.G.K. (1979) Phytochemistry,
    18, 609-610.
9. Anderson, D.M.W., Farquhar, J.G.K. and McNab, C.G.A. (1983)
    Phytochemistry, 11, 2481-2484.
10. Anderson, D.M.W., Hendrie, A., and Munro, A.C. (1972)
    Phytochemistry, 11, 733-736.
11. Anderson, D.M.W. and Herbich, M.A. (1963) J. Chem. Soc.,
    1-6.
12. Anderson, D.M.W. and Karamalla, K.A. (1966) J. Chem. Soc.
    (C), 762-764.
13. Anderson, D.M.W. and Hendrie, A. (1971) Carbohyd. Res., 20,
    259-268.
14. Anderson, D.M.W. and Stoddart, J.F. (1966) Carbohyd. Res.,
    2, 104-114.
15. Anderson, D.M.W. and Cree, G.M. (1968) Carbohyd. Res., 6,
    385-403.
16. Vandevelde, M.C. and Fenyo, J.C. (1985) Carbohyd. Polymers,
    5, 251-273.
17. Anderson, D.M.W. (1986) in Gums and Stabilisers for the Food
    Industry, 3, (eds. Phillips, G.O., Wedlock, D.J. and
    Williams, P.A.) pp. 79-86, Elsevier Applied Science
    Publishers, London, U.K.
18. Anderson, D.M.W., Bell, P.C., and McDougal, F.J. (1986)
    Food Additives and Contaminants, 3, 305-312.
19. Anderson, D.M.W., Howlett, J.F. and McNab, C.G.A. (1987)
    Phytochemistry, 26, 837-839.
20. Anderson, D.M.W., Bell, P.C., Gill, M.C.L., McDougal, F.J.
    and McNab, C.G.A. (1986) Phytochemistry, 25, 247-249.
21. Anderson, D.M.W., Howlett, J.F. and McNab, C.G.A. (1986)
    Food Hydrocolloids, 1, 95-99.
22. Anderson, D.M.W., Howlett, J.F. and McNab, C.G.A. (1987)
    Phytochemistry, 26, 309-311.

23. Anderson, D.M.W, McDougal, F.J. and McNab, C.G.A. (1987) Food Hydrocolloids, 1, 243-246.
24. Anderson, D.M.W. (1986) Food Additives and Contaminants, 3, 231-234.
25. Goldstein, I.J., Hay, G.W., Lewis, B.A. and Smith, F. (1965) Methods in Carbohydrate Chemistry, 5, 361-364.
26. Anderson, D.M.W. (1986) Food Additives and Contaminants, 3, 123-132.
27. Anderson, D.M.W. and McDougal, F.J. (1987) Food Additives and Contaminants, 4, 125-132.
28. Anderson, D.M.W. and Yin, X.S., studies in progress.
29. Anderson, D.M.W. and McDougal, F.J. (1987) Food Additives and Contaminants (in press).
30. Gammon, D.W., Stephen, A.M. and Churms, S.C. (1986) Carbohyd. Res., 158, 157-171.
31. Anderson, D.M.W. (1977), Process Biochemistry, 12 (10), 24-29.

Studies of biopolymer gels by compressive de-swelling

A.Lips, P.M.Hart and A.H.Clark

Unilever Research, Colworth House, Sharnbrook, Bedford MK44 1LQ, UK

ABSTRACT

An apparatus has been developed which permits the mechanical application of pressure to aqueous biopolymer gels in the range of osmotic stress $7.10^3 - 4.2 \times 10^5 Nm^{-2}$. Both kinetic and equilibrium studies of de-swelling have been performed. The former indicate that the dependence on time of fluid expressed does not conform to a simple diffusion model and is consistent with three exponential processes. Re-swelling studies on compressed gels indicate a complete quantitative recovery of both size and shape for agar but not for gelatin. This suggests that the crosslinks in agar gels are effectively permanent.

Equilibrium measurements on agar gels typically required 10^6 s; de-swelling equilibria could not be achieved with gelatin gels. A 5% w/w agar reference gel was compressed at different pressures to yield a range of equilibrium states. The equilibrium values of pressure and concentration were fitted to the Flory-Erman model for covalently crosslinked networks; the fitting parameters were the shear modulus of the uncompressed gel, the polymer solvent Flory - Huggins interaction parameters X_1, X_2 and a coefficient A determining the contribution to the elastic free energy from entropic volume restriction of the network crosslinks. The inferred value of $X_1 (> .5)$ suggests that water is a poor solvent for agar. However, spontaneous precipitation of agar appears to be prevented by the elastic contribution from entropic volume restriction acting in opposition to the Flory - Huggins mixing and the elastic network retraction terms.

1. INTRODUCTION

The energetics and dynamics of physical crosslinks in non-covalently crosslinked gels may dictate a range of behaviour between the extremes of (i) complete reversibility of crosslinks and therefore equilibrium control of their density and (ii) effective permanence of the crosslinks ("frozen-in" during gel formation) with the gel responding to

measurements of material lost from the sample may be recorded
for up to three months.

3. RESULTS AND DISCUSSION

The photographs in Figure 2 indicate that compressively
de-swollen (30 psi for 1 month) agar gels fully recover their
original shape and dimensions on subsequent re-swelling.   It
appears therefore that the important crosslinks in agar are
formed at elevated temperatures and that under ambient
conditions these links are effectively permanent and no
longer under equilibrium control.   Lack of equilibrium
control may also explain the fact that fully set agar gels,
irrespective of concentration, do not swell or shrink
appreciably when placed in contact with excess solvent at
room temperature.

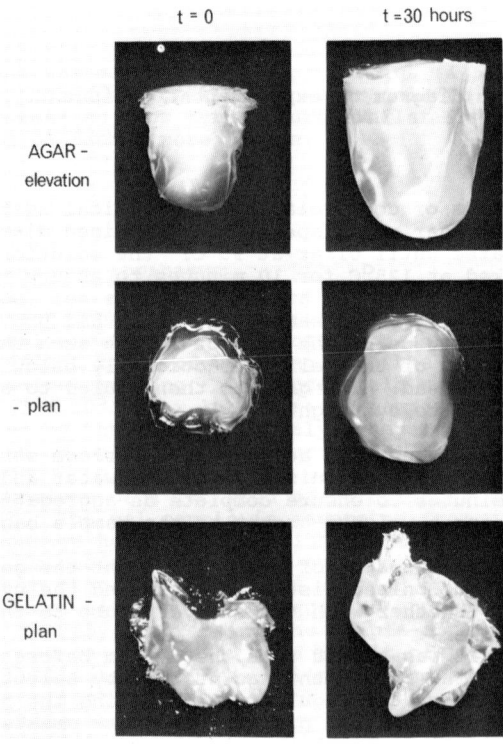

2.   Photographs of de-swollen agar and gelatin gels before
      and after re-swelling.

Figure 2 also shows that gelatin gels do not retain memory of their original shape on prolonged de-swelling. A significant degree of impermanence in the crosslinks is thus indicated. Consistent with this conclusion are the results of rheological studies on gelatin gels indicating significant creep on long time scales (7) and also less direct studies based on the comparison of the rheology of de-swollen gels and freshly prepared gels both at the same concentration (8). Measurements of de-swelling kinetics, displayed in logarithmic time t in Figure 3, reinforce the difference in the nature of the crosslinks between the two types of gel. For agar it is possible to measure an equilibrium response to the applied pressure within a time scale of ca. 1 month. On the basis of Figure 3, gelatin gels would require several months to reach a plateau limit. It is not feasible to maintain gelatin free from bacteriological contamination in the apparatus over such long times.

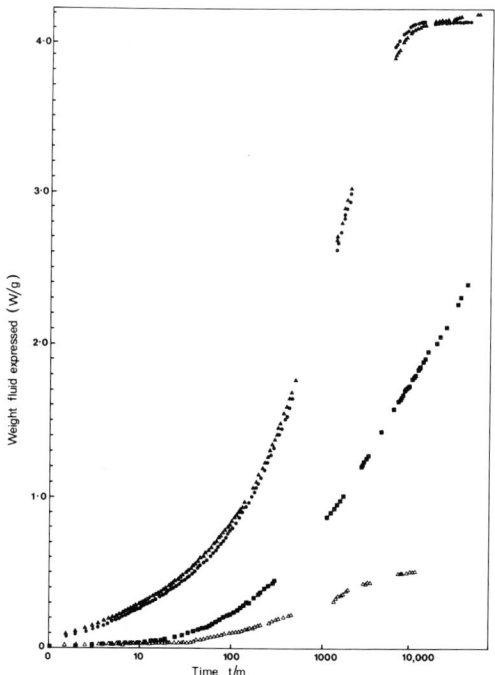

3. Weight fluid expressed (W) versus log time data for (a) agar gel (5% w/w) at 30psi (▲ ●), (b) agar gel (5% w/w) at 2.5 psi (△), and (c) gelatin gel (5% w/w) at 30 psi (■)

TABLE 1

| Process | 30 psi Analysis | | 2.5 psi Analysis | |
	Amplitude (g)	Time (min)	Amplitude (g)	Time (min)
1	0.39 ± 0.02 (9.5%)	10 ± 1	0.016 ± 0.003 (3.1%)	0.5 ± 1.0
2	1.3 ± 0.1 (31.0%)	339 ± 4.0	0.15 ± 0.02 (29.1%)	171 ± 40
3	2.5 ± 0.1 (59.5%)	2472 ± 150	0.35 ± 0.02 (67.8%)	2034 ± 200

Table 1. Results of least squares fitting three exponential
processes to agar gel expression data (weight of fluid
expressed - versus - time).   Analyses are given for 30psi
and 2.5psi data sets. (see also Figs. 3 & 4)

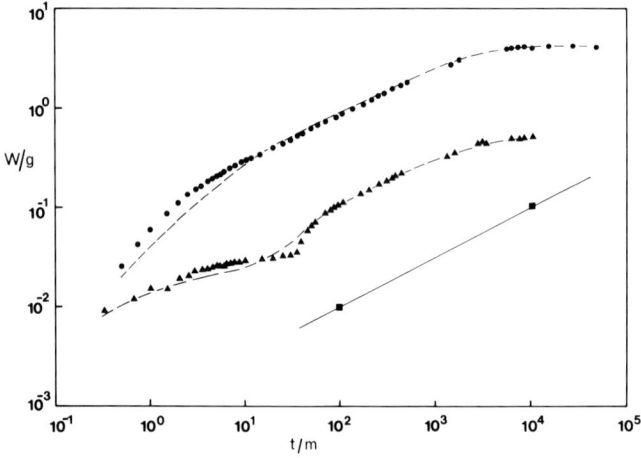

4.   Data sets (a) and (b) from Fig 3   re-plotted in log-log
form (● 30psi and ▲ 2.5psi). Best fit three-process
exponential models are indicated by broken lines (see also
Table 1). $t^{1/2}$ slope is indicated by straight line (■).

If diffusion control applies (9) we would expect a linear
dependence of eluted solvent weight W on $t^{1/2}$.   The double
logarithmic plots of log W vs. log t for agar in Figure 4
are not linear and would only conform to a $t^{1/2}$ relationship
over restricted time intervals.   In fact the full kinetic
profiles are consistent with three exponential processes the
amplitudes and characteristic time constants of which
(determined by least squares analysis) are summarised in
Table 1. Their physical significance is not yet clear and in
an effort to explain them we are currently examining the
relevance of models for plug expression/consolidation (10).
In the following we concentrate on de-swelling equilibria
which we have shown can be measured for agar. Figure 5
depicts the dependence of equilibrium agar concentration C_e
on applied pressure P. The establishment of these
equilibrium states involved continuous kinetic measurement at
each pressure until a true plateau response of W against log
t (Figure 3) was reached.
Complete reversibility could be demonstrated in that
de-swollen gels recovered solvent quantitatively, according
to Figure 5, on partial or full release of pressure and the
equilibria were insensitive to previous de-swelling history.

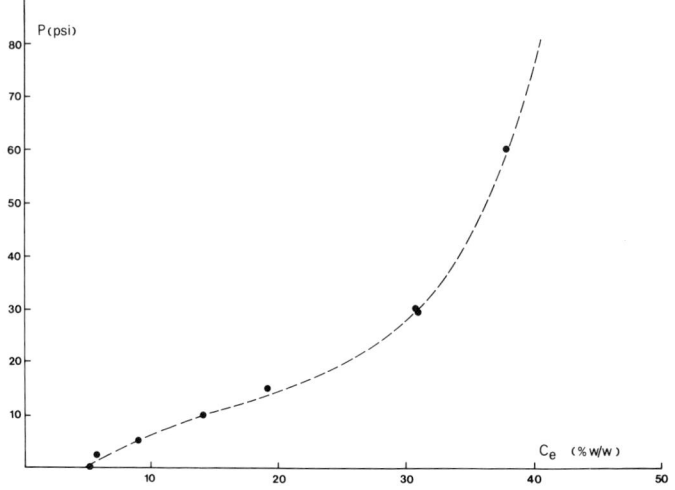

5.   Equilibrium pressure (P) - versus - concentration (C_e)
data for compressed agar gels (●). Best fit theoretical
model (see text and table 2) is indicated by broken line.

Since, in view of the preceding evidence, the crosslinks in
agar can be considered to be effectively permanent it is not
unreasonable to attempt to model the data in Figure 5 on the
basis of the theories of swelling established for chemically
crosslinked gels. We follow the most recent development of
the Flory theory by Erman and Flory (11). The reduced
chemical potential $\Delta\widetilde{\mu}_1$ for the solvent in a nonionic polymer
network may be written

$$\Delta\widetilde{\mu}_1 = (\mu_1 - \mu_1{}^0)/RT = \ln(1-v_2) + v_2 + Xv_2{}^2 + G^0 V_1((v_2/v_2{}^0)^{1/3} - 0.5 (v_2/v_2{}^0))/RT \quad [1]$$

where the first three terms constitute the familiar
Flory-Huggins expression for the mixing of polymer segments
of volume fraction v_2 with solvent. We anticipate that X,
the parameter characterising the interaction between solvent
and polymer, is dependent on composition and can be expressed
empirically as $X = X_1 + X_2 v_2$. G^0 is the elastic modulus of
the gel in its reference state $v_2{}^0$ at which the crosslinks
were introduced and V_1 is the molar volume of the solvent.
The elastic contribution to the solvent chemical potential,
as written, refers to a perfect tetrafunctional affine
network (11,12). There are two opposing effects which are
both entropic in origin; $G^0 V_1(v_2/v_2{}^0)^{1/3}/RT$ represents the
contribution from the segments between crosslinks and $-G^0$
$V_1(v_2/v_2{}^0)/2RT$ the effect of restricting crosslinks to a
given volume. It should be noted, however, that for the
equivalent phantom network (11,12), in which the constraint
on fluctuations of crosslink junctions is relaxed, G^0 is
replaced by $G^0/2$; also since there is no entropy change on
deformation associated with volume restriction of crosslinks
in the phantom network case only the first contribution
mentioned above applies. For real networks, therefore, we
can expect an empirical elastic contribution of the form
$G^0 V_1((v_2/v_2{}^0)^{1/3} - A(v_2/v_2{}^0))/RT$ where the parameter A
reflects such factors as functionality of crosslinks, degree
of perfection of network (viz: cycle rank (11,12)) and local
constraints on network junctions. Since for an isotropic
pressure P applied on the gel $\Delta\widetilde{\mu}_1 = -PV_1/RT$ at equilibrium
with V_1 again the molar volume of the solvent, we may then
write

$$P = -RT(\ln(1-v_2) + v_2 + (X_1 + X_2 v_2)v_2{}^2)/V_1 - G^0((v_2/v_2{}^0)^{1/3} - A(v_2/v_2{}^0)) \quad [2]$$

P, v_2 and $v_2{}^0$ being determined by measurement (Figure 5).
The value 0.65 ml g^{-1} is typical for the partial specific
volume of polysaccharides as can, for example, be inferred
from X-ray scattering literature. This value was taken to
estimate v_2 from the equilibrium polymer concentration C_e;
the reference state v_2^0 corresponds to $C_e = 5$ (%w/w), the
concentration at which the agar gel was formed. A standard

least squares procedure was employed to fit equation [2] to
the data points in Figure 5. Four variable parameters G^o,
X_1, X_2 and A were employed. The optimum values obtained for
these are shown in Table 2 and the corresponding fit to the
experimental data is indicated in Figure 5. It is
gratifying to find that the value determined for G^o is in
satisfactory agreement with shear modulus results typically
measured for 5% agar gels (13) particularly when sources of
uncertainty (see footnote to Table 2) are taken into account
and   it should be noted that, whilst constraining the value
of X_2 in the least squares refinement to zero produces a less
satisfactory overall fit, similar values for A and G^o are
determined.
In the light of this last observation and the fact that G^o is
in reasonable agreement with experiment our fit is primarily
determined by the two variable parameters, X_1 and A, and it
is interesting that the inferred value for X_1 = .6 i.e. >0.5
suggests that water is a poor solvent for agar.

TABLE 2

Results of fitting equn (1) to equilibrium pressure (P) -
concentration data (C_e) for compressed 5.0% w/w agar gels
(Four parameter and three parameter least-squares fits).
See also Fig. 5.

G^o (dyne cm^{-2})	X_1	X_2	A
full refinement $0.8 \pm 0.5 \ 10^6$	0.60 ± 0.06	0.2 ± 0.1	1.2 ± 0.2
$X_2 = 0$ $1.4 \pm 0.5 \ 10^6$	0.68 ± 0.02	0.0(fixed)	1.2 ± 0.2

NOTE: The experimental G^o result for 5.0% agar gel from ref
(4) is 2.4 10^6 dyne cm^{-2}. The somewhat lower value
determined here is probably a consequence of assuming the
classical rubber theory front factor of unity when deriving
equation 1 in the text rather than the 'generalised' front
factor 'a' suggested in Ref. 2.

The inferred magnitude of A, suggesting affine behaviour,
needs special discussion. According to theory it should vary
from 0 to 1 in inverse proportion to junction functionality
(12). In view of the likely participation of agar chains in
several junction zones per chain and the fact that such zones
may consist of bundles of chains (4) a low functionality
description seems unlikely. Therefore some other explanation
must be sought for the unexpectedly large value of A inferred
here. Perhaps relevant is the fact that junction zones in

biopolymer gels cannot be idealised as point crosslinks and the entropy change of restricting such zones may need to take account of significant contributions from excluded volume. In most of the literature concerned with the swelling of gels that are covalently crosslinked in the melt the restriction term has not been emphasised primarily because it is relatively unimportant. However, in the present case, this restriction term is much more important because the reference state is that of a dilute aqueous gel (5% as initially prepared) and compression is the mode of deformation.

Normally in a compression experiment in a good solvent (X<.5) the resistance to network collapse is provided by a net negative contribution to the solvent chemical potential from the Flory-Huggins mixing expression ($\ln (1 - v_2) + v_2 + Xv_2^2$) which opposes the positive contribution from the contractive elastic network term $G^O(v_2/v_2^0)^{1/3}$ and any externally applied pressure. In the present case, because X >0.5, both the mixing expression and the contractive network contribution should facilitate network collapse; it is then only the anomalous restriction term $-G^O A(v_2/v_2^0)$ which can provide a balance. This would imply that it is the volume restriction of crosslinks which prevents the total collapse of the network and expression of solvent. Some limited syneresis is actually a characteristic of agar gels.

4.   REFERENCES

1.   J.R. Hermans, J. Polym Sci., Part A, 3, 1859 (1965).

2.   A.H. Clark & S.B. Ross-Murphy, Brit. Polymer J.,17, 164 (1985).

3.   D. Oakenfull, J. Food Sci. 49, 1103 (1984).

4.   A.H. Clark & S.B. Ross-Murphy, Adv. Polymer Sci, 83, 57, (1987).

5.   I.C.M. Dea, A.A. McKinnon, D.A. Rees., J. Mol. Biol, 68, 153, (1972).

6.   M. Zrinyi and F. Horkay., J. Polymer Sci, Polymer Phys Ed., 20, 815, (1972).

7.   H.J. Poole, Trans Faraday Soc., 21, 114 (1925).

8.   H. McEvoy, S.B. Ross-Murphy and A.H. Clark, Polymer, 26, 1493 (1985).

9.   A.M. Hecht, E. Geissler, J. Chem Phys., 73, 4077, (1980).

10. Filtration,Principles and Practices, Part 1., Edited by Clyde Orr, Chapter 5, p.361, Marcel Decker Inc., New York and Basel (1977).

11. B. Erman and P.J. Flory, Macromolecules, 19, 2342, (1986).

12. P.J. Flory, Proc. R. Soc. London A., 351, 351, (1976).

13. A.H. Clark, R.K. Richardson, S.B. Ross-Murphy and J.M. Stubbs, Macromolecules, 16, 1367, (1983).

Enzymic solutions to polysaccharide related industrial problems

Barry V.McCleary,[+*] Joan Harrington[*] and Helen Allen[*]

+Biocon Biochemicals Ltd, Kilnagleary, Carrigaline, Co. Cork, Ireland
*Biological and Chemical Research Institute, NSW Department of Agriculture, Rydalmere, NSW 2116, Australia

ABSTRACT

Enzymes find widespread use in the brewing, baking, fruit-juice and starch-syrup industries. Some of these applications are discussed in the present communication with particular emphasis on the use of enzymes in wheat-processing, bread making and production of low-carbohydrate beer. The development of specific assays for particular polysaccharides and polysaccharide degrading enzymes is also described.

KEYWORDS

Pentosan, arabinoxylan, barley β-glucan, galactomannan, starch, pectin, transglucosidase.

INTRODUCTION

A range of microbial and plant-derived enzymes are used in the processing of cereals and fruits. In a number of these processes, polysaccharides are partially or totally depolymerised, or otherwise modified. In the processing of cereals, the major polysaccharide substrate is starch and the nature of this, and of the starch degrading enzymes employed, are usually well understood. However, in some applications different commercial enzyme preparations with the same levels of what is considered to be the required activity, can behave quite differently. This may be due either to differences in specificity (i.e. substrate sub-site binding requirements) or to the presence of minor, but functionally significant side-activities. The identification of a particular side-activity may first require the separation and

characterisation of its respective substrate. Alternatively, one
may employ highly purified enzymes to help identify the specific
polysaccharide or its functional significance in a particular
process. Thus, to determine the relative importance of
wheat-flour pentosan (arabinoxylan), (1→3)(1→4)-β-D-glucan and
glucomannan in dough development and on bread properties, we have
used highly purified xylanase, (1→3)(1→4)-β-D-glucanase
(lichenase) and (1→4)-β-D-mannanase, respectively, to
depolymerise these polymers in situ and thus to destroy their
functional properties. Alternatively, the purified enzymes can
be used to develop specific quantitative tests for particular
polysaccharides.

In the current article, I will describe some general aspects
of the use of enzymes in the processing of polysaccharide
containing materials with particular reference to work underway
in our laboratory. I will also describe novel test kits we have
developed for the measurement of plant polysaccharides and
polysaccharide degrading enzymes.

ARABINOXYLANS

Introduction
Arabinoxylans (pentosans) are constituents of the endosperm
cell-walls of most cereals. They are readily fractionated into
soluble and insoluble components(1), but the chemical/physical
basis of this difference in solubility is not clear. For
wheat-flour arabinoxylans, there is considerable evidence
suggesting that insolubility may be due to chemical cross-linking
through ferulic acid residues. Water soluble pentosans form
highly viscous aqueous solutions and it has been estimated(2)and
subsequently demonstrated(3)that more than 20% of the water in
wheat-flour dough is associated with pentosans. Water-soluble
pentosan in dough absorbs eleven times its weight of water and
water-insoluble pentosan absorbs ten times its weight(4).

Through their water-binding and viscosity building properties,
arabinoxylans are of functional significance in the separation of
starch and gluten during wheat-flour processing(5). They also
affect dough development and bread properties and can act as
antinutritional factors when they are present in excess
quantities in the diet of chickens (e.g. when rye-flour forms a
major part of the diet).

Wheat-flour fractionation
The water requirement in the separation of starch and gluten
from wheat-flour slurries has recently been reduced to two cubic
metres per tonne of flour. However, concentration of waste water
is a problem due to the presence of high concentrations of
starch, and especially arabinoxylan. Increasing the total solids
concentration to only 15% can give solution viscosities of
greater than 1000 cp, and the viscosity increases exponentially
with concentration(5). Treatment of the waste water with
α-amylase and a xylanase preparation followed by heating to 85°C
in a jet cooker gave depolymerisation of starch and pentosan and
denaturation of protein which was removed on a whirlpool
separator. The liquor could then be readily concentrated to 60%

total solids in a steam cooker.

An alternative approach to this problem might be the use of pentosanase (xylanase) enzyme in the original wash water. Depolymerisation of pentosan should allow easier separation of starch and gluten, resulting in less loss of small starch granules (B-fraction <10μm) in the wash water. For such treatment to be practically viable it would be essential that the xylanase preparation be devoid, or essentially devoid of α-amylase and particularly, proteases. Even minor proteolysis of gluten could lead to loss of its "vital" viscoelastic properties.

Dough development and baking

It is generally accepted that wheat-flour arabinoxylan plays an important functional role in dough development and that it significantly affects such bread properties as crumb and crust structure and loaf volume(1,6). However, conflicting results have been reported by different authors as to the functional significance of various minor flour components, including arabinoxylan. These differences can be attributed mainly to the types of experiments employed in such studies. These experiments were usually based on the aqueous fractionation of flour components followed by reconstitution of various fractions, or, they involved the addition of particular fractions to whole flour, followed by baking trials. A major problem with these approaches is that minor components such as fermentable sugars or lipids can be lost, or alternatively, the functional properties can be altered or lost during extraction and fractionation.

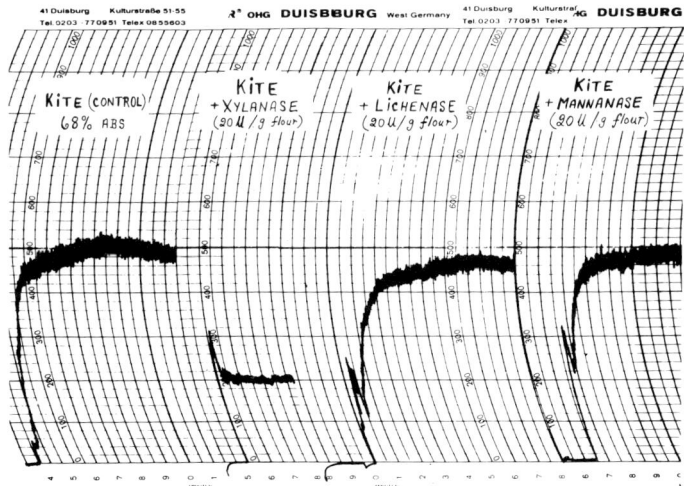

Fig. 1. Effect of β-xylanase, lichenase or β-mannanase on the strength of wheat-flour doughs. Control doughs contained 10g flour and 6.8mL of water. Treated doughs contained 6.8mL of water containing 200U of enzyme.

In our studies(3,7)we adopted a different approach. Instead
of fractionating the flour components, we used highly purified
endo-depolymerases to specifically depolymerise and destroy the
functional properties of particular polysaccharides in the dough
mixture. The enzymes employed were lichenase [endo-(1→3)(1→4)-
β-D-glucanase] from Bacillus subtilis, xylanase from Aspergillus
niger and β-mannanase from Aspergillus niger. Each of the
enzymes was devoid of the other activities and contamination with
protease and starch degrading activities was less than one part
in 500,000 (by activity).

The effect of excess quantities of each of these enzymes
(200U/10g of wheat flour) on Farinograph dough strength is shown
in Fig. 1. Addition of xylanase resulted in an almost immediate
loss of dough strength. The Farinograph reading fell from 500 to
200 Brabender Units (BU) and remained at this value over an
extended period. Lichenase caused a detectable, but less
significant decrease in dough strength and β-mannanase had a very
minor effect. The proportional decrease in dough strength is
approximately correlated to the percentage weight of each
component in the flour.

The properties of the dough after treatment with xylanase or
lichenase can be restored completely by the addition of given
amounts of guar flour. The amount of guar flour required is
approximately equal to the amount of arabinoxylan or
mixed-linkage β-glucan present in the flour. Thus, for xylanase
treated flour with a known arabinoxylan content of 2-3%,
0.25-0.30g of guar flour per 10 grams of wheat flour (i.e.
2.4-2.9%) is required to restore the original properties to the
dough and bread derived from the flour. We have termed the
amount of guar flour required to restore the dough strength of
xylanase treated flour to 500BU, as the Guar Adjustment Factor
(GAF). This factor may prove useful in defining differences
observed in the behaviour of flours from different wheat
varieties. A significant and consistent difference in GAF has
been observed in doughs produced from flours of "hard" and "soft"
wheat varieties.

MIXED-LINKAGE BETA-GLUCANS

Introduction
      Barley endosperm, cell-wall β-glucan [(1→3)(1→4)-β-D-glucan]
can seriously affect the malting quality of barley grain and the
brewhouse performance of the derived malt and of unmalted barley
(used as adjunct)(8). Barley β-glucan represents only 2.5-5.5% of
whole barley-flour weight, but through its limited solubility,
its viscosity building properties and its tendency to precipitate
from solution as gelatinous masses, it can seriously reduce the
rate of wort and beer filtration, and can lead to hazes,
precipitates and gels in the final product. This polymer is
depolymerised by malt β-glucanase, which is synthesized during
the malting process. Alternatively, it can be depolymerised by
the addition of microbial enzymes to the mash, the wort or the
final beer. Enzymes employed include fungal cellulase-type
preparations and bacterial (1→3)(1→4)-β-D-glucanase (lichenase).

Measurement of barley β-glucan and malt β-glucanase

The major single reason why β-glucan problems have persisted as long as they have in the malting and brewing industries is that the methods available for the measurement of this component and of the enzymes responsible for its breakdown in vivo (malt β-glucanase) were tedious and unreliable. Consequently, a major aim of our research in this area was to develop new, reliable techniques for the measurement of these components.

The method developed for the measurement of barley β-glucan employs highly purified bacterial β-glucanase (lichenase) and fungal β-glucosidase(9). The β-glucan is degraded to oligosaccharide fragments with the former enzyme, and the oligosaccharides are hydrolysed to free D-glucose by β-glucosidase. Glucose is specifically measured with glucose oxidase/peroxidase reagent. The method is both specific and accurate and has been adopted as the Australian standard method by the Royal Australian Chemical Institute.

Malt β-glucanase is specifically measured using a novel substrate, namely a chemically modified and dyed barley β-glucan(10). The principle of this assay procedure is outlined in Fig. 2. The method has been evaluated by eighteen laboratories throughout Australia and in Europe, and for five malt samples, coefficient of variation values of less than 6% were obtained.

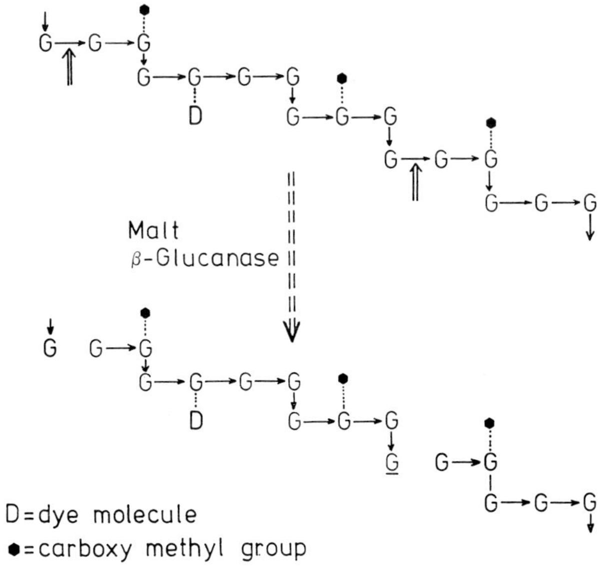

Fig. 2. Theoretical basis of the malt β-glucanase assay procedure.

The β-glucan and malt β-glucanase methods described here allow the maltster and the brewer to monitor these components at all stages throughout the conversion of barley to beer, thus assisting them in optimising their processes. In situations where high levels of β-glucan persist in the wort or beer, microbial β-glucanases can be added (if legislation allows). These enzymes can also be added to chicken diets which contain high levels of barley flour. When present at high levels, the β-glucans can reduce feed intake and lead to "sticky-droppings". β-Glucanase alleviates these problems by destroying the viscosity-building and gelling properties of the glucan.

GALACTOMANNANS

Introduction
     The use of enzymes for the modification of galactomannans is well documented(11). The functional properties are increased by controlled treatment with α-galactosidase to remove a proportion of the D-galactosyl residues. Galactomannans with solution and interaction properties similar to those of locust-bean gum can be prepared from guar or lucerne-seed galactomannans by reducing the D-galactose content to about 20-23%. Viscosity-related processing problems or feed-related problems can be resolved by treatment of galactomannan containing materials with β-mannanase. The enzyme finds application in the coffee industry and in the treatment of the high protein meal obtained on the milling of guar seed. Incomplete separation of the splits (galactomannan containing endosperm material) from the proteinaceous cotyledon/seed coat material, leads to problems with the digestion of the latter material by chickens. In coffee manufacture, β-mannanase can be used to reduce the viscosity of coffee extracts, allowing these extracts to be evaporated to a higher solids content by traditional evaporation before going to the spray- or freeze-dryer(12).

PECTIC SUBSTANCES

Introduction
     The pectic substances include all polygalacturonic acid-containing materials(13). In the pectinic acids, all the carboxyl groups are in the free acid form. Pectins are partially esterified polygalacturonic acids. The ester groups are generally methyl, but rapeseed and sugar beet pectins are also esterified with acetic acid groups. Other constituents commonly present in pectic polysaccharides are D-galactosyl, L-arabinosyl and, in some cases, D-apiosyl units. Lesser amounts of L-rhamnosyl and D-xylosyl and traces of L-fucosyl, units are also present(11).

Fruit juice manufacture
     Commercial enzyme preparations containing pectin lyase, pectate lyase, endo-polygalacturonanase and pectin esterase find widespread application in the processing of fruit and vegetables.

The action of these on pectin and pectate is summarised in
Fig. 3. The type of enzyme employed is dictated by the nature of
the pectic substance present, the final product desired and the
stage of processing at which the enzyme is employed. The
beneficial effects of these enzymes in fruit processing is well
understood and well documented. A less well understood problem

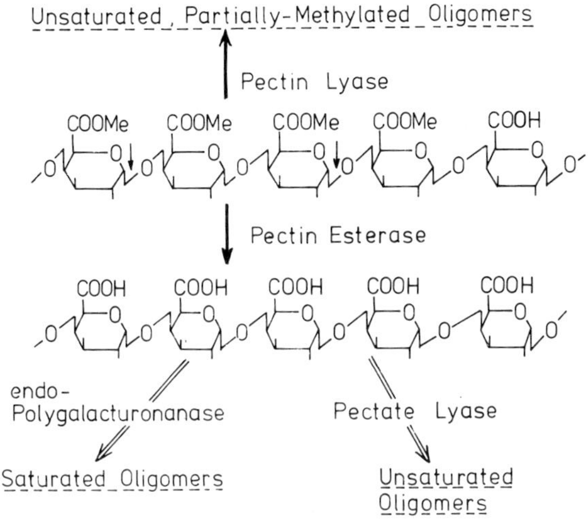

Fig. 3.   Enzymic hydrolysis of pectin and pectate.

is that associated with the formation of precipitates in fruit
juice concentrates. These precipitates are now known to be
arabinan in nature, and it is considered that inclusion of the
enzyme endo-arabinanase in the pectinase preparation may resolve
the problem. Arabinans are composed of a backbone of (1→5)-
α-linked L-arabinosyl residues to which (1→3)-α-linked
L-arabinosyl groups are attached (Fig. 4). The arabinan from
sugar beet is highly substituted by (1→3)-α-L-arabinosyl branch
units and consequently is hydrolysed to an extent of only 3% by
endo-α-L-arabinofurananase. On treatment of the beet arabinan
with α-L-arabinofuranosidase, a polymer of (1→5)-α-linked
L-arabinose could be precipitated from solution. This was
hydrolysed by endo-α-L-arabinofurananase to an extent of 23%,
with release of a series of L-arabino-oligosaccharides initially,
and, on extended incubation, of L-arabinose and (1→5)-
α-L-arabinobiose(14). Partially debranched arabinan was
hydrolysed by endo-α-L-arabinofurananase at 16 times the rate for
native arabinan(15).
    The fact that arabinans from different sources such as apple
pectin, pear pectin or beet pectin show different degrees of
susceptibility to hydrolysis by endo-α-L-arabinofurananase could
be due to differences in the degrees of substitution of the
(1→5)-α-L-arabinan mainchain by branch units.

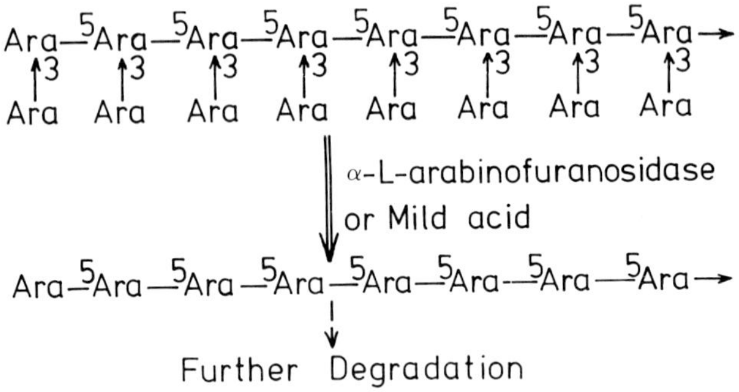
Ara—5Ara—5Ara—5Ara—5Ara—5Ara—5Ara—5Ara→
↑3 ↑3 ↑3 ↑3 ↑3 ↑3 ↑3 ↑3
Ara Ara Ara Ara Ara Ara Ara Ara
α-L-arabinofuranosidase
or Mild acid
Ara—5Ara—5Ara—5Ara—5Ara—5Ara—5Ara—5Ara→
Further Degradation

Fig. 4. Debranching of arabinan by acid or α-L-arabinofuranosidase.

The problems caused by precipitation of α-L-arabinan can most probably be avoided by including sufficient α-L-arabinofuranosidase and α-L-arabinofurananase in the enzyme mixture; or alternatively, by ensuring that the preparation contains minimal quantities of α-L-arabinofuranosidase. The arabinan precipitates may be due to association of non-modified, highly branched arabinan released from the pectic substance, or more likely, to the association of partially or highly, debranched fragments produced on heating the juice concentrate under conditions of low pH, or by the action of α-L-arabinofuranosidase on the released "native" arabinan.

Coffee processing

In the depulping of coffee beans, traditional processes have relied on enzymes produced during natural fermentation. However, this process is subject to many uncontrollable outside influences and thus can give variable results. In contrast, depulping and mucilage removal with industrial pectinases is more controllable giving clean, white and shiny beans with a higher weight recovery and less risk of tainting or of quality loss through microbial spoilage(12).

STARCH

Dough development and bread properties

Starch is the major carbohydrate reserve in cereals and is vital in the baking and brewing processes(13). Wheat flour for baking purposes must contain low, but measurable levels of α-amylase. α-Amylase is required for the degradation of starch to fermentable sugars during the dough-development stage. At these temperatures (~30°C), only damaged starch granules are attacked. If excess levels of cereal α-amylases are present in flours (i.e. flours from weather damaged wheat), then excessive starch dextrinisation occurs during the early stages of baking (i.e. after intact starch granules have gelatinised) leading to "sticky-crumb" problems. If flours are deficient in α-amylase, then it is advisable to use a fungal α-amylase (<u>Aspergillus niger</u>) supplement, as this enzyme attacks damaged starch granules but is rapidly inactivated at temperatures below that required for starch gelatinisation(13).

<u>Wort and beer production</u>
    In the production of wort in beer manufacture, the combined action of malt α- and β-amylase give essentially complete degradation of starch to maltose and glucose. However, a low concentration of higher degree of polymerisation (d.p.) dextrins (possibly branched dextrins) are also present. These higher d.p. dextrins are not fermented by yeast. To produce low-carbohydrate (Light) beers, the wort is treated with amyloglucosidase which gives essentially complete conversion of the dextrins and maltose to glucose (Fig. 5). The incomplete hydrolysis of dextrins to glucose is usually associated with contamination of the amyloglucosidase with another enzyme termed transglucosidase.

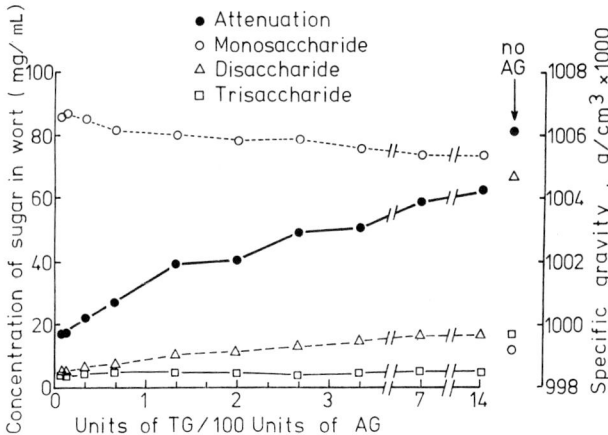

Fig. 5. Treatment of wort with amyloglucosidase to increase fermentability and produce low carbohydrate beer: the effect of transglucosidase contamination of amyloglucosidase. Untreated wort is shown as "no AG". Specific gravities (attenuations) were determined on the final beer.

Transglucosidase is an α-glucosidase which has very high
transferase activity(16). When incubated with maltose, glucose
from the reducing end of the molecule is released into solution,
whereas the non-reducing D-glucosyl group is preferentially
transferred to the 6-hydroxyl of glucose (yielding isomaltose) or
to the 6-hydroxyl on the non-reducing terminal D-glucosyl group
of maltose (yielding panose). The isomaltose and panose can
subsequently be hydrolysed by transglucosidase, but are only
slowly hydrolysed by amyloglucosidase. Consequently, they
accumulate in the wort, and because they cannot be fermented by
yeast, they carry through into the beer, resulting in beer with
increased specific gravities (attenuation values).

Starch syrup manufacture
    A number of processes based on acid and/or enzyme treatment
are available (Scheme 1)(13). In acid processes, a starch slurry
of about 35-40% is acidified to about pH 2 with hydrochloric
acid. This is then passed through a converter at a steam
pressure of 30 pounds per square inch. The starch is gelatinised
and depolymerised to a predetermined level and the reaction is
terminated by adjusting the pH to 4-5 with alkali. The syrup is
then clarified by filtration and/or centrifugation, concentrated
to 60% dry matter and deodourised and decolourised by treatment
with powdered or granular carbon and passed through ion-exchange
resin to remove minerals and proteins. In syrup manufacture from
wheat starch, contaminating arabinoxylan (pentosan) is converted
to sparingly soluble xylan by preferential removal of the acid-
labile α-L-arabinofuranosyl branch residues, and this can cause
filtration problems. This problem can be resolved by treating
the syrup with a suitable xylanase preparation after the pH has
been adjusted back to pH 4-5.
    Acid/enzyme processes involve only limited hydrolysis by acid
with negligable production of dextrose. The syrup can then be
treated with a range of enzymes to yield various end products.
Thus, β-amylase treatment, or preferably β-amylase plus

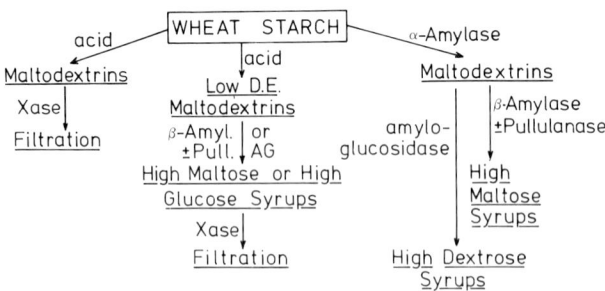

Scheme 1.   Enzymic and acid treatment of wheat starch. Xase is
             xylanase; β-Amyl. is β-amylase; Pull. is pullulanase;
             and AG is amyloglucosidase.

pullulanase, yields high maltose syrups. In enzyme/enzyme processes, the starch is cooked and preliminary depolymerisation is achieved with a temperature stable α-amylase. The syrup produced can be modified with a single enzyme or a mixture, to give high maltose, or high fermentable sugar content. High dextrose syrups are also produced using amyloglucosidase (glucoamylase). This enzyme readily cleaves the α-1,4-linkages in amylose and amylopectin, but acts on the α-1,6-linkages in amylopectin at a reduced rate. However, an amyloglucosidase present in <u>Cladosporium resinae</u> culture broths(17)readily cleaves the α-1,6-linkages giving rapid and essentially complete hydrolysis of starch to glucose.

<u>Measurement of α-amylase</u>
α-Amylase is vitally important in baking, brewing and numerous other processes. However, techniques available for the quantification of this activity are tedious and not readily automated. In response to this need, we(18)developed an assay employing a defined oligosaccharide substrate, namely blocked p-nitrophenyl maltoheptaoside. The principle of the assay procedure is outlined in Fig. 6. Basically, the substrate is

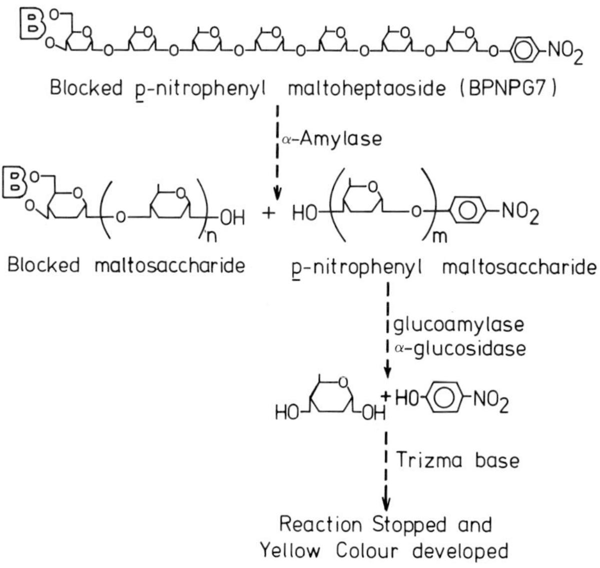

Blocked p-nitrophenyl maltoheptaoside (BPNPG7)

α-Amylase

Blocked maltosaccharide   p-nitrophenyl maltosaccharide

glucoamylase
α-glucosidase

Trizma base

Reaction Stopped and
Yellow Colour developed

Fig. 6. Theoretical basis of the α-amylase assay procedure.

prepared in the presence of excess quantities of glucoamylase and
α-glucosidase. Both of these enzymes are exo-enzymes and thus
have no action on the native substrate due to the presence of the
blocking group on the non-reducing terminal D-glucosyl group.
When the oligomer is cleaved by endo-acting α-amylase, the
nitrophenyl oligosaccharide released is rapidly and completely
hydrolysed to free D-glucose and p-nitrophenol. The reaction is
terminated and the p-nitrophenol colour is developed on addition
of Trizma base (pH~10). The assay is simple and accurate and can
be readily automated. For wheat flour samples, correlation
coefficients of 0.988, 0.971 and 0.991 were obtained when the
method was compared to the Phadebas assay, Liquefaction Number
(calculated from Hagberg Falling Number units) or Farrand method,
respectively.

ACKNOWLEDGEMENTS

    The authors would like to acknowledge the expert technical
assistance of Mrs. E. Nurthen, Ms. I. Shameer, and
Ms. H. Sheehan. We also thank Ms. Anne Payne for assistance in
the preparation of this manuscript.

REFERENCES

1. Shelton, D.R. and D'Appolonia, B.L., Cereal Food World, 30
    (1985)437-442.
2. Bushuk, W., Bakers Digest, 40 (1966)38-40
3. McCleary, B.V., Int.J. Biol. Macromol. 8(1986)349-354.
4. Kulp, K.,Cereal Sci. Today, 13(1986)414-417
5. Wieg, A.J., Starch, 36(1984)135-139.
6. Hoseney, R.C., Food Technol. 38(1984)114-117
7. McCleary, B.V., Gibson, T.S., Allen, H. and Gams, T.C.,
    Starch, 38,(1986)433-437.
8. Bamforth, C.W., Brewers Digest, 57(1982)22-35
9. McCleary, B.V., and Glennie-Holmes, M., J. Inst. Brew.
    91(1985)285-295
10. McCleary, B.V. and Shameer, I., J. Inst. Brew. 93(1987)87-90.
11. McCleary, B.V. and Matheson, N.K., Adv. Carbohydr. Chem.
    Biochem., 44(1987)147-276.
12. Ehlers, G., Food Review, Aug/Sept. 1984, 37-38.
13. Pomeranz, Y., (1985) Functional Properties of Food
    Components, Academic Press Inc., New York.
14. Tagawa, K. and Kaji, A., Carbohydr. Res. 11(1969)293-301.
15. L. Weinstein and P. Albersheim, Plant Physiol
    63(1979)425-432.
16. Pazur, J.H. and Ando, T., Arch. Biochem. Biophys.
    93(1961)43-49.
17. McCleary, B.V. and Anderson, M.A., Carbohydr. Res.
    86(1980)77-96.
18. McCleary, B.V. and Sheehan, H., J. Cereal Sci. (1988) in
    press.

Solution conformations of the polysaccharide gellan

V.Crescenzi and M.Dentini

Department of Chemistry, University of Rome, Rome, Italy

Abstract

The study deals with the influence exerted by guanidinium hydrochloride, urea, and sucrose, respectively, on the stability of the ordered chain conformation that can be assumed by gellan chains in dilute aqueous solution. Guanidinium hydrochloride and sucrose stabilize the ordered chain state of gellan, though with different mechanisms, and eventually lead to gel formation. In contrast, urea acts as a typical denaturing agent favouring the randomly coiled state of gellan. Much less pronounced appears the influence of the same cosolutes on the solution properties of branched polysaccharides structurally related to gellan.

Introduction

Gellan belongs to a novel group of structurally related anionic bacterial polysaccharides that have recently resulted in academic and industrial interest because of their unique rheological and/or gelling properties. The repeating units of these biopolymers (1-2) all contain a linear tetrasaccharide with the structure shown in Fig. 1 (acyl substituents omitted).

The nature of the side-chains and of the fourth sugar residue (α-L-Sugp) in the backbone however varies as indicated in Table I.

Comparative studies carried out on gellan, welan, and rhamsan clearly show that the dilute aqueous solution properties of such structurally similar polysaccharides are, in fact, profoundly different(3,4).

V.Crescenzi and M.Dentini

TABLE 1

Polysaccharide	R_1	R_2	α-L-Sugp
gellan[a]	H	H	rha
welan[b]	H	α-L-rhap(1-> or α-L-manp(1->	rha
rhamsan[c]	β-D-glcp 1 ↓ 6 α-D-glcp(1->6)	H	rha
S-88[d]	H	α-L-rhap(1->	rha/man
S-198[e]	α-L-rha(1->4)[*]	H	rha/man

a) from Pseudomonas elodea; b) Alcaligenes ATCC 31555;
c) AlcaligenesATCC 31961 ; d) Alcaligenes 31554 ;
e) Alcaligenes ATCC 31853; *) Contained in only ca.50% of
   the repeat units

->3)-β-D-glcp-(1,4)-β-D-glcAp-(1,4)-β-D-glcp-(1,4)-α-L-sugp(1->
    6/4   3
    |
    R_1 R_2

Fig.1 :Backbone repeating unit in common to the five polysaccharides
indicated in Table I.

In particular, our data demonstrate that while welan and rhamsan exhibit solution properties nearly unperturbed by changes in ionic strength, temperature or pH, to the contrary gellan , in dilute aqueous solution (ca. 0.08 %w/V),upon addition of univalent salts (0.01-0.10M) undergoes a cooperative, order->disorder conformational transition accompanied by abrupt changes in chiroptical properties and characterized (at 25°) by :
1) an enthalpy change of ca.- 6 kJ/mole of repeating unit ; 2) an increase in intrinsic viscosity by a factor ca. 3.

Moreover, such transition, which can be attributed to an "expanded coil-> helix" transition (5), is thermally reversible with no hysteresis. At higher polymer and added salt concentrations gellan solutions eventually turn into stable gels.

The stiff, helical conformation adopted by gellan under the experimental conditions mentioned above (i.e. in solution) must be prerequisite to the formation, at higher polymer and salt concentrations, of the "junction - zones" characteristic of the gel phase. A study of the main factors governing the stability of the gellan ordered conformation in aqueous media is therefore important also for an understanding of its performance   as a gelling hydrocolloid.

The following results deal with the influence of cosolutes such as guanidinium hydrochloride, urea, and sucrose, respectively, on the solution behavior of gellan. For comparison purposes a few data concerning other polysaccharides of Table I are also briefly reported.

Results
a) Gellan / Guanidinium HCl

As experienced with different 1:1 valent salts (5), addition at 25° of guanidinium hydrochloride (gu.HCl) to a dilute (0.08 % w/V) gellan aqueous solution brings about a change in optical activity that can be traced to a conformational change of the polysaccharide (Fig. 2).

This change is thermally reversible with no detectable hysteresis ,(Fig. 3), the midpoint, transition temperatures being correlated with gu.HCl concentration as found with other simple salts. In conclusion, for concentrations up to ca. 50 mM, guHCl behaves just as any other 1:1 valent electrolyte and thus is able to trigger the coil->helix transition of gellan. However, interesting enough, guHCl at concentrations higher than ca. 0.1M induces gel formation : indeed, in comparison to all other 1:1 valent salts considered so far (5), guHCl appears to be the strongest "gel promoter" for aqueous gellan.

b) Gellan / Urea
Urea is found to act as a typical "denaturing" agent towards gellan ordered

V.Crescenzi and M.Dentini

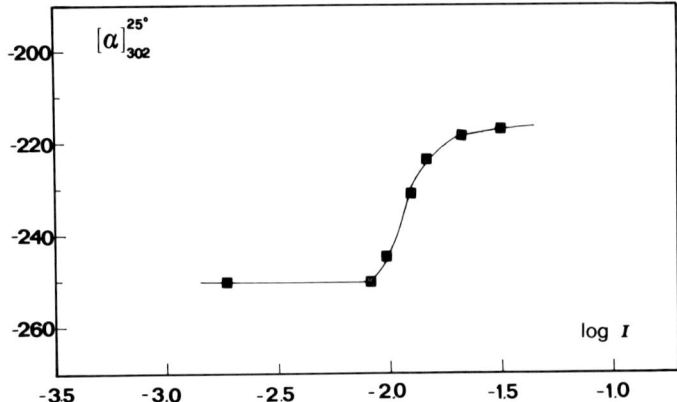

Fig.2 :Dependence of gellan (0.08% w/V) optical activity at 25° on
guanidinium hydrochloride concentration. I is the total ionic strength.

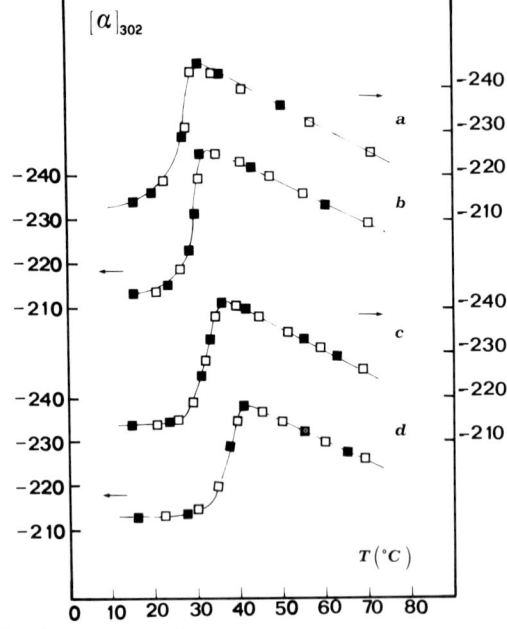

Fig.3: Melting of gellan ordered conformation at
different guanidinium hydrochloride
concentrations(Mx10^2) : a) 1.4 ; b) 2.0 , c)
2.5, d) 2.75

chain state, as demonstrated by the thermal-transition data collected in Fig. 4. In agreement with the optical activity data, the results of viscosity measurements show that at 25° in 5M urea (and 75 mM Me4NCl) the conformational state of gellan completely reverts back to a randomly coiled one having an intrinsic viscosity nearly three times lower than that of the ordered state (in 75mM Me4NCl).

c) Gellan / Sucrose

In contrast to urea, sucrose confers stability towards the ordered chain state of gellan in dilute aqueous solution. In fact, data of Fig. 5 clearly demonstrate that gellan intrinsic viscosity is markedly enhanced upon addition of sucrose increasing sigmoidally with sucrose concentration.

A series of similar experiments carried out using welan, rhamsan, and S-88 (see Fig.1 and Table I) show that all these polysaccharides exhibit equilibrium and transport properties affected very little by changing the chemical and physical variables which so profoundly influence the solution behavior of gellan.

Discussion

Results presented above lead, in our opinion, to the following conclusions:

a) Introduction of a sugar side-chain (see Table I and Fig.1) onto the common tetrasaccharide repeating unit always markedly reduces the "effective conformational space" otherwise available to the latter (that is to gellan). In fact, only gellan among the five structurally related polysaccharides of Table I would be able to enjoy a relative conformational freedom and display chain flexibility. Thus, the stereoregular, linear chains of gellan are sensitive to variations in ionic strength or temperature and to the addition of non-ionic cosolutes, and react by changing conformation.

b) simple salts, including gu.HCl, stabilize selectively the ordered chain state of gellan in dilute aqueous solutions and room temperature. Concentrations of univalent salts higher than ca. 0.2M (0.1M, in the case of gu.HCl) promote gel formation.

c) urea destabilizes the helical conformation of gellan, probably *via* direct interactions with the polysaccharide , favouring a coiled state of the chains: and this is what has also been found in the case of different polysaccharides (6) with the notable exception of xanthan (7).

To the contrary, sucrose makes the helical state of gellan more stable and, as found with other polysaccharides, creates conditions favourable to the onset of gelation(8).

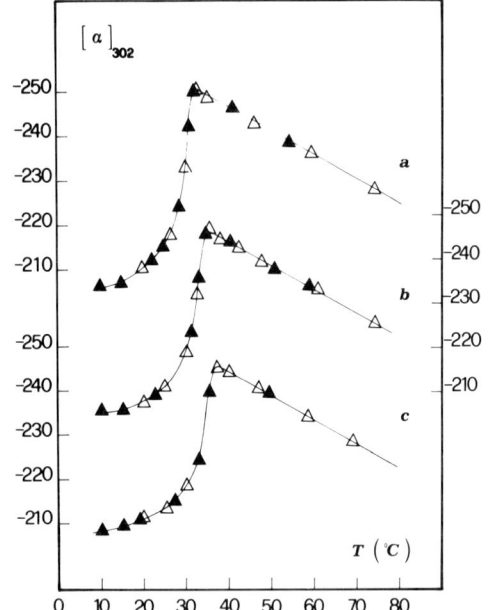

Fig.4. Melting curves for gellan in 70mM Me4NCl
and different urea concentrations (a 0.15M,
b 1.0M , c. 0.5M)

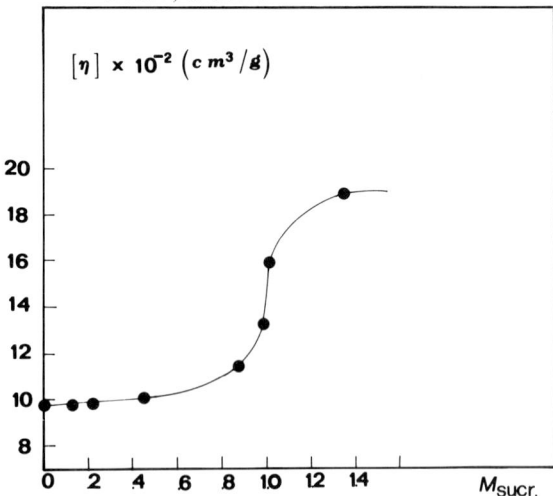

Fig.5 .Intrinsic viscosity of gellan in 6.3mM Me4NCl as a function of
sucrose concentration at 25°

Acknowledgements

The polysaccharide samples were kindly provided by Dr. R.Moorhouse (Kelco Co., San Diego, California).This work has been carried out with financial support of the Italian Ministero Pubblica Istruzione, Rome, Italy.

References

1) P-E. Jansson, B.Lindberg, P.A.Sandford ; Carbohydr. Res., <u>124</u>, 135 (1983)
M.A.O'Neill, R.R. Selvendran, V.J.Morris ;Carbohydr. Res., <u>124</u>, 123 (1983)
R. Moorhouse ; Am. Chem. Soc. Meeting, Denver, USA, 1987
2) T.A.Chowdhury, B.Lindberg, U.Lindquist ; Carbohydr. Res., 161, 127 (1987)
3) V.Crescenzi, M.Dentini, T.Coviello, R.Rizzo ; Carbohydr. Res., <u>149</u>, 425 (1986)
4) V.Crescenzi, M.Dentini, I.C.M.Dea ; Carbohydr. Res., <u>160</u>, 283 (1986).
5) V.Crescenzi, M.Dentini, T.Coviello, A.Cesàro, S.Paoletti, F.Delben; Gazz. Chim. Ital., in press.
6) M.Watase, K.Nishinari ; Food Hydrocoll.,<u>1</u>, 25 (1986)
7) S.A.Frangou, E.R.Morris, D.A.Rees, R.K.Richardson,S.B.Ross-Murphy ; J.Polymer Sci., Polymer Letter Ed., <u>20</u>, 531 (1982).
8) S.M.Fiszman, E.Costell, L.Duràn ; Food Hydrocoll., <u>1</u>, 113 (1986)

Conformational studies of α-(1→4) glucans in solid and solution states by NMR spectroscopy

Michael J.Gidley

Unilever Research Laboratory, Colworth House, Sharnbrook, Bedford MK44 1LQ, UK

ABSTRACT

A number of experimental approaches to the determination of polysaccharide conformations are discussed with examples drawn from studies of α-(1→4) glucans. A recently introduced 2D NMR experiment is shown to provide information on the average values of the two conformation-determining torsion angles, ϕ and ψ. The range of conformations present in frozen solutions can be assessed by ^{13}C cross polarisation and magic angle spinning NMR, and may be characterised by direct comparison with spectra of solid materials having well defined conformations. The molecular structures present in amylose gels are probed and found to be of two types, namely motionally-restricted double helices and more mobile regions containing all energetically-allowed conformations.

INTRODUCTION

The determination of polysaccharide solution conformation is important in studies of the molecular origins of the properties of aqueous polysaccharide systems. NMR spectroscopy has played an increasingly important role in carbohydrate conformation studies in recent years due to the advent of higher magnetic field strengths and the use of a range of new multiple pulse experiments. Many of these studies have focussed on the oligosaccharide chains of glycoproteins (1,2) with repeating polysaccharides receiving less attention (3,4).

In this report, I describe two approaches to the NMR study of polysaccharide solution conformation using α-(1→4) glucans (Fig.1) as a model system. In the first approach, high resolution solution state NMR spectroscopy is applied to oligosaccharides in conjunction with a chiroptical test of the suitability of specific oligosaccharides as conformational models of related polysaccharides. In the second approach, frozen polysaccharide solutions are studied by high resolution solid state ^{13}C (CP/MAS) NMR spectroscopy

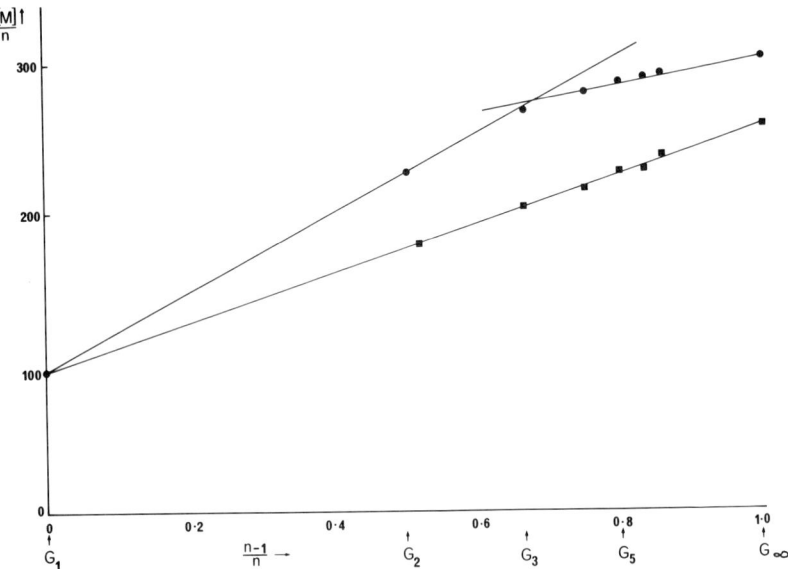

Figure 2.   Freudenberg plots for α-(1 → 4) glucan oligomers
in water (●) and dimethyl sulphoxide (■). The molar rotation
at 589nm [M] divided by the number of glucose residues,n, is
plotted against (n-1)/n. Optical rotation measurements in
dimethyl sulphoxide solution have been corrected for
refractive index differences to allow a direct comparison
with aqueous solutions (Dintzis,F.R. and Tobin,R.(1969)
Biopolymers 7,581-593)

Karplus relationship which describes the variations of
three-bond coupling constants to torsion angles is not
accurately known (7); and secondly, the single values derived
for ϕ and ψ would represent averages of all conformations
present. If the individual conformers are not known, it has
been shown that such average structures have no physical
meaning (16). Clearly, it is important to determine not only
averaged structural parameters but also the range of
conformers present.

^{13}C CP/MAS NMR STUDIES OF FROZEN SOLUTIONS

   Averaged structural information is derived from solution
NMR studies as conformational fluctuations occur on a
timescale faster than that of the NMR experiment. If
solutions are frozen, however, then individual conformers
will be trapped and could be probed by ^{13}C cross polarisation

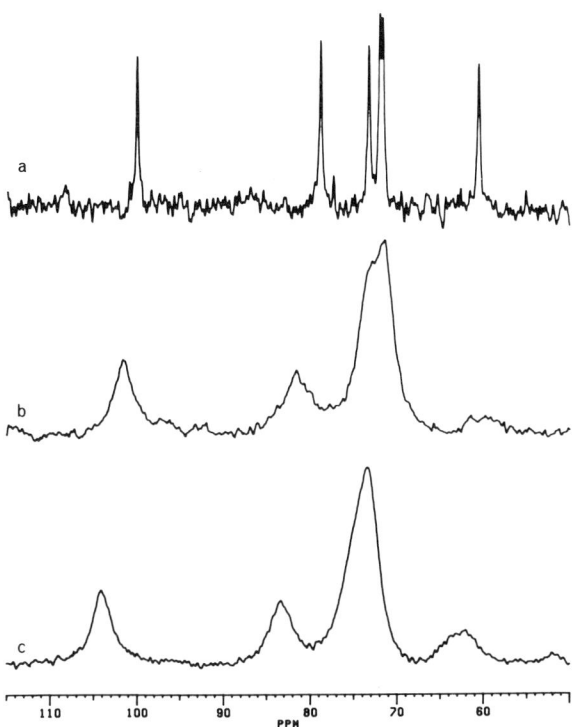

Figure 3 (a).Single pulse ^{13}C spectrum of a 10% w/v solution
of amylose in dimethyl sulphoxide at 303K   (b).^{13}C CP/MAS
spectrum of the same solution at 238K.(c).^{13}C CP/MAS spectrum
of solid V-type amylose hydrate

and magic angle spinning (CP/MAS) NMR spectroscopy. By
analysing a range of solid α-(1→ 4) glucans, we have
recently shown that C-1 and C-4 chemical shifts are primarily
determined by conformational features (17). Spectra obtained
from frozen solutions could therefore be compared with solid
samples having characteristic conformations (as determined by
X-ray diffraction). Figure 3 shows spectra of DMSO solutions
of amylose recorded at 303K (Fig. 3a) and 238K (Fig. 3b).
The frozen solution is seen to have broader spectral features
than those of the ambient solution and all signals show
significant chemical shift differences. Comparison with
solid V-type amylose (Fig. 3c) shows that frozen DMSO
solutions (Fig. 3b) have very similar chemical shifts and
therefore similar conformations to the V-type structures
characterised by X-ray diffraction (18). The chemical shift

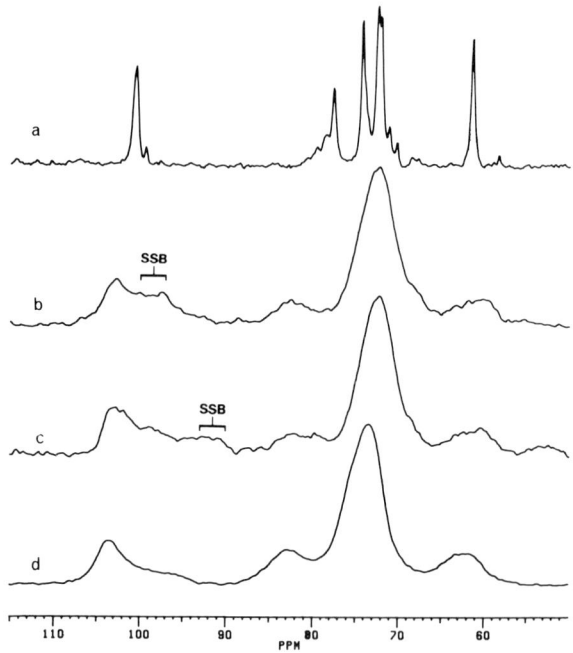

Figure 4. (a) Single pulse ^{13}C spectrum of a 15% w/v aqueous solution of waxy maize β-limit dextrin at 303K (b) ^{13}C CP/MAS spectrum of the same solution at 233K (c) as spectrum (b) with slower spinning speed to show effect of spinning side bands, SSB (d) ^{13}C CP/MAS spectrum of solid amorphous starch

differences between ambient and frozen solutions are being further investigated.

Frozen aqueous solutions of polymeric α-(1→4) glucans show a much wider range of C-1 and C-4 chemical shifts (Fig. 4b) with weighted average values being similar to those observed at ambient temperature (Fig. 4a). Solid amorphous α-(1→4) glucans have very similar spectral features (Fig. 4d) to those of frozen solutions (Fig. 4b) suggesting that aqueous solutions contain all energetically-allowed glycosidic conformations (17). Although it could be argued that conformations characterised in frozen solution are not necessarily representative of those present at ambient temperatures, the fact that weighted average low temperature chemical shifts coincide with observed ambient shifts for aqueous solutions adds some confidence to the approach adopted. Further work is required to elucidate the nature of the conformational origin of chemical shift effects to expand on recent initial studies in this area (17,20).

MOLECULAR STRUCTURES IN AMYLOSE GELS

Gelation results in a decrease in mobility of a significant fraction of a polysaccharide. This enhanced chain rigidity is reflected by a decrease of T_2 values and sufficient broadening of ^{1}H and ^{13}C high resolution signals so that at least part of the signal becomes indistinguishable from the baseline. It might be expected that signals from rigid regions of the polysaccharide gel may be sufficiently 'solid-like' to be detectable in a ^{13}C CP/MAS experiment. This has indeed been shown to be the case for pustulan gels (21).

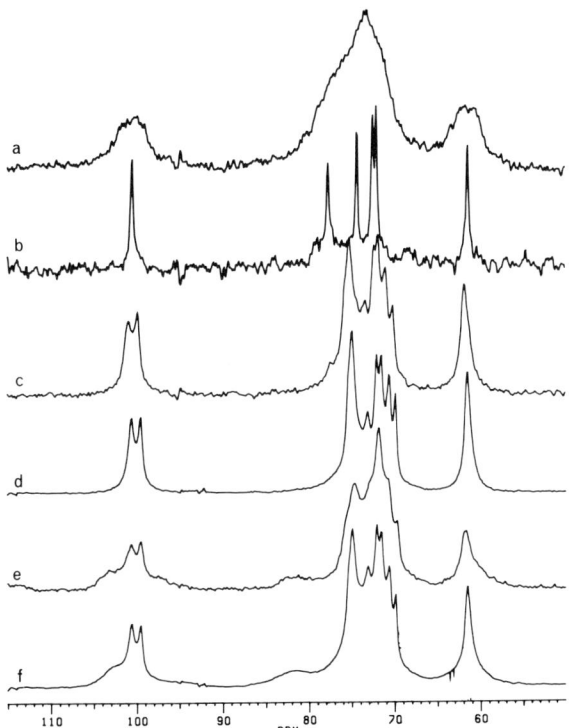

Figure 5.(a) Single pulse ^{13}C spectrum of a 10% w/v aqueous potato amylose gel at 303K (b) as (a) with 500Hz magic angle spinning (c) ^{13}C CP/MAS spectrum of the same sample (d) ^{13}C CP/MAS spectrum of highly crystalline B-type amylose (e) ^{13}C CP/MAS spectrum of the same gel as a-c at 233K (f) Simulation of spectrum (e) by addition of 67% of Fig.5c spectrum and 33% of Fig.4c spectrum

We have examined 10% w/v aqueous gels of both natural (polydisperse) and synthetic monodisperse (22) amyloses of DP300 and 2800 by NMR techniques. Samples in a MAS rotor were analysed using single pulse excitation i.e. analogous to a conventional high resolution experiment, and gave a broad but recognisable spectrum (Fig. 5a). The broadness was shown to be due to residual chemical shift anisotropy rather than T_2 relaxation or chemical shift dispersion effects as magic angle spinning at speeds comparable to the observed linewidths (200 - 500 Hz) resulted in a well-resolved spectrum essentially identical to that found for aqueous amylose solutions (Fig. 5b). A CP/MAS experiment on the same sample resulted in a dramatically different spectrum (Fig.5c) which was identified as being typical (17,19) of B-type (double-helical) solid material (Fig. 5d). Two NMR experiments have therefore shown that 'mobile' and 'immobile' regions in amylose gels have distinct conformations ie. solution-like and double-helical respectively. If the solution-like spectrum obtained by single pulse excitation (Fig. 5b) reflects motionally-averaged conformations, the range of such conformations could be assessed by a CP/MAS experiment on a frozen gel. The spectrum thus obtained (Fig. 5e) shows features due to both B-type double helices and amorphous $\alpha-(1 \rightarrow 4)$ glucan (Fig. 4c) suggesting that the same range of conformation is present in the mobile gel component as is found in aqueous solutions. By computer addition of appropriate proportions of spectra characteristic of amorphous (Fig. 4c) and B-type double helical (Fig. 5d) conformations, the CP/MAS spectrum of the frozen gel can be simulated accurately (Fig. 5f). This therefore suggests that in the ambient gel, the two conformational types which can be separately characterised (Fig. 5b,c) are the only ones present.

The relative proportions of double helix and amorphous conformations in gels can be estimated by simulation of CP/MAS spectra of frozen gels (Fig. 5e,f). As the mobilities of these two conformations appear to be very different, it might be expected that ^1H T_2 measurements of ambient gels could also resolve and quantify two components. In line with this expectation, two polysaccharide components are readily distinguished in amylose gels by virtue of their different relaxation timescales ($\sim$10 μsec and $\sim$1msec for the rigid and mobile components respectively). Quantification of these two components agrees well with analysis of CP/MAS spectra of frozen gels (Table 2).

Table 2   QUANTIFICATION OF MOLECULAR ORDERING IN 10% AMYLOSE
          GELS

Amylose	% B-type[a]	% rigid component[b]
Potato (polydisperse)	67 ± 2-3	70 ± 2-3
Synthetic (DP300)	83 ± 2-3	88 ± 2-3
Synthetic (DP2800)	67 ± 2-3	72 ± 2-3

a. from ^{13}C CP/MAS measurements   b. from ^{1}H T_2 measurements

These experiments therefore provide a complete
quantitative description of the molecular conformations
present in amylose gels and suggest that the rigid
cross-links in the gel are due to double helices and that the
more mobile inter-helical regions exhibit the full range of
energetically-accessible conformations.

REFERENCES

1. Homans,S.W.,Dwek,R.A.,Boyd,J.,Mahmoudian,M.,Richards,W.G.
    and Rademacher,T.W.(1986) Biochemistry,25,6342-6350
2. Brisson,J.R and Carver,J.P.(1983)Biochemistry,
    22,3671-3680 and 3680-3686
3. Casu,B.(1985) in Polysaccharides-Topics in Structure and
    Morphology, (ed. Atkins,E.D.T.)pp1-40. Macmillan Press,
    Basingstoke, UK.
4. Gorin,P.A.J.(1981)Adv.Carbohydr.Chem.Biochem., 38,13-104
5. Rees,D.A.,Morris,E.R.,Thom,D. and Madden,J.K.(1982) in
    The Polysaccharides, Vol.1 (ed.Aspinall,G.O.)pp195-290.
    Academic Press, New York,USA.
6. Dais,P.,Shing,T.K.M and Perlin,A.S.(1984)J.Am.Chem.Soc.,
    106,3082-3089
7. Hamer,G.K.,Balza,F.,Cyr,N.and Perlin,A.S.(1978)
    Can.J.Chem., 56,3109-3116
8. Lemieux,R.U. and Bock,K.(1983) Arch.Biochem.Biophys.
    221,125-134
9. Shashkov,A.S.,Lipkind,G.M. and Kochetkov,N.K.(1986)
    Carbohydr.Res.147,175-182
10. Morris,G.A. and Hall,L.D.(1982) Can.J.Chem.60,2431-2441
11. Barker,R. and Serianni,A.S.(1986)Acc.Chem.Res.19,307-313
12. Gidley,M.J. and Bociek,S.M.(1985)J.Chem.Soc.Chem.Commun.
    220-222
13. Bax,A. and Freeman,R.(1982)J.Am.Chem.Soc.104,1099-1100
14. Rees,D.A.(1970)J.Chem.Soc.(B),877-884
15. Freudenberg,K.,Freidrich,K.,Bumann,I. and Soff,K.(1932)
    Justus Liebigs Ann.Chem.494,41-62
16. Jardetzky,O.(1980)Biochim.Biophys.Acta 621,227-232
17. Gidley,M.J. and Bociek,S.M.(1987) submitted for
    publication

18. Rappenecker,G.and Zugenmaier,P.(1981) Carbohydr.Res.
    89,11-19
19. Gidley,M.J. and Bociek,S.M.(1985)J.Am.Chem.Soc. 107,
    7040-7044
20. Veregin,R.P.,Fyfe,C.A.,Marchessault,R.H. and
    Taylor,M.G.(1987)Carbohydr.Res. 160,41-56
21. Stipanovic,A.J.,Giammatteo,P.J. and Robie,S.B.(1985)
    Biopolymers 24, 2333-2343
22. Gidley,M.J.,Bulpin,P.V. and Kay,S.(1985) in Gums and
    Stabilisers for the Food Industry -3(eds.Phillips,G.O.,
    Wedlock,D.J. and Williams,P.A.)pp 167-176 Elsevier,
    London,UK

Fourier transform infrared spectroscopy and biopolymer functionality

R.H.Wilson, B.J.Goodfellow and P.S.Belton

AFRC Institute of Food Research, Norwich Laboratory, Colney Lane, Norwich NR4 7UA, UK

ABSTRACT

The technique of Fourier transform infrared spectroscopy, in conjunction with new methods of sample presentation, has been used to study biopolymer functionality in aqueous systems. Changes in the spectra of kappa and iota carrageenan gels have been observed with both counterion and temperature. Most of these differences arise from conformational changes of the polysaccharide backbone but there is also evidence of direct ion-sulphate interaction in certain ion forms, consistent with previous NMR results. An assignment of the region 1400-800 cm^{-1} has been made for the carrageenan gels. In waxy maize starch gels changes in the infrared spectrum are observed during storage. These have been related to the development of crystallinity in the material. The time course of the observed FTIR measurements during retrogradation is very similar to those produced from other techniques, such as shear modulus.

INTRODUCTION

The technique of Fourier transform infrared (FTIR) spectroscopy in combination with new methods of sample presentation such as attenuated total reflectance (ATR) now allow the acquisition of high quality spectra from previously quite intractable materials and, in particular, aqueous solutions. In this paper the potential of the technique is illustrated for the study of food polysaccharides. Attention has been concentrated on the region 1400 - 800 cm^{-1}, where a number of bands which have been found to be very sensitive to biopolymer conformation are seen[1]. Most of these arise from highly coupled C-C and C-O stretching modes of the polymer backbone[2]. However, in the case of carrageenan, bands arising from the sulphate groups also appear in this region[1].

R.H.Wilson et al.

The appearance of sulphate bands makes FTIR potentially very attractive for the study of the role of counterions in carrageenans, for while a technique such as NMR line width(3,4) may be used to monitor ion mobility it offers no indication as to the precise origin of such effects. FTIR can be used to simultaneously study changes in both the polysaccharide backbone and sulphate groups with changing counterion type and temperature.

The second polysaccharide studied is waxy maize starch. Changes in the spectrum of starches have beeen observed during both the gelatinisation and retrogradation processes(5). Waxy maize starch was found to exhibit very pronounced changes during storage and therefore provides the best system for a comparative study with the more usual methods of following retrogradation such as shear modulus measurements or X-ray diffraction.

EXPERIMENTAL

All FTIR measurements were carried out on a Digilab FTS60 spectrometer operating at 4 cm^{-1} resolution and equipped with a TGS detector. 128 or 64 interferograms were co-added before Fourier transformation. Triangular apodization was employed.

Spectra of the various pure ion forms of kappa and iota carrageenan were prepared as decribed previously(6,7). Gels (2%) were prepared by heating the required amount of pure ion form carrageenan with water in a sealed digestion bomb at 363 K for two hours.

Variable temperature carrageenan spectra (128 scans) were obtained using a Specac heatable transmission cell fitted with ZnSe windows and a 50μm spacer. The gels were directly injected into the cells which were preheated at 353 K. FTIR spectra were then obtained at various temperatures during the controlled cooling of the gel. A spectrum of water at each temperature was also obtained as a background. The absence of hysteresis was demonstrated by the identical spectra produced when the cooled samples were re-heated. Variable temperature spectra were also obtained using a heatable ATR cell. This was an adaptation of a Spectra-Tech continuously variable angle ATR cell with a ZnSe crystal (45°, 50 x 20 x 3 mm parallelogram). The normal solid sample holder plates were replaced with ones containing a shallow sample well and equipped with 12V soldering iron heating elements and a thermocouple. When the cell was assembled the sample was sealed within the shallow well by the crystal. The temperature was controlled with a Specac controller. Spectra produced were identical to those using transmission techniques. However, the latter method was found to be manipulatively much easier, with the problems associated with injecting a gel into a conventional transmission cell eliminated.

Ageing studies of waxy maize starch gel were carried out in a Spectra-Tech CIRCLE cell with ZnSe crystal. The sealed cell was stored at 274 K between FTIR measurements (64 scans) which were made regularly over a period of three weeks. The empty cell was used as background and a spectrum of water was also acquired. The water spectrum was digitally subtracted from all starch spectra. Measurements of shear modulus were also made at intervals using a Rank Brothers pulse shearometer(8).

RESULTS AND DISCUSSION

A representative spectrum of a carrageenan (potassium Kappa carrageenan) at 298 K is shown in figure 1. In this figure are also given the tentative assignments for the bands in the region(1). Assignment of the S-O stretching modes is made largely by comparison with similar sulphated polysaccharides. The band assigned to S-O symmetric stretch at 1090 cm^{-1} is seen as a shoulder in both kappa and iota carrageenans only in ion forms for which NMR(3,4) gives evidence of ion-polymer interaction. The S-O assymetric stretch is affected to a small extent but there is a possibility that this band is superimposed upon a second, weaker mode that may actually be undergoing change. This type of band is seen in spectra of some unsulphated polysaccharides, presumably arising from another C-O related mode.

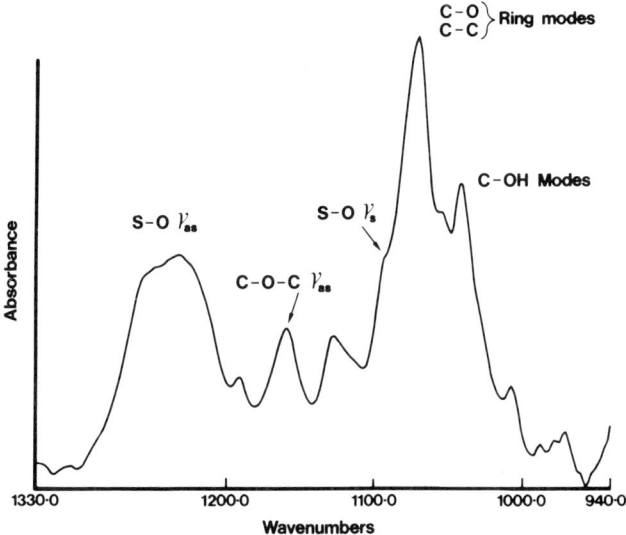

Figure 1.   Infrared spectrum of potassium kappa carrageenan gel.

Apart from changes in the sulphate bands the spectra of the different ion forms of kappa carrageenan show relatively minor differences. Sodium kappa carrageenan has a weaker 1040 cm^{-1} band (C-OH mode) than potassium or caesium. Presumably these differences arise from ion induced conformational changes. These may be brought about by solvent mediated effects, as a result of the high local counterion concentrations predicted for polyelectrolytes(9). In the iota carrageenan spectra the 1090 cm^{-1} S-O mode is more pronounced, consistent with the increased sulphate content. Generally there is a stronger counterion effect in this system than in the kappa series(1). If the observed changes at 1090 cm^{-1} were due to conformational changes on gelling then it would be expected that similar changes would be seen in sodium iota carrageenan. Although this is gelled there is no appearance of the 1090 cm^{-1} band, consistent with the NMR observations(3,4).

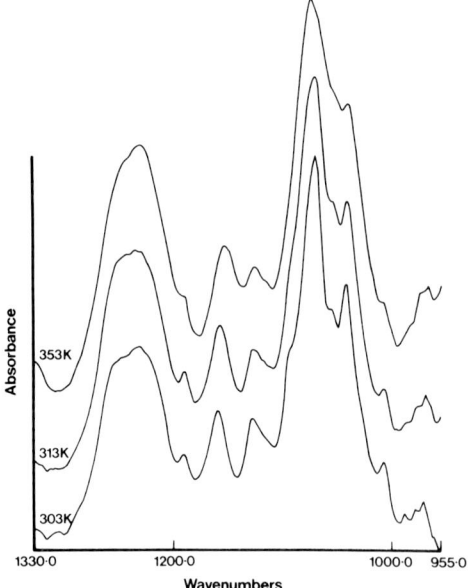

Figure 2.

Variation of infrared spectrum of potassium kappa carrageenan with temperature.

The variable temperature cycle of potassium kappa carrageenan is illustrated in figure 2. During cooling of the sample from 353 K to 298 K a number of effects can be seen. There are changes in the intensity of some bands, particularly, the band at 1090 cm^{-1} does not appear until the approximate temperature of the sol - gel transition (about 313 K) is reached. As this band has been assigned to a sulphate mode(1) its development implies increasing ion-sulphate interaction during gel formation. This result is consistent with the chemical shift observed by NMR(10). The S-O assymmetric stretch is also affected in that during cooling the initial broad band becomes a doublet. Other band changes can be assigned to conformational changes presumably associated with helix formation. There is also a definite band narrowing in the region 1100-1000 cm^{-1} during gel formation. These sharp lines are associated with a more highly ordered structure in the gel state. There is no apparent development of a band at 1090 cm^{-1} in sodium kappa carrageenan which does not gel and for which NMR(3,4) indicates no strong ion-polymer interactions.

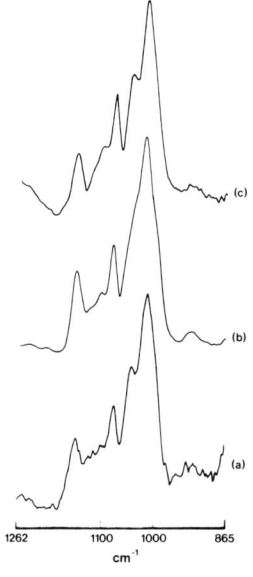

Figure 3.

Infrared spectra of waxy
starch in water
(a) before gelatinisation
(b) after gelatinisation
(c) after storage of gel
    at 274K for 21 days.

Reproduced with
permission from (2)

A spectrum of waxy maize starch (20%) in water is shown in figure 3. A number of bands can be seen with the most intense at 1019 cm^{-1} and a weaker band at 1046 cm^{-1}. There is also a weak shoulder on the low frequency side at about 1000 cm^{-1}. During gelatinisation at 363 K both this shoulder and the 1046 cm^{-1} band are lost. However, during storage at 274 K a shoulder gradually grows on the high frequency side of the 1019 cm^{-1} band whilst the 1046 cm^{-1} band is gradually reformed. The development of this band occurs by a gradual line narrowing of the broad band in the gelatinised material, centred at 1019 cm^{-1}. The 1046 cm^{-1} band has been related to the degree of crystallinity of the material by spectroscopic investigation of highly crystalline model compounds(11). The general line narrowing occurs as the initial disordered gel gives way to a more ordered helical state with a resultant decrease in conformational environments and hence hydrogen-bond energies.

The extent of the growth of the 1046 cm^{-1} band can be quantified by taking the ratio of the intensity of the band at 1019 cm^{-1} to that at the frequency (1037 cm^{-1}) at which the minimum of the valley between the 1019 and 1046 cm^{-1} bands occurs in the fully retrograded material. When this ratio, R, and measurements of shear modulus are made on a similar sample and are plotted against storage time then very similar time course curves result (figure 4). This indicates a direct correlation between a spectroscopic measurement and a functional property. This opens up the possibility that FTIR can be used to rapidly measure the extent of retrogradation of starch materials not only in model systems but also in real foods.

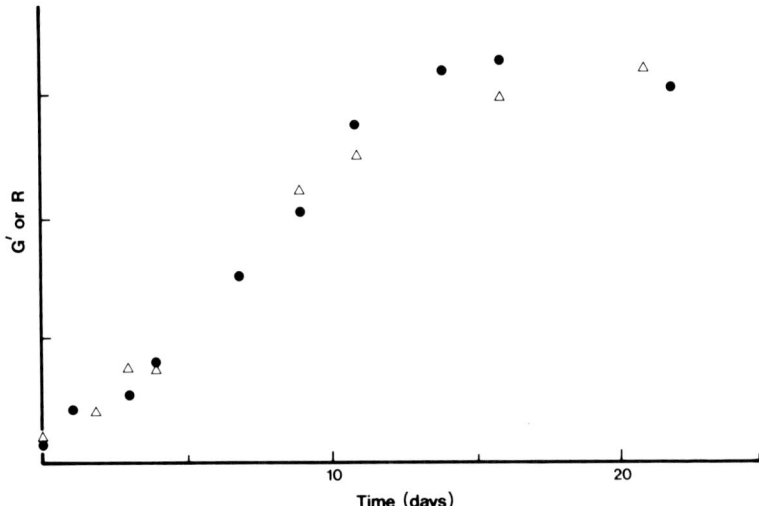

Figure 4. Time course of FTIR (●) and shear modulus (▲) measurements for
waxy maize starch retrogradation.

Reproduced with permission from (2).

   In conclusion, FTIR would appear to offer great potential for the study
of gelation and retrogradation in food biopolymers. The ability to deal
with wet and intractable systems should allow a very wide range of
applicability including intact foodstuffs.

REFERENCES

1. Belton, P.S., Wilson, R.H. and Chenery, D.H., (1986) Int. J. Biol.
    Macromol., 8, 247-251.

2. Tipson, R.S. and Parker, F.S., (1980) in The Carbohydrates, (eds.
    Pigman, W. and Horton, D.) vol. 1B, pp1394-1436, Academic Press, London.

3. Belton, P.S., Chilvers, G.R., Morris, V.J. and Tanner, S.F., (1984)
    Int. J. Biol. Macromol., 6, 303-308.

4. Belton, P.S., Morris, V.J. and Tanner, S.F., (1985) Int. J. Biol.
    Macromol., 7, 53-56.

5. Wilson, R.H., Kalichevsky, M.T., Ring, S.G. and Belton, P.S. Carbohydr,
    Res., accepted for publication. (UK 3432).

6. Morris, V.J. and Chilvers, G.R., (1981) J. Sci. Food Agric., 32,
    1235-1241.

7. Morris, V.J. and Belton, P.S., (1982) Prog. Food Nutr. Sci., 6, 55-66.

8. Ring, S.G. and Stainsby, G., (1985), J. Sci. Food Agric., 36, 607-613.

9. Manning, G.S., (1978) Q. Rev. Biophys. 11, 179-246.

10. Belton, P.S., Morris, V.J. and Tanner, S.F., (1986) Macromolecules, 19, 1618-1621.

11. Wilson, R.H., Kalichevsky, M.T., Ring S.G and Belton, P.S., unpublished results.

samples, such as the chemical composition already mentioned, molecular weight and molecular distribution, pH and ionic strength of solvent, as well the methods of extraction and the compositions of extracting solvents. Here, the effects of extraction procedures using dilute Na_2CO_3, EDTA and CDTA, on intrinsic viscosities of alginate samples were studied. The results of pre-treating the weeds with aqueous formaldehyde prior to extraction with sodium carbonate solutions are also reported.

MATERIALS AND METHODS

The alginate was extracted from brown seaweeds collected from Likas Bay, Kota Kinabalu, Sabah, Malaysia. All the weeds were washed thoroughly with tap water and finally with distilled water and dried in the air. The dried weeds were ground and used for various extraction methods.

Extraction with Na_2CO_3, di-sodium ethylenediamine tetraacetic acid (EDTA) and di-sodium 1,2-di-aminocyclohexane-N,N,N,N-tetraacetic acid (CDTA).
Suitable amounts of the ground weeds were extracted with aqueous solutions of sodium carbonate (pH 12), EDTA (pH 7.5 and 9.0) and CDTA (pH 7.5) respectively. The concentration of the sequestrant in all cases were 0.1 M and the pH was adjusted using buffer solutions. The extraction was performed at room temperature with continuous stirring. Samples were taken out from extraction reservoirs at suitable intervals and centrifuged at 5000 rpm for 15 minutes. The alginate was precipitated from the viscous solution using 3 volume of ethanol. The precipitate washed thoroughly with ethanol, then oven dried at 40°C.
For determining the percentage yield, suitable amount of the ground weeds were extracted for 24 hours, centrifuged at 5000 rpm and the alginate precipitated from the supernatant using 3 volumes of ethanol, washed thoroughly with ethanol and oven dried at 40°C to a constant weight.

Pre-treatment with formaldehyde prior to extraction with Na_2CO_3 and EDTA.
The ground weeds were treated with 1% and 10% formaldehyde, respectively. The amount of formaldehyde used was just sufficient to wet the weeds. The extract was filtered and the supernatant was discarded. The residues was air dried and extracted as mentioned above using 0.1 M Na_2CO_3 or 0.1 M EDTA.

Seaweeds extraction at different pH values and the effect of pH on intrinsic viscosity.
Seaweeds treated with 10% formaldehyde and samples without any pre-treatment were used. Samples were extracted with 0.1 M EDTA at pH 6,7,8,9, 10, 11, 12 and 13, respectively. The extraction time was 6 hours for each batch. After the extraction was completed, the viscous solution was centrifuged and the alginate precipitated with 3 volume of ethanol. The precipitate was oven dried at 40°C. This samples was used to find the relationship between intrinsic viscosity and pH during extraction.

Intrinsic viscosity
      Intrinsic viscosities for all samples were determined in Cannon-
Ubbelohde, semi-micro dilution viscometer size 75. The solvent
used was 0.2 M NaCl and the alginate samples were dialysed (dialysis bags, Medicell International Ltd.) to equilibrium against
0.2 M NaCl for at least 48 hours with constant stirring. Final
concentration of alginates were determined using the phenol-sulphuric acid reaction (6). The intrinsic viscosities were calculated
as usual by plotting reduced viscosity, η_{sp}/C and inherent viscosity, $\ln \eta_r/C$ against C and the limiting intercept at C = 0 will
give the intrinsic viscosity.

Mannuronate/guluronate ratios and block composition.
      The block composition was determined by using a partial hetero-
geneous hydrolysis technique (4) and the M/G ratio determination
was performed by a modified version of a total hydrolysis technique
(2,7). D-mannuronolactone was used as standard for both mannuronic acid and guluronic acid in phenol-sulphuric acid assay (6)
with a division factor of 1.7 used to allow for the greater colour
development by guluronic acid.

Determination of Mark-Houwink parameters.
      This was performed by plotting log [η] versus log $\bar{M}_w$ for dif-
ferent molecular weight sample obtain by fractionation or auto-
claving. The exponent 'a' for each batch was obtained from the
slope of log [η] versus log $\bar{M}_w$. Intrinsic viscosity for alginates from Laminaria hyperborea, Sargassum sp. 2 and Fucus sp.
were determined at 0.05 M, 0.10 M, 0.20 M and 0.50 M NaCl, respectively, and the intrinsic viscosities at infinite ionic strength
were obtained from a plot of [η] versus $1/\sqrt{I}$. $\bar{M}_w$ were determined by laser light scattering, sedimentation-diffusion (7) and
viscometry techniques. The plots of log [η] versus log $\bar{M}_w$ resulted in straight line for all the samples studied.

RESULTS AND DISCUSSION

Percentage yield of Na-alginate.
      Table 1 gave the percentage yield in different sequestring
agents used for isolating alginate. It shows that EDTA and CDTA
gave a higher percentage of alginate at a shorter period (6 hours)
compared to sodium carbonate extraction, although the percentage
of alginates are not significantly different at longer times of
extraction.

Table 1 : EFFECT OF Na_2CO_3, EDTA AND CDTA ON YIELD OF ALGINATE
           FROM SARGASSUM sp. 2

| Time of extraction | Percentage of alginate | | |
(hours)	Na_2CO_3(0.1 M) pH 12	EDTA (0.1 M) pH 7.5	CDTA (0.1 M) pH 7.5
6	20	34	38
24	38	40	42
48	40	42	43

Mannuronate to guluronate ratio
    Mannuronate to guluronate ratios are given in Table 2. Sargassum
sp. 2 gave the lowest M/G ratio with the highest composition of
guluronate blocks (GG blocks) and is identical to the M/G ratio
from Laminaria hyperborea samples. Fucus sp. on the other hand,
gave the highest M/G ratio and mannuronate blocks (MM blocks)
while alginate from Ascophyllum nodosum contained the highest
alternating blocks.

Table 2 :   M/G RATIO AND BLOCKS COMPOSITION FOR ALGINATE FROM
            VARIOUS SPECIES.

Source of alginate	Block composition			M/G	M/G
	-MGMG-	-MMMM-	-GGGG-	a	b
Sargassum sp. 1	36	24	40	0.72	0.76
Sargassum sp. 2	33	20	47	0.57	0.60
Fucus sp.	26	60	14	2.70	2.72
Ascophyllum nodosum	61	25	14	1.25	1.52
Laminaria hyperborea	32	14	54	0.43	0.60

Note: a - partial heterogeneous hydrolysis technique.
      b - complete hydrolysis.

Effect of extraction with Na$_2$CO$_3$, EDTA and CDTA on [η].
    Figure 1 gave the intrinsic viscosity as a function of extrac-
tion times. EDTA and CDTA were used because they are better se-
questring agents for calcium ions, compared to carbonate. It clearly shows that the intrinsic viscosity increases with time for
all the sequestrant used, up to a maximum intrinsic viscosity.
CDTA gives the highest intrinsic viscosity at 20.8 dL/g for extraction of 8 hours or longer carried out under neutral conditions. While for samples extracted in 0.1 M Na$_2$CO$_3$ (pH 12) and
EDTA (pH 9) the intrinsic viscosity drops. This is most significant especially for Na$_2$CO$_3$ extraction. This clearly demonstrates
that alginate solution are fairly stable under neutral conditions
and depolymerisation occurs under alkaline condition.

Effect of formaldehyde pre-treatment on intrinsic viscosity.
    Table 3 gives the results of formaldehyde treatment on intrin-
sic viscosity. The results shows that pre-treatment with formaldehyde prior to normal extraction using 0.1 M sodium carbonate
resulted in higher intrinsic viscosity. The samples pre-treated
with 10% formaldehyde gives higher intrinsic viscosity and a
whiter product. While Na-alginates not subjected to pre-treatment
with formaldehyde are dark browm.
    Pre-treated weeds with formaldehyde gives alginate with almost
constant intrinsic viscosity as shown in Fig. 1 even they are extracted with 0.1 M Na$_2$CO$_3$ at prolonged times.

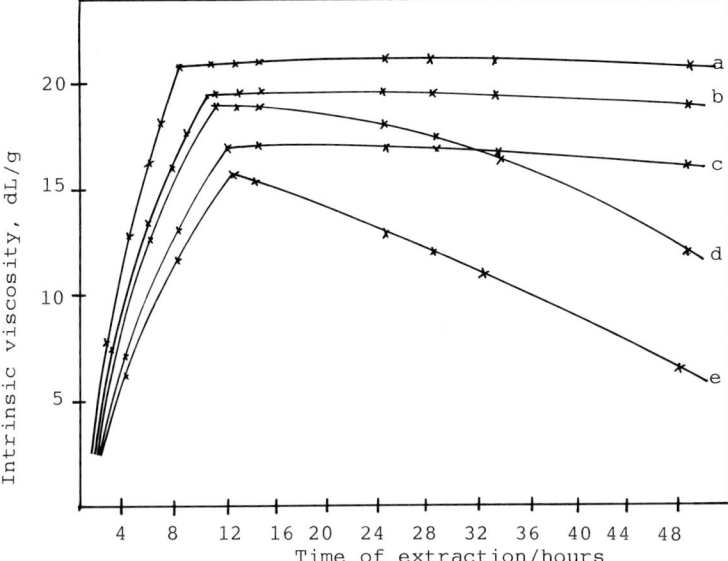

Fig. 1 : Intrinsic viscosity as a function of extraction times, seaweed extracted in (a) 0.1 M
CDTA, pH 7.5, (b) 0.1 M EDTA, pH 7.5 (c)
pretreated with HCHO, extracted in 0.1 M
Na_2CO_3 (d) 0.1 M EDTA, pH 9 and (e) 0.1 M
Na_2CO_3 pH 12.

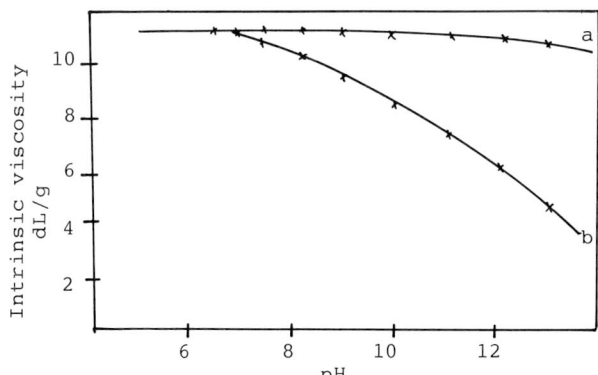

Fig. 2 : Intrinsic viscosity as a function of pH during extraction, (a) seaweed pretreated with
HCHO, extracted in 0.1 M EDTA, (b) extracted
in 0.1 M EDTA. Extraction time was 6 hours.

Table 3 : EFFECT OF FORMALDEHYDE PRE-TREATMENT ON INTRINSIC
          VISCOSITY (ALGINATE FROM SARGASSUM sp.1)

Time of soaking the weeds in formaldehyde before extracting with 0.1 M Na_2CO_3	Intrinsic viscosity, mL/g	
	1% HCHO	10% HCHO
0 hours	675	675
12 hours	742	862
24 hours	820	981
36 hours	845	1025

Results in Fig. 1 demonstrate that the reduction in $[\eta]$ after achieving a maximum $[\eta]$ in 0.1 M Na_2CO_3 is not entirely caused by alkaline degradation but may be caused by phenolic compounds present in the brown seaweeds. The phenolic compounds may contribute to oxidative-reductive depolymerisation. Sargassum sp. was extracted at different pH using di-sodium EDTA as calcium ion sequestrant and the effect on $[\eta]$ is given in Fig. 2. With the normal extraction procedure the $[\eta]$ decreases with increasing pH, but after treating the weeds with formaldehyde, the $[\eta]$ is almost constant up to pH 12, then there was a slight reduction in $[\eta]$ probably due to alkaline degradation.

Effect of M/G ratio on Mark-Houwink parameters.
    Table 4 gives the effect of M/G ratio on the Mark-Houwink parameters. It clearly shows that the 'a' values decrease with increasing ionic strength. The M/G ratio has a certain effect on the 'a' values. High guluronate samples such as from Sargassum sp.2 (M/G ratio 0.57) have higher values of 'a' compared to high mannuronate samples such as from Fucus sp. (M/G ratio of 2.7).
    The values of 'a' in the Mark-Houwink equation ($[\eta]$ = KM^a) is usually taken as an indication of the molecular stiffness or persistence length. According to the most commonly applied statistical and hydrodynamical treatment of polymers (8) a value of 'a' between 0.5 - 0.8 indicate a randomly coiled molecule, values between 1.0 - 1.2 indicates a stiff coil while value of 'a' equal to 1.8 indicate a rigid rod. Intermediate types of behaviour are possible; thus a polymer may show partial free flowing properties or have a configuration which is somewhere between that of flexible coil and a stiff rod. According to hydrodynamic theories, a high value of 'a' obtained at infinite ionic strength indicates that the uncharged molecules is very extended in aqueous solution.
    The result shows that the guluronate rich fractions are stiffer than high mannuronate fractions because low M/G ratio samples give higher 'a'values compared to high mannuronate samples. This might be the result of either stronger carboxylate-carboxylate repulsion between adjacent α -linked guluronic acid residues or the decreased flexibility of this glycosidic linkage due to more extensive steric restrictions than in the case of β -linked mannuronic acid.
    In Table 4 the intrinsic viscosities are expressed in dL/g.

Table 4 : EFFECT OF M/G RATIO ON MARK-HOUWINK PARAMETERS.

Concentration of NaCl (M)	High guluronate samples	
0.05	$4.15 \times 10^{-5} \bar{M}_w^{1.21}$	$3.72 \times 10^{-5} \bar{M}_w^{1.23}$
0.10	$6.39 \times 10^{-5} \bar{M}_w^{0.91}$	$5.24 \times 10^{-5} \bar{M}_w^{0.93}$
0.20	$11.6 \times 10^{-5} \bar{M}_w^{0.86}$	$8.04 \times 10^{-5} \bar{M}_w^{0.88}$
0.50	$13.22 \times 10^{-5} \bar{M}_w^{0.84}$	$10.57 \times 10^{-5} \bar{M}_w^{0.85}$
∞	$15.53 \times 10^{-5} \bar{M}_w^{0.82}$	$13.91 \times 10^{-5} \bar{M}_w^{0.82}$
Samples	Laminaria hyperborea	Sargassum sp.2
M/G ratio	0.60	0.57

Concentration of NaCl (M)	High mannuronate samples	
0.05	$6.11 \times 10^{-5} \bar{M}_w^{1.15}$	$6.92 \times 10^{-5} \bar{M}_w^{1.10}$
0.10	$9.27 \times 10^{-5} \bar{M}_w^{1.08}$	$10.55 \times 10^{-5} \bar{M}_w^{1.00}$
0.20	$15.10 \times 10^{-5} \bar{M}_w^{0.84}$	$16.27 \times 10^{-5} \bar{M}_w^{0.82}$
0.50	$17.33 \times 10^{-5} \bar{M}_w^{0.82}$	$18.14 \times 10^{-5} \bar{M}_w^{0.79}$
∞	$18.95 \times 10^{-5} \bar{M}_w^{0.79}$	$19.06 \times 10^{-5} \bar{M}_w^{0.75}$
Samples	Ascophyllum nodosum	Fucus spp.
M/G ratio	1.52	2.70

CONCLUSIONS

Extraction of weeds by using EDTA and CDTA can result in a higher percentage of alginate and higher viscosity. Prolonged extraction in sodium carbonate solution can cause significant decreases in [η] and can be overcome by pre-treating the weeds with formaldehyde. Formaldehyde will react with phenolic compounds usually present in most brown seaweeds and form insoluble products and the depolymerisation by phenolic compounds on alginate can be minimised.

The Mark-Houwink parameters obtained in this study clearly shows that the high guluronate samples are stiffer than high mannuronate samples.

ACKNOWLEDGEMENTS

Gratitude is extended to The British Council for providing an air ticket for Mr. Fasihuddin B. Ahmad to attend this conference and to The North East Wales Institute and The Committee for the Gums and Stabilisers Conference for their financial assistance.

REFERENCES

1. Chapman, V.J. and Chapman, D.J. (1980), Seaweeds and their Users, Third Edition, pp 194-196, Chapman and Hall, London & New York.
2. Haug, A. and Larsen, B. (1962), Acta Chem. Scand. 16, 1908-1918.
3. Haug, A., Larsen, B. and Smidsrod, O. (1967), Acta Chem. Scand. 21, 691-704.
4. Haug, A., Larsen, B. and Smidsrod, O. (1974), Carbohydrate Res. 32. 217-225.
5. Haug, A., Myklestad, S., Larsen, B. and Smidsrod, O. (1967), Acta Chem. Scand. 21, 768-772.
6. Dubois, M., Gilles, K.A., Hamilton, J.K., Ribers, P.A. and Smith, F. (1956), Anal. Chem. 28, 350-360.
7. Wedlock, D.J., Fasihuddin, B.A. and Phillips, G.O. (1986), Int. J. Biol. Macromol. 8, 57-61.
8. Mitchell, J.R. (1979) in Polysaccharides in Food, (Eds. Blanshard, J.M.V. and Mitchell, J.R.), pp 55-71, Butterworths, London.

Kinetics of conformational ordering of gelatin and related oligopeptides

D.M.Goodall, S.J.McBurney, T.A.Tancock and P.D.Bailey

Chemistry Department, University of York, York YO1 5DD, UK

ABSTRACT

A simple mechanism of a reversible transition between random coil and a triple helix has been used to model the kinetics of thermally-induced renaturation and denaturation of rat-tail tendon gelatin, a collagen-related oligopeptide (Pro-Pro-Gly)$_n$, and the collagen-derived peptide fragment α1-CB2. Initial results are reported on the kinetics of ordering of α-pig skin gelatin induced by a salt dilution jump.

INTRODUCTION

Although there have been many studies of the conformational ordering of gelatin, using quenching to initiate renaturation, the detailed mechanism is not fully understood(1). A variety of inter-mediates have been postulated(2,3), both parallel(4) and sequential(5) mechanisms have been proposed, and normally no account has been taken of the kinetic influence of the reverse reaction. Synthetic oligopeptides(6,7) and protein fragments(7,8) have been used as model compounds for collagen renaturation and denaturation. In this study we have tested a simple mechanistic scheme to simulate reaction progress curves for gelatin, collagen fragments and synthetic oligopeptides, drawing analogies with conformational transitions in related biopolymer systems(9).

Conformational transitions in gelatin may be driven by changing the salt concentration as well as by changing the temperature, since the transition midpoint temperature, T_m, is a function of salt concentration. We shall report preliminary work on the kinetics of conformational ordering of α-pig skin gelatin induced by a salt dilution jump.

COMPUTER SIMULATIONS

The aim of this work was to test a very simple mechanism for the renaturation and denaturation of collagen-related molecules. The kinetic scheme used was that of three coils (C) in equilibrium with a triple helix (TH), with a third order renaturation rate constant, k_3, and a first order denaturation rate constant, k_{-1}.

$$3C \underset{k_{-1}}{\overset{k_3}{\rightleftharpoons}} TH \qquad\qquad (1)$$

Such a scheme has previously been used with collagen-related oligo-peptides (6) and collagen-derived peptide fragments (8). Reaction was induced by rapid heating or cooling perturbations (upward or downward T jumps), and relaxation followed by optical rotation. Because an analytical solution cannot be obtained for reaction (1), kinetic treatments are normally restricted to initial rate studies (4a,8) or small perturbation relaxation analysis (7). We have followed Piez and Sherman (8) in using computer simulation. The simple mechanism (1) was fitted to literature data using the program Simula (Shell Thornton Research), with k_3 and k_{-1} as variables and known optical rotation for the helix and coil forms.

$(Pro-Pro-Gly)_{15}$

Oligopeptides in the family $(Pro-Pro-Gly)_n$ (n = 10-15) have been investigated kinetically by Sutoh and Noda (6). Their experimental data was presented in plots of log $(k_f c_o^2)$ versus $1/T$, where k_f is the third order renaturation rate constant and c_o the initial peptide concentration. We have chosen to replot as $k_f c_o^2$ versus T in Figure 1, by analogy with previous work on polysaccharides (9) and polynucleotides (10,11).

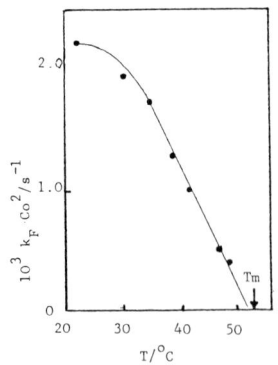

Figure 1. Rate constant for helix formation of $(Pro-Pro-Gly)_{15}$ as a function of temperature. Experimental data from Sutoh and Noda (6).

It is evident that for $(Pro-Pro-Gly)_{15}$ the forward rate extrapolates to zero at 52° C, which is within experimental error equal to the transition midpoint $T_m = 53^\circ$ C established from equilibrium studies(12).

The behaviour of Figure 1 is similar to that of other biopolymers with regularly repeating residue structures (e.g. poly AU (10), iota and kappa carrageenan (9)). Kinetics are interpreted in terms of slow nucleation followed by steady state growth. At T_m there is no driving force for propagation of the boundary between helix and coil regions, so the rate drops to zero.

For lower oligomers the kinetic T_m appears to be somewhat higher than the equilibrium T_m. It would be of interest to study a wider range of oligomers to establish the effect on k_f of chain length variation. For kappa carrageenan, the kinetics of conformational ordering are the same both for high chain length and hydrolysed samples(13).

α1-CB2

Piez and Sherman (8) have made a detailed study of the helix-coil transition in the 36-residue peptide α1-CB2 obtained from cyanogen bromide digests of α1-chain collagen. The equilibrium position and initial rate of renaturation varied with peptide concentration as expected for reaction (1).

Typical simulations of their reaction progress curves are shown in Figure 2, taking data for 0.96 mM α1-CB2 at 10^0 C. Rate constants found to provide the best fit for renaturation were $k_3 = 60$ dm^6 mol^{-2} s^{-1}, $k_{-1} = 1.8 \times 10^{-5}$ s^{-1}, and for denaturation $k_3 = 100$ dm^6 mol^{-2} s^{-1}, $k^{-1} = 2.4 \times 10^{-5}$ s^{-1}.

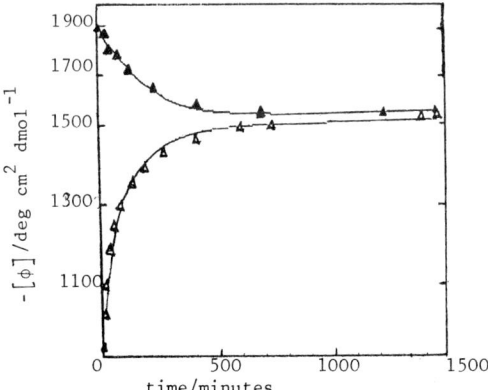

Figure 2.   Simulation of relaxations following temperature jumps on the collagen fragment α1—CB2, fitted to experimental data of Piez and Sherman (8) : T jumps 5 to 10^0 C (▲) and 27 to 10^0 C (▲).

Table 1. RATE CONSTANTS FROM SIMULATIONS OF RENATURATION AND
          DENATURATION KINETICS OF COLLAGEN-RELATED MOLECULES
          USING A SIMPLE RANDOM COIL TO TRIPLE HELIX MODEL

Material	$T/^{\circ}C$	$k_3/dm^6\ mol^{-2}\ s^{-1}$	k_{-1}/s^{-1}
α1-CB2	10	$(1.0 \pm 0.6) \times 10^2$	$(1.8 \pm 0.5) \times 10^{-5}$
	5	$(1.9 \pm 1.2) \times 10^2$	$(1.0 \pm 0.1) \times 10^{-6}$
$(Pro-Pro-Gly)_{15}$	15	1.7×10^3	$< 1 \times 10^{-7}$
Rat-tail tendon collagen	40	5.8×10^5	6.8×10^{-4}
	15	5.0×10^4	3.8×10^{-6}

Rate constants from the simulations are collected in Table 1. All
values of k_3 are per mole of material in the triple-helix form,
with the rate equation as in (8)

$$\frac{d\xi}{dt} = \frac{d[TH]}{dt} = \frac{1}{3}\frac{d[C]}{dt} = k_3[C]^3 - k_{-1}[TH] \qquad (2)$$

$d\xi/dt$ is the rate of reaction defined with correct regard for the
stoichiometric coefficients (14), in contrast to reference (6) where
$k_f = 3k_3$. Rate constants per mole of helix residue, k_3', may
be obtained using

$$k_3' = k_3/n^2 \qquad (3)$$

where n is the number of residues per chain.

For α1-CB2 data is presented as mean and standard deviation by
averaging best-fit rate constants from simulations at one temperature
covering reactions starting from both extremes of order and disorder,
and variations in peptide concentration. It should be noted that

- The simple reaction scheme (1) gives a fair qualitative fit

- Overall reaction progress curves can be simulated, whereas rate
  constants obtained by Piez and Sherman (8) gave a good account
  only of initial rates of renaturation.

- the temperature dependence of k_{-1} is far greater than that
  of k_3

Because k_3 from initial rates (8) is greater than k_3 from the overall reaction curve, and our simulations suggest a systematic trend for k_3 to be higher from denaturation than from renaturation, it is evident that reaction (1) is a simplification of the mechanism of the order-disorder transition for α1-CB2. As has previously been pointed out (8), this could be due to heterogeneity in strand alignment.

Rat-tail tendon collagen

   The seminal paper by Flory and Weaver (2) introduced the widely used equation for the variation of renaturation rate with degree of undercooling below the collagen melting temperature. Their mechanistic scheme was however unrealistic, insofar as it suggested one-third order denaturation kinetics to accompany first order renaturation. Figure 3 shows that the reaction scheme (1) can fit Flory and Weaver's original data, giving rate constants summarised in Table 1.

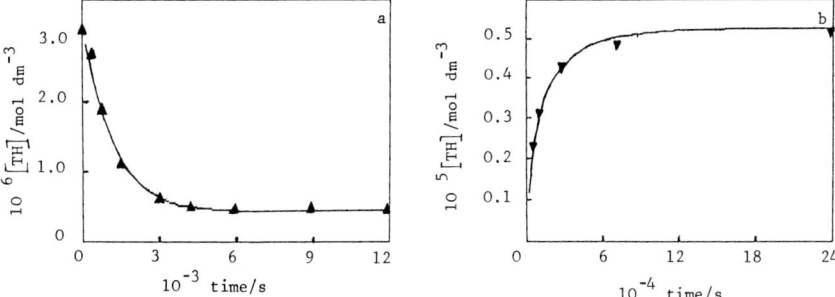

Figure 3. Simulation of relaxations following temperature jumps on rat tail collagen, fitted to experimental data of Flory and Weaver (2) : denaturation at 40° C ($\blacktriangle$) and renaturation at 10° C ($\blacktriangledown$).

   It can be shown that the temperature dependence of k_3 is in fair agreement with the behaviour expected from a renaturation rate constant which incorporates nucleation and growth contributions with known enthalpy change for growth of a helix residue.

SALT-DILUTION JUMPS

   The aim of this part of the work was to study the kinetics of renaturation of a gelatin sample following a salt dilution jump. The transition midpoint temperature, T_m, of the conformational transition has been shown to vary with the nature and concentration of added salt, with Ca^{2+} and SCN^- ions having the greatest effect (15). On decreasing the salt concentration both T_m and the amplitude of the optical rotation transition,

$\Delta[\alpha]_T$, were found to increase. The temperature perturbation needed
to induce a thermally-driven transition normally requires several
seconds, whereas the salt concentration can be changed within
milliseconds using a stopped-flow apparatus (16). Thus any rapid
structural changes masked in the dead time of a quenching experiment
might show up in a salt-dilution jump experiment.

α-Gelatin was prepared from a commercial sample of acid-processed
pig skin gelatin by ammonium sulphate precipitation. Samples left in
$0.7M$ $CaCl_2$ for 24 hours at 10° C were diluted twofold in the
polarimetric stopped-flow apparatus (16). No evidence was obtained
to suggest that a fast optical rotation change had been missed in
previous polarimetric studies. Figure 4 shows the reaction progress
curve on a longer timescale, 1–10,000 minutes, followed on a
polarimeter (Perkin Elmer 141) following a ten-fold dilution jump
from a starting concentration $0.7M$ $CaCl_2$, 1 mg ml^{-1} α gelatin.
The biphasic nature of the reaction accords with previous results on
α-pig skin gelatin following thermal quenching (17). The initial
phase is likely to correspond to chain association, and the secondary
phase to strand alignment in annealing or aggregation.

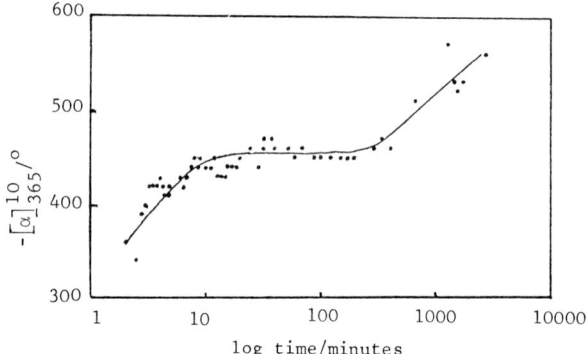

Figure 4. Optical rotation change following a salt dilution jump of
α pig skin gelatin at 10° C. Initial concentrations
$0.7M$ $CaCl_2$, 1 mg ml^{-1} gelatin ; final concentrations
$0.07M$ $CaCl_2$, 0.1 mg ml^{-1} gelatin.

CONCLUSION

Renaturation of gelatin has been studied kinetically using a salt
dilution jump and gives results analogous to those obtained by thermal
quenching. A simple mechanistic scheme of a random coil to triple-
helix transition has been shown using computer-modelling to give a
reasonable fit to reaction progress curves for both natural and
synthetic collagen-related peptides as well as for the native material.

ACKNOWLEDGEMENT We thank the A.F.R.C. and Rowntree Mackintosh plc for
support through the Cooperative Studentship Award Scheme to S.J.M. and
T.A.T.

REFERENCES

1. Godard, P., Biebuyck, J.J., Barriat, P.A., Naveau, H. and Mercier, J.P. (1980) Makromol. Chem., 181, 2009-2018.
2. Flory, P.J. and Weaver, E.S. (1960) J. Amer. Chem. Soc., 82, 4518-25.
3. Finer, E.G., Franks, F., Philips, M.C. and Suggett, A. (1975) Biopolymers, 14, 1995-2005. Eagland, D., Pilling, G., Suggett, A. and Wheeler, R.G. (1974) Faraday Disc. Chem. Soc., 57, 181.
4. a) Harrington, W.F. and Rao, N.V. (1970) Biochemistry, 9, 3714-3724.
   b) Harrington, W.F. and Karr, G.M. (1970) Biochemistry, 9, 3725-3733.
   c) Harrington, W.F. and Hauschka, P.V. (1970) Biochemistry, 9, 3734-3763.
5. Yuan, L. and Veis, A. (1973) Biophys. Chem., 1, 117-124.
6. Sutoh, K. and Noda, H. (1974) Biopolymers, 13, 2477-2488.
7. Weidner, H., Engel, J. and Fietzik, P. (1974) Proc. 2nd Int. Symp. Poly(amino acids), Polypeptides and Proteins, Rehovot, pp. 419-435.
8. Piez, K.A. and Sherman, M.R. (1970) Biochemistry, 9, 4134-4140.
9. Goodall, D.M. and Norton, I.T. (1987) Acc. Chem. Res., 20, 59-65.
10. Blake, R.D., Klotz, L.C. and Fresco, J.R. (1968) J. Amer. Chem. Soc., 90, 3556-3562.
11. Lee, C.H. and Wetmur, J.G. (1972) Biopolymers, 11, 549-561.
12. Sutoh, K. and Noda, H. (1974) Biopolymers, 13, 2391-2404.
13. Austen, K.R.J., Goodall, D.M., Lloyd, D.K. and Norton, I.T. manuscript in preparation.
14. Atkins, P.W., Physical Chemistry, 3rd edn., Oxford University Press, Oxford, U.K.
15. von Hippel, P.H. and Wong, K-Y, (1962) Biochemistry, 1, 664-674.
16. Goodall, D.M. and Lloyd, D.K. (1986) Gums and Stabilizers for the food industry 3, (eds. Phillips, G.O., Wedlock, D.J. and Williams, P.A.) pp. 497-500.
17. Veis, A. and Schnell, J. (1967) Symposium on Fibrous Proteins Australia 1967, (ed. Crewther, W.G.) pp. 193-204.

Melting behaviour of gelatin gels: origin and control

J.P.Busnel, S.M.Clegg[+] and E.R.Morris[+]

Laboratoire de Physico-Chemie Macromoléculaire, Université du Maine, Route de Laval, 72017 Le Mans Cedex, France
[+]Department of Food Research and Technology, Cranfield Institute of Technology, Silsoe College, Silsoe, Bedford MK45 4DT, UK

ABSTRACT

Gelatin samples held at moderate temperatures (15 - 30°C) before quenching to low temperature (5°C) give stronger gels than those cooled directly to 5°C and show progressive development of a gel fraction of enhanced thermal stability, leading eventually to bimodal melting. We attribute this behaviour to selective formation at the holding temperature of helices longer than those formed at lower temperature, and present evidence from 'initial slope' kinetics and Monte Carlo simulation to suggest that the higher-melting species are predominantly functional, inter-molecular helices rather than 'wasted' intramolecular double-hairpin structures.

INTRODUCTION

It is well known in the Industry that the thermal history of gelatin gels can have a profound effect on their physical properties (1,2). In particular, gels held for some time at a moderate temperature before final quenching to a lower temperature are more rigid than those quenched directly to the lower temperature (3,4). We now report a detailed study of this behaviour, and offer an interpretation that may have much wider implications for understanding and controlling the formation and melting of gelatin gels.

MATERIALS AND METHODS

The gelatin used was a first-extract limed ossein sample kindly supplied by Rousselot. Solutions were prepared at 45°C, and sodium azide (0.02%) was incorporated to inhibit bacteria. Differential scanning calorimetry (DSC) studies were carried out on a Setaram microcalorimeter using a sample volume of ∿1ml and a scan rate of 0.1 degrees per minute. Optical rotation was measured at 365 nm on a Perkin-Elmer 241 polarimeter using thermostatted cells of pathlength 1 cm or 10 cm as appropriate. The fraction of residues in the ordered, helical conformation (f_H) was calculated using the observed optical rotation of the disordered sol state (above 40°C) and the literative value for fully-ordered native collagen (5). Gel rigidity (storage

modulus, G') was measured at a frequency of 0.5Hz on a Sangamo
Viscoelastic Analyser using cone and plate geometry of cone
angle 2 degrees and diameter 5 cm.

THERMAL HISTORY AND GEL MELTING

   Figure 1 shows DSC melting endotherms for the same gelatin
solution gelled under different thermal conditions.   For each
family of curves illustrated, the 'reference' state is a gel
prepared by direct quenching of the hot (45°C) solution to a
fixed temperature of 5°C for 16 hours, by which time further
conformational ordering, as judged by optical rotation, is
negligible over the time period of the DSC measurement (∿7 hours).
The other traces in each family show the effect of holding the
sample for varying times at a higher temperature (T_O) before
again cooling to 5°C for 16h.   In each case, as the length of
time at the holding temperature is increased the melting
endotherms show progressive development of a higher-melting peak
and eventually assume a distinct bimodal character, with the
families of curves for each holding temperature crossing at a
single isosbestic point.   In all cases, however, the overall
enthalpy change is essentially constant (≈ 3.2 kJ mol^{-1}) and the
net helix fraction at 5°C is also virtually unchanged (at ∿75%).

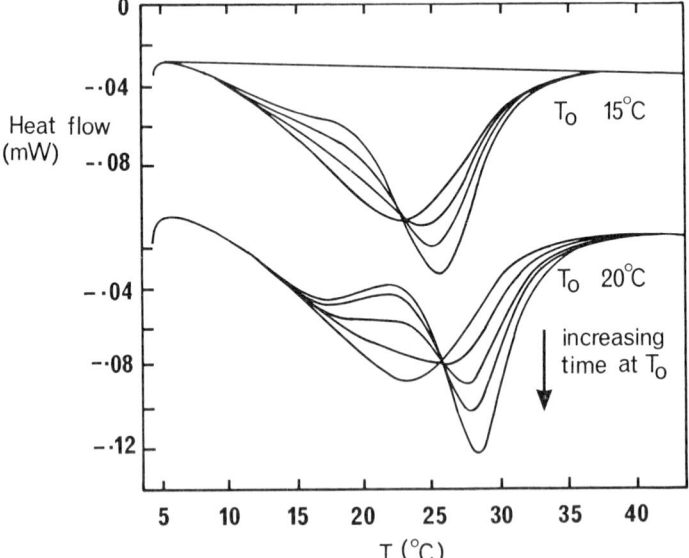

Figure 1.   DSC melting endotherms for gels (20 mg ml^{-1}) held
(upper curves) at 15°C for 0, 1.5, 7 and 24 hours and (lower
curves) at 20°C for 0, 2, 7, 24 and 72 hours before final quenching for 16 hours at 5°C in all cases.   Similar families of curves
were obtained using holding temperatures of 25°C and 30°C.

As illustrated in Fig. 2a, however, the rigidity (G') of gels held at higher temperature before final aging at 5°C is higher than that of the same sample cooled directly to 5°C, and remains higher throughout the melting process.   In the case of samples held for sufficiently long at higher temperature to give obvious bimodal melting behaviour by DSC, the temperature-course of loss of rigidity on heating (dG'/dT) also shows bimodal character, (Fig. 2b) with peaks that correspond closely to the DSC minima.

To investigate this behaviour further we have used optical rotation (α) to monitor the temperature-course of helix melting (i.e. local conformational change).   As illustrated in Fig. 3 the rate of decrease in helix fraction with increasing temperature (dα/dT) is closely superimposable on the overall melting profile from DSC, both for samples quenched directly to 5°C and for those first held at higher temperature (T_O) to give bimodal melting.   Figure 3 also shows the temperature-course of helix-melting when the sample is held for the same length of time at T_O, but then melted directly from this temperature without first cooling to 5°C.   Under these conditions the temperature-depend-ence of dα/dT is unimodal, joins the bimodal melting-curve at the isosbestic point illustrated in Fig. 1, and follows the same temperature-course thereafter.

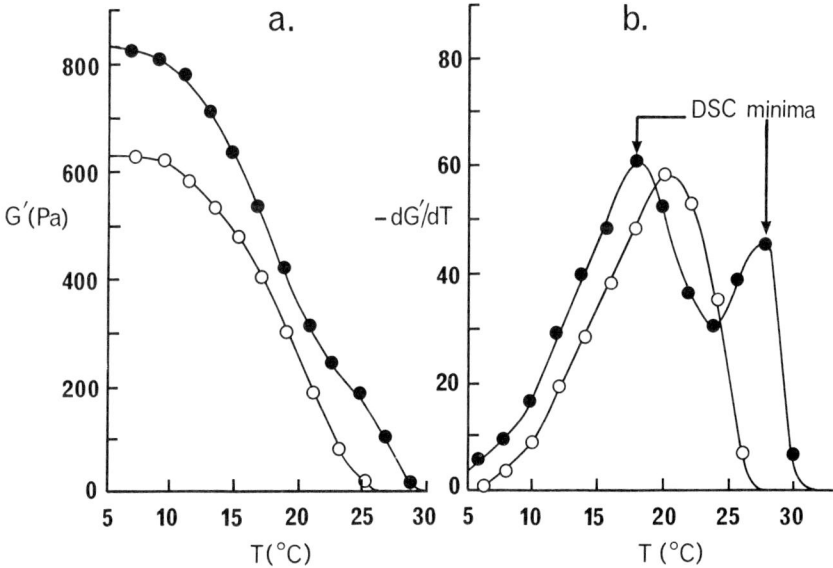

Figure 2.   Temperature-dependence of rigidity (G', 0.5 Hz) for gels (20 mg ml^{-1}) quenched to 5°C for 16 hours, either directly (o) or after holding for 24 hours at 20°C (●);   the final helix fraction is virtually identical in both cases ($f_H \cong 0.75$). a) absolute values of G'   b) rate of change of G' with increasing temperature (dG'/dT).

It therefore appears that the bimodal melting behaviour
characterised independently by DSC, optical rotation and gel
rigidity has its origin in the selective formation at T_O of a
population of helices of greater thermal stability than those
formed at lower temperature (5°C), which then remains virtually
unchanged when the temperature is subsequently decreased.
Quantitatively, the fractional amount by which the mid-point
melting temperature (T_m) exceeds the temperature at which the
helices are formed (T_O) decreases linearly with increasing T_O
(Fig. 4a) and extrapolates to zero at $T_O \cong 36$°C, which probably
corresponds to the melting point of the parent collagen (6). At
each value of T_O the helix fraction increases linearly with the
logarithm of holding time throughout the timescale of likely
practical importance, as illustrated in Fig. 4b. Operationally,
therefore, we can predict both the amount (f_H) and thermal
stability (T_m) of the high-melting species formed after a known
time at a specific holding temperature (T_O). The origin of the
enhanced stability, however, is less clear-cut.

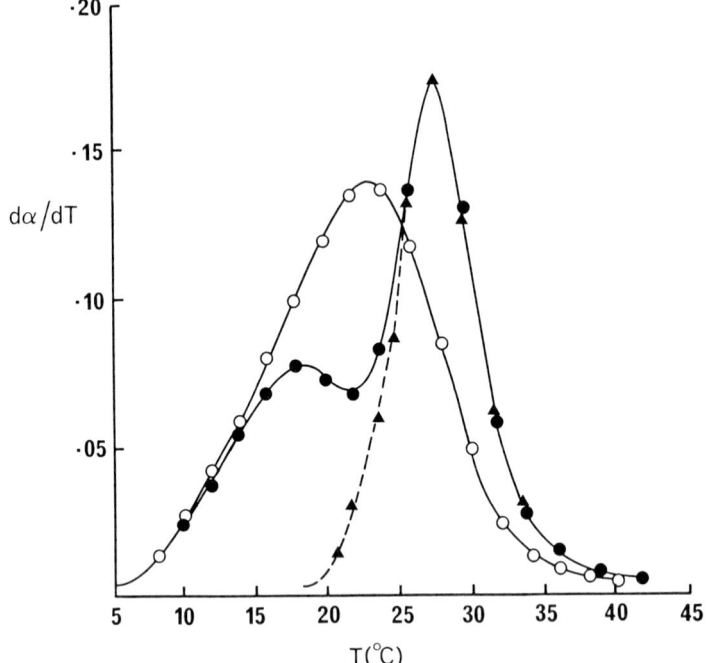

Figure 3. Thermal melting of gels (20 mg ml^{-1}) quenched to 5°C
for 16 hours directly (o) or after holding for 24 hours at 20°C
(●); symbols show local conformational change by optical rotation
($d\alpha/dT$) and solid lines show overall enthalpy change (Fig. 1)
by DSC. Also shown (-▲-) is the optical rotation change for the
same sample melted directly after holding at 20°C for 24 hours.

The close superposition (Fig. 3) of the temperature-course of the enthalpy change monitored by DSC and the conformational change by optical rotation argues against stabilisation of helices by aggregation unless the enthalpy change associated with the aggregation was undetectably small. A more likely interpretation is that the minimum critical sequence length for helix formation increases with increasing temperature, so that the helices formed at the higher temperature (T_O) are longer than those formed on subsequent cooling and therefore melt at higher temperature (7). However, since holding at higher temperature has little if any effect on the final overall helix content, longer helices would of necessity imply fewer helices and therefore, at first sight, a weaker network rather than the more rigid structure observed experimentally (e.g. Fig. 2). We now present evidence from other lines of investigation which offers a unifying interpretation of these apparently conflicting observations.

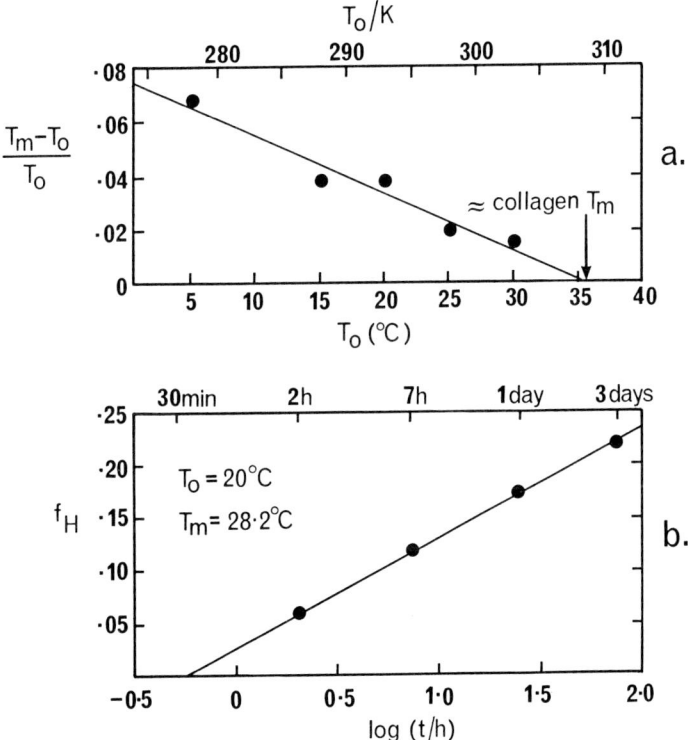

Figure 4. (a) Dependence of mid-point melting temperature(T_m) on setting temperature (T_O).
(b) Increase in helix fraction (f_H) with time (t) at fixed temperature (T_O, illustrated for T_O = 20°C).

'INITIAL SLOPE' KINETICS

    There have been numerous studies of the dynamics of conform-
ational ordering in collagen/gelatin systems, and many different
and often conflicting, mechanisms have been proposed. (8).   It
is well established, however, that the kinetics of the sustained,
gradual increase in helix fraction (and gel strength) at long
times after quenching are particularly complex.   In the very
early stages of reaction, however, before the fraction of
disordered sequences has decreased significantly from 100%,
helix content increases linearly with time, and thus the initial
slope of reaction progress curves such as that shown in Fig. 5
provides a direct measure of the rate of helix formation under
these simple conditions.   By using known changes in UV absorption
to monitor helix formation at low concentrations and optical
rotation at higher concentrations (9), we have determined initial
reaction rates over a very wide range of gelatin concentrations
$(0.05 - 32$ mg ml^{-1}).

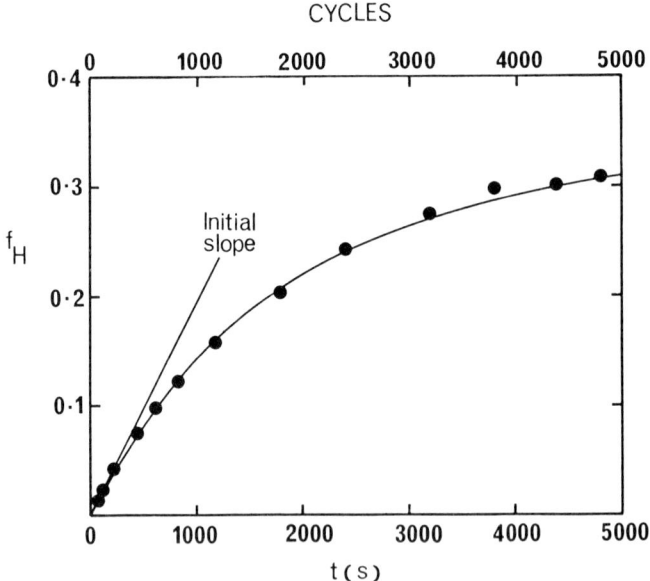

Figure 5.   Comparison of observed reaction progress curve (solid
line) for helix formation in gelatin (2 mg ml^{-1}; 10°C) with
values of helix fraction from the Monte-Carlo simulation
described in the text, using A = 5, B = 11.5 and p(inter)= 0.5
(i.e. equal probability of inter- and intra-molecular reaction).
Determination of rate of conformational ordering in the early
stages of reaction using the initial slope (S) of the reaction
progress curve is also illustrated.

At concentrations below ~ 0.5 mg ml^{-1} the initial slope
($S = df_H/dt$) remains constant, as shown in Fig. 6; thus since
$f_H = [\text{helix}]/c$, the absolute rate of helix formation is directly
proportional to concentration (c), and so the reaction displays
first-order kinetics, consistent with an intramolecular process.
At higher concentrations, however, the value of S increases
substantially (by almost an order of magnitude over the
concentration range studied), indicating the onset of a second
mechanism of conformational ordering. To examine the kinetics
of this second process we have subtracted the constant value of
S at low concentrations (S_1) from the experimental values of S
at higher concentrations to obtain the contribution ($S_2 = S-S_1$)
from the second process. As shown in Fig. 6, the slope of log
S_2 vs log c is very close to unity (~ 1.02) and thus the
additional process at higher concentration is second-order,
consistent with the involvement of two chains in the rate-limiting step to triple-helix formation.

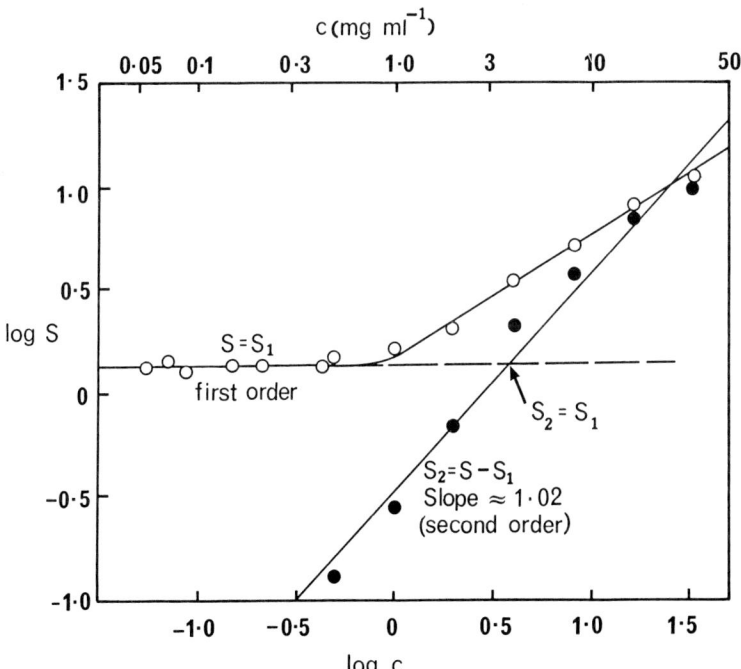

Figure 6. Concentration-dependence of 'initial slope' kinetics
for renaturation of gelatin (10°C). Open symbols: observed slope
($S = df_H/dt$); filled symbols: $S_2 = S - S_1$ where S_1 is the
constant value of S at low concentration.

PROPOSED MODEL FOR GELATIN RENATURATION

To interpret this behaviour we propose the following set of 'minimum assumptions' illustrated schematically in Fig. 7.

1) Helix formation is initiated at a β-bend or similar metastable 'kink' which brings two chain segments into close spatial proximity.

2) These may then form a helix nucleus by collision with a third strand.

3) At low concentration this is most likely to be a segment of the same chain while at higher concentration an intermolecular collision is more probable.

4) Nucleation is rate-limiting, so that the intramolecular process shows first-order kinetics and the intermolecular process is second-order.

5) Subsequent helix growth is rapid, and proceeds as far as is geometrically possible (i.e. until reaching the end of one of the participating strands or until, as illustrated in Fig. 7, an intramolecular loop is fully utilised).

Qualitatively this model offers a simple explanation of the higher rigidity of gels whose thermal history has included a holding period at higher temperature. Since intramolecular helices involve three strands from the same chain their maximum possible length will be one third of the chainlength while for intermolecular helices, where only two of the participating strands are from the same chain, the maximum length will be half the chainlength. We would therefore expect the requirement for progressively longer helices with increasing temperature (6) to favour formation of intermolecular helices (which will of course contribute to the strength of the gel network) in preference to shorter, intramolecular structures (which are 'wasted').

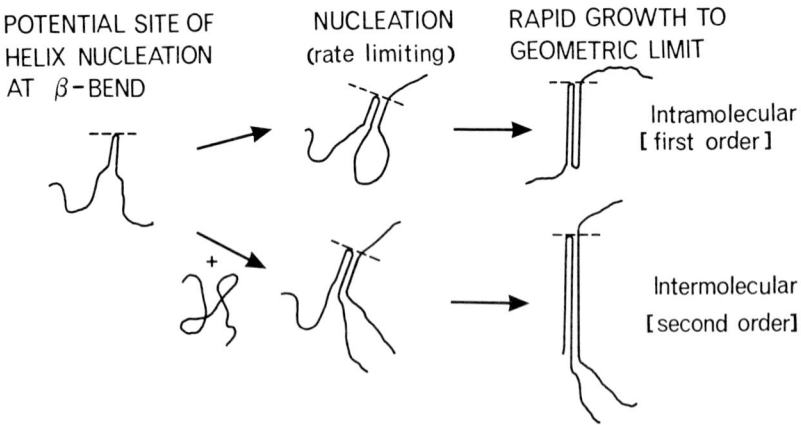

Figure 7. Schematic illustration of proposed kinetic model.

MONTE-CARLO SIMULATION

As a first step in making this argument more quantitative we
have carried out a preliminary computer simulation of the renaturation process based on the model proposed above. Briefly, we
first generate a large population of chains; in the results
reported here we have used 1000 chains with a gaussian length-
distribution centred at 0.8 of the maximum length. The probability that helix formation is intermolecular rather than intramolecular [p(inter)] is chosen to simulate concentration. In
each cycle of the program a chain is selected at random (with
the probability of selection directly proportional to the chainlength). If it is disordered (as it of course would be at the
start of the simulation)then a point within it is chosen at
random to simulate the position of the β-bend shown in Fig. 7.
A second point is chosen at random, either within the same chain,
to simulate intramolecular nucleation, or within a second,
randomly selected disordered sequence, to simulate intermolecular
nucleation; the intermolecular option is chosen if p(inter)
exceeds a random number generated between 0 and 1. In either
case the second point is regarded as the position at which the
third strand joins the β-bend to form the helix nucleus (Fig. 7),
and the maximum potential helix length (L) is calculated. The
stability of the potential helix is then assessed by calculating
the probability of unwinding, p(un) = A exp(-BL), where A and B
have pre-selected values chosen to simulate temperature, as
discussed later. If p(un) exceeds a random number generated
between 0 and 1, nucleation is considered to have failed and the
program continues to the next cycle, leaving the chain(s) disordered. Otherwise the three participating strands of the helix
are stored as ordered sequences, and any residual disordered
loops or tails are considered as additional candidate sequences
for helix formation in later cycles of the program.
So far we have considered the case where the sequence selected
randomly at the beginning of the cycle is disordered. If,
however, it is already ordered (which becomes progressively more
likely as the simulation proceeds) then the stability test
described above is applied. If p(un) is less than the random
number generated the program passes on to the next cycle leaving
the helix intact; otherwise the chosen sequence and its two
helix partners are considered to have become disordered and are
combined with any contiguous disordered sequences in the same
chain(s). The number and length-distribution of inter- and
intramolecular helices may be printed out at any stage, and
iteration continued with the same values of A and B (to simulate
longer times at the same temperature) or with different values
(to simulate temperature change).
In the present preliminary studies a combination of values for
A and B were chosen by trial and error to give a reasonable
quantitative match (Fig. 5) to experimental reaction progress
curves at 10°C (A = 5; B = 11.5); A was then held constant and
B increased or decreased to simulate lower or higher temperatures
respectively (i.e. as B increases the probability that a helix of
a given length will unwind decreases). By comparison of the
helix fraction attained after a large number of iterations (e.g.
10,000) with experimental values at long times (where in both

cases the rate of change in f_H is several orders of magnitude
lower than the initial rate) we have chosen B = 75 to simulate
the final quench temperature in our DSC experiments (5°C), and
B = 15 to simulate a typical holding temperature (∿20°C).

Figure 8 shows the length-distribution of helical sequences
obtained after 0, 1000 and 5000 cycles at, A = 5, B = 15 followed
by iteration at A = 5, B = 75 until the overall helix fraction
had attained an essentially constant value ∿72% in all cases).
The results from this simulation parallel closely the experimental
observations by optical rotation and DSC (Figs. 1 and 3).   When
direct cooling to 5°C is simulated (iteration at B = 75 only) the
distribution is monodisperse and centred at low helix length
(i.e. low melting);   prior iteration at B = 15 (to simulate
holding at 20°C) shifts the populations of helix lengths to
higher values (i.e. higher melting) and eventually leads to a
bimodal distribution, with curves crossing at an isosbestic point
as observed experimentally.

Although the relative probabilities of inter- and intramolecular helix formation were set to be equal [p(inter) = 0.5]
the proportion of intermolecular helices after iteration at B = 15
(to simulate holding at 20°C) was about 75% of the total, with a
length-distribution centred at much higher values (∿32% of the
maximum chainlength, in comparison with ∿18% for intramolecular).

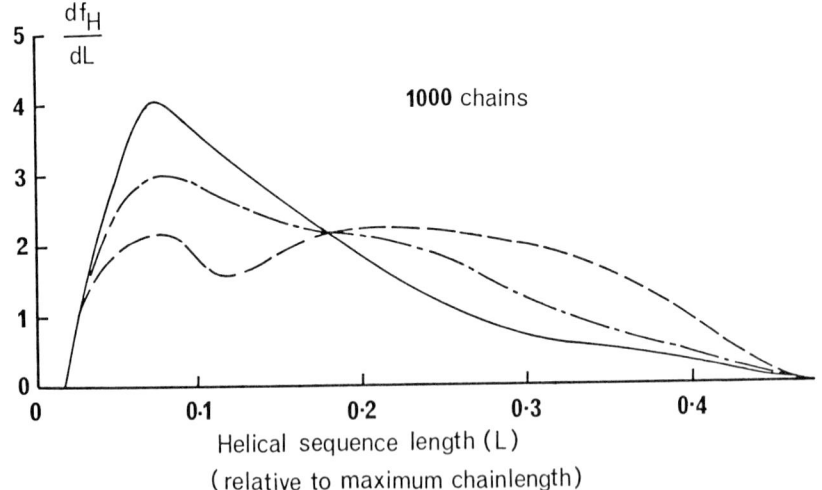

Figure 8.   Monte-Carlo simulation of gelatin renaturation.
Results are shown for the length-distribution of helical
sequences after 0 (——), 1000 (-·-) and 5000 (---) cycles at
A = 5, B = 15 followed by iteration to essentially constant helix
fraction (f_H ≅ 0.72 in all cases) at A = 5, B = 75, with p(inter)
set at 0.5 throughout.

On subsequent further iteration at B = 75 (to simulate
cooling to 5°C) this population of long (i.e. high-melting)
intermolecular helices remains virtually unchanged but is
augmented by additional shorter helices (inter- and intramolecular), leading to the bimodal length-distribution shown in
Fig. 8.

In summary, therefore, we propose that from simple geometric
and statistical considerations intermolecular triple helices
requiring a maximum of two strands from a single chain will have
a considerably higher average length and therefore a greater
probability of formation and survival, particularly at higher
temperatures, than intramolecular helices where a single chain
provides all three strands, and that this simple proposal
provides a unified interpretation of both the greater thermal
stability and higher rigidity of gels whose thermal history
includes a holding period at higher temperature.

ACKNOWLEDGEMENTS

We thank Drs. S.B. Ross-Murphy, I.T. Norton and D.M. Goodall
for helpful discussions, and the AFRC for the award of a Cooperative Research Studentship to S.M. Clegg.

REFERENCES

1. Stainsby, G. (1977) in The Science and Technology of Gelatin,
   (eds. Ward, A.G. and Courts, A.) pp 179-207. Academic Press,
   London, UK.
2. Ledward, D.A. (1986) in Functional Properties of Food Macromolecules, (eds. Mitchell, J.R. and Ledward, D.A.) pp 171-201.
   Elsevier, London, UK.
3. te Nijenhuis, K. (1981) Colloid Polym. Sci., 259, 522-535.
4. te Nijenhuis, K. (1981) Colloid Polym. Sci., 259, 1017-1026.
5. Djabourov, M. and Papon, P. (1983) Polymer, 24, 539-635.
6. Flory, P.J. and Weaver, E.S. (1960) J. Amer. Chem. Soc., 82,
   4518-4525.
7. Cantor, C.P. and Schimmel, P.R. (1980) Biophysical Chemistry,
   Vol. III, Chapter 23. W.H. Freeman, San Francisco, USA.
8. Goodall, D.M., McBurney, S.J., Tancock, T.A. and Bailey, P.D.
   (1987) This Volume, preceding paper.
9. Busnel, J.P., Morris, E.R. and Ross-Murphy, S.B.,
   in preparation.

Part 2

GELATION AND RHEOLOGICAL PROPERTIES

Gelation of ionic polysaccharides

Marguerite Rinaudo

Centre de Recherches sur les Macromolécules Végétales, CNRS, BP 68, 38402 Saint-Martin d'Hères Cedex, France

ABSTRACT

The paper gives some general information on non-covalent gels formed from ionic polysaccharides. The thermoreversible gels of K-carrageenan and gellan are described. Then ionic gels crosslinked by divalent counterions are discussed ; the examples of alginate but mainly of pectins are given.

INTRODUCTION

Polysaccharide gelation takes an important place in the food industry. There is a large originality of these polymers compared with the synthetic ones ; the native polysaccharides are often stereoregular at least on part of the chains. Then cooperative interactions may occur and by example physical gels may be formed ; some systems will be discussed in the following.

In that respect, it seems that two types of non-covalent gels are formed with biopolymers :

- thermoreversible gels formed in well defined thermodynamic conditions (temperature, ionic concentration...). It is the case of K-carrageenan, gellan... (but also of agarose or curdlan which are neutral polysaccharides...) ;

- ionic gels crosslinked by calcium counterions in direct relation with the chemical structure of the polymers ; this is the case for pectins with low degree of esterification or alginates.

The mechanisms of gelation and the role of the structure of the polymers will be developed.

THERMOREVERSIBLE GELS

Many systems are now recognized to form thermoreversible gels ; the sol-gel transition is, in each case, controlled by the thermodynamic properties of the systems. Increasing the Flory-Huggins parameter χ by increasing the ionic concentration or decreasing the temperature, leads to the phase transition. Usually, it must be considered that due to the stereoregularity of the polysaccharides, cooperative interchain interactions may be established leading to a stable network at least under given experimental conditions.

In addition, the sol-gel transition is often directly related to a conformational transition (coil state-helical conformation). This means that the thermodynamic behaviour must be described in the sol state by a model of a chain and in the gel state by a model of a rod-like molecule.
The examples of K-carrageenan and gellan gum will be discussed.

Gellan gum has been studied extensively in recent years; it is a heteropolysaccharide with a low charge parameter (one carboxylic group per four saccharide units) ; the chemical structure is recalled in Figure 1a (1,2). The polymer is acetylated on the C6 position of one of the D-glucose residues in the native polysaccharide. Deacetylation of the polymer considerably changes its properties in solution and its gelation ability (1).

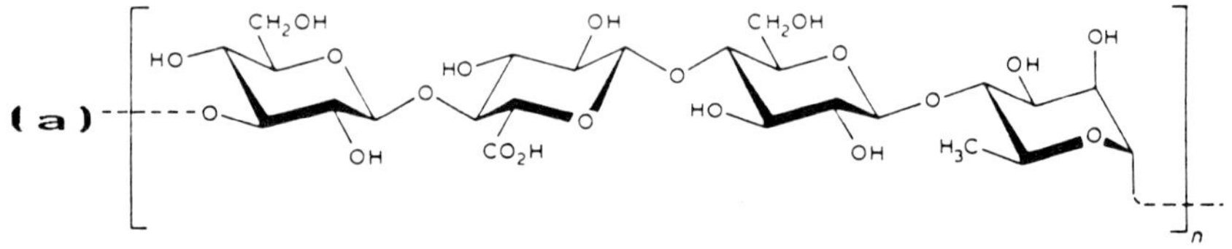

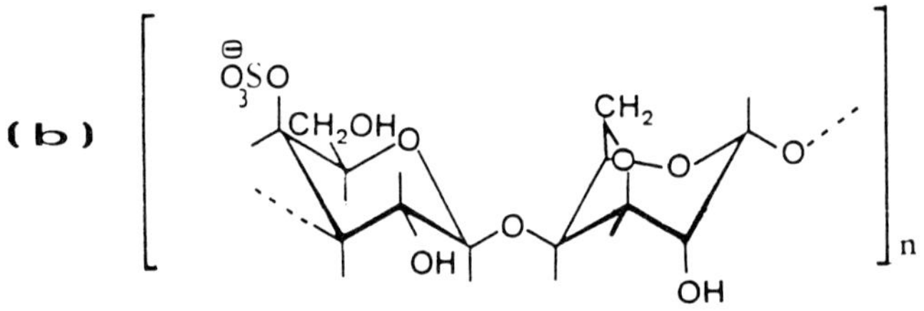

Figure 1. Chemical structure of gellan (a) and K-carrageenan (b).

The deacetylated form of   gellan gives a stiff and brittle gel in the presence of external salt or when the temperature decreases. It is clear that the deacetylation favours packing of the chains and the formation of the gel. A gel-sol transition in the presence of salts (NaCl for example) is associated with a conformational transition demonstrated from optical rotation. Until now, no definitive conclusion is given for the conformations on both sides of the transition. The results on the ordered conformation investigated in the solid state suggest the existence of a single contracted helix or an extended double helix (3,4). Both these structures have a charge density twice that of a single chain but presently no evidence of a change in the charge parameter during the conformational transition was obtained ;   the transition from double helix to single chain is not proven (5). But, the gel seems to be formed by aggregation of these stiff double helices ; it presents a large hysteresis between setting and melting temperatures at least in the conditions used by Moorhouse (1). In that respect, the behaviour presents many analogies to agarose or agarose-like polymers.

Kappa carrageenan has been studied extensively in our laboratory by C. Rochas (6-8). The chemical structure is given in Figure 1b. From thermodynamic measurements, it was demonstrated that helical dimers are formed ; they may then form aggregates which gel under well-defined conditions of temperature and salt concentration described by a phase diagram (Figure 2). The limitation of the different domains is dependent on the nature of the counterions, on the temperature and on the ionic concentration (taking into account the free counterions from the polyelectrolyte). The sol-gel transition always coincides with a conformational transition of the polymer from the coil state to a helical dimer conformation ; these two transitions are much dependent on the nature of the counterions with the most effective K^+, Rb^+, Cs^+ ; K^+ was proved to form ion pairs, decreasing the net charge of the polymer and its solubility and then to favour dimer formation and phase separation. The anhydrogalactose units present in kappa carrageenan but also in the iota carrageenan structure are essential for gelation.

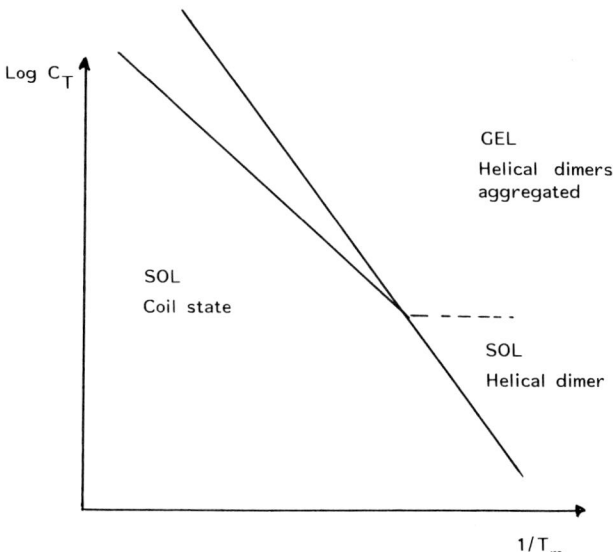

Figure 2. Schematic phase diagram for K-carrageenan. C_T is the total ionic concentration and T_m the temperature of conformational transition.

In contrast to covalent gels obtained by crosslinking of synthetic ionic polymers, when these polysaccharide gels are formed their degree of swelling is only slightly affected by an increase of external salt concentration or by addition of a non-solvent (9). The deswelling effect is less when the charge density is lower going in the order agarose < K-carrageenan < i-carrageenan.

Other systems based on ionic polysaccharides are now obtained when two different polymers are mixed. This was previously discussed by Morris (10,11). An example is the gelation obtained when xanthan gum, an anionic polymer, is mixed with guar gum (neutral polymer). Cooperative interactions are established preferentially when the xanthan is in the disordered conformation.

It may be predicted that many other thermoreversible gelling systems will be found in the near future.

CALCIUM INDUCED GELATION

In collaboration with J.F. Thibault we investigated the role of the degree of esterification and of the distribution pattern of the carboxylic groups along the chains upon gelation. A commercial apple pectin was purified and progressively deesterified by (1) cold sodium hydroxide leading to a random distribution of free carboxylic groups and (2) an orange pectin esterase leading to a blockwise distribution of the free carboxylic groups. Then samples with different average charge densities corresponding to different degrees of esterification were obtained (12,13) (Figure 3a).

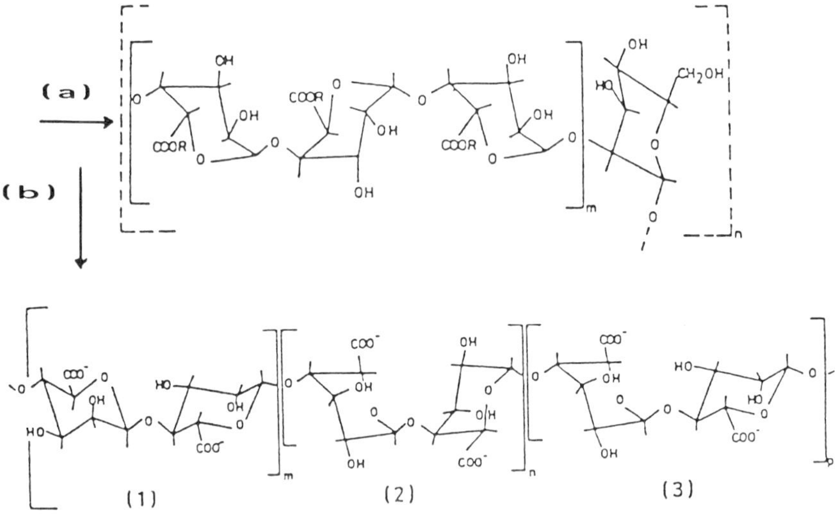

Figure 3. Chemical structure of pectin (a) (with R = -H or -CH$_3$) and alginate (b) (poly β-D-mannuronate (1) and poly α-L guluronate (2) blocks, alternating sequence (3)).

First of all, a large modification of the behaviour in the presence of Ca^{+2} counterions exists in solution around an average degree of esterification (DE) of 50 % ; when DE is lower than 50 %, aggregation is obtained in the presence of excess calcium. This is demonstrated from thermodynamics or light scattering (see figure 4). In the same range of DE a transition is observed in the activity coefficient of calcium.

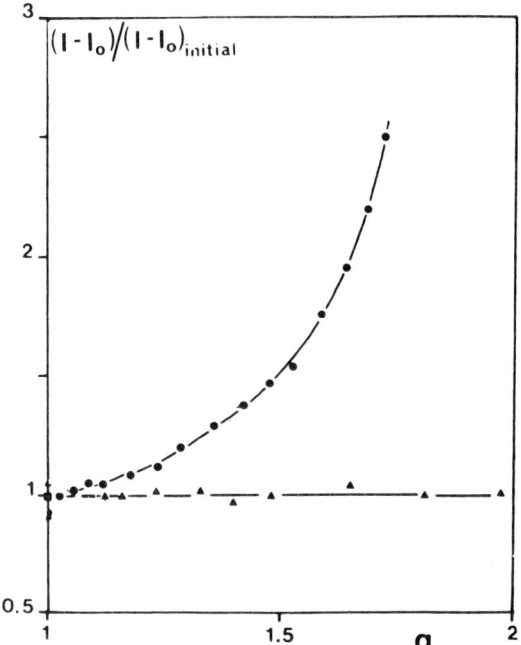

Figure 4. Change in scattered light upon addition of calcium chloride (●) and potassium chloride (▲) on calcium pectinate (DE<2 % ; polymer concentration 0.2 g/l).

The role of the counterions is of first importance ; if the role of a series of divalent counterions (Ba, Sr, Ca, Mg) is investigated, the originality of Mg^{+2} is pointed (12,14) : for low DE pectins, the activity coefficient for Mg^{+2} is twice that of Ca^{+2} ; our interpretation is that, in dilute solution, the interaction of Mg^{+2} is as predicted from the electrostatic theory of polyelectrolytes. In contrast the activity of calcium is in agreement with the formation of a dimer accompanied with a conformational change demonstrated by circular dichroïsm (15).

The following coefficients of transport determined from conductivity in dilute solution were obtained :

	$f\ Mg^{+2}$	$f\ Ca^{+2}$	
Experimental values	0.280	0.110	
Theoretical prediction	0.276	0.276	Single chain
		0.138	dimer

This table demonstrates clearly the role of the nature of the counterions.

If the solution of polymer is concentrated or in the presence of an excess of divalent electrolyte, a gel is formed by aggregation of dimers in the presence of calcium but never with Mg^{+2}.

Finally, the role of the pattern of distribution of the free carboxylic groups is shown in Figure 5 ; for the same average charge parameter corresponding to an average degree of esterification DE = 30 %, the polymer with blockwise distribution forms much more aggregates than a random distribution ; this means that a given minimum length of the carboxylic blocks is necessary to form a permanent junction. The minimum length estimated in the literature   is around 20. The model adopted usually for this very specific calcium binding is that of the egg box model (16).

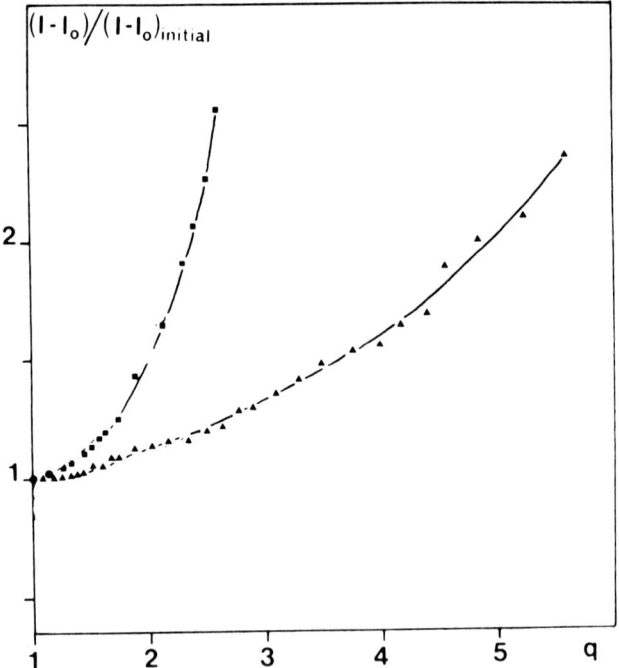

Figure 5. Change in scattered light upon addition of calcium chloride on calcium pectinate (DE 30 % ; polymer concentration : 0.2 g/l)) ▲ random ; ● blockwise.

The gelation of alginate in the presence of calcium (Fig.3,b) is analogous to that of pectins with low degree of esterification. In this case, the junctions are made of blocks of guluronic acid (configuration of L-guluronic acid in alginate is the mirror image of the D-galacturonic acid in pectin except at carbon 3);

the properties of a sample of alginate is not only controlled by the ratio of mannuronic acid/guluronic acid but also by the length and distribution of the blocks in the molecules.

The behaviour of these gels is very different from that of thermoreversible gels described previously ; the junctions are stable against temperature and only exchangeable at least partially by excess of monovalent counterions.

CONCLUSION

Some ionic polysaccharides are known to form non covalent network structure with original physical properties. They are usually more rigid that the covalent gels obtained with synthetic polymers.
Two type of gels can be described from single polymer gels :

- thermoreversible gels from low charge density polymers which adopt a stiff conformation (often an helical dimer or a double helix). Junctions are formed by aggregation (or pseudo crystallisation) of these stiff chains. The phase separation is controlled by the thermodynamics of the system (temperature, ionic concentration, solubility of the polymer...) ;

- ionic gels formed with polymers having a high charge density and a large stereoregularity in which the junctions are stabilised by cooperative calcium association. This interaction is dependent on the configuration of the uronic acids and on the nature of the counterion with high selectivity.

Today, multicomponent gels are developed and these mixed gels are found to allow control of the physical properties of the networks formed.

REFERENCES

1. Moorhouse, R., Colegrove, G.T., Sandford, P.A., Baird, J.K. and Kang, K.S. (1981) in "Solution Properties of Polysaccharides"; Ed. D.A. Brant, ACS Symposium Series 150, 111-124.
2. Crescenzi V. and Dentini, M. (1987), Carbohydr. Res, 160, 283-302.
3. Caroll, V., Miles, M.J. and Morris, V.J. (1982), Int. J. Biol. Macromol., 4, 432-433
4. Upstill, C., Atkins, E.D.T. and Attwell, P.T. (1986) Int. J. Biol. Macromol., 8, 275-288.
5. Rinaudo, M. and Milas, M., Communication "International Symposium on electrical interactions in complex fluids", June 1987, Colmar, France.
6. Rochas, C. and Rinaudo, M. (1982), Carbohydr. Res. 105, 269-272.
7. Rinaudo, M., Rochas, C. and Michels B. (1983), J. Chim. Phys., 80, 305-308.
8. Rochas, C. and Rinaudo, M. (1984), Biopolymers 23, 735-745.
9. Rinaudo, M. and Landry, S., Polymer Bull. (to be published)
10. Morris, V.J. (1985) in "Gums and Stabilisers for The Food Industry", vol. 3, Ed. Phillips, G.O., Wedlock, D.J., Williams, P.A., Publ. Elsevier, 87-99.

11. Cairns, P., Miles, M.J., Morris, V.J. and Brownsey, G. (1987), Carbohydr. Res., 160, 411-423.
12. Thibault, J.F. and Rinaudo, M. (1985), Biopolymers, 24, 2131-2143.
13. Thibault, J.F. and Rinaudo, M. (1986), 25, 455-468.
14. Thibault, J.F. and Rinaudo, M. (1985), British Polymer J., 17, 181-184.
15. Thibault, J.F. and Rinaudo, M. (1986) in "Chemistry and Function of Pectins", Edit. Fishman, M.L. and Jen, J.J., ACS Symposium series 310, 61-72.
16. Grant, G.T., Morris, E.R., Rees, D.A. and Smith, P.J.C. and Thom, D., (1973), Febs Lett., 32, 195-198.

Size and stability of the junction zones in gels of iota and kappa carrageenan

David G.Oakenfull and Alan Scott

CSIRO Division of Food Research, PO Box 52, North Ryde, NSW 2113, Australia

ABSTRACT

Information about junction zones – the regions of polymer where the molecules interact to form a gel network – can be derived from the kinetics of gelation and from the relationship between shear modulus and concentration for dilute weak gels. These measurements give (1) the average number of segments of polymer that are involved and (2) the size and thermodynamic stability of the junction zones. We have used these methods to study the gelation of κ- and ι-carrageenan. In carrageenan gels association is believed to occur through the formation of intermolecular double helices and subsequent association of these helices promoted by potassium ions. Our results support this two-step mechanism and confirm that the junction zones are complex. In gels of κ-carrageenan the average junction zone is an aggregate of about six double helices and in gels of ι-carrageenan the junction zones involve fewer polysaccharide segments with 2-3 associated helices. The junction zones in κ-carrageenan gels are also larger in terms of number average molecular weight and have greater thermodynamic stability than those in ι-carrageenan gels. ι-carrageenan appears to have too high a charge density for extensive aggregation, accounting for the softer texture of ι-carrageenan gels compared with κ-carrageenan.

INTRODUCTION

The carrageenans are alternating copolymers of β-D-galactose and 3,6-anhydro-α-D-galactose, linked (1→3) and (1→4) and designated ι or κ according to the relative number of sulphate ester substituents (1), as shown in Fig. 1. The linkage pattern introduces a twist into the molecule giving rise to the possibility of helical structures (2). Evidence primarily from spectroscopic studies suggests that gelation

involves association of the polysaccharide chains through the
formation of intermolecular helices. Mechanisms have been
proposed (3,4) along the lines shown in Fig. 2. Carrageenan
chains associate by the formation of intermolecular double
helices to form 'ordered domains'. Gelation then occurs with
the subsequent aggregation of these domains to form a molecular
network. We have derived further information about the
gelation mechanism from studying the kinetics of the process
and from measurements of shear modulus.

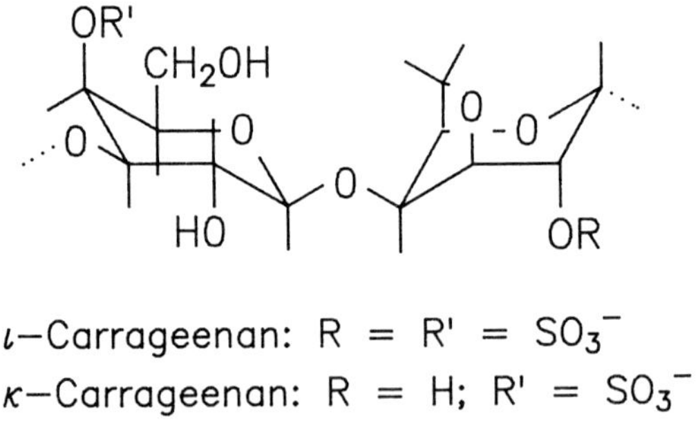

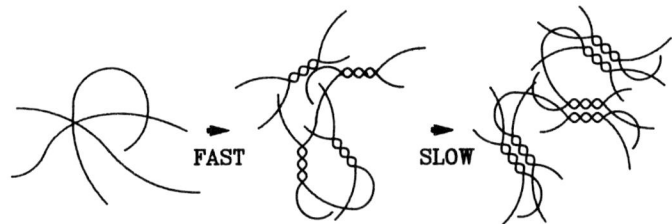

ι–Carrageenan: R = R' = SO_3^-

κ–Carrageenan: R = H; R' = SO_3^-

Figure 1. Idealised disaccharide repeating structure of κ- and
ι-carrageenan.

![schematic representation with FAST and SLOW arrows]

Figure 2. Schematic representation of the 'domain model' for
gelation of the carrageenans. The first step is the formation
of double helices which then aggregate further to form a gel
network.

Kinetics

The time required to form a gel of small but precisely
predetermined rigidity is a useful measure of the rate of
gelation (5,6), the setting being inversely proportional to the
initial rate. Because measurements are confined to the very
early stages of the gelation process, it is reasonable to
assume that each potential cross-linking locus (L) along a

polysaccharide chain acts as an independent species in solution, so that if n of these form a junction zone (J) then,

$$nL \rightarrow J \tag{1}$$

and the rate of gelation,

$$v = kc^n \tag{2}$$

where k is the rate constant and c is the concentration of polysaccharide. Thus the slope of log(rate of gelation) vs log(concentration) gives the 'reaction order' which is effectively the number of polysaccharide chains involved in the formation of a junction zone (6).

Shear Modulus

From the theory of rubber elasticity, an equation can be derived (7) relating shear modulus (G) to the concentration by weight (c) of the gel-forming polymer:

$$G = (RTc/M) \cdot (M[J] - c)/(M_J - c) \tag{4}$$

M is the number average molecular weight of the polymer, M_J the number average molecular weight of the junction zones and the quantity [J] is effectively the 'molar concentration' of junction zones. The derivation again relies on the assumption that the gels are very weak, with polymer concentrations close to the gel threshold. The polymer chains linking junction zones would then be long enough to approach Gaussian behaviour (8). As in the kinetic approach, this condition also makes it reasonable to assume that the cross-linking loci act as independent species in solution and that the formation of junction zones is an equilibrium process subject to the law of mass action.

It then follows that if n is again the number of cross-linking loci that form a junction zone and K_J is their association constant, that

$$K_J = [J]M_J^n\{n(c - M_J[J])\}^{-n} \tag{5}$$

By using numerical methods, equations 4 and 5 can be combined, eliminating the quantity [J] and giving a relationship between shear modulus and concentration in terms of M, M_J, K_J and n. Values of these can then be obtained which best fit experimental data of G vs c. Thus we have two independent estimates of the number of polysaccharide chains involved in a junction zone (n), as well as information about the size and thermodynamic stability of junction zones.

Our results confirm that the junction zones in carrageenan gels involve a large number of polysaccharide chains. There are substantial differences between the ι and κ forms, reflecting their different degrees of sulphation. A preliminary account of the kinetic part of this work has appeared previously (6).

EXPERIMENTAL

Materials

The ι and κ-carrageenan were Sigma products purified as their potassium salts by methods described previously (9,10). All solutions were prepared in deionised water with 0.02 M potassium chloride.

Gelation Kinetics

A very simple procedure gave the time required for a solution to form a gel with a predetermined small rigidity, without in any way disturbing the sample mechanically. A fixed weight (10 g) of solution was placed in each of a series of flat-bottomed cylindrical vials (diameter 2.74 cm) held in a thermostatted waterbath (298 K). These were then inverted sequentially and the time required to form a gel just strong enough to remain held in position was recorded. The reciprocal of the setting time was then proportional to the rate of gelation – a procedure equivalent to estimating a rate constant by the initial rate method (5,6).

Shear Modulus

Measurements of shear modulus were made by the method recently developed by Oakenfull, Parker and Tanner (11). The gels were formed in cylindrical dishes (radius 27 mm) and subjected to a small compression with a flat-ended cylindrical probe (radius 4.5 mm). An analytical electronic balance was used to determine the applied force. A plot of force *vs* extent of penetration gives the apparent Young's modulus of the sample which can be converted into the absolute shear modulus by multiplying by a factor of 0.0715 calculated from the geometry of the cylinder and probe.

RESULTS AND DISCUSSION

(1) Gelation Kinetics

Evidence from light scattering has been taken to suggest that in carrageenan gels the junction zones involve a large number of polysaccharide chain segments, possibly as many as ten (3). Our kinetic results support this suggestion. In Fig. 3 we show a logarithmic plot of gelation rate against the weight concentration of polysaccharide. The slopes of the

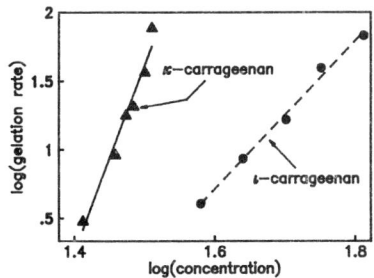

Figure 3. Logarithmic plot of gelation rate (100/t, t in minutes) against the weight concentration of polysaccharide (g/kg). The slopes of the lines indicate the number of segments of polysaccharide chain forming a junction zone.

lines, indicating the average number of segments of polysaccharide chain per junction zones are, for κ-carrageenan 12.5 ± 0.9 and for ι-carrageenan 4.5 ± 0.6. Thus in gels of κ-carrageenan the average junction zone is an aggregate of about 6 double helices but in gels of ι-carrageenan the junction zones are considerably less complex with 2-3 associated double helices.

(2) Shear Modulus

        In Fig. 4 we show shear modulus vs concentration for κ and ι-carrageenan gels at concentrations close to the gel threshold.

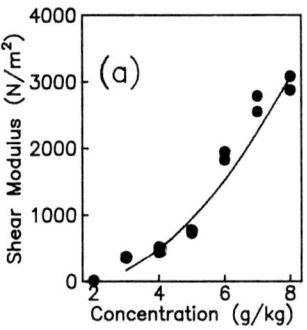 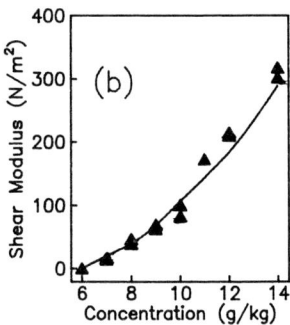

Figure 4.   Plots of shear modulus vs concentration for potassium salts of (a) κ-carrageenan and (b) ι-carrageenan at 298 K. The curves were calculated from equations 4 and 5 using the values for K, M_J, K_J and n given in Table I.

        Using equations 4 and 5, we calculated from these data values for number average molecular weight (M), number average molecular weight of the junction zones (M_J), association constant for junction zone formation (K_J) and the number of segments of polysaccharide chain which form a junction zone (n).   These are given in Table I.   There are very obvious differences between the results for the two carrageenans. κ-carrageenan has larger junction zones (M_J) than ι-carrageenan by a factor of more than two. κ-carrageenan's junction zones are also the more thermodynamically stable with a free energy of formation (ΔG_J) of -133 kJ/mole compared with -15.6 kJ/mole. The values obtained for n were in each case half those obtained from the kinetic measurements.   This apparent discrepancy can be explained in terms of the mechanism shown in Fig. 2 and tends to further support that mechanism.

Table I.   SIZE AND THERMODYNAMIC STABILITY OF THE JUNCTION
ZONES IN GELS OF κ- AND ι-CARRAGEENAN CALCULATED FROM SHEAR
MODULUS DATA AT 298 K (EQUATIONS 4 AND 5).

	κ-carrageenan	ι-carrageenan
M	105 000	178 000
M_J	8 900	4 100
n	6.4	2.1
K_J [a]	1.65×10^{23}	554
ΔG^o_J [b]	-133	-15.6
No. of monomer units per junction zone	37	14
ΔG^o_J per monomer unit [b]	-3.58	-1.12

[a] In units of mole fraction.   [b] kJ/mole.

Norton *et al.* (12) have shown that the first step in the
gelation process, dimerisation through the formation of
helices, occurs very rapidly.   Thus the aggregating unit, the
cross-linking locus, in the derivation of equations 3 and 4, is
a helix which already involves two polysaccharide chains.   The
values of n from shear modulus data should therefore be doubled
to give the actual numbers of polysaccharide chain segments per
junction zone.   This then gives 12.8 for κ-carrageenan and 4.2
for ι-carrageenan, almost identical to the values from gelation
kinetics.

In Table I we have included the contribution per monomer
unit to the free energy of formation of junction zones for the
two carrageenans.   In the case of κ-carrageenan the value is
numerically   more   than   double   that   for   ι-carrageenan.
Presumably this is a consequence of the different degrees of
sulphation of the two carrageenans.   Aggregation is promoted by
the   potassium   counterions   which   form   ion-pairs   with   the
sulphate groups (4).   Potassium (and also rubidium) interact
specifically   with   the   carrageenans   (9,10),   ionic   radius
apparently being the critical factor.   The role of the ion

seems to be that of relieving electrostatic repulsion through ion-pair formation while at the same time allowing the polysaccharide chains to be in suitable proximity for the formation of strong intermolecular hydrogen bonds. Our results suggest that in ι-carrageenan the higher charge density (one sulphate group per monosaccharide unit) causes the free energy of association to be less negative and so insufficient to overcome the considerable entropic barrier to aggregation of segments from a large number of polysaccharide chains. Thus the thermodynamic driving force for association is much less, explaining why ι-carrageenan forms softer, weaker gels than κ-carrageenan at equivalent concentrations.

REFERENCES

1. Rees, D. (1969) Adv. Carbohydr. Chem. Biochem., 24, 267-332.
2. Rees, D.A. (1977) Polysaccharide Shapes, 80 pages, Chapman and Hall, London, UK.
3. Morris, E.R., Rees, D.A. and Robinson, C. (1980) J. Mol. Biol., 138, 349-362.
4. Rochas, C. and Rinaudo, M. (1984) Biopolymers, 23, 735-745.
5. Oakenfull, D. and Scott, A. (1986) in Gums and Stabilisers for the Food Industry 3, (ed. Phillips, G.O., Wedlock, D.J. and Williams, P.A.) pp 465-475, Elsevier, London, UK.
6. Morris, V.J. and Oakenfull, D. (1987) Chem. Ind., 201-202.
7. Oakenfull, D. (1984) J. Food Sci., 49, 1103-1104 & 1110.
8. Mitchell, J.R. and Blanshard, J.M.V. (1979) in Food Texture and Rheology, (ed. Sherman, P.) pp 425-435. Academic Press, London, UK.
9. Morris, V.J and Belton, P.S. (1982) Prog. Food Nutr. Sci., 6, 55-66.
10. Belton, P.S., Morris V.J. and Tanner S.F., (1985) Int. J. Biol. Macromol., 7, 53-62.
11. Oakenfull, D.G., Parker, N.S. and Tanner, R.I., these Proceedings.
12. Norton, I.T., Goodall, D.M., Morris, E.R. and Rees, D.A. (1983) J. Chem. Soc., Faraday Trans. 1, 79, 2489-2500.

Electrochemical approach to studies of electrostatic interactions and ion binding in concentrated biopolymer systems

P.M.Hart, A.H.Clark and A.Lips

Unilever Research, Colworth House, Sharnbrook, Bedford MK44 1LQ, UK

ABSTRACT

D.G. Hall has developed a comprehensive thermodynamic theory for solutions of polyelectrolytes, surfactants and other charged colloids.(J.Chem.Soc.,Faraday Trans I, 1981,<u>77</u>,1121). Electrochemical methodology based on this treatment can provide information both on ion binding and the electrostatic contribution to the Donnan osmotic pressure which may be regarded as a measure of the electrostatic interaction between the polyelectrolyte molecules. The theory is purported to be generally applicable, it does not depend on assumptions regarding the geometry of the charged species, and can be more successful than Manning theory. Moreover, the application to concentrated charged systems is straightforward and the theory incorporates excluded volume effects.

Application to biopolymer systems has so far been limited. We have therefore evaluated the method in the context of caseinate solutions which have already been extensively characterised by other approaches. When applied to concentrated caseinate solutions, the method can yield the established values for calcium binding, the voluminosity of the casein micelles, and the magnitude of electrostatic interactions.

1. INTRODUCTION

The Poisson-Boltzmann (P-B) equation is central to most theoretical descriptions of electrostatic interactions between charged species, be they colloidal particles, polyelectrolytes, surfactant micelles or small ions. In its linearised form, which is mathematically more tractable, the equation is the basis of the Debye Huckel theory of strong electrolytes and of early descriptions of electrostatic interactions in site-specific ion binding (1-3). It is employed also in the recent Manning theory for polyelectrolyte solutions (4) where the key notion is that of counterion condensation on the polyelectrolyte to reduce a high primary surface charge to an effective critical charge per unit length of polyion; the

135

distribution of uncondensed counterions and coions is then
describable by a linearised form of the P-B equation. In
general, mathematical problems limit the application of the P-B
equation to systems with relatively simple, well-defined
geometries. The charged colloid entities of typical biopolymer
systems, however, are invariably complicated and moreover
polydisperse in size and geometry. The only simple prospect of
modelling such systems is limited to the description of
suitable averages rather than the complete spatial distribution
of the various charged constituents (which in principle is
within the scope of P-B theory). Such an approach has recently
been developed by Hall (5-8). It is based on thermodynamic
arguments involving the concept of the Donnan equilibrium
between a solution of charged colloid plus electrolyte and a
solution of electrolyte alone. The treatment has been very
successful in the description of micellar solutions of ionic
surfactants and of the characteristic behaviour of dilute
polyelectrolyte solutions improving on Manning theory.
Moreover it has provided a comprehensive thermodynamic
framework for the measurement and interpretation of the binding
of charged, mutually interacting ligands (eg. ionic
surfactants) to macromolecules (8).

The requirement of equal chemical potentials in the
system under study (^{S}) and the Donnan solution in osmotic
equilibrium (^{D}), in respect of any electrically neutral
combination of diffusible coions (_) and counterions (+), is
expressed in the Hall theory by

$$(c_{z_+}^D)^{\nu_+} \quad (c_{z_-}^D)^{\nu_-} \quad (\gamma_\pm^D)^{\nu_+ + \nu_-} = \quad <c_{z_+}^S>^{\nu_+} \quad <c_{z_-}^S>^{\nu_-} \quad (\gamma_\pm^S)^{\nu_+ + \nu_-} \quad [1]$$

where C denotes ionic concentration, $\gamma_\pm$ mean activity
coefficient and ν the stoichiometric coefficients for the
neutral electrolyte dissociating according to
$A\nu_+ \ B\nu_- \rightarrow \nu_+ A^{z+} + \nu_- B^{z-}$ with z representing the number of charges
of the ion and $\nu_+ \ z_+ + \nu_- \ z_- = 0$
By diffusible ions are meant all ionic species other than the
charged macroions. The average concentration of a particular
species of diffusible ion in the system $<c^S>$ is taken to be its
total concentration minus those fractions which are either
specifically bound or electrostatically condensed on the
macroions. The Debye Huckel equation is prescribed for
estimating the mean activity coefficient, the calculation
extending to all diffusible ions but not the macroion. By
introducing the concept of an effective degree of dissociation
for the macroion, Hall was able to provide a predictive
capability for relations of the above type which moreover can
take account of excluded volume effects.

From an experimental viewpoint, the chemical potentials
of electrically neutral combinations of diffusible coions and
counterions are readily measured by electromotive-force (e.m.f)
methods. The preferred approach is to use cells with
electrodes reversible to counterion and coion in which liquid
junction effects cancel and which therefore permit reliable

measurements in concentrated colloids circumventing the
problems (viz: suspension effect(9)) and interpretation of the
more commonly employed single ion activity measurements. If
such back-to-back measurements are performed for all the
neutral combinations of diffusible ions (ie. neutral salts) in
the system, the average concentration of the ions in both
system and Donnan solution can be inferred by reference to the
known activity products for standard solutions of the salts.
In such calculations, the reasonable assumption is made that
the coions do not bind specifically to the macroion. If the
total concentrations of the ions in the system are known it is
then straightforward to determine the site-specific (excluding
double layer effects) binding of counterions to the macroion.
The treatment of Hall in addition provides a simple expression
for estimating the electrostatic contribution to the Donnan
osmotic pressure which may be regarded as a measure of an
average electrostatic interaction between the macroions.

The procedure outlined above differs from the more
frequently used thermodynamic methods for measuring ion binding
namely equilibrium dialysis and single ion electrode studies.
Both of these pose difficulties of interpretation, and it is
then usual practice to quote values of "apparent" binding
without correction for the "Donnan effect". By contrast the
method of Hall purports to measure directly the site specific
counterion binding and furthermore the magnitude of the
electrostatic contribution to the Donnan osmotic pressure.
The latter affords a direct insight into the nature of the
interactions between polyelectrolyte molecules. By the Hall
procedure, these can be studied conveniently in very
concentrated polyelectrolyte solutions, and this is the major
objective of this study.

It is necessary to comment further on the formulation of
equation [1] in terms of the average concentrations of the
mobile ions and therefore its similarity with the classical
Donnan expression (10). Experimental evidence (11,12) and
theoretical arguments (13) indicate that the use of average
concentrations is not always justified and that in general the
details of the distribution of ions should be considered.
However, following Overbeek (13), we will show that, provided
the distribution of the diffusible ions in the double layers
surrounding the macroions can be represented by a linearised
form of the P-B equation, then equation [1] can be used with
confidence in the prediction of counterion binding. As
indicated for example by electrokinetic methods many
polyelectrolyte systems have sufficiently small potentials to
warrant the use of a linearised P-B expression. Moreover, to
the extent that the Manning theory has found wide application,
the assumption of low potentials in the double layer
surrounding the macroion is not unreasonable.
Extensive knowledge exists already on the binding of
calcium to caseinate (14-17), on the (low) electrokinetic
potentials of the macroions in such systems(18), their tendency
to micellise, and the nature of the micellar structures(19,20).

In view of this background, caseinate solutions provide useful
systems for evaluating the scope and applicablilty of the Hall
thermodynamic approach to concentrated biopolymer systems.

2. DETAILS OF APPROACH

Chemical analysis on the chosen commercial sample of
sodium caseinate (ex. DeMelkindustrie Veghel) indicated the
presence of significant levels of sodium (1.4%w/w), calcium
(0.08%w/w) and potassium ions(0.0063%w/w) with chloride ion
(0.23%w/w) the only important diffusible anion.
The natural pH of solutions of the caseinate (as supplied)
was in the range 6.5 to 7, and to such unbuffered solutions
were added known amounts of calcium chloride. E.m.f.
measurements were performed at $22^{\circ}C$ involving reversible
electrodes for sodium,calcium, potassium and hydrogen ions each
sharing a reversible chloride electrode as the common reference
electrode.   Calibration plots were constructed for standard
solutions of the various chloride salts. Conformity to near
Nernstian behaviour in the performance of each of the
electrodes could be demonstrated and it was ascertained that
cross selectivity to cations was not significant for any of the
electrodes under the conditions of study. The potentiometric
measurements were carried out with an Orion EA 940 ion analyser
with high impedance detection for both the monitor and
reference electrodes. The back-to-back e.m.f. measurements
scale with the logarithm of the activity products
$(c_{z_+})^{\nu_+}$ $(c_{z_-})^{\nu_-}$ $(\gamma^{\pm})^{\nu_+ + \nu_-}$ which are readily quantified through
measurements on standard solutions of the various salts.
For the Donnan solution the concentrations of the ionic
species Na^+, Ca^{++}, K^+ and H^+ and Cl^- are then specified by four
such measurements together with the condition of electrical
neutrality, $[Na^+]^D + 2[Ca^{++}]^D + [H^+]^D + [K^+]^D = [Cl^-]^D$. Numerical
solution of these equations is straightforward; an initial
estimate is obtained by assuming that all the activity
coefficients are unity. Next the Davies extension of the
Debye-Huckel equation is used to obtain an improved estimate of
activity coefficients

$$\log_{10} \; \gamma_i = - \frac{0.509 \; z_i 2 \sqrt{I}}{1 + \sqrt{I}} + 0.13 \; z_i^2 \; I \qquad [2]$$

in terms of I ($=1/2 \; \Sigma z_i^2 c_i$) the ionic strength calculated on
the basis of the first estimate of the ionic concentrations.
Iteration readily yields convergent solutions for the ionic
concentrations and the single ion activity coefficients γ_+^D
and γ_-D $[\gamma_+ \nu_+ \gamma_- \nu_- = \gamma_\pm \nu_+ \nu_-]$.

The diffusible ions of the systems under study $<c^S>$ are
not themselves subject to electrical neutrality, rather their
net charge $[<Na^+>^S + 2<Ca^{++}>^S + <K^+>^S + <H^+>^S - <Cl^->^S]$
balances the negative charge Z of the polyions. Since for pH>5
caseinate is substantially negatively charged it is not
unreasonable to assume that the chloride ions are coions and as
such unbound, and their average concentrations can be taken

therefore as known. Average concentrations of free counterions and their activity coefficients are then estimated simply from measured activity products, again following a simple iterative procedure based on [2]. Since total concentrations of diffusible ions are known the levels of bound counterions can be inferred.

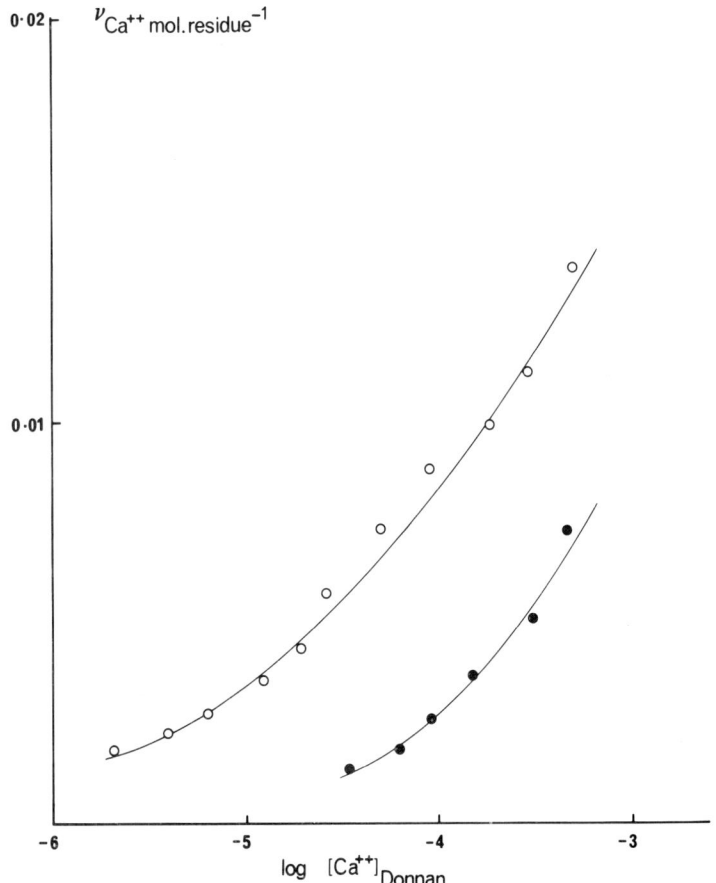

1. Calcium binding (calcium ions per amino acid residue) to caseinate at 0.45% w/w vs. the logarithm of the concentration of calcium ions in the Donnan solution; total sodium concentrations are: O - 2.83×10^{-3} mol.dm^{-3} and $\bullet$ - 1.28×10^{-2} mol. dm^{-3}.

3. DILUTE CASEINATE SOLUTIONS

The concentration of caseinate selected was 0.45% w/w, a
level representative of previous binding studies(14-17). Fig.1
shows two binding isotherms obtained by the present method. To
facilitate comparison with previous studies, the levels of
calcium binding are expressed in numbers of ions per amino acid
residue taking an average molecular weight of 112 for the
latter based on the known sequence data for α, β and κ-casein
(21-23) and the established ratio of these components in
caseinate (24). The concentration of calcium ions chosen as the
abscissa in Fig.1 is that of the Donnan solution (rather than
of the system). This choice provides a common reference and
parallels the quantity measured by equilibrium dialysis. The
isotherms are in reasonable agreement with previous work
(14-17) showing the expected levels of calcium binding and the
sensitivity of binding to ionic strength.

4. CONCENTRATED CASEINATE SOLUTIONS

Reproducible e.m.f. measurements could be obtained with
systems as concentrated as 15%w/w. Interpretation of the data
in terms of the above considerations alone, however, yielded
unrealistic (negative) binding of sodium ions. Inclusion of
the concept of excluded volume can remove this discrepancy.
The effective concentrations of the diffusible ions are taken
as greater than the nominal values because a fraction of the
water is inaccessible to the ions by a combination of hydration
of the caseinate and micellisation. We represent the excluded
volume effect in terms of V(= volume of water inaccesible to
ions/g of caseinate) and this is expected to be of similar
order to the range of values, 2 - 5 ml.g^{-1}, observed in studies
of casein voluminosity (18-20). In the analysis V is treated as
a variable parameter and binding isotherms are generated for
calcium and sodium ions. It is known (15) that sodium ions have
ca. 1/100th of the affinity of calcium ions for binding to
casein. Moreover the intrinsic binding constants to organic
phosphate groups (which are the predominant substrates in ion
binding to casein) suggest that the binding of sodium ions is
too weak to be easily measured in the dilute caseinate
solution. However at 15% and 8% the local concentrations of
sodium ions are high and significant fractions can be expected
to be bound. In fact prior to the addition of calcium chloride
mainly sodium ions are bound. Optimisation in the choice of
values of V was dictated by the requirements of consistency
with the known binding affinities for sodium and calcium ions
in dilute systems where excluded volume effects are of no
consequence. We find that a model of initial ion exchange (two
sodium ions for one calcium ion) followed by more extensive
calcium binding past the stage of complete desorption of sodium
ions can meet these criteria. In particular realistic values
are found for the amounts of calcium bound (Figs. 2 & 3) and
corresponding results for V are similar to estimates of
voluminosity quoted for casein micelles.

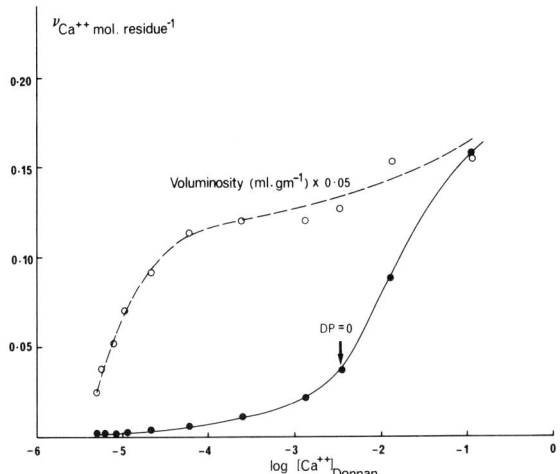

2.   Calcium binding (calcium ions per amino acid residue) to caseinate at 8% w/w vs. the logarithm of the concentration of calcium ions in the Donnan solution (●). Voluminosity estimates are also shown (○). Arrow indicates point at which Donnan potential is zero.

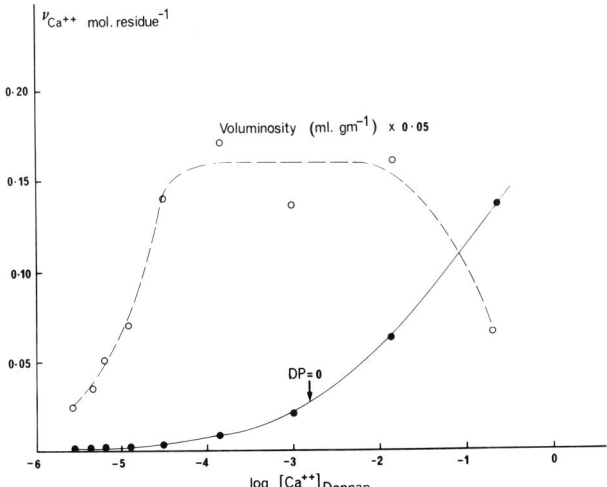

3.   Calcium binding (calcium ions per amino acid residue) to caseinate at 15% w/w vs. the logarithm of the concentration of calcium ions in the Donnan solution (●). Voluminosity estimates are also shown (○). Arrow indicates point at which Donnan potential is zero.

5. LOCAL VS AVERAGE CONCENTRATIONS

As indicated in the introductory section some justification is
required for the use of average concentrations (and implicit
average electrical potentials) in the Hall treatment.   The
rigorous procedure would be to relate local concentrations to
local potentials and take averages of local concentrations.
This of course presupposes some knowledge of the geometry of
the charged macroions and their mutual configurations.
The Boltzmann distribution requires that

$$C_j{}^L = C^D{}_j \exp (-z_j e\, \psi^L/kT) \qquad\qquad [3]$$

where $C_j{}^L$ is the local concentration of ionic species j and
charge number z_j where e is the electronic charge ψ^L the local
electrical potential and $C_j{}^D$ the concentration in the Donnan
solution; k is the Boltzmann constant and T temperature.   We
are interested in the average concentrations given by

$$<C_j> = C_j{}^D < \exp (- z_j e\, \psi^L /kT)> \qquad\qquad [4]$$

If $z_j e\, \psi^L/kT < 1$ everywhere , $\exp (-z_j e\, \psi^L/kT) \simeq 1 - z_j e \psi^L/kT$,
and

$$<C_j> = C_j{}^D \exp (- z_j e <\psi>/kT) \qquad\qquad [5]$$

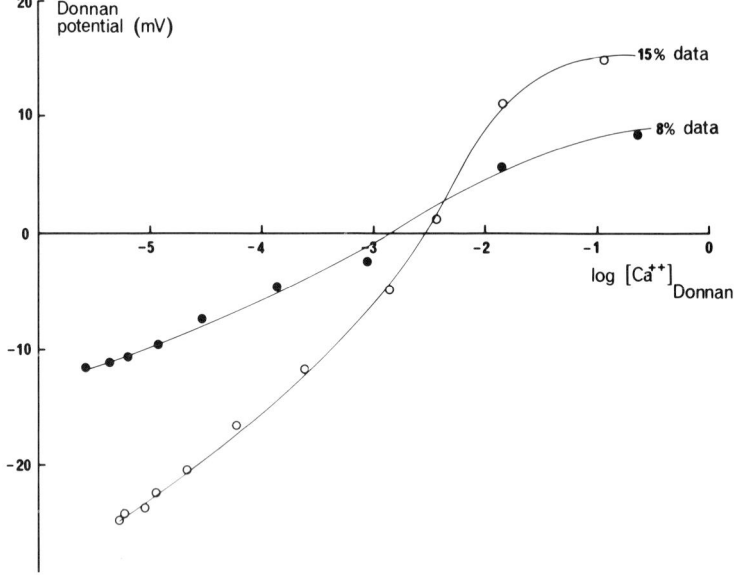

4.   Inferred Donnan potentials vs. the logarithm of the
concentration of calcium ions in the Donnan solution for
concentrated caseinate solutions.

where $<\psi>$ is now an average potential. Therefore, in the linear P-B regime, at positions where local potentials ψ^L equal the average potential $<\psi>$, the local concentrations of both coions and counterions are equal to their average concentrations. The use of equation [1] is thus justified and can be considered to relate to positions in the system where local potentials have the numerical value of the average potential $<\psi>$.

The contribution of the diffusible ions to the osmotic pressure π_{el} is found by averaging the local values over all configurations, viz:

$$\pi_{el} = RT< \sum_j c^D_j \, [\exp \, (\, - z_j e \, \psi^L/kT) - 1]> \qquad [6]$$

and if it can be assumed that all configurations are equally probable then

$$\pi_{el} = RT \sum_j c_j^D \, [\exp \, (\, -z_j e < \psi >/kT) -1] \qquad [7]$$

where $<\psi>$ is determined by equation [5].

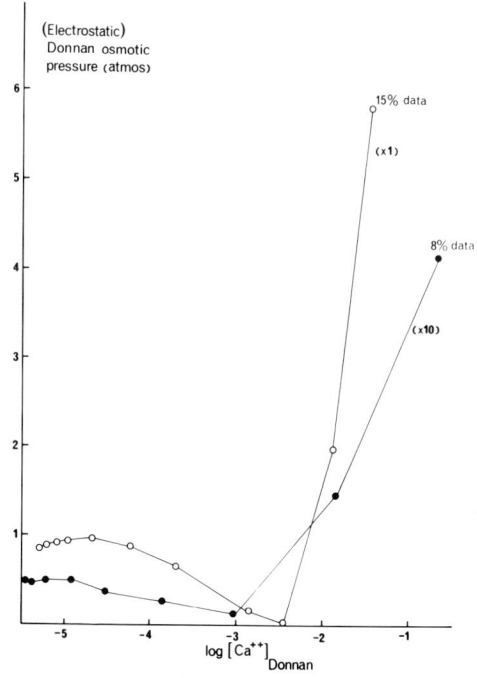

5. Electrostatic contribution to the Donnan osmotic pressure vs. the logarithm of the concentration of calcium ions in the Donnan solution for concentrated caseinate solutions.

In data analysis we have applied [7] & [5], where necessary with concentrations replaced by activities, and have therefore assumed that electrostatic repulsion between the caseinate molecules is insufficient to create lattice-like order. Fig.4 shows the average (Donnan) potentials so inferred for both the 8% and 15% caseinate systems.

Both the negative potentials and the osmotic pressure term π_{el} (in Fig.5) decrease with calcium addition and vanish at the point of precipitation.
It is gratifying that the inferred Donnan potentials are within the range expected from electrokinetic studies (18).

In conclusion, the application of the Hall approach seems promising and merits further evaluation.

Acknowledgement

We are indebted to Prof. Denver G.Hall from Unilever Research Port Sunlight Laboratory, for his valuable guidance and stimulating discussions.

REFERENCES

1.   K. Linderstrom-Lang, C.R. Trav.Lab. Carlsberg ,Ser Clim.,
     1924,15,No.7
2.   C. Tanford and J.G. Kirkwood, J.Am.Chem.Soc. 1957,79,5333
3.   T.L. Hill, Arch.Biochem.Biophysics,1955,57,299
4.   G.S. Manning, J.Chem.Phys., 1969,51,524,3249
5.   D.G. Hall, J. Chem.Soc. Faraday Trans.1,1981,77,1121-1156
6.   D.G. Hall,Colloids and Surfaces,1982,4,367
7.   D.G. Hall, J.Chem Soc.Faraday Trans.I,1984,80,1193
8.   D.G. Hall, J.Chem Soc., Faraday Trans I,1985,81,885
9.   H. Pallman, Kolloidchem. Beihefte, 1930,30,334
10. F.G. Donnan, Z. Electrochem, 1911,17,572
11. H. Hammerstein, Biochem Z., 1924,147,481
12. F.W. Klaarenbeek, Over Donnen evenwichten bij solen van
     Arabische Gom, Thesis, Utrecht, 1946
13. J.Th.G. Overbeek, In Kruyt H.R. (Ed.) Colloid Science,
     Vol.1, Elsevier, Amsterdam.,1952,191
     Also J.Th.G. Overbeek, Prog.Biophys.Chem., 1956,6,57
14. D.F. Waugh, C.W. Slattery and L.K. Creamer, Biochemistry,
     1971,10 817
15. C.W. Slattery and D.F. Waugh, Biochemical
     Chemistry,1973,1,104
16. D.G. Dalgleish and T.G. Parker, J. of Dairy Res. 1980.
     47,113
17. T.G. Parker and D.G. Dalgleish, J.of Dairy Research,
     1981,48,71
18. D.G. Dalgleish, J. of Dairy Research, 1984,51,425
19. T.A.J. Payens, J. of Dairy Research, 1979,46,291

20. P. Walstra,J. of Dairy Research,1979,46,317
21. J.C. Mercier, F. Grosslande & B. Ribadeau-Dumas,
     Eur.J.Biochem,1971,23,42
22. B.Ribadeau-Dumas, G. Brignon, F. Grosslande and
     J.C. Mercier, Eur.J. Biochem,1972,25,505
23. J.C. Mercier, G. Brignon & B. Ribardeau-Dumas, Eur.J.
     Biochem,.1973,35,222
24. A.H. Clark and S.B. Ross-Murphy, Advances in Polymer
     Science, 1987,83,57-192

Studies on the gelation of the microbial polysaccharides XM6 and gellan gum

H.D.Chapman, G.R.Chilvers, M.J.Miles and V.J.Morris

AFRC Institute of Food Research, Norwich Laboratory, Colney Lane, Norwich NR4 7UA, UK

ABSTRACT

XM6 and gellan gum are examples of gelling bacterial polysaccharides. The regular chemical repeat units of these polymers make them ideal models for investigating polysaccharide gelation. A range of experimental techniques has been employed to characterise the molecules in solution and the mode of association in the gel state. Static light scattering studies on dilute solutions of the polysaccharides have been used to investigate their conformation in solution and the mode of association upon gelation.

KEYWORDS: XM6, Gellan gum, bacterial polysaccharides, gelation, light scattering.

INTRODUCTION

XM6 is the name given (1) to the extracellular polysaccharide produced by the bacterium Enterobacter (NCIB 11870). The polymer is a branched anionic heteropolysaccharide with the chemical repeat unit (2) shown in fig. 1a. The polymer disperses in water to produce a viscous liquid. At sufficiently high polymer concentrations and ionic strength,XM6 samples form thermoreversible gels (1). The melting-point of the gels increases with increasing ionic strength (1). The low polymer concentrations required for gelation, coupled with the ability to adjust the melting-point of the gel to room temperature, have suggested (1) potential food applications in areas where the 'melt-in-mouth' characteristic of gelatin may be an important attribute. X-ray diffraction studies of aligned fibres prepared from XM6 gels show highly crystalline patterns consistent with strong polymer-polymer association and gelation (3,4). Analysis of the X-ray data (4), coupled with model-building calculations (5), suggest an 8_3 double helix, sections of which can crystallise in either an orthorhombic or tetragonal unit cell. Melting of the gels is accompanied by a sharp change in optical rotation which has been attributed to a conformational transition from an ordered 'helical' structure to a disordered

'random-coil' structure (1).   This article describes light
scattering studies on dilute solutions of XM6.   Conditions have
been chosen to permit examination of the proposed 'ordered' and
'disordered' conformations.

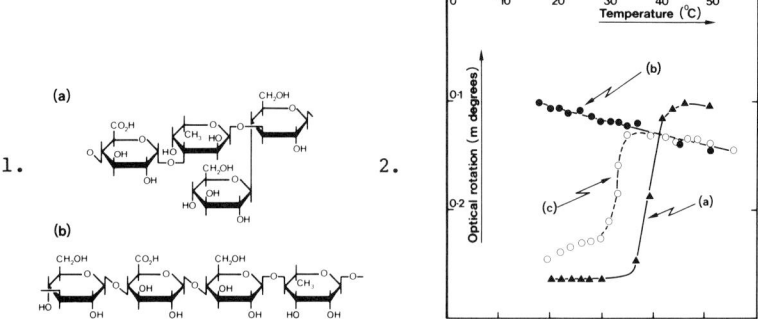

Fig. 1.Chemical repeat units of (a) XM6, (b) deesterified gellan
     gum.

Fig. 2.Optical rotation versus temperature plots for (a) 2% Na
     XM6 in the presence of 0.5M NaCl, (b) 2% TMA XM6 in water,
     (c) 2% TMA XM6 in 0.5M TMACl. Cell length 10cms, wavelength
     589nm.
     Gellan gum is the extracellular polysaccharide produced by
Pseudomonas elodea (6).   The polymer is a linear anionic
heteropolysaccharide (7,8).   In the absence of base labile
substituents the chemical repeat unit is the tetrasaccharide
(7,8) shown in fig. 1b.   The native polymer is esterified on the
(1→3) linked glucose residue (9).   Esterification amounts to
incomplete C2 substitution with L-glycerate and ~50% C6
substitution with acetate (9).   The commercial gelling agent
'Gelrite' is completely deesterified.   Gellan gum is an
extremely good gelling agent.   Suggested industrial uses include
a broad spectrum gelling agent (10-12), a microbial growth
medium or a medium for plant tissue culture (10,13-16),
deodorant gels (17) and microencapsulation (18,19).   Potential
food applications have been studied (11,12).   On the basis of
successful toxicity studies Kelco have announced the intention
to seek approval for the use of gellan gum as a food additive
(11,18).   Esterification inhibits gelation of gellan gum and
cations promote intermolecular association and gelation
(10,12,20).   The native polymer produces soft elastic gels (10).
Progressive deesterification results in increasingly brittle
gels (10).   X-ray diffraction studies (21-24) of fibres prepared
by stretching native gellan gum gels show good polymer alignment
but poor crystallinity.   Progressive deesterification results in
increased crystallinity but no detectable change in the type or
dimensions of the helix.   Gellan gum forms a three-fold helix
with a pitch of 2.82nm.   Alternative stereochemically acceptable
models (24) are a right-handed 3_1 contracted single helix and a

left-handed 3_2 extended double helix. Modelling studies (25)
have, at present, failed to establish which of these types of
helix are present within the junction zones of the gel. The
role played by the ester substituents is not completely
understood. Acetylation at C6 of the (1→3) linked glucose does
not disrupt formation of either three-fold helix (25) but the
recently discovered L-glycerate may disrupt helix formation and
could be important in limiting crystallisation. Gelation is
sensitive to the type of counterion and the ionic strength of
the medium (12,20). Increasing the ionic strength increases the
melting-point of the gels. Recently Crescenzi and coworkers
(26,27) have suggested that gellan gum may undergo a reversible
'helix-coil' transition in dilute solution. These authors
suggest that the ordered and disordered forms of the
polysaccharide may be isolated by controlling the ionic strength
of the medium. This article describes light scattering studies
on gellan gum solutions. Conditions have been chosen to permit
examination of the proposed 'ordered' and 'disordered'
conformations.

MATERIALS AND METHODS

Samples of XM6 were a gift from Dr. I.T. Taylor (ICI,
Biological Products Business). Large scale production,
isolation and clarification have been described by O'Neill and
coworkers (2). An additional clarification step involved
filtering hot (90°C) aqueous XM6 dispersions through 8 μm
Millipore filters. Pure salt forms of XM6 were prepared by ion-
exchanging the polymer into the acid form and then neutralising
solutions with the appropriate hydroxide. Gelrite, the
potassium salt of fully deesterified gellan gum was purchased
from Kelco-AIL. Pure salt forms were prepared by precipitating
'gellanic acid' at pH ⩽3.0 and then redissolving the acid form
by addition of the appropriate hydroxide. Samples of XM6 and
gellan gum were stored as freeze dried powders.

Static light-scattering studies were made using a modified
Malvern 4300 spectrometer system. The light source was a
vertically polarised He-Cd (wavelength λ = 441.6nm) laser and
data were collected over the angular range 30-130°. The
apparatus was calibrated with benzene and the Rayleigh ratio for
benzene was calculated to be 64.23 x 10^{-6}cm^{-1}. Samples of XM6
and gellan gum were cleaned by filtration through 3μm Millipore
filters. Refractive indices of solutions were measured using an
Abbé refractometer. Specific refractive index increments
(dn/dc) were measured using a Chromatix KMX-16 laser
differential refractometer. (dn/dc) was found to be 0.156cm^{3}g^{-1}
and 0.136cm^{3}g^{-1} for XM6 and gellan at λ = 632.8nm. These values
were corrected as described by Huglin (28) to give (dn/dc) =
0.157cm^{3}g^{-1} and 0.142cm^{3}g^{-1} for XM6 and gellan at λ = 441.6nm.
Partial specific volumes were determined using an Anton Parr DMA
602 density measuring cell and a DMA 60 meter. Optical rotation
was measured using an AA-100 polarimeter (Optical Activity Ltd).

RESULTS AND DISCUSSION

Fig. 2 shows the optical rotation change observed upon

melting a 2% Na XM6 gel prepared in the presence of 0.5M NaCl.
A similar optical rotation transition was observed (fig. 2) for
tetramethylammonium (TMA) XM6 prepared in the presence of 0.5M
TMACl.   In the presence of TMA cations the transition
temperature was lowered and the samples were viscoelastic rather
than elastic gels.   In the absence of added electrolyte TMA XM6
shows a monotonic variation of optical rotation with temperature
(fig. 2).   Nisbet et al. (1) report that the melting-point of
the XM6 gels increases with increasing ionic strength but is
independent of polymer concentration.   These authors suggest
that the melting results from a helix-coil transition.   If this
assumption is correct then it should be possible to isolate the
'coil' and 'helical' conformation in dilute solution.

     Static light scattering studies were made on Na XM6 and TMA
XM6 solutions in the absence of added electrolyte.   A typical
Zimm plot for Na XM6 is shown in fig. 3.   The experimental
conditions correspond to the proposed disordered 'random coil'
state of the polysaccharide.   Values of the measured molecular
weights $\langle M \rangle_W$, radii of gyration $\langle R_g^2 \rangle_Z^{\frac{1}{2}}$, and the second osmotic
virial coefficients A_2 are listed in table I.   The high radii of
gyration make it difficult to carry out the extrapolation to
zero angle.   Thus the agreement between results obtained for the
different salt forms may be regarded as satisfactory.   Data
determined for TMA XM6 in 0.5M TMACl are also shown in table I.

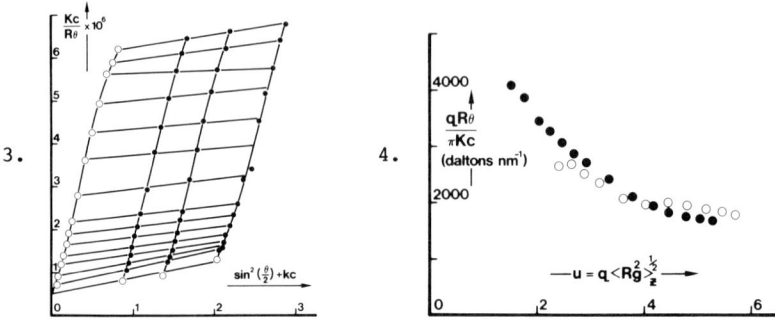

3.

4.

Fig. 3.Zimm plot for Na XM6 in water.

Fig. 4.Holtzer plot of the 'zero-concentration' data for (●) Na
XM6 in water and (○) TMA XM6 in 0.5M TMACl.
These results were obtained under conditions favouring the
ordered 'helical' state of the polysaccharide.   Increasing the
ionic strength does not lead to an increase in molecular weight
ruling out the possibility of a coil to double helix transition.
Since the radius of gyration remains unchanged upon increasing
the ionic strength three possible situations could arise:
firstly the polymer could remain in the coil state, secondly the
polymer could remain in an ordered helical state and thirdly the
polymer could undergo a coil-to-single helix transition for
which the decrease in contour length, arising from an increase
in mass per unit length, is exactly offset by the increase in

the Kuhn statistical segment length of the polymer. Under favourable conditions the mass per unit length of the polymer may be estimated from a Holtzer plot (29) of the 'zero concentration' scattering data. A Holtzer plot for Na XM6 in water is shown in fig. 4. The high angle 'zero concentration' data for TMA XM6 in 0.5M TMACl is also shown in fig. 4. The Holtzer plot for a semi-flexible coil should rise from zero at U=0, pass through a maximum at $U \geq 1.4$ and then tend to a plateau value, corresponding to the mass per unit length, at high U values (29). Clearly the polymers behave as semi-flexible coils with the same mass per unit length under high and low salt conditions. This rules out the possibility of a coil-to-helix transition. It remains to be determined whether the polymer is in the coil or helical form. The present data suggest a plateau value corresponding to <1400 daltons nm^{-1}. The expected mass per unit length for the polyanion in the 'coil' form would be ~490 daltons nm^{-1}. An eight-fold helix with a pitch of 4.95nm would have a mass per unit length of ~1040 daltons nm^{-1} for the ordered polyanion. Thus the present experimental data coupled with X-ray diffraction studies favour a solution of 8_3 double helices.

The change in optical rotation is too small to measure directly at the low polymer concentrations employed for light scattering studies. At higher polymer concentrations the optical rotation transition has only been observed under conditions where there is clear evidence for intermolecular association and gelation. The present data favour association and crystallisation of segments of 8_3 double helices. Quantitative analysis of the changes in the circular dichroism band of the uronic acid could be used to assess whether changes in the environment of the carboxyl group are solely responsible for the optical rotation transition.

TABLE I: Light scattering data on XM6 solutions

Sample	Solvent	$<M>_W$ daltons x 10^{-6}	$<R_g^2>_Z^{\frac{1}{2}}$
Na XM6	H_2O	2.0 ± 0.3	153 ± 7
TMA XM6	H_2O	1.8 ± 0.3	160 ± 8
TMA XM6	0.5M TMACl	1.4 ± 0.2	165 ± 8

The remaining data have been obtained on solutions of gellan gum. Recently Crescenzi et al. (26,27) have claimed to have observed a thermally-reversible ordered 'helix'-disordered 'coil' transition in dilute gellan gum solutions. In the presence of electrolyte they observed (26,27) a sharp decrease in intrinsic

viscosity and a sharp change in optical rotation on heating
gellan gum solutions. These transitions were found to be
thermally reversible. The transition temperature was found to
decrease with decreasing ionic strength and, at sufficiently low
ionic strength, the transition temperature was less than ambient
temperature and the sample remained in the disordered state.

TABLE II: Light scattering data on gellan gum solutions

Sample	Solvent	$\langle M \rangle_W \times 10^{-6}$	$\langle R_g^2 \rangle_z^{\frac{1}{2}}$ (nm) Zimm	(nm) Kratky	$\langle Rg \rangle_{vis.}$ Viscosity
Native gellan	DMSO	2.2	97	–	90
Gelrite (TMA)	H_2O	2.1	125	130	–
Gelrite (TMA)	0.5M TMA	2.0	142	138	–

Static light scattering studies were performed on
samples of Na gellan and TMA gellan samples in the absence of
added electrolyte. The Na gellan samples were found to be
heavily aggregated and it was not possible to obtain true
solutions. Such aggregation probably accounts for the high
relaxation times observed in previous electric birefringence
studies on gellan samples (21,22) and these data can not be
taken to reflect the intrinsic rigidity of the polysaccharide
structure. Fig. 5 shows a Zimm plot for aqueous TMA gellan.
These conditions correspond to the proposed disordered state.
The calculated molecular weight and radius of gyration are
reported in table II. The data obtained for TMA gellan in 0.5M
TMACl are shown in fig. 6. These conditions correspond to the
proposed 'ordered' state of the polysaccharide. There is
evidence at low polymer concentrations and low scattering angles
for polymer aggregation. Extrapolation of the higher angle data
by the Zimm method has been used to estimate the molecular
weight and radius of gyration of the unaggregated molecules
(table II). Once again the high radii of gyration coupled with
aggregation at high salt concentrations make it difficult to
obtain very accurate estimates of molecular weight. The absence
of any change in molecular weight argues against a coil-double
helix transition. The absence of any change in molecular weight
or radius of gyration for the unaggregated species suggest that
the increase in intrinsic viscosity reported by Crescenzi et al.
(26,27) must reflect the intermolecular aggregation observed in
the present light scattering studies. It remains to be
established whether the unaggregated species in high or low salt
regimes is a helix or coil. The data shown in fig. 4a and table
I are consistent with those reported previously for gellan gum

solutions in dimethyl sulphoxide (21,30). Fig. 7 shows a Kratky plot (29) of the 'zero concentration' scattering data for TMA gellan in the absence of added electrolyte. The plateau region is indicative of a Gaussian chain. The fact that the plateau extends to $U = U^* = 5$ means (29) that the Kuhn statistical segment length l_K will be less than $12\langle R_g^2\rangle_z^{\frac{1}{2}}/(\Pi U^*) = 95$nm. In the absence of any linear increase of the Kratky plot at high U it is impossible to provide a better estimate of l_K or determine the mass per unit length. The plateau value provides an alternative estimate of the radius of gyration. The plateau height is equal to $2\langle M\rangle_W/\langle R_g^2\rangle_z$ giving $\langle R_g^2\rangle_z^{\frac{1}{2}} = 122$nm. This is probably a more accurate estimate of $\langle R_g^2\rangle_z^{\frac{1}{2}}$ because the plateau is easier to define than the slope of the 'zero concentration' data in the Zimm plot. Values of $\langle R_g^2\rangle_z^{\frac{1}{2}}$ obtained from the Kratky plots under low and high salt conditions are listed in table II. The stiffness of the polymer may be assessed by calculating the characteristic ratio C_∞. This compares the square of the measured radius of gyration with the square of the value calculated for a freely jointed chain of n sugar residues each of length l. For gellan n may be taken to be the number of sugar residues in the polymer backbone. If M_S is the molecular weight of the repeat of the polyanion then $n=4\langle M\rangle_W/M_S$. Taking

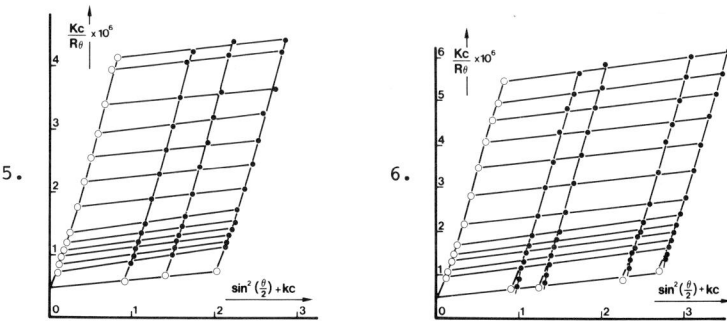

Fig. 5.Zimm plot for TMA gellan in water.

Fig. 6.Zimm plot for TMA gellan in 0.5M TMACl.

$M_S = 643$ daltons, $l = 0.437$nm, $\langle M\rangle_W = 2 \times 10^6$ daltons and $\langle R_g^2\rangle_z^{\frac{1}{2}} = 122$nm gives $C_\infty = 6\langle R_g^2\rangle_z/(nl^2) = 38$. Talashek and Brant (31) have recently calculated a value of $C_\infty = 17.9$ for high molecular weight gellan in the disordered coil state. Unless the sample polydispersity is such that $\langle M\rangle_z/\langle M\rangle_W \gtrsim 2$ then the present data favour a stiffer ordered helical structure in dilute solution. In previous studies of gellan gum solutions in dimethyl sulphoxide (21,30) there was found to be good agreement between the values of the radius of gyration calculated from intrinsic viscosity data and measured by light scattering. In view of the different polydisperse averages measured by the two techniques this result argues against a broad molecular weight

distribution. The present studies favour a model for gelation
involving side-by-side association and crystallisation of
segments of the gellan helices. The order-disorder transition
reported by Crescenzi et al. (26,27) appears to involve melting
and setting of gel precursors.

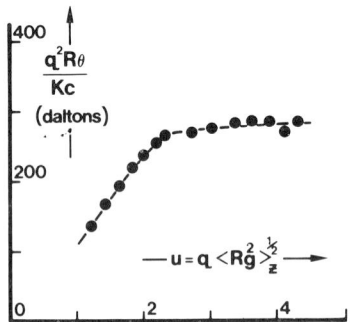

Fig. 7.Kratky plot for TMA gellan in water.

REFERENCES

1.   Nisbet, B.A., Sutherland, I.W., Bradshaw, I.J., Kerr, M.,
      Morris, E.R. and Shepperson, W.A. (1984) Carb. Polym., 4,
      377.
2.   O'Neill, M.A., Morris, V.J., Selvendran, R.R., Sutherland,
      I.W. and Taylor, I.T. (1986) Carb. Res., 148, 63.
3.   Morris, V.J. and Miles, M.J. (1986) Int. J. Biol.
      Macromol., 8, 342.
4.   Atkins, E.D.T., Attwool, P.T., Miles, M.J., Morris, V.J.,
      O'Neill, M.A. and Sutherland, I.W. (1987) Int. J. Biol.
      Macromol. (in press).
5.   Atkins, E.D.T., Attwool, P.T., Sanson, C.E. and Veluraja,
      K. (unpublished work).
6.   Kang, K.S. and Veeder, G.T. (1982) U.S. Patent 4, 326, 053.
7.   O'Neill, M.A., Selvendran, R.R. and Morris, V.J. (1983)
      Carb. Res., 124, 123.
8.   Jansson, P.E., Lindberg, B. and Sandford, P.A. (1983) Carb.
      Res., 124, 135.
9.   Kuo, M-S., Mort, A.J. and Dell, A. (1986) Carb. Res., 156,
      173.
10.   Moorhouse, R., Colegrave, G.T., Sandford, P.A., Baird, J.
      and Kang, K.S. (1981) In 'Solution Properties of
      Polysaccharides" ed. D.A. Brant, ACS Washington, ACS Symp.
      Ser., 150, 111.
11.   Sanderson, G.R. and Clark, R.C. (1983) Food Technol., 37,
      63.

12. Sanderson, G.R. and Clark, R.C. (1984) In 'Gums and Stabilisers for the Food Industry. 2. Application of Hydrocolloids' eds. G.O. Phillips, D.J. Wedlock and P.A. Williams, Pergammon Press, Oxford, p.201.
13. Cottrell, I.W. (1980) In 'Industrial Potential of Fungal and Bacterial Polysaccharides', ACS Washington, ACS Symp. Ser. 126, 251.
14. Harris, J.E. (1985) Appl. Environ. Microbiol., 50, 1107.
15. Lin, C.C. and Casida Jnr., L.E. (1984) Appl. Environ. Microbiol., 47, 427.
16. Shungu, D., Valiant, M., Tutlane, V., Weinburg, E., Weisburger, B., Koupal, L., Gadebusch, H. and Stapley, E. (1984) Appl. Environ. Microbiol., 46, 840.
17. Sandford, P.A., Cottrell, I.W. and Pettitt, D.J. (1984) Pure Appl. Chem., 56, 879.
18. Pettitt, D.J. (1986) In 'Gums and Stabilisers for the Food Industry 3' eds. G.O. Phillips, D.J. Wedlock and P.A. Williams, Elsevier Appl. Sci., London, p.451.
19. Chilvers, G.R. and Morris, V.J. (1987) Carb. Polym. (in press).
20. Morris, V.J. (1987) In 'Food Technology - 1' eds. R.D. King and P.S.J. Cheetham, Elsevier Appl. Sci., London, Chap. 5.
21. Miles, M.J., Morris, V.J. and O'Neill, M.A. (1984) In 'Gums and Stabilisers for the Food Industry. 2. Application of Hydrocolloids', eds. G.O. Phillips, D.J. Wedlock and P.A. Williams, Pergammon Press, Oxford, p. 485.
22. Carroll, V., Chilvers, G.R., Franklin, D., Miles, M.J., Morris, V.J. and Ring, S.G. (1983) Carb. Res., 114, 181.
23. Carroll, V., Miles, M.J. and Morris, V.J. (1982) Int. J. Biol. Macromol., 4, 432.
24. Attwool, P.T., Atkins, E.D.T., Upstill, C., Miles, M.J. and Morris, V.J. (1986) In 'Gums and Stabilisers for the Food Industry. 3.', eds. G.O. Phillips, D.J. Wedlock and P.A. Williams, Elsevier Appl. Sci., London, p. 135.
25. Upstill, C., Atkins, E.D.T. and Attwool, P.T. (1986) Int. J. Biol. Macromol., 8, 275.
26. Crescenzi, V., Dentini, M., Coviello, T. and Rizzo, R. (1986) Carb. Res., 149, 425.
27. Crescenzi, V., Dentini, M. and Dea, I.C.M. (1987) Carb. Res., 160, 283.
28. Huglin, M.B. (1972) In 'Light Scattering from Polymer Solutions', ed. M.B. Huglin, Academic Press, London and New York, p. 165.
29. Schmidt, M., Paradossi, G. and Burchard, W. (1985) Makromol. Chem. Rapid Commun., 6, 767.
30. Brownsey, G.J., Chilvers, G.R., I'Anson, K. and Morris, V.J. (1984) Int. J. Biol. Macromol., 6, 211.
31. Talashek, T.A. and Brant, D.A. (1987) Carb. Res., 160, 303.

Studies on the mechanism of gelation for xanthan – galactomannan and xanthan – glucomannan mixed gels

G.J.Brownsey, P.Cairns, M.J.Miles and V.J.Morris

AFRC Institute of Food Research, Norwich Laboratory, Colney Lane, Norwich NR4 7UA, UK

ABSTRACT

Studies have been made to investigate the molecular basis of the synergistic behaviour of xanthan-galactomannan and xanthan-glucomannan gels. X-ray diffraction studies of aligned fibres prepared from the mixed gels provide evidence for molecular binding between xanthan and the glucomannan konjac mannan and also between xanthan and the galactomannans carob gum and tara gum. Rheological measurements reveal the need to denature the xanthan helix in order to obtain inter-molecular binding and gelation. Present data favour a binding between the cellulosic backbone of xanthan and the stereochemically compatible backbones of the galactomannans and the glucomannan.

KEYWORDS: Xanthan, Carob gum, Tara gum, Konjac mannan, gels, gelation, X-ray diffraction.

INTRODUCTION

Xanthan gum is a bacterial polysaccharide produced commercially from Xanthomonas campestris. Carob gum and tara gum are galactomannans which are extracted respectively from the seeds of the trees Ceratonia siliqua and Caesalpinia spinosa. Xanthan-carob and xanthan-tara mixtures form thermo-reversible gels under conditions for which neither xanthan nor the galactomannans alone will gel (1-3). Dea and coworkers (2,3) have attributed gelation to an intermolecular binding between the two polysaccharides. These authors favoured a binding between the xanthan helix and regions of the galactomannan backbone depleted in galactose substituents. Recent X-ray diffraction studies (4-6) have confirmed the existence of xanthan-galactomannan binding but suggest an interaction between the stereochemically compatible cellulosic backbone of xanthan and the mannan backbone of the galactomannan. The X-ray diffraction data will be described and used to discuss the mechanism of gelation.

Konnyaku (a gel of konjac mannan) is a popular food in Japan. Konjac mannan is a glucomannan extracted from the tubers of

Amorphophallus konjac. Sols of konjac mannan will gel when
heated at alkaline pH (7). The alkali treatment is believed to
deacetylate the polysaccharide. However, the role played by
acetyl groups in inhibiting gelation and the molecular basis for
gelation are still incompletely understood. X-ray diffraction
studies of native konjac mannan will be described and used to
discuss the molecular basis of gelation.

There is experimental evidence for the growth of deacetylated
glucomannan crystals at the surface of cellulose fibrils (8).
The similar crystal structures reported for deesterified
glucomannans and cellulose suggest the possibility of intermolecular binding and gelation of xanthan-glucomannan mixtures.
Since acetylation of the glucomannan inhibits gelation, and in
order to avoid complications due to the gelation of the
glucomannan alone, studies have been confined to xanthan-native
konjac mannan mixtures. Rheological data will be presented to
demonstrate experimental conditions under which such mixtures
will gel. These data together with X-ray diffraction studies of
mixed gels have been used to investigate the molecular basis for
gelation.

MATERIALS AND METHODS

Samples of xanthan and carob gum were purchased from Sigma
Chemicals. The sample of tara gum was a gift from Dr. C. Blood.
Konjac flour was purchased from Senn Chemicals. The flour was
dispersed in water, heated to 90°C, cooled to room temperature
and left to stand overnight. The dispersion was filtered
through a glass sinter to remove insoluble matter.

Fibres for X-ray diffraction studies of gels were prepared by
cutting the gels into strips which were then stretched under
conditions of controlled relative humidity and temperature.
Samples which did not gel were poured onto Teflon or glass
substrates and partially dried to form films. Oriented fibres
were prepared from these films. X-ray data were recorded
photographically. The interior of the camera was flushed with
helium to reduce background scattering. Cu K_α radiation
(wavelength 0.154nm) was used and calcite was dusted onto the
fibres for calibration purposes. Creep compliance studies were
performed using an Instron 3250. Optical rotation was measured
using an AA-100 polarimeter.

RESULTS AND DISCUSSION

Xanthan-galactomannan gels

Typical x-ray fibre patterns obtained for xanthan, carob and
tara are shown in fig. la-c. X-ray diffraction patterns
obtained for fibres prepared from carob-xanthan and tara-xanthan
mixed gels are shown in figs. 1 d,e. It is clear from fig. 1
that the fibres prepared from the mixed gels yield new X-ray
fibre patterns providing experimental evidence for xanthan-
galactomannan binding in these two systems. Intermolecular
binding between xanthan and certain galactomannans was first
suggested by Dea and coworkers. These authors (2,3) suggested a
cooperative interaction in which the helical structure of the

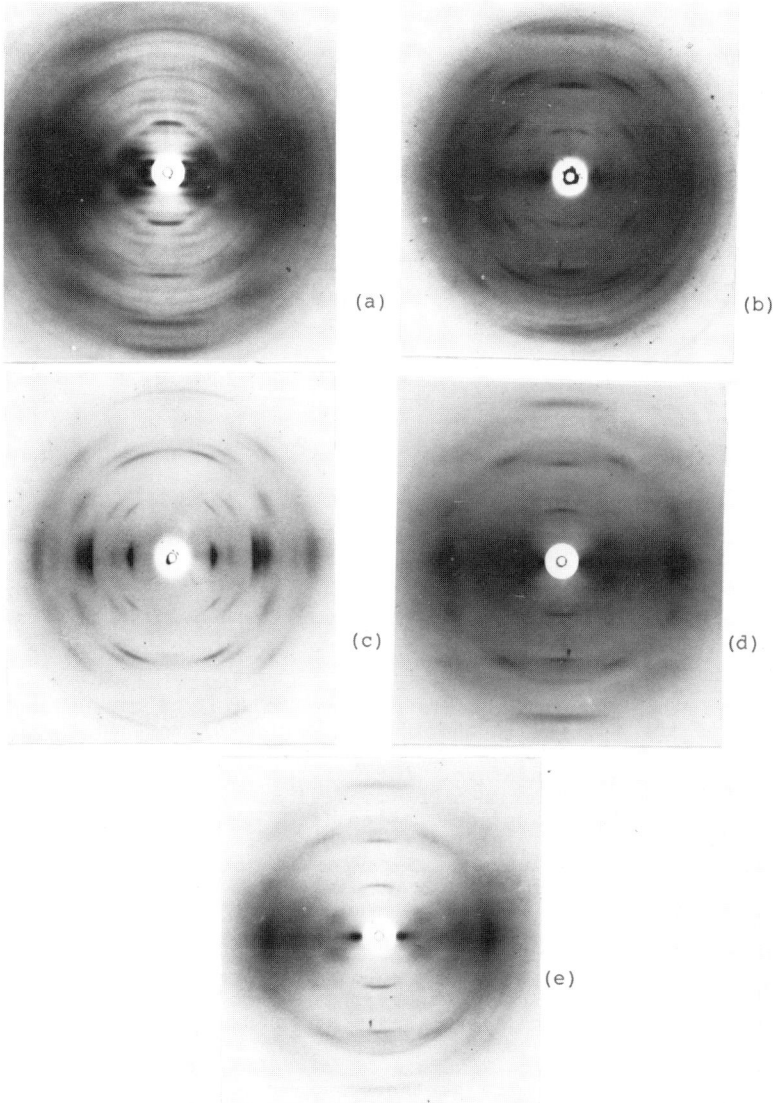

Fig. 1. X-ray fibre diffraction patterns obtained at a
wavelength of 0.154nm and a relative humidity of ~98%.
(a) xanthan, stretched ~300%, (b) carob, stretched 300%
- stored ~3 years at 20°C, (c) tara, stretched 200%,
(d) carob(65%)-xanthan(35%) gel, stretched 300%, (e)
tara(55%)-xanthan(45%) gel, stretched 300%.

xanthan molecule is retained and binds to regions of the galactomannan backbone depleted in galactose. The stoichiometry of the suggested interaction is undefined and there is no obvious steric compatibility between the mannan backbone of the galactomannan and the xanthan helix. This makes it difficult to construct a 3D model of the proposed interaction and assess its merits in terms of the measured X-ray fibre patterns (figs. 1d,e).

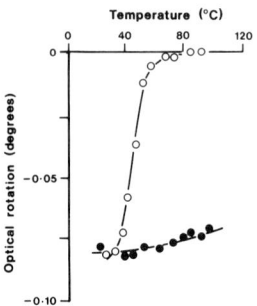

Fig. 2. Optical rotation measured for a 0.5% Na xanthan solution at a wavelength of 589nm. o - xanthan in water, • - xanthan in 0.5M CaCl$_2$

The following mixing experiments were carried out to test whether the xanthan helix is retained within the mixed junction zone of the two polysaccharides. Samples of xanthan and carob were separately dispersed in water at 95°C and each sample cooled to room temperature. Optical rotation studies, which are sensitive to the helix-coil transition, were used to monitor helix formation in xanthan samples (fig. 2). Thorough mixing of the xanthan (helical form) with carob at room temperature (25°C) did not lead to gelation. X-ray diffraction patterns obtained for fibres prepared from such mixtures were poor but showed reflections characteristic of xanthan alone. The absence of reflections attributable to pure carob is not surprising because crystalline carob patterns of the type shown in fig. 1 are only obtained upon annealing or prolonged storage of carob fibres. When the non-gelling xanthan-carob samples were heated to 95°C and then cooled to room temperature they gelled. Fibres prepared from such gels gave the new X-ray diffraction pattern shown in fig. 1d. Heating is clearly necessary to induce gelation. To determine whether heating merely enhanced mixing or was necessary to denature the xanthan helix, a third experiment was performed in which xanthan and carob were mixed at high temperatures, but with xanthan in the helical form. This was achieved by adding calcium chloride to raise the helix-coil transition temperature above 100°C (fig. 2). When such mixtures were heated and recooled they did not gel.

These experiments suggest molecular binding between the denatured cellulosic backbone of xanthan and segments of galactomannan chain. Intermolecular binding is attributed to the stereochemical compatibility between the mannan backbone of the galactomannan and cellulosic backbone of xanthan which permits co-crystallisation. The reflections present in the diffraction pattern which are attributable to the mixed junction zones alone are equivalent to a galactomannan lattice for which only 0kl reflections are allowed. This situation could arise if in the a-direction the crystal is small or imperfect. It is possible to envisage molecular mechanisms (4,5) by which xanthan molecules could poison growth of galactomannan crystals in the a-direction. However, the resultant laminate structures should give rise to streaking of reflections on all layer lines. This was not observed experimentally suggesting that the a-direction is aperiodic. A simple model (4-6) would be to envisage a random co-crystallisation of xanthan and the galactomannan leading to a galactomannan lattice threaded with xanthan molecules. The trisaccharide side-chains of xanthan are larger than the galactose side-chains of the galactomannans and may act as defects leading to aperiodicity in the a-direction.

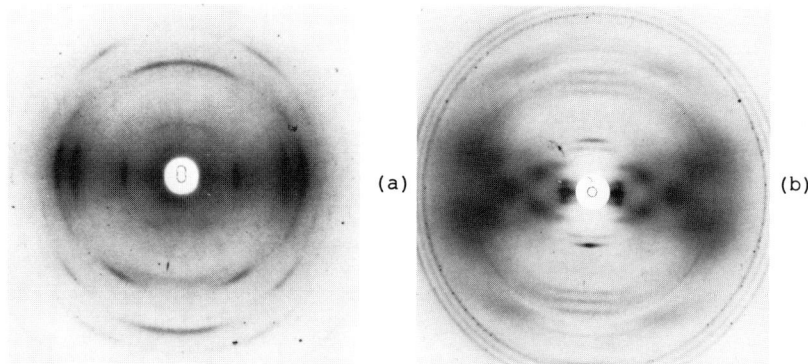

Fig. 3. X-ray fibre diffraction patterns obtained at a
       wavelength of 0.154nm and a relative humidity of ~98%.
       (a) native konjac mannan, stretched 100%, (b) xanthan
       (50%)-native konjac mannan (50%), stretched 200%

Native konjac mannan

Fig. 3a shows an X-ray diffraction pattern obtained from fibres prepared from native 'acetylated' glucomannan. The patterns obtained are highly crystalline and may be indexed onto a mannan II lattice. Deacetylated konjac mannan is reported (8) to crystallise in the mannan II lattice. Fig. 3a clearly demonstrates that the acetyl substituents do not prevent sterically molecular association and crystallisation.

Acetylation does not preclude gelation and presumably
deesterification upon alkali treatment results in an increased
tendency to crystallise and gel. The position of the ester
substituents is unknown. A possible explanation for a decrease
in polymer solubility is that deesterification increases the
number of potential inter-molecular hydrogen bonds.

Xanthan-konjac mannan gels

There is evidence for the affinity of mannan-based and
cellulose-based polysaccharide structures. Chanzy et al. (8)
report evidence for the growth of glucomannan crystals on the
surface of cellulose fibrils. Rydholm (9) reports
reprecipitation of glucomannan on wood pulp during pulping of
softwood. Such observations suggest a possible intermolecular
binding between glucomannans and the cellulosic backbone of
xanthan. Initially gelation of xanthan-konjac mannan mixtures
was used as an indirect indicator of intermolecular binding. To
eliminate complications arising from the gelation of the
glucomannan component alone studies were made on xanthan-native
(acetylated) konjac mannan. Samples of xanthan and konjac
mannan were separately dispersed in water at 95°C and each
sample cooled to room temperature. Neither sample at 1 or 2%
total polymer concentration gelled on cooling to room
temperature. Mixtures of xanthan and native konjac mannan (1:1)
at a total polymer concentration of 1% were prepared in various
ways. Optical rotation was used to monitor the conformation of
the xanthan molecules at concentrations, temperatures and ionic
strengths equivalent to that within the mixture (fig. 2).
Simple creep compliance tests (fig. 4) were used to distinguish
between gels and viscous liquids. The samples were subjected to

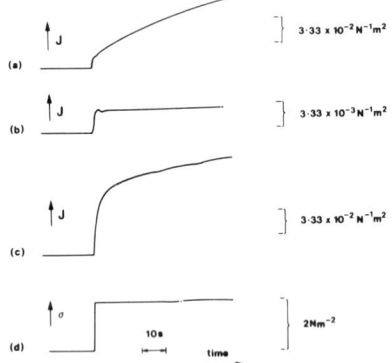

Fig. 4. Creep compliance (J) studies on xanthan-native konjac
mannan mixtures. Total polymer concentration 1%. (a)
mixtures prepared at 25°C, (b) mixtures heated to 95°C
and recooled to room temperature (25°C), (c) mixtures
in the presence of 0.5M $CaCl_2$ heated to 95°C and then
recooled to room temperature, (d) applied stress profile
(σ). The compliance values are calculated from the
maximum stress in the parallel plate configuration.

a step-wise increase in stress (fig. 4d). Mixtures prepared at
room temperature (fig. 4a) with xanthan in the helical state
(fig. 2) did not gel. Mixtures prepared at 95°C (fig. 4b) under
conditions which lead to denaturation of the xanthan helix (fig.
2) did gel on cooling to room temperature. In the presence of
added calcium chloride the helix-coil transition temperature for
xanthan can be raised above 100°C (fig. 2). Mixtures of xanthan
and native konjac mannan prepared at 95°C (fig. 4c), with
xanthan in the helical form, did not gel upon recooling to room
temperature. These data suggest intermolecular binding between
the glucomannan and the denatured cellulosic xanthan backbone.
X-ray diffraction studies of fibres prepared from oriented gels
provide a direct method for probing inter-molecular binding.
Fig. 3b shows an X-ray fibre diffraction pattern for a xanthan-
native konjac mannan gel. The patterns obtained for the mixed
gels are similar to those obtained for xanthan alone. The most
obvious difference is the different layer-line spacing. For
example, the layer lines corresponding to $\ell = 1,2,3$ indicate a
helical pitch of 5.6nm compared to that of 4.7nm observed for
xanthan alone. The new X-ray diffraction patterns obtained
provide direct evidence for inter-molecular binding but
interpretation of such patterns is complicated by the complex
and possibly irregular structure of the glucomannan (10).

REFERENCES

1. Morris, E.R., Rees, D.A., Young, G., Walkinshaw, M.D. and
    Darke, A. (1977) J. Mol. Biol., 110, 1.

2. Dea, I.C.M. and Morrison, A. (1975) Adv. Carb. Chem.
    Biochem., 31, 241.

3. Dea, I.C.M., Morris, E.R., Rees, D.A., Welsh, E.J., Barnes,
    H.A. and Price, J. (1977) Carbohydr. Res., 57, 249.

4. Cairns, P., Miles, M.J. and Morris, V.J. (1986) Nature,
    322, 89.

5. Cairns, P., Miles, M.J., Morris, V.J. (1987) Carbohydr.
    Res., 160, 411.

6. Morris, V.J. and Miles, M.J. (1986) Int. J. Biol.
    Macromol., 8, 342.

7. Maekaji, K. (1974) Agr. Biol. Chem., 38, 315.

8. Chanzy, H.D., Grosrenaud, A., Joseleau, J.P., Dube, M. and
    Marchessault, R.H. (1982) Biopolymers, 21, 301.

9. Rydhdm, S. (1965) "Pulping processes", Interscience, New
    York.

10. Takahashi, R., Kusakabe, I., Kusama, S., Sakurai, Y.,
    Murakami, K., Maekawa, A. and Suzuki, T. (1984) Agric.
    Biol. Chem., 48, 2943.

Viscoelastic response of xanthan gum/guar gum blends

R.C.Clark

Kelco Division of Merck & Co. Inc., San Diego, CA, USA

ABSTRACT

The viscoelastic properties of xanthan gum/guar blends have been measured using various blend ratios, ion levels and total gum levels. These data show that the interaction has its maximum "synergism" at blend ratios of roughly 3:2 (xanthan gum:guar gum) in the absence of added ions. This optimal gum ratio appears constant with changes in total gum level. Addition of either sodium ions (at levels $> 10^{-3}$ molar) or calcium ions (at levels $> 5 \times 10^{-4}$ molar) diminishes the interaction significantly.

INTRODUCTION

It has been known for some time that xanthan gum and various galactomannan polysaccharides such as guar gum, locust bean gum and tara gum can interact in a "synergistic" manner. It is also known that galactomannans with a relatively unsubstituted mannose backbone can interact with xanthan gum to form soft, elastic gels. Mixing xanthan gum with galactomannans having a more even and/or higher level of substitution results in a viscosity that is higher than that of either polymer under certain conditions {1} {2} {3}.

This interaction is important to the food industry as it is used in applications that require lower cost or a lower level of stability than provided by xanthan gum alone {4}. Examples include dairy products, sauces, soups, frozen foods and certain salad dressings.

Since certain of these applications depend upon specific rheological properties, additional information above and beyond the normal steady shear viscosity measurements is desirable. For example, a relatively high elastic modulus is required to stabilize an emulsion effectively. On the other hand, a high degree of elasticity is *not* desired in a soup or most sauces. Since all foods contain a variety of ions, it is also necessary to know how the xanthan–galactomannan interaction changes upon the addition of certain ions.

MATERIALS AND METHODS

The xanthan gum and guar gum samples were commercial samples used without further purification or modification. The xanthan gum sample was KELTROL T (KELTROL is a Trademark of Merck & Co., Inc. [Rahway, New Jersey], Kelco Division, USA) and SUPERCOL U (SUPERCOL is a Trademark of Aqualon Co., Wilmington, Deleware) was used as the source of guar.

To allow for the moisture inherent in all polysaccharide samples, solutions were prepared on a moisture corrected basis. Moisture contents were measured by using a microwave analyzer. Values were taken immediately before preparing a solution and ranged from 9% to 11% by weight. No correction was made for actual polysaccharide content in either the guar gum or xanthan gum samples.

Mixing was carried out with a 3 blade propeller type mixer operating at a speed just below the point where air entrainment occurs (600 to 1200 rpm). The appropriate amount of gum was added to the water phase (containing added salts, if needed) with the usual care to avoid lumping. As the sample thickened, the stirrer speed was increased to ensure good mixing. Although no evidence of unhydrated gum particles could be seen after 20 to 30 minutes of mixing, a 2 hour stirring time was used for all samples. The solutions were then allowed to remain at room temperature (22°) for 20 to 24 hours before being measured.

Rheological measurement of the solutions was carried out with a Rheometrics Fluids Rheometer (RFR–7800) using Couette fixtures having a R_i/R_0=0.967 and L_b=35.5 mm. The instrument was equipped with a dual 10/100 gm cm transducer so a wide range of torque values from about 5×10^{-3} to 10^{+2} gm cm were able to be measured accurately. Samples were loaded into the instrument and allowed to "rest" for 5 minutes before being measured. A specially constructed cap was placed over the top of the Couette cup to minimize sample evaporation. Except as specially noted on the figures, all viscoelastic measurements were taken using a 10% strain level over a frequency range from 0.1 to 63 rad/sec. Instrument response at frequencies greater than this was deemed not acceptable.

For readers not familiar with viscoelastic measurements, the following brief background is given (see {5} for additional background). Nearly all materials have a certain degree of solid–like character (elasticity) and a certain degree of liquid–like character (viscosity). In other words, most materials are *viscoelastic* in nature. One common method of measuring this involves applying a sinusoidally varying **strain** or deformation to a sample and recording the resulting **stress** or force. A perfectly elastic material will show no phase shift between the applied strain wave and the resulting stress wave. The ratio between these two wave amplitudes can be used to calculate the elastic modulus of the sample. On the other hand, a perfectly viscous material will show an exact 90° phase shift between the two wave forms. Here, the ratio of the two wave amplitudes is termed the viscous modulus. All materials have a phase shift between 0° (elastic) and 90° (viscous) which can be determined along with the magnitude of the resulting stress.

RESULTS AND DISCUSSION

The results of this investigation are divided into three sections. First, the basic rheological properties of xanthan gum, xanthan gum/guar gum blends and guar gum are described (figures 1-3). Next, the effect of various blend ratios on xanthan gum/guar gum blends is described for low gum levels (figures 4 & 5) and for higher levels (figures 6 & 7). Finally, the effects of increasing levels of sodium ions (figures 8 & 9) and calcium ions (figures 10 & 11) are covered.

Section One

The viscoelastic response of 0.5% xanthan gum prepared in 0.01 molar NaCl at a 10% strain level is shown in figure 1. It is seen that the fluid is primarily elastic in its response since the G' value (elastic modulus) is higher than the G" value (viscous modulus) at all frequencies. In rheological terms, these xanthan solutions are considered to be gel–like since the slopes of the moduli curves are relatively flat with respect to frequency and G' is higher than G". The quantity labeled as "Eta *" (η*) is the complex viscosity. It is calculated by dividing the vector sum of G' and G" by the deformation frequency in radians per second. For most materials, η* correlates with the steady shear viscosity, *eta* (η). However, in the case of highly elastic fluids like xanthan gum solutions, η* can be several percent larger than η, especially at low imposed strain levels. This is thought to be due to the large amount of "structure" or association that exists between chains in xanthan gum solutions. Taken together, these data confirm what is well known about xanthan gum solutions. They are gel–like in nature and have a pseudoplastic (or "shear thinning") rheological flow profile.

Figure 1. Rheological Properties of 0.5% Xanthan Gum in 0.01 Molar NaCl

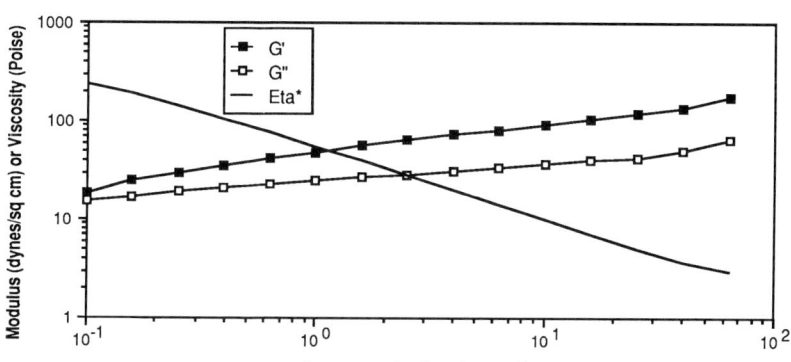

Figure 2 shows the data for a solution of 0.5% xanthan gum/guar gum (1:1) prepared in 0.01 molar NaCl. Qualitatively, these data are similar to the xanthan gum data in figure 1. This sample is also gel–like although the G' value is closer to the G" value at all frequencies. This indicates that the strength of the chain–chain associations is diminished for the blend. Such blends are also pseudoplastic in nature but the blend complex viscosity (@ 0.1 rad/sec) is about 85 Poise compared to about 230 Poise for the unblended xanthan sample. Xanthan gum/guar gum blends do not always have a higher viscosity than unblended xanthan.

Figure 2. Rheological Properties of 0.5% Xanthan Gum/Guar Gum (1:1) Blend in 0.01M NaCl

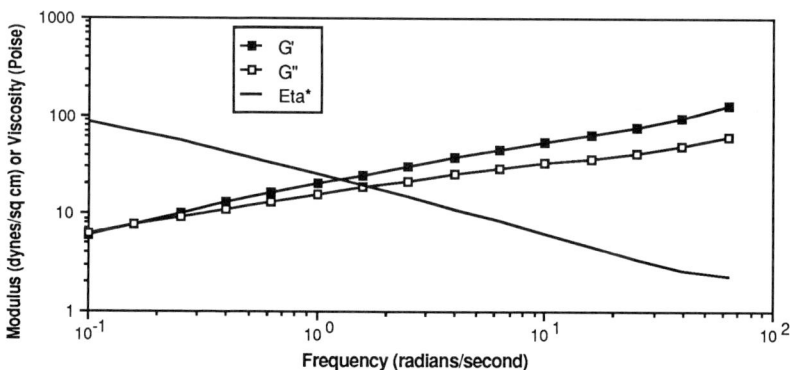

The properties of the unblended guar gum solution shown in figure 3 are different. This sample is not gel–like since the moduli curves have noticeably more slope and G' < G" for most frequencies. The sample is also less pseudoplastic and there are indications of a lower Newtonian region at low frequencies. If a sample of guar is made up to the same steady shear viscosity value (η) as xanthan, it will be less elastic or gel–like. Solutions of guar gum alone will not have the suspending ability of xanthan gum or xanthan gum/guar gum blends. This property results from the elasticity that is present in the apparently liquid xanthan gum sample.

R.C.Clark

Figure 3. Rheological Properties of 0.5% Guar Gum in 0.01M NaCl

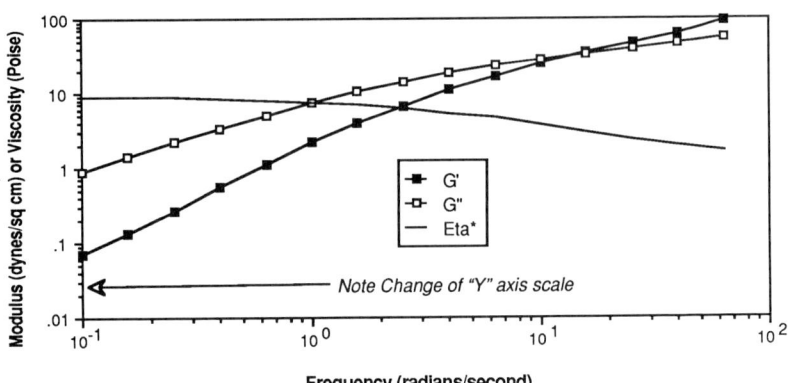

Frequency (radians/second)

Section Two

The data in figure 4 show how various blend ratios affect the complex viscosity and phase angle of xanthan gum/guar gum blends at 0.1% total concentration in deionized (DI) water. The maximum for eta * occurs at a blend ratio of about 3:2 (xanthan gum:guar gum) which agrees closely with previous findings under similar conditions {6}. It is also interesting to note that the solution is most elastic at this blend ratio, i.e., has the smallest phase angle. This indicates that this is the point at which the associations between chains are the strongest. It also confirms that xanthan gum/guar gum blends build viscosity mostly by increased elasticity. In deionized water, there is what would be termed a "synergistic" interaction, that is, the viscosity of the blend at a 3:2 ratio is greater than either blend component.

It is noteworthy that unblended xanthan gum is not very elastic at this concentration indicating a low degree of chain association.

Figure 4. Rheological Properties of 0.1% Xanthan Gum/Guar Gum (1:1) in DI Water

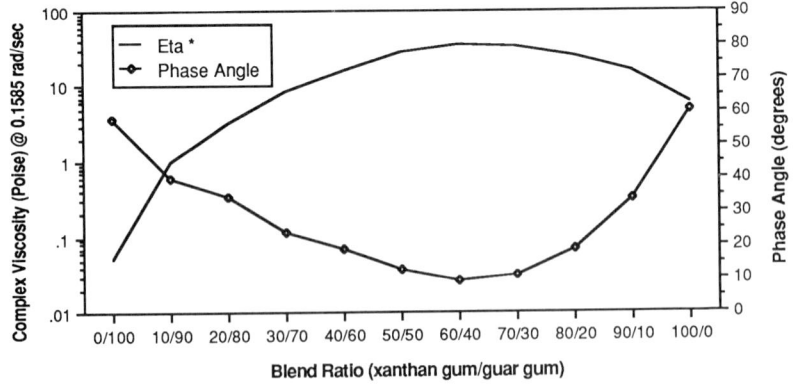

Blend Ratio (xanthan gum/guar gum)

The effect of hydrating the xanthan gum/guar gum solution in 0.5% NaCl (0.0856 molar) is shown in figure 5. These data indicate that the interaction between the two hydrocolloids decreases significantly in the presence of sodium ions. All blend ratios have less viscosity than the unblended xanthan gum. There does seem to be a small amount of interaction present at blend ratios of about 1:9 (xanthan/guar) since the viscosity increases markedly at that point and the sample becomes more elastic. However these data are suspect since they were collected at the torque threshold of the rheometer.

Figure 5. Rheological Properties of 0.1 % Xanthan Gum/Guar Gum (1:1) in 0.5% NaCl

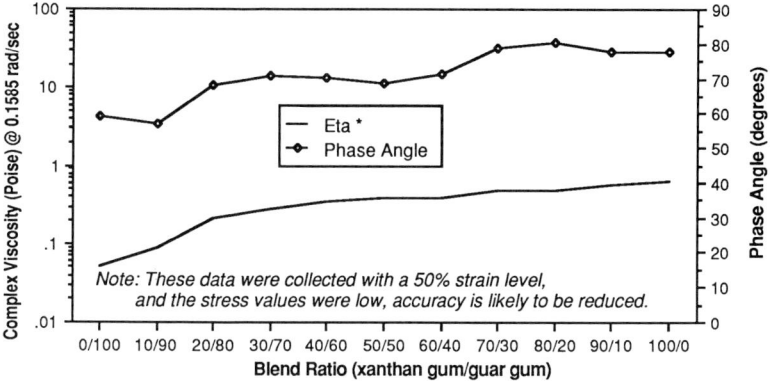

Figure 6 gives data for various blend ratios at 0.5% total gum hydrated in DI water. As was observed with the samples made at 0.1% in DI water, there is a clear indication of an interaction with a maximum at around a 3:2 blend ratio. At this ratio the sample was also the most elastic (minimum phase angle) indicating the highest degree of chain to chain association. Figures 4&6, show the optimum blend ratio in DI water is *independent* of gum concentration.

Figure 6. Rheological Properties of 0.5% Xanthan Gum/Guar Gum (1:1) in DI Water

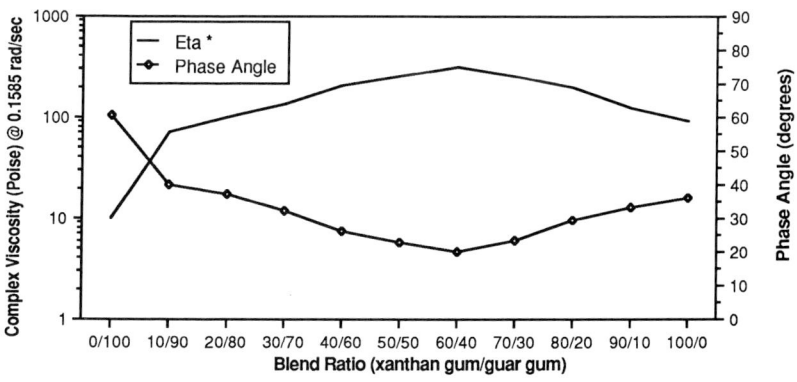

When the 0.5% solutions were prepared in 0.5% NaCl the interaction was again diminished (figure 7). The highest viscosity was obtained with the unblended xanthan gum. Again there is a disproportionate increase in viscosity at blend ratios around 1:4. The maximum elasticity (minimum phase angle) occurs at a blend ratio of 3:7. This is not the point of maximum viscosity, however. There does seem to be a degree of chain to chain association when sodium ions are present but this is diminished as compared to the situation in DI water.

Figure 7. Rheological Properties of 0.5 % Xanthan Gum/Guar Gum (1:1) in 0.5% NaCl

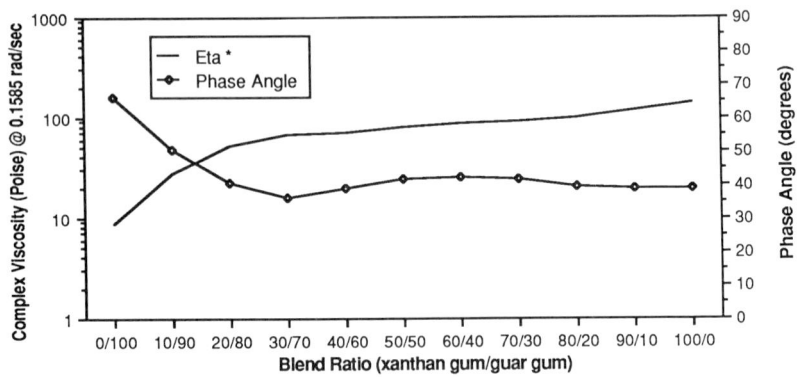

Section Three
The previous data show the xanthan gum/guar gum interaction is affected by added ions. In the following figures, the reasons behind this are examined.

Figure 8 shows how the rheological properties (elastic modulus (G'), viscous modulus (G") and phase angle) are affected by changes in sodium ion concentration (added as NaCl). Elastic modulus and phase angle data both indicate a point of maximum chain to chain interaction at a sodium concentration of about 1×10^{-3} molar.

Figure 8. Effect of Sodium Level on 0.5% Xanthan Gum Rheological Properties

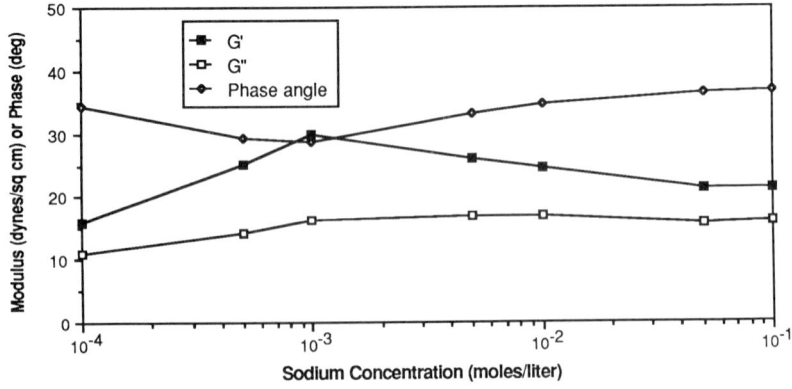

In figure 9 it is seen that the interaction between xanthan and guar starts to decrease when sodium levels increase above 10^{-3} molar sodium. Above this level, the elastic modulus falls sharply and the phase angle goes from 28° to over 45° (indicating a diminished association between the polymer chains). This change occurs at the same ion level as the viscosity decrease in the unblended xanthan gum, it appears that the xanthan conformation change may be responsible for most of the decrease.

Measured, but not shown here, were data on the effect of sodium ions on guar gum. These data show no ionic effect in the range tested as would be expected from a non–ionic gum.

Figure 9. Effect of Sodium Level on 0.5% Xanthan Gum/Guar Gum (1:1) Rheological Properties

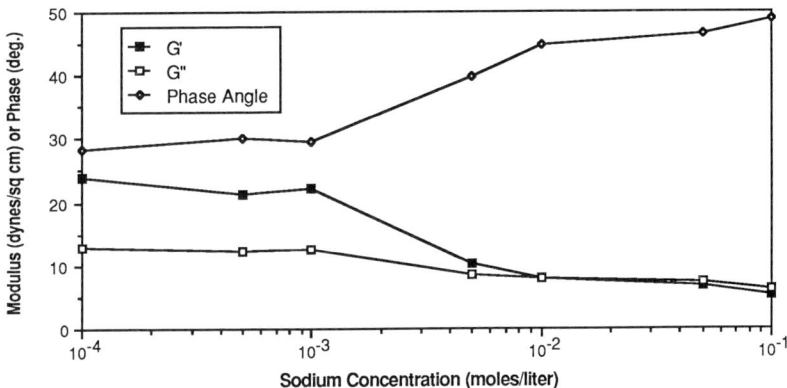

Data for xanthan gum solutions containing calcium ions (figure 10) show a trend similar to that seen for sodium ions. The main difference between Na and Ca is the ion level at which maximum association occurs. From these data, this appears to be 5 x 10^{-4} molar calcium as CaCl$_2$. At this concentration, the phase angle is at a minimum and G' is at a maximum. Higher levels of calcium ions diminish the interactions within the xanthan gum solutions.

Figure 10. Effect of Calcium Level on 0.5% Xanthan Gum Properties

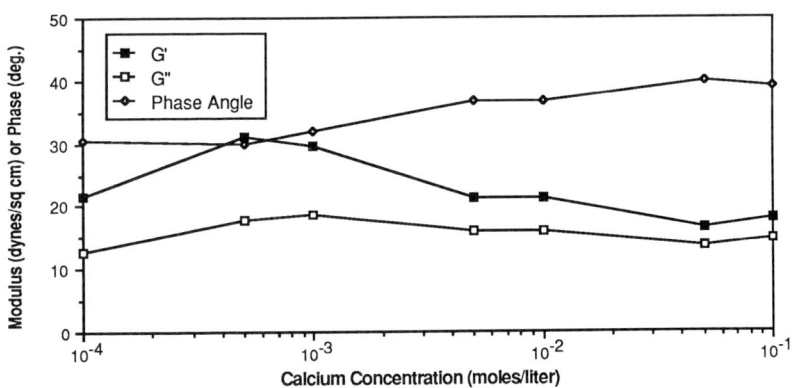

With the information on the effect of calcium on xanthan (figure 10) it is not surprising to find the same trend for the effect of added Ca on xanthan gum/guar gum blends (figure 11). At calcium ion concentrations of about 5 x 10^{-4} molar, the blend interaction begins to decrease. The moduli decrease more and the phase angle increases more with continued addition of calcium ions, indicating that calcium is more effective than sodium at inhibiting the xanthan gum/guar gum chain to chain associations.

Figure 11. Effect of Calcium Level on 0.5% Xanthan Gum/Guar Gum (1:1) Rheological Properties

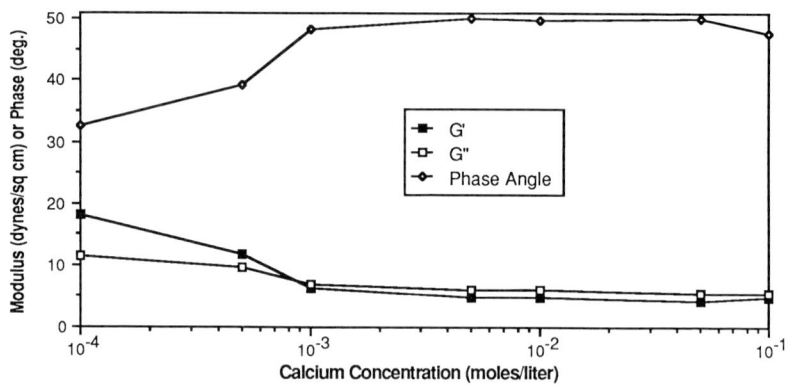

Calcium Concentration (moles/liter)

CONCLUSIONS
- Under certain conditions, the rheological properties of xanthan/guar blends are similar to unblended xanthan gum. These blend rheological properties are different from unblended guar.
- In deionized water the optimum blend ratio is about 3:2 (xanthan gum/guar gum).
- Above about 1 x 10^{-3} molar Na or 5 x 10^{-4} molar Ca^{+2}, the interactions between xanthan gum and guar gum are greatly diminished.
- The conformation of the xanthan gum molecule is known to change due to charge screening at concentrations similar to those mentioned above. It therefore seems likely that it is the conformation of the xanthan molecule which largely controls the interaction changes at different ion levels.

REFERENCES

1 Tako, M. and Nakamura, S. (1986) FEBS Letters, 204 (1), 33–36.
2 McCleary, B. V. and Neukom, H. (1982) Prog. Fd. Nutr. Sci. 6, 109–118.
3 McCleary, B. V. (1984) Carbohydrate Polymers, 4, 253–270.
4 Collyer, S. G. (1984) in Gums and Stabilizers for the Food Industry 2 Application of Hydrocolloids, (ed. G. O. Phillips, D. J. Wedlock and P. A. Williams) pp 349–355. Pergamon Press,Oxford, UK.
5 Morris, E. R. (1984) in Gums and Stabilizers for the Food Industry 2 Application of Hydrocolloids, (ed. G. O. Phillips, D. J. Wedlock and P. A. Williams) pp 57–78. Pergamon Press,Oxford, UK.
6 Tako, M. and Nakamura S. (1985) Carbohydrate Research, 138, 207–213.

Copyright © 1987 Merck & Co., Inc. (Rahway, New Jersey), Kelco Division, USA All Rights Reserved

Sidechain – mainchain interactions in bacterial polysaccharides

G.Robinson, C.E.Manning, E.R.Morris and I.C.M.Dea[+]

Department of Food Research and Technology, Cranfield Institute of Technology, Silsoe College, Silsoe, Bedford MK45 4DT, UK
[+]Leatherhead Food R.A., Randalls Road, Leatherhead, Surrey KT22 7RY, UK

ABSTRACT

The specific role of sidechains in directing and controlling the conformation and functional interactions of bacterial poly- saccharides is illustrated for the gellan/welan family, poly- saccharide XM-6, xanthan and rhizobium capsular polysaccharide, and contrasted with the less-specific solubilising influence typical of branched polysaccharides of plant origin.

A   →3)-β-D-Glc-(1→4)-β-D-GlcA-(1→4)-β-D-Glc-(1→4)-α-L-Rha-(1→

B   →3)-β-D-Glc-(1→4)-β-D-GlcA-(1→4)-β-D-Glc-(1→4)-α-L-Rha-(1→
                                                     3
                                                    ↑
                                                    1   β-D-Gal
                                        α-L-Rha   1
                                                          ↓
                                                          4

C   →3)-β-D-Glc-(1→4)-α-D-GlcA-(1→3)-α-L-Fuc-(1→   β-D-Gal
                4   1
               ↑   ↓
               1   6
     β-D-Glc   D   →3)-α-D-Man-(1→3)-β-D-Gal-(1→4)-α-D-Glc-(1→
                                                                  2
                                                                 ↑
                                                                 1
                                                          α-D-Gal

(All residues are in the pyranose ring form)

Figure 1. Repeating sequences of A) gellan, B) welan, C) poly- saccharide XM-6 and D) rhizobium capsular polysaccharide.

G.Robinson et al.

INTRODUCTION

In plant polysaccharides the usual effect of sidechains linked covalently to the polymer backbone is to confer enhanced solubility. Two examples of this type of behaviour are discussed in detail elsewhere in this volume: in the galactomannan series (which includes locust bean gum, guar gum and gum tara) galactose sidechains act as 'spacers' which inhibit ordered close-packing of the parent mannan in the solid state (1) and therefore make it easier for chains to dissociate and go into solution; similarly in tamarind galactoxyloglucan (2) the cellulose backbone is solubilised by sidechains attached to approximately two-thirds of the glucosyl residues (a degree of substitution close to that in guar gum).

In both these cases the sidechains are 1,6-linked to the polymer backbone, and this is a common, though by no means exclusive, structural motif in branched polysaccharides of plant origin. Since the 1,6 linkage involves three covalent bonds (rather than two in the case of attachment through a hydroxyl group of the sugar ring) this mode of linkage confers an additional drive to solubility through greater loss of conformational entropy on adoption of a packed, solid-state structure (3).

Sidechains in bacterial polysaccharides can have a similar solubilising effect: for example, the triple-helix of curdlan (whose near-insolubility is reflected in the formation of stiff, opaque gels) is rendered soluble by substitution of every third residue in the 1,3-β-D-glucan backbone with 1,6-linked β-D-glucosyl sidechains which inhibit packing without affecting the ordered structure (4). In other cases, however, they can have a quite different role, by modifying or restricting the conformational options of the polymer backbone and promoting the formation of ordered structures with different backbone geometry from that adopted when sidechains are absent. We report here some recent studies of structure/conformation/function relationships in bacterial polysaccharide systems, and offer some tentative proposals on the functional role of sidechains.

MATERIALS AND METHODS

Gellan (ATCC 31461), deacylated form, and welan (ATCC 31555) were kindly supplied by Kelco Division of Merck & Co. Inc., and polysaccharide XM-6 (NCIB 11870) and capsular polysaccharide from Rhizobium trifolii (strain TA-1) were generously donated by Dr. I.W. Sutherland and Dr. L.P.T.M. Zevenhuizen, respectively. Gellan, welan and XM-6 were converted to the sodium salt form by ion exchange, and solutions were dialysed exhaustively against the appropriate salt solution prior to measurement.

DSC measurements were carried out on a Setaram microcalorimeter using a sample volume of ∿1 ml and scan rate of 0.1 degrees per minute. Rheological measurements were made on a Sangamo Viscoelastic Analyser using a cone and plate geometry of cone angle 2° and diameter 5 cm. Optical rotation was measured on a Perkin Elmer 241 polarimeter. Intrinsic viscosity was determined under 'zero shear' conditions on a couette viscometer.

XANTHAN

The extracellular polysaccharide from Xanthomonas campestris
(xanthan) has been investigated in far greater detail than any
other microbial polysaccharide. The β-1,4-glucan (cellulose)
backbone is substituted at O(3) of alternate residues by charged
trisaccharide sidechains which solubilise the polymer. The first
residue in each sidechain, D-mannose, is linked diaxially
through adjacent hydroxyl groups (to the backbone at O(1) and to
the rest of the sidechain at O(2)) giving rise to a tight 'hair-
pin bend' which allows the sidechains to fold down compactly
against the main chain, leading to formation of a 5-fold ordered
structure rather than the 2-fold conformation adopted by unsubstituted cellulose.

Solid-state x-ray fibre-diffraction evidence is almost equally
consistent with two sterically-feasible candidate structures: an
extended single helix, and a co-axial double helix in which the
individual strands would have very similar geometry to that in
the single-stranded model (5,6). The conformational transition
(7) observed in solution from the high-temperature/low-salt disordered form of the polymer to the low-temperature/high-salt
ordered form follows first-order kinetics, favouring the single-
helix proposal, but light scattering studies show a mass per
unit length for the ordered form twice that expected for a single
chain, and this has been interpreted as evidence for a double-
helix. However, on cooling from high temperature, the conformational change (as monitored by techniques such as optical rotation or differential scanning calorimetry) is virtually complete
before the onset of molecular weight increase (by light scattering) is observed (7), suggesting side-by-side dimerisation of
single helices rather than formation of a co-axial double helix.
Similarly, when cadoxen is used to destabilise the ordered
structure (8), the observed decrease in molecular weight (although
again interpreted as unravelling of a double helix) occurs
(Fig. 2) at substantially lower concentrations of cadoxen than
the conformational transition, consistent with dissociation and
subsequent 'melting' of single helices.

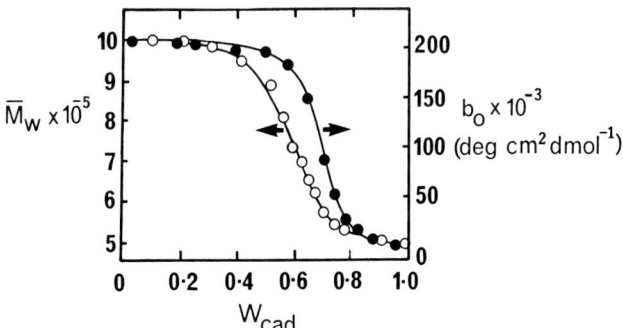

Figure 2. Molecular weight and optical activity (Moffitt parameter, b_0) of xanthan with varying weight-fraction of cadoxen
(W_{cad}). From (8) with permission.

Although solutions of xanthan in the ordered conformation
show rheological properties analogous to those of gels (9), they
cannot support their own weight under gravity (and therefore
flow). Xanthan does, however, form self-supporting mixed gels
with locust bean gum and other structurally-related plant polysaccharides such as gum tara and konjac mannan, and we have
previously interpreted this behaviour in terms of association of
the xanthan ordered structure to unsubstituted or sparingly substituted regions of the plant polysaccharide backbone (10).
Recent x-ray studies of dried and oriented mixed gels of xanthan
and locust bean gum (11) do indeed show diffraction intensities
indicative of specific interaction between the two polymers, but
it has been suggested that the interacting regions of xanthan are
in the 2_1 (cellulose) conformation, to match the corresponding
2_1 conformation of the mannan chain, rather than in the 5_4
ordered form normally imposed by the sidechains (5,6).
This is at first sight an attractive proposal, since it does
provide a mechanism for an exact, repeating match of interacting
sequences. However, the evidence on which it is primarily based
(formation of cohesive gels only when xanthan is cooled from the
disordered form in the presence of the co-synergist, but not when
the two polymers are mixed together under conditions where the
xanthan is already ordered) seems far from conclusive, since the
mixing process could, in itself, destroy the gel network as it
forms (as happens, for example, when calcium salts are mixed
directly with solutions of alginate, or when carrageenan or
furcellaran solutions are stirred during cooling to the ordered
form). It is also difficult to reconcile with the ability of
galactomannans to stabilise the ordered structure of xanthan to
higher temperatures (10). However, more definitive comparison
of the two proposed models clearly requires further experiment,
such as chemical, enzymic or genetic modification of the xanthan
sidechains to a form in which the 5_4 conformation is no longer
adopted, so that by one model the interaction would be eliminated
and by the other model enhanced.

GELLAN

Gellan gum (Fig. 1) is the parent-member of a family of
structurally-related novel bacterial polysaccharides whose
functional properties and potential practical applications
have been reviewed recently (12). Figure 3 shows thermal
transitions of gellan observed by differential scanning calorimetry (DSC) on heating and cooling under gelling (added salt)
and non-gelling (salt-free) conditions. In the absence of salt
there is a single, fully-reversible transition, with no detectable thermal hysteresis, which follows exactly the same temperature-course as the conformational change observed by optical
rotation. Under gelling conditions, in the presence of added
salt, the transition is not only shifted to higher temperature,
as would be expected from simple polyelectrolyte considerations,
but is much larger, shows pronounced hysteresis between heating
and cooling, and in the heating direction occurs in two distinct
stages.

We interpret these results as follows: i) the single process observed under non-gelling conditions is a simple conformational transition from a high-temperature coil state to a low-temperature double-helix form, as characterised by x-ray fibre-diffraction in the solid state (13); ii) gel formation requires further intermolecular association by cation-mediated helix-helix aggregation, as in carrageenan (14), and this aggregation process contributes the additional enthalpy change observed under gelling conditions: iii) in the cooling direction helices aggregate as soon as they form, so that the two processes appear as a single thermal transition, but in the heating direction disaggregation occurs first (giving rise to the first step in the DSC melting curve) and then the individual helices melt (giving the second step). Consistent with this interpretation, the first melting step occurs over the same temperature range as loss of macroscopic gel structure, and the enthalpy change in the second step is the same as that of the single process observed under non-gelling conditions.

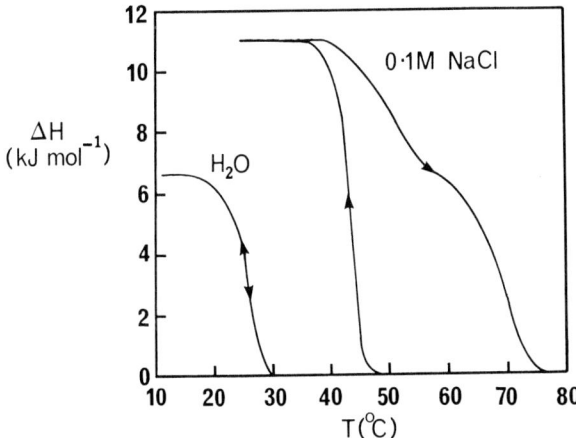

Figure 3.   Thermal transitions (16) of gellan (1% w/v) on heating and cooling in water and in 0.1 M NaCl.

WELAN

Welan gum has the same backbone structure as gellan, but each tetrasaccharide repeat unit of the main chain carries a 1,3-linked α-L-rhannose or α-L-mannose sidechain (Fig. 1). We have found no evidence of a conformational transition in welan under any accessible conditions of temperature and ionic strength, consistent with studies elsewhere (15). Thus the molecule is either a 'random coil' which does not order under hydrated conditions, or it is locked in a stable, ordered structure which is very resistant to denaturation. Our results strongly favour the latter proposal.

Firstly, solutions show rheological properties (Fig. 5) quite different from those of conformationally-disordered polysaccharides, but similar to xanthan 'weak gels' (9): solid-like response (G') exceeds viscous flow (G") with little frequency-dependence in either modulus, and small-deformation dynamic viscosity ($\eta^\star$) is higher than steady-shear viscosity (η) at equivalent values of frequency (ω) and shear-rate ($\dot{\gamma}$). Secondly, welan solutions show (16) no detectable high-resolution NMR signal, irrespective of temperature, consistent with the extreme line-broadening characteristic of conformationally-rigid chains (17). Thirdly, in contrast to conformationally-disordered polyelectrolytes where intramolecular electrostatic repulsion between chain segments is progressively screened), the intrinsic viscosity of welan (Fig.4) shows almost no variation with ionic strength, indicating a rigid ordered conformation.

It therefore appears that, in welan, attachment of a sidechain to the gellan backbone stabilises the ordered structure, but inhibits the aggregation process necessary for gelation of the parent molecule. The position of sidechain attachment, shielding the glucuronate carboxy group which is presumably involved in cation-induced gelation of gellan, supports this interpretation, but final verification must clearly await x-ray evidence of whether or not the ordered structure of the welan backbone is indeed the same as that already characterised for gellan.

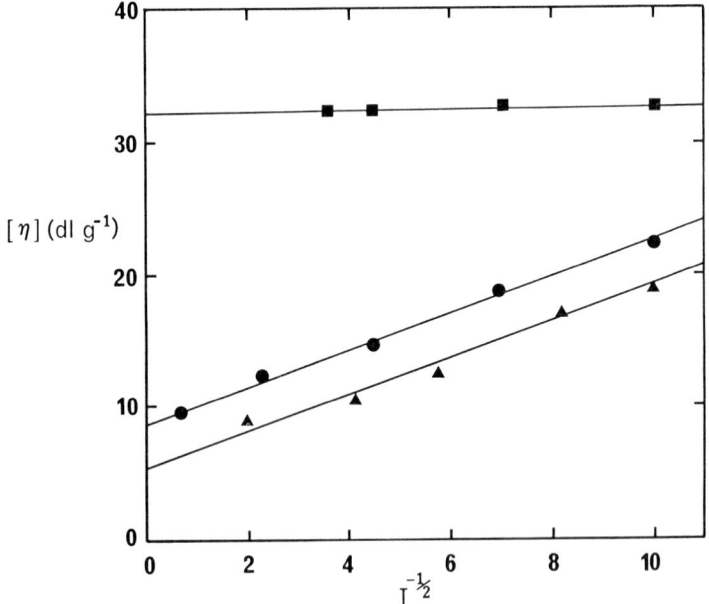

Figure 4. Variation of intrinsic viscosity ([η]) with ionic strength (I) for welan (■), polysaccharide XM-6 (▲), and (18) alginate (●) of molecular weight $\cong$ 5 x 10^5.

POLYSACCHARIDE XM-6

Figure 4 also shows the ionic-strength dependence of intrinsic viscosity for another recently isolated bacterial polysaccharide XM-6 (Fig. 1) which, like gellan, forms thermally-reversible cation-mediated gels at low polymer concentration (19). X-ray evidence (20) indicates an 8-fold double helix structure in the solid state, and it is suggested elsewhere in this volume (21) that this structure persists under hydrated conditions at both high and low temperature, and that gel formation on cooling occurs by cation-induced association of pre-existing helices, rather than by a coil-helix transition as suggested previously (19). However, the intrinsic viscosity results in Fig. 4, obtained at a temperature (30°C) just above completion (19) of gel-melting (as monitored by G') and chiroptical change (optical rotation), show a degree of flexibility comparable to that of disordered polysaccharides such as sodium alginate rather than ordered species such as xanthan or, as proposed here, welan. The rheological properties of XM-6 in the sol state (fig. 5) are also typical of a disordered polysaccharide ($\eta = \eta^*$; G" > G'; both moduli increase steeply with increasing frequency) and the high resolution NMR spectrum (19) shows linewidths similar to those for other polysaccharides in the disordered form.

Since XM-6 is a polyelectrolyte, adoption of a highly-expanded stiffened chain-conformation under salt-free conditions is to be expected, and may explain why little, if any, further increase in radius of gyration is observed (21) on going from this state to the salt-induced ordered form.

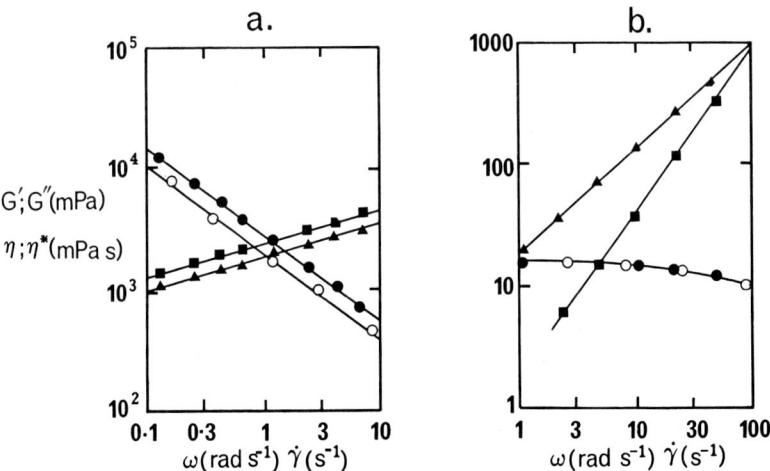

Figure 5. Frequency (ω) dependence of G' (■), G" (▲) and η^* (●) and shear-rate ($\dot{\gamma}$) dependence of η (o) for a) welan (Na⁺ salt form; 0.5% w/v; 1M NaCl; 25°C) and b) polysaccharide XM-6 (Na⁺ salt form; 0.3% w/v; 0.2M NaCl; 37°C).

RHIZOBIUM CAPSULAR POLYSACCHARIDE

The gel-forming neutral capsular polysaccharide from
Rhizobium trifolii has an unusual primary structure (Fig. 1) in
which a single residue in the backbone trisaccharide repeating
sequence carries two sidechains, one a disaccharide attached
through a flexible 1,6 linkage and the other a 1,2-linked monosaccharide in a highly restricted steric location. As illustrated in Fig. 6, the temperature-course of conformational change
associated with gel formation and melting is also unusual (22).

In the cooling direction conformational ordering, as monitored
by optical rotation, follows the sigmoidal form observed for
many gelling polysaccharides, but in the initial stages of heating there is an anomalous further 'dip' in optical rotation before
the expected reversal to the high-temperature value, and over the
same temperature-range there is a pronounced 'plateau' in the
decrease in gel rigidity (G'). Both these unusual features are
progressively diminished, and ultimately eliminated, with decreasing concentration, suggesting that they may be associated with
an aggregation process, but as yet we have no satisfactory
detailed explanation to offer for their molecular origin. What
is clear, however, is that the sidechains have a crucial role in
directing the conformational behaviour of the main chain, since
when they are removed (by Smith degradation) gelation is
eliminated and, as illustrated in Fig. 6, the unsubstituted backbone no longer adopts the ordered conformation.

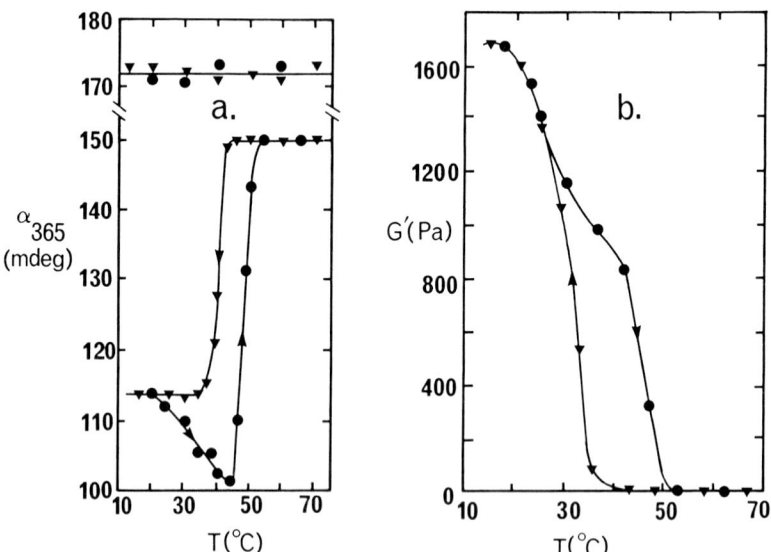

Figure 6. Temperature-course of a) optical rotation and b) gel
rigidity for rhizobium capsular polysaccharide on heating (●)
and cooling (▼). The upper curve in (a) is for the non-gelling,
debranched material. From (22) with permission.

ACKNOWLEDGEMENTS

We thank Dr. I.W. Sutherland and Dr. L.P.T.M. Zevenhuizen for polysaccharide samples, Dr. M.J. Gidley for help and advice, and Kelco Division of Merck for studentship support to C.E. Manning.

REFERENCES

1. Winter, W.T., Food Hydrocolloids, in press.
2. Reid, J.S.G., Edwards, M. and Dea, I.C.M., this volume.
3. Rees, D.A. and Scott, W.E. (1971) J. Chem. Soc. B, 469-479.
4. Kashiwagi, Y., Norisuye, T. and Fujita, H. (1981) Macromolecules, 14, 1220-1225.
5. Moorhouse, R., Walkinshaw, M.D. and Arnott, S. (1977) ACS Symp. ser., 45, 90-102.
6. Okuyama, K., Arnott, S., Moorhouse, R., Walkinshaw, M.D., Atkins, E.D.T. and Wolf-Ullish, C. (1980) ACS Symp. ser., 141, 411-427.
7. Norton, I.T., Goodall, D.M., Frangou, S.A., Morris, E.R. and Rees, D.A. (1984) J. Mol. Biol., 175, 371-394.
8. Kitagawa, H., Sato, T., Norisuye, T. and Fujita, H. (1985) Carbohydr. Polym., 5, 407-422.
9. Morris, E.R. in Gums and Stabilisers for the Food Industry 2 Application of Hydrocolloids (eds. Phillips, G.O., Wedlock, D.J. and Williams, P.A.) pp 57-78.
10. Dea, I.C.M., Morris, E.R., Rees, D.A., Welsh, E.J., Barnes, H.A. and Price, J. (1977) Carbohydr. Res., 57, 249-272.
11. Cairns, P., Miles, M.J. and Morris, V.J. (1986) Nature, 322, 89-90.
12. Pettitt, D.J. (1986) in Gums and Stabilisers for the Food Industry 3 (eds. Phillips, G.O., Wedlock, D.J. and Williams, P.A.) pp 451-463.
13. Chandrasekaran, R., Millane, R.P. and Arnott, S., this volume.
14. Morris, E.R., Rees, D.A. and Robinson, G. (1980) J. Mol. Biol., 138, 349-362.
15. Crescenzi, V., Dentini, M. and Dea, I.C.M. (1986) Carbohydr. Res., 160, 283-302.
16. Robinson, G., Morris, E.R. and Gidley, M.J., Carbohydr. Polym., in preparation.
17. Bryce, T.A., McKinnon, A.A., Morris, E.R., Rees, D.A. and Thom, D. (1974) Faraday Discuss. Chem. Soc., 57, 221-229.
18. Smidsrød, O. and Haug, A. (1971) Biopolymers, 10, 1213-1227.
19. Nisbet, B.A., Sutherland, I.W., Bradshaw, I.J., Kerr, M., Morris, E.R. and Shepperson, W.A. (1984) Carbohydr. Polym., 4, 377-394.
20. Morris, V.J. and Miles, M.J. (1986) Int. J. Biol. Macromol., 8, 342-348.
21. Chapman, H.D., Chilvers, G.R., Miles, M.J. and Morris, V.J., this volume.
22. Gidley, M.J., Dea, I.C.M., Eggleston, G. and Morris, E.R. (1987) Carbohydr. Res., 160, 381-396.

Molecular structures of gellan and other industrially important gel forming polysaccharides

R.Chandrasekaran, R.P.Millane and Struther Arnott

Whistler Center for Carbohydrate Research, Purdue University, West Lafayette, IN 47907, USA

ABSTRACT

Several microbial and algal polysaccharides are extensively used in major industries because of their gel forming properties. Both gellan, a polysaccharide from the bacterium Auromonas elodea and the capsular polysaccharide from Rhizobium trifolii are recently developed polymers and are potentially useful because of their gelling characteristics at very low concentrations. Carrageenans are structural polysaccharides from the red algae Rhodophyceae and are extensively used in the food industry as gelling and structuring agents. We have used x-ray diffraction analysis of oriented fibers to determine the molecular structures of these polysaccharides and rationalized some of their physical properties. Gellan forms a double-helix containing two parallel, half staggered, 3-fold left-handed chains of pitch 5.64nm. The Rhizobium polysaccharide is either a 2-fold single helix of pitch 1.96nm or a 4-fold, half-staggered, parallel double helix of pitch 3.93nm. Iota-carrageenan also forms a half-staggered parallel double helix made up of 3-fold right-handed chains of pitch 2.66nm. On the other hand, the kappa-carrageenan (which is chemically identical to iota but its anhydrogalactose residues are not sulfated) structure that best fits the diffraction data is a non-half-staggered parallel double helix containing right-handed 3-fold chains of pitch 2.5nm. Double helical junction zones are implicated for the gel formation in all these polysaccharides.

INTRODUCTION

A large number of polysaccharides either secreted by bacteria or produced by algae have already found a common place commercially or have the potential for commercialization in the near future. Some of these polysaccharides exhibit large viscosities at very low concentrations in solution and, in addition, can form gels of varying strength and texture. Consequently, such polymers have important applications in the food industry. As part of a structural study of gel-forming polysaccharides, we have examined four different polymers. The main experimental probe is x-ray diffraction analysis of oriented fibers and the aim is to investigate the molecular basis of gel formation in each case.

Gellan is a linear, extracellular, anionic polysaccharide secreted by the microorganism Auromonas elodea. It has the tetrasaccharide (A-B-C-D) repeating unit (1,2)

→3)-β-D-Glc-(1→4)-β-D-GlcA-(1→4)-β-D-Glc-(1→4)-α-L-Rha-(1→.

Approximately 6% (by weight) of the native material is O-acetylated and it forms weak gels. Upon deacetylation, the aqueous solutions are highly viscous even at concentrations as low as 0.04% and form stiff, brittle gels.

The capsular polysaccharides from many strains of the Rhizobium bacteria, all of which are involved in the nitrogen fixation process in plants, also have gelling properties. The capsular polysaccharide from Rhizobium trifolii (strain TA-1) is neutral and has a doubly branched hexasaccharide as the repeating unit (3,4) shown below:

```
            β-D-Gal-(4←1)-β-D-Gal
                 1
                 ↓
                 6
+4)-α-D-Glc-(1→3)-α-D-Man-(1→3)-β-D-Gal-(1→
                 2
                 ↑
                 1
             α-D-Gal
```

The course of the gel-sol transition, in the temperature range 20-90°C, has been followed by measuring the variation in optical rotation and it suggests that the mechanism of gel formation involves conversion from a conformationally mobile random coil in solution, to a rigid, ordered conformation in the gel slate (4). The transition midpoint temperature is approximately 49°C on heating and 42°C on cooling.

Carrageenans are gel forming polysaccharides belonging to a family of the marine red algae Rhodophyceae. They are extracted commercially from seaweeds and are extensively used in the food industry for gelling, thickening, bodying and emulsion stabilization in both water-based and milk-based systems. Two principal gelling fractions called iota- and kappa-carrageenan, are built up of alternating β-D-galactose and 3,6-anhydro-α-D-galactose units, with different degrees of O-sulfate group substitutions. The disaccharide repeating unit of iota-carrageenan is

→3)-β-D-Gal-4-sulfate-(1→4)-3,6-anhydro-α-D-Gal-2-sulfate-(1→.

The repeating unit of kappa-carrageenan is the same as in iota except that its anhydrogalactose is not sulfated. Both fractions form thermally reversible gels upon heating and cooling of aqueous solutions. The proposed gelling mechanism (5) is as follows: The polymer chains in solution exist as random coils at temperatures above the melting point of the gel. On cooling, the molecules form double helical junction zones and a three-dimensional polymer network builds up. The junction zones alone do not result in gel formation. Rather, they lead to non-interacting polymer domains via the cross-linking of up to approximately 20 chains. Gelation occurs following the aggregation of the double helical junction zones. Iota-carrageenan in the presence of Li$^+$, for example, forms non-aggregating double helices and hence does not gel (6).

Using x-ray diffraction techniques, we have characterized the ordered conformations of the molecular helices of these polysaccharides in oriented fibers and films so as to understand their physical properties.

EXPERIMENTS

Polysaccharide samples suitable for x-ray studies were prepared as fibers or films that were stretched to give well oriented, and in favorable cases polycrystalline, specimens. For fiber preparation, drops of concentrated polysaccharide solution (≥ 5 mg/ml) were placed between two glass rods in the fiber puller, allowed to dry slowly and stretched to at least twice the original length. Films were prepared by casting the drops on a teflon block and allowed to dry overnight in a desiccator over silica gel. Strips of film were cut and hung vertically from support rods. Small weights (5-10 g) were suspended from the lower end of the strip to induce stretching and hence orientation in the film specimens. In all cases, polycrystallinity was achieved during the process of drying and/or stretching by maintaining an appropriate relative humidity and temperature surrounding the specimen. X-ray diffraction patterns were obtained using pinhole focusing, flat film cameras and nickel filtered CuKα radiation. The lithium salt of gellan gives diffraction patterns containing sharp Bragg reflections (Fig. 1a) as does the calcium salt of iota-carrageenan (Fig. 1c). On the other hand, the Rhizobium capsular polysaccharide (Fig. 1b) and the potassium salt of kappa-carrageenan (Fig. 1d) give patterns which show mainly continuous intensities on layer lines indicating that their molecules are reasonably well oriented but that there is no long range lateral organization.

The layer line spacings and overall intensity distribution provide important details such as the pitch and symmetry of the helix, on the basis of which possible molecular models can be constructed. In all recent studies, an Optronics P-1000 rotating drum microdensitometer was used to digitize the diffraction patterns and the intensities were determined using a computerized analysis (7,8).

STRUCTURE ANALYSIS

The Linked-Atom Least-Squares (LALS) refinement system (9) was used to produce stereochemically satisfactory helical models of the polysaccharide chains consistent with the observed pitch and symmetries. Models containing left- and right-handed single helical chains, and parallel and anti-parallel double helices of either chirality were examined where appropriate. In general, bond lengths, bond angles and pyranose conformations were fixed at standard values (10). Glycosidic bridge and side chain conformation angles and, the position of the second chain for the double helix, were variable parameters. These were adjusted so as to minimize (in a least-squares sense) the expression:

$$\Omega = \Sigma \, \omega_m ({}_oF_m - F_m)^2 + \Sigma e_i \Delta\theta_i^2 + \Sigma c_j \Delta d_j^2 + \Sigma \lambda_h G_h$$

$$= \qquad X \qquad + \quad E \quad + \quad C \quad + \quad L$$

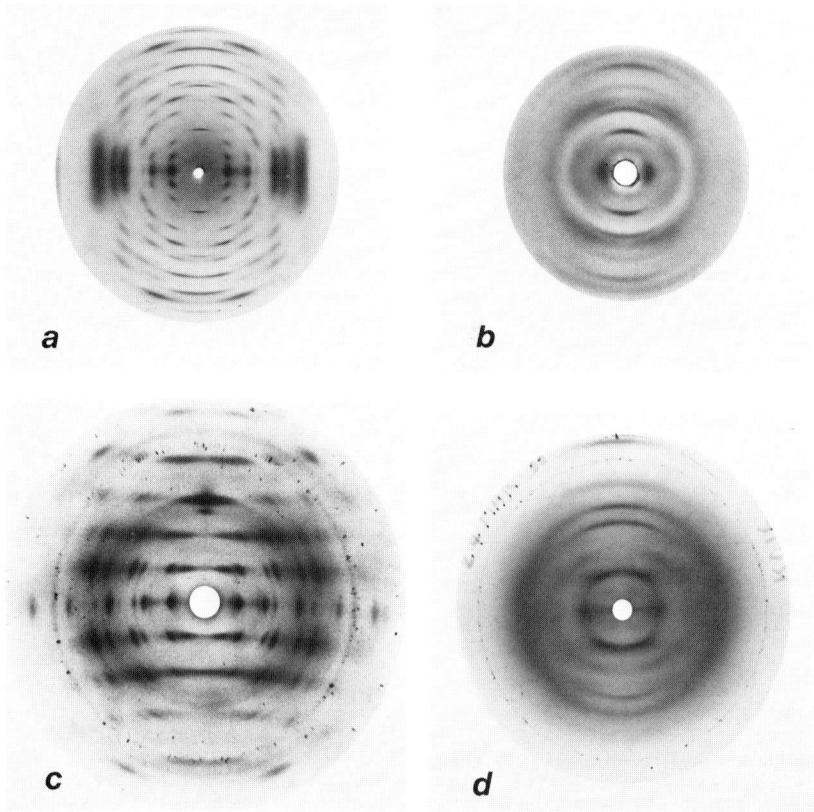

Figure 1.   X-ray fiber diffraction patterns from (a) an oriented and
polycrystalline sample of the lithium salt of gellan, (b) an oriented sample
of the capsular polysaccharide from <u>Rhizobium</u> <u>trifolii</u>, (c) an oriented and
polycrystalline sample of the calcium salt of iota-carrageenan and (d) an
oriented sample of the potassium salt of kappa-carrageenan.

The agreement between the observed ($_oF_m$) and calculated (F_m) structure
amplitudes (Bragg and/or continuous) is optimized by the first term X.   The
second term E ensures that the varied conformation and bond angles of the
molecule are within their expected domains. Both non-bonded and hydrogen bond
interactions within, and between, molecules are optimized by the third term
C.   The last term L imposes ring closure and helix connectivity constraints.
Hamilton's significance test (11) on the basis of Ω, X or C was used to
compare competing models.

DETAILS OF MOLECULAR MORPHOLOGIES

Gellan

All the Bragg spots in Fig. 1a can be indexed on the basis of a trigonal unit cell with dimensions $a = b = 1.56(3)$nm and $c = 2.82(5)$nm. A previous x-ray study (12), however, did not provide a molecular model compatible with the x-ray data. Meridional spots on the $\ell = 3n$ layer lines indicate that the molecule has $3n$-fold helix symmetry. Both single helices of pitch 2.82nm and half-staggered double helices of pitch 5.64nm of either chirality have been carefully examined. Of the six possible models listed in Table I, only two - the 3-fold left-handed single helix (Model 2) and the 3-fold double helix (Model 4) also with left-handed chains - are stereochemically satisfactory. Structure factor calculations with two molecules in the unit cell led to crystallographic R values of 0.7 and 0.4 for Models 2 and 4 respectively, indicating the clear superiority of the double helix model.

Table I. CHARACTERISTICS OF THE GELLAN POLYANION MODELS

Model	Helix	Type	Chirality of chain	Pitch (nm)	Steric feasibility
1	3-fold single	--	right	2.82	SC
2	3-fold single	--	left	2.82	Good
3	3-fold double	HS	right	5.64	SC
4	3-fold double	HS	left	5.64	Good
5	6-fold double	HS	right	5.64	SC
6	6-fold double	HS	left	5.64	SC

HS = parallel, half-staggered, SC = some short contacts.

Further refinement of the crystal structure (13) resulted in a final R value of 0.22 and 0.24 corresponding to antiparallel and parallel packing of two double helices in the unit cell.

The gellan double helix, shown in Fig. 2a, is fairly slim. The polysaccharide chain contains three intramolecular hydrogen bonds per tetrasaccharide repeat, all of them involving the glucuronide residue. The double helix is stabilized by interchain OH•••O hydrogen bonds at every carboxylate group. The donor and acceptor groups are in the interior of the molecule and are located close to the helix axis at radial distances between 0.23 and 0.31nm. As a donor, O6C forms bifurcated hydrogen bonds with O61B and O62B; similarly O3D hydrogen bonds to O2B. Taking the antiparallel packing as being the best representative of the crystal structure (Fig. 3), it is found that most of the intermolecular hydrogen bonds involve either the free hydroxyls of rhamnose or the hydroxymethyl groups of glucose A residues which are located farthest from the helix axis. Each molecule in the unit cell interacts with its three nearest neighbors, laterally separated by 0.9nm from it, as shown in Fig. 3.

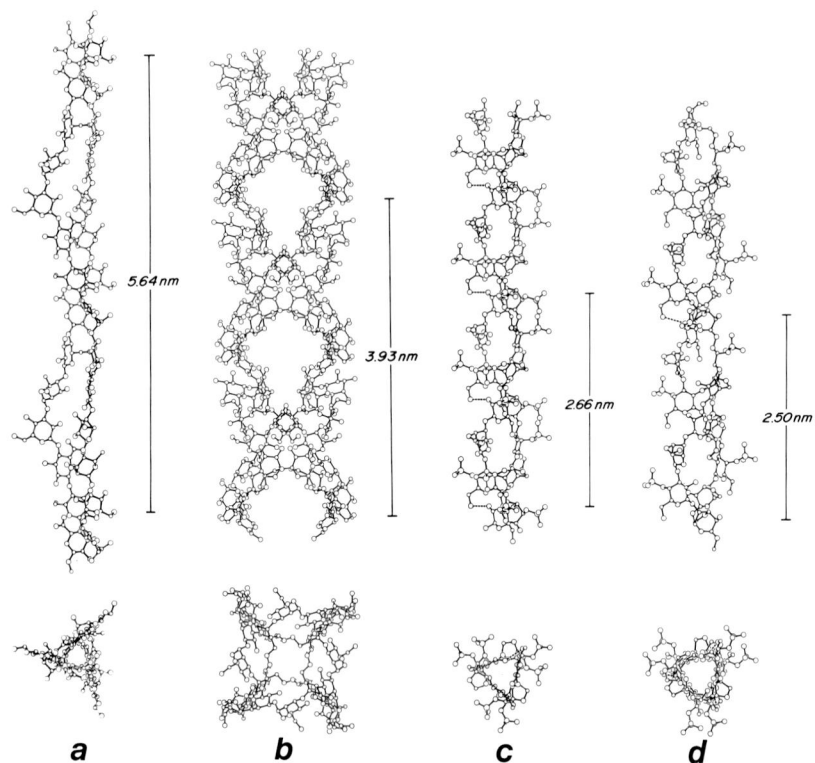

Figure 2.   Two mutually perpendicular views of the double helical structures
of (a) gellan, (b) the capsular polysaccharide from Rhizobium trifolii, (c)
iota-carrageenan and (d) kappa-carrageenan.   The interchain hydrogen bonds in
carrageenans are shown by dashed lines.

Capsular polysaccharide from Rhizobium trifolii

The diffraction pattern (Fig. 1b) shows that the molecules are reasonably
well oriented and there is only short range lateral organization in the
fiber.   The layer line spacings show that the c-repeat of the molecule is
1.96nm and the second layer meridional spot shows a 2n-fold helical
symmetry.   The strong equatorial spot suggests a molecular diameter of the
order of 1.8nm.

Consistent with the observed layer line spacings and helix symmetry, there
are five different molecular models (Table II). The first is a 2-fold helix
of pitch 1.96nm (Model 1).   Two such chains can form either parallel (Model 2)
or antiparallel (Model 3) co-axial 2-fold double helices.   The two chains in
Model 2 must be non-half-staggered or else the odd layer lines on the

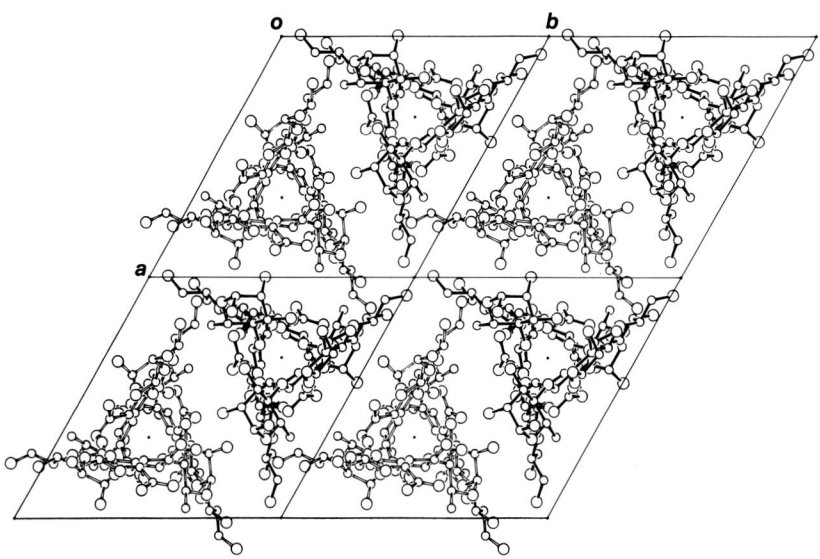

Figure 3. Helix axis projection of the gellan crystal structure corresponding to the antiparallel packing arrangement.

diffraction pattern would be extinguished. Models in which conformationally identical 4-fold polysaccharide chains of pitch 3.93nm form parallel, half-staggered, double helices are also compatible with the diffraction pattern. Both right- (Model 4) and left-handed (Model 5) helices are possible candidates.

Table   II.   CHARACTERISTICS   OF   THE   MOLECULAR   MODELS   OF   THE   CAPSULAR POLYSACCHARIDE FROM RHIZOBIUM TRIFOLII.

Model	Helix	Type	Chirality of chain	Pitch (nm)	Steric feasibility
1	2-fold single	--	--	1.96	Good
2	2-fold double	P,NHS	--	1.96	SC
3	2-fold double	AP	--	1.96	SC
4	4-fold double	P,HS	right	3.93	Good
5	4-fold double	P,HS	left	3.93	Good

P = parallel, AP = antiparallel, HS = half-staggered, NHS = non-half-staggered, SC = some short contacts.

Examination of each of the models and the influence of the three possible staggered conformations about the (1→6) linkage reveals that the 2-fold single helix (Model 1) and 4-fold double helices (Models 4 and 5) are stereochemically satisfactory (14). In the gauche plus orientation about the (1→6) linkage, the disaccharide side chain is fully extended and protrudes away from the backbone. In the trans orientation, the side chain folds on the backbone but the gauche minus orientation is sterically unacceptable. The 4-fold parallel double helix, of either chirality, is free from steric compression and a representative structure is shown in Fig. 2b.

Iota-carrageenan

Samples of the divalent salts (e.g. Ca^{2+}, Sr^{2+}) of iota-carrageenan can be prepared in the form of well oriented and polycrystalline fibers which give good diffraction patterns as in Fig. 1c. The Bragg spots can be indexed on the basis of a trigonal unit cell of dimensions a = b = 1.37nm and c = 1.33nm. Detailed analysis (15) shows that the molecule is a double helix (Fig. 2c) made up of two identical, parallel, right-handed, 3-fold helices of pitch 2.66nm which are translated from each other along their common helix axis by exactly half the pitch. There are no intrachain hydrogen bonds but the double helix is stabilized by hydrogen bonds between the hydroxyl groups O6 and O2 of galactose residues in different chains. These groups are buried in the interior of the helix. As can be seen in Fig. 2c, the sulfate groups are on the periphery of the helix and are probably involved in intermolecular interactions through cation bridges.

Kappa-carrageenan

Reasonably oriented but non-crystalline fibers can be prepared from samples of the potassium salt of kappa-carrageenan. Consequently, the diffraction pattern (Fig. 1d) contains continuous intensities on layer lines. Meridional diffracted intensity on the 6th and 9th layer lines indicates that the molecule forms a 3-fold helix and the layer line spacings correspond to a molecular pitch of 2.5nm. Both left- and right-handed single helices are free from steric compression but their fits to the x-ray data are rather poor. Half-staggered, parallel double helices as in iota-carrageenan are incompatible with the layer line spacings. On the other hand, double helices containing right-handed 3-fold chains in which the position and orientation of the second chain are offset from the half-staggered position by small amounts, provide a greatly improved x-ray agreement (16). Coaxial, antiparallel double helices are also sterically acceptable and their x-ray agreements are marginally inferior to those for non-half-staggered parallel double helices. The best model (Fig. 2d) is a parallel double helix similar to iota-carrageenan but one chain is offset from the half-staggered position by a 0.1nm translation along, and 28° rotation about, the helix axis. The crystallographic R value for this model is 0.23.

Despite the lack of sulfation on the anhydrogalactose residues and the shorter pitch compared to iota-carrageenan, kappa-carrageenan is conformationally very similar to iota-carrageenan. Since the chains are not half-staggered, there are only half the number of interchain O6•••O2 hydrogen bonds between galactose residues on opposite chains in kappa- (Fig. 2d) compared to iota-carrageenan (Fig. 2c).

CONCLUSION

X-ray diffraction analysis of gellan, the capsular polysaccharide from Rhizobium trifolii, and iota- and kappa-carrageenan reveal that they can all form double helices. Recent developments in intensity measurement and structure refinement procedures have enabled us to determine the molecular architectures of these gel-forming polysaccharides that are widely used in, or potentially useful to, the food industry. The carboxyl groups in gellan, the side chains in the Rhizobium trifolii capsular polysaccharide and the sulfate groups in carrageenan seem to influence the physical properties of these polymers. The crystal structures provide information on the possible intermolecular interactions, that may mimic the aggregation of junction zones during gel formation.

ACKNOWLEDGEMENTS

We thank Deb Zerth for word processing the manuscript, Robert Wariberg for photography and NSF for financial support (Grants 8512599 and 8606942).

REFERENCES

1. O'Neill, M.A., Selvendran, R.R. and Morris, V.J. (1983) Carbohyd. Res., 124, 123-134.
2. Jansson, P.E., Lindberg, B. and Sanford, P.A. (1983) Carbohyd. Res., 124, 135-139.
3. Zevenhuizen, L.P.T.M. and van Neerven, A.R.W. (1983) Carbohyd. Res., 124, 166-171.
4. Gidley, M.J., Dea, I.C.M., Eggleston, G. and Morris, E.R. (1987) Carbohyd. Res., 160, 381-396.
5. Rees, D.A. (1969) Adv. Carbohyd. Chem. Biochem., 24, 267-332.
6. Morris, E.R., Gidley, M.J., Murray, E.J., Powell, D.A. and Rees, D.A. (1980) Int. J. Biol. Macromol., 2, 327-337.
7. Millane, R.P. and Arnott, S. (1985) J. Appl. Crystallogr., 18, 419-423.
8. Millane, R.P. and Arnott, S. (1985) J. Macromol. Sci. Phys., B24, 193-227.
9. Smith, P.J.C. and Arnott, S. (1978) Acta Crystallogr., A34, 3-11.
10. Arnott, S. and Scott, W.E. (1972) J. Chem. Soc. Perkin Trans II, 324-335.
11. Hamilton, W.C. (1965) Acta Crystallogr., 18, 502-510.
12. Upstill, C., Atkins, E.D.T. and Attwool, P.T. (1986) Int. J. Biol. Macromol., 8, 275-288.
13. Chandrasekaran, R., Millane, R.P., Arnott, S. and Atkins, E.D.T. (1987) Carbohyd. Res. (submitted).
14. Chandrasekaran, R., Millane, R.P., Walker, J.K., Arnott, S. and Dea, I.C.M. (1987) in Recent Developments in Industrial Polysaccharides, (ed. Crescenzi, V., Dea, I.C.M. and Stivala, S.S.) In press. Gordon and Breach Science, New York, USA.
15. Arnott, S., Scott, W.E., Rees, D.A. and McNab, C.G.A. (1974) J. Mol. Biol., 90, 253-257.
16. Millane, R.P., Chandrasekaran, R., Arnott, S. and Dea, I.C.M. (1987) Carbohyd. Res. (submitted).

Effect of polysaccharide thickeners on organoleptic attributes

Z.V.Baines and E.R.Morris

Department of Food Research and Technology, Cranfield Institute of Technology, Silsoe College, Silsoe, Bedford MK45 4DT, UK

ABSTRACT

Although the 'thickness', 'stickiness' and 'sliminess' of polysaccharide systems are perceived subjectively as separate, independent attributes, the numerical values assigned to them by sensory panels (using magnitude estimation against a fixed control) are virtually identical. Rotational viscosity at a fixed shear-rate of ~ 50 s^{-1} (as suggested twenty years ago by Dr. F.W. Wood) correlates well with all three attributes for normal solutions (e.g. samples thickened with disordered, 'random coil' polysaccharides) but seriously underestimates the perceived texture of 'weak gels' (e.g. samples thickened with xanthan), where the tenuous network structure is disrupted in making the measurement. Equivalent non-destructive oscillatory measurements at ~ 50 rad s^{-1}, however, give a reliable objective index of thickness, stickiness and sliminess in both cases. Contrary to previous suggestions there is no direct correlation between sliminess and degree of shear thinning. 'Flavour release' and perceived sweetness are unaffected by disordered polysaccharides at concentrations (c) below the coil-overlap value (c*) but decrease sharply at higher concentrations, with both attributes following the same generalised dependence on c/c*, irrespective of polysaccharide type and molecular weight. Over the concentration range of practical importance (up to $\sim 1\%$ w/v) xanthan shows no detectable suppression of perceived flavour or taste.

INTRODUCTION

While the ultimate criterion of the 'eating quality' of food products must, of course, be subjective assessment (e.g. by expert tasters or 'naive' consumer panels) there are obvious advantages in convenience, precision and long-term reproducibility from developing objective instrumental measurements that provide a reliable, predictive index of perceived organoleptic attributes. We report here some recent evidence that the relationships between objective rheological properties of polysaccharide thickeners and their effect on perceived 'in-mouth' texture and texture-related attributes such as 'flavour release' may be simpler and more readily quantifiable than previously supposed.

PERCEIVED THICKNESS AND OBJECTIVE VISCOSITY

For samples that have a single, unambiguous viscosity (Newtonian solutions such as sugar syrups) there is a direct linear
relationship (equation [1]) between the logarithm of viscosity
(η) and the logarithm of perceived 'thickness' (T) assessed by a
sensory panel using the technique of magnitude estimation against
a fixed control (1).

$$\log T = 0.22 \log \eta + 1.69 \qquad (r^2 = 0.98) \qquad [1]$$

$$\text{i.e.} \quad T = 1.69 \, \eta^{0.22} \qquad\qquad\qquad [2]$$

Thus the relationship between objective stimulus (η) and subjective response (T) follows a 'power law' dependence (equation [2])
of the general form proposed by Stevens (2) for many other
aspects of sensory perception (loudness of sound, brightness of
light, etc.). The pre-exponential parameter (i.e. the constant,
1.69, in equation [1]) depends on the 'thickness' of the control
sample and has therefore no fundamental significance, but the
exponent (0.22) is independent of the choice of control and shows
the change in objective viscosity required to produce a specific
change in perceived thickness.

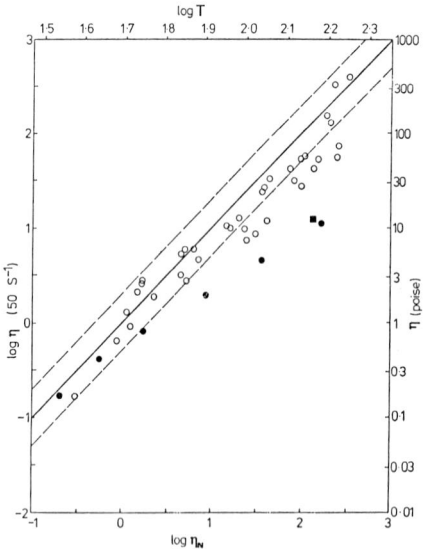

Figure 1. Comparison of 'equivalent Newtonian viscosity' (η_N)
calculated from panel scores for thickness (T) by equation [1]
with rotational viscosity at 50 s^{-1}, as suggested by Wood (5).
The solid line corresponds to perfect agreement and the dashed
lines indicate the confidence limit (p = 0.95) on panel scores.
Filled symbols show 'weak gels'. From Cutler, Morris and Taylor
(1), with permission.

For most polysaccharide samples, however, viscosity does not have a single, fixed value, but decreases with increasing shear rate ('shear thinning'), often by several orders of magnitude over the range of shear-rates of practical significance. The choice of appropriate measuring conditions is therefore a critical issue in comparing the objective viscosity of polysacc- haride systems with their perceived texture.

In previous studies (1,3,4) we have shown that for most samples rotational viscosity measured at a fixed shear rate of 50 s^{-1}, as suggested by Wood (5), and 'mouthfeel' viscosity derived using the shear-stress/shear-strain conditions proposed by Shama and Sherman (6) both give excellent agreement with subjective values of 'equivalent Newtonian viscosity' (1,3) calculated from panel scores for thickness (T) by equation [1]. However, as illustrated in Fig. 1, the perceived viscosity of certain samples, particularly those incorporating xanthan as thickener, is seriously underestimated.

DYNAMIC VISCOSITY AND 'WEAK GELS'

Xanthan is unusual in having a rigid, ordered chain-conform- ation in solution, which leads to the formation of a tenuous intermolecular network (7,8). One consequence of this 'weak gel' character (9) is that the viscosity of xanthan solutions measured under small-deformation (oscillatory) conditions where the net- work remains intact (dynamic viscosity, η^*) is substantially higher than conventional shear viscosity (η) measured under large- deformation conditions (e.g. in a rotational viscometer) where the network is broken down. In the absence of any such network (e.g. solutions of conformationally-disordered polysaccharides, or Newtonian syrups), by contrast, η and η^* are the same when measured at equivalent values of shear rate (s^{-1}) and frequency (rad s^{-1}).

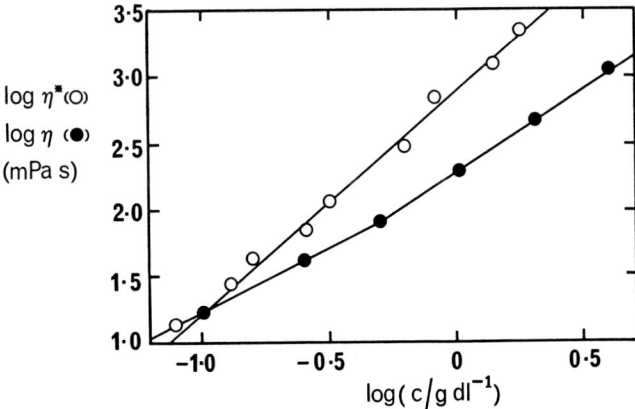

Figure 2. Comparison of dynamic viscosity (η^*) at 50 rad s^{-1} and steady-shear viscosity (η) at 50 s^{-1} for xanthan 'weak gels'.

The divergence between the small-deformation dynamic
viscosity of xanthan solutions measured at 50 rad s^{-1} and their
large-deformation shear viscosity at 50 s^{-1} increases with
increasing concentration, as shown in Fig. 2.   This behaviour
parallels closely the increasing divergence observed (Fig. 1)
between the rotational viscosities of 'weak gels' and the values
anticipated from panel scores for thickness, suggesting that the
higher values obtained under small deformation might be a more
appropriate index of 'in-mouth' texture.
   As illustrated in Fig. 3 there is indeed an excellent correlation (r^2 = 0.95) between perceived thickness (T) and dynamic
viscosity (η*) at 50 rad s^{-1}, and in particular results for
'random coil' polysaccharide solutions and for weak gels fall
smoothly within the same distribution (10).   We therefore conclude
that conventional viscosity measurements at a fixed shear-rate of
50 s^{-1}, as proposed twenty years ago by Wood (5), provide a
reliable objective measure of the perceived thickness of normal
fluids (where η = η*) but underestimate the thickness of 'weak
gels' (where η < η*), while equivalent small-deformation oscillatory measurements of η* at 50 rad s^{-1} correlate directly with
perceived thickness for both types of system.

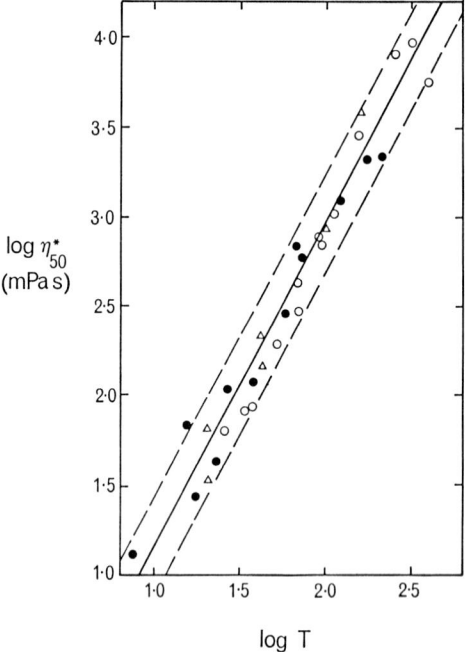

log T

Figure 3.   Correlation (10) of the perceived thickness (T) of
samples thickened with guar gum (o), pre-gelatinised starch ($\triangle$),
and xanthan ($\bullet$) with dynamic viscosity (η*) at 50 rad s^{-1}. Dashed
lines show the confidence limit (p = 0.95) on panel scores.

'STICKINESS', 'SLIMINESS' AND SHEAR THINNING

In a previous investigation of the organoleptic properties of disordered polysaccharides (11) a close 1:1 relationship was reported between 'thickness' and 'stickiness', with panel scores not only being highly correlated ($r^2 = 0.98$) but having virtually the same numerical values for both. As illustrated in Fig. 4, panel scores for 'sliminess' are also essentially identical to those for 'thickness' (or 'stickiness'), both for 'normal' solutions of disordered polysaccharides and for xanthan 'weak gels'. Thus although the thickness, stickiness and sliminess of thickened polysaccharide systems are perceived subjectively as separate organoleptic attributes, they are quantitatively indistinguishable. In particular this implies that the direct correlation shown in Fig. 3 between objective values of dynamic viscosity (at 50 rad s^{-1}) and subjective 'thickness' applies with equal validity to 'stickiness' and 'sliminess', in apparent conflict with the long-held view (12,13) that the 'sliminess' of polysaccharide solutions is inversely related to their degree of shear-thinning.

The relationship between shear-thinning behaviour and sliminess was first proposed in 1962 by Szczesniak and Farkas (12) who used a range of polysaccharide samples standardised to a fixed viscosity (1200 mPas) at low shear-rate (0.5 RPM on a Brookfield viscometer), and found that the samples that showed the greatest drop in viscosity with increasing shear-rate (e.g. xanthan) were perceived as least slimy while those closest to Newtonian behaviour were rated the slimiest.

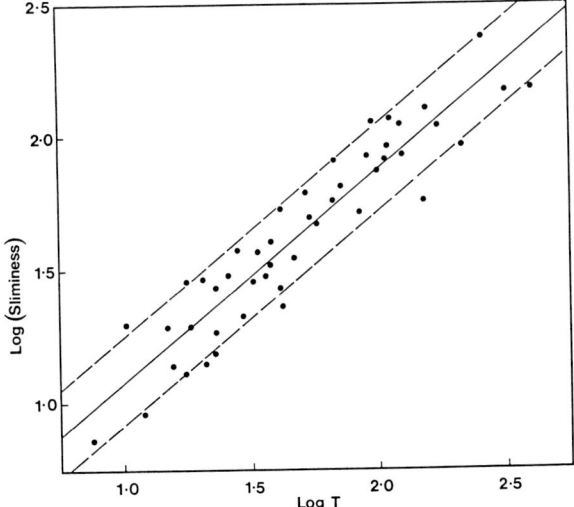

Figure 4. Correlation (10) of the perceived 'sliminess' and perceived thickness (T) of samples as in Fig. 3. Dashed lines show the confidence limit (p = 0.95) on panel scores.

Unfortunately, a consequence of this experimental design is
that, as illustrated in Fig. 5a, the samples with the most
pronounced shear-thinning behaviour will also have the lowest
viscosities at higher shear rates, such as 50 s^{-1}. By removing
the restriction of constant viscosity at low-shear rate, however
the two effects can be studied independently, as in the investigation (10) from which results are shown in Figs. 3 and 4, and
it is then evident that the objective determinant of perceived
sliminess (and perceived 'thickness' and 'stickiness') is
viscosity at 50 s^{-1} (in steady-shear) or 50 rad s^{-1} (under
oscillation), while the degree of shear-thinning has no direct
influence on perceived texture, as illustrated schematically in
Fig. 5b.

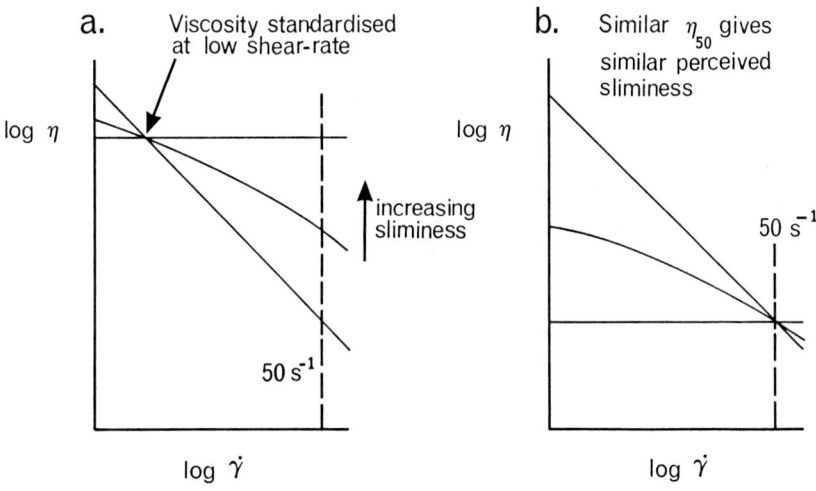

Figure 5.   Objective correlate of perceived 'sliminess'.
      (a)   Samples standardised to constant viscosity at low
             shear rate: η(50 s^{-1}) decreases with degree of shear-
             thinning, with consequent decrease in sliminess.
      (b)   Similar values of η(50 s^{-1}) give similar 'sliminess',
             irrespective of degree of shear-thinning.

TASTE INTENSITY AND 'FLAVOUR RELEASE'

Suppression of the perceived intensity of flavour and taste in
thickened systems is clearly an issue of considerable technical
and commercial significance in product formulation, but there is
very little systematic understanding of the processes involved.
We now report preliminary results from studies of taste intensity
and 'flavour release' in polysaccharide model systems, which
indicate some striking generalities of behaviour.

Figure 6 shows sensory panel scores (14) for the perceived thickness, sweetness and flavour intensity of a range of solutions all incorporating the same objective concentration of sucrose and flavouring, but thickened with different concentrations of a single 'random coil' polysaccharide sample. The sharp break in the concentration-dependence of perceived thickness corresponds to the onset of coil-overlap and entanglement (the c^* transition) and is paralleled by a corresponding change (4) in concentration-dependence of objective viscosity. Below this point panel scores for flavour and taste (sweetness) remain essentially constant, but decrease steeply at higher concentrations of polysaccharide, with both attributes showing the same concentration-dependence.

Quantitatively, the reduction in perceived intensity of flavour and taste with increasing polymer concentration (c) can be fitted with reasonable precision by equation [3]:

$$S = S_0/[1 + (c/c_{\frac{1}{2}})^p]$$ [3]

where S denotes the sensory panel score, S_0 is the constant value of S at low polymer concentrations, $c_{\frac{1}{2}}$ is the concentration at which $S = S_0/2$ and p is the absolute value of the terminal slope of log S vs. log c at high concentration.

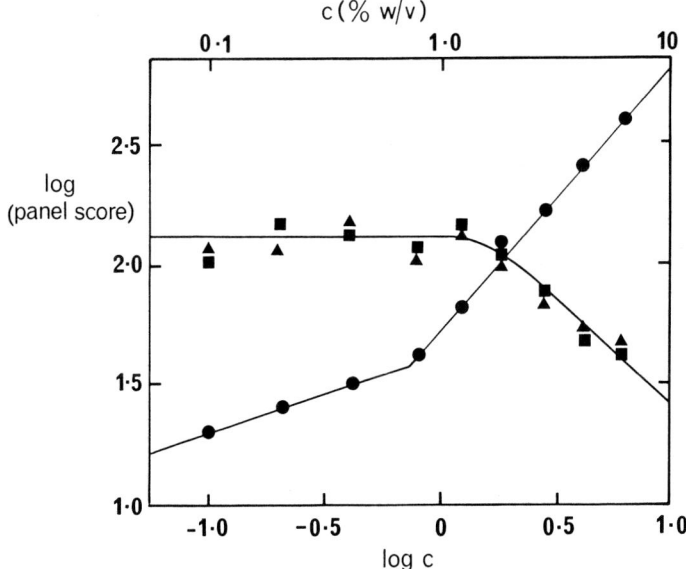

Figure 6. Concentration-dependence (14) of perceived thickness (●), sweetness (■) and flavour intensity (▲) for solutions thickened with a single sample of guar gum (MW ≅ 5×10^5).

Role of the xanthan structure on the rheological properties in aqueous solutions

Françoise Callet, Michel Milas and Bernard Tinland

Centre de Recherches sur les Macromolécules Végétales, BP 68, 38402 Saint-Martin d'Hères Cedex, France

ABSTRACT

Contrary to work published previously, no evidence of dissociation of the xanthan double-helix was found although similar measurements have been done. An attempt to explain these discrepancies is given. Thus, the xanthan exists in solution under two different ordered conformations (native and renatured) without change in molecular weight. It is shown by intrinsic viscosity that the rheological properties are dependent on these conformations. For xanthan samples with different acetyl and pyruvate contents we have shown that the substituents have no influence on the viscosity even in the semi-dilute regime. We conclude that the intrinsic viscosity of xanthan is the parameter which defines most accurately, the specific viscosity and the yield value, independently of the nature of the ordered conformation and acetyl and pyruvate contents.

INTRODUCTION

In the ordered conformation, xanthan gum can be characterized by a native and a renatured form without change in molecular weight (1,2). Most of the work on the xanthan conformation have been unaware of these two ordered structures. Independently of the existence of these two states, several groups (3,5) concluded that xanthan takes on an ordered double-stranded helix in contrast with other workers (6-8). The same disagreement exists on the contribution of acetyl and pyruvate substituents to the rheological properties of xanthan solutions (9-12).

In this work, we have tried to take stock of these questions and to show the influence of the different xanthan conformations on the rheological properties.

MATERIAL AND METHODS

Native xanthan : the unpasteurized culture-broths were provided by Shell (Sittingbourne U.K.), samples A and B, and Rhône-Poulenc (Melle), sample C. The xanthans were recovered following the procedure described previously (1). To preserve the native conformation, the solutions were then prepared by dissolving the powder in 0.01 M NaCl. To get the renatured conformation, the solutions were heated for 2 minutes at 80°C. Renatured xanthan can also be obtained by dissolving xanthan powder in pure water. NaCl is added just before the measurements. (1,2). Deacetylated xanthan, pyruvate-free xanthan and acetyl and pyruvate free xanthan were obtained as described elsewhere (13). Samples with lower molecular weights were prepared by sonication of the xanthan solutions (1 g/L, 0.01M NaCl) at 0°C with a Branson sonifier (Mod. 12) for various times (1 to 45 mn.). The solutions were filtered through Millipore filters (0.2 μm) and precipitated as described previously (1). The solvent cadoxen was prepared in the same way as described by Sato et al. (3). The solutions were prepared by mixing weighed amounts of Na salt xanthan with the cadoxen. Nevertheless, in order to reduce the dissolving time of xanthan in cadoxen, for measurements before 10 h, the xanthan was first dissolved in water at high concentration (1 % w/w) then diluted at 0.4 g/L with cadoxen. In these conditions the final cadoxen percentage is 96 %. The xanthan-cadoxen solutions (C = 0.4 g/L) were stored at 20°C.

Acetyl and pyruvate contents were estimated by [1]H n.m.r. (1) and expressed as the fraction of side chains substituted by these groups. Viscosity measurements were carried out at 25°C in the lower Newtonian regime, using a low-shear 30 from Contraves. Yield values were obtained with a Carrimed controlled stress rheometer equipped with a cone (diameter 4.0 centimeters, angle 4.0 degrees). Intensities of light scattered from 0.01 M aqueous NaCl solutions and cadoxen solutions of Na xanthan at 25°C were measured with a Fica 5000 light scattering photometer in an angular range from 22.5 to 150° at 546 nm and with a low-angle laser light-scattering apparatus (Chromatix KMX 6) at 6°.

Optical clarification of xanthan solutions was achieved either by centrifuging the solution,4 h at 20 000 g (for measurements with the Fica apparatus or by filtration through 0.2 millipore filters for the Chromatix apparatus. The range of xanthan concentrations used was from 0.4 to 0.1 g/L with the Fica and 0.1 to 0.025 g/L with the Chromatix. No difference in $\overline{M}_w$ was found by these two methods. The specific refractive index increments used were 0.155 (mL/g) in 0.01 M NaCl (1) and 0.164 (mL/g) in cadoxen (3).

XANTHAN CONFORMATION

We have shown previously (1,2) that xanthan gum can exist under two ordered conformations. Table I gives the properties of the native (N) and renatured (R) forms. With the same $\overline{M}_w$, the renatured form is more viscous than the native one. But the difference between these two forms depends on the samples. The ratio η_R / η_N at 1g/L and $[\eta]_R / [\eta]_N$ can be larger than 30 and 2.5 respectively. The renatured conformation is more rigid than the native one (1). In particular conditions, Lecourtier et al. (4) found that it was possible to dissociate the double helix to a single one. We have not observed this dissociation, but in some cases have observed degradation (1,14).

Dissociation of the double stranded xanthan had also been suggested by Sato et al. (3) using cadoxen as solvent. We have repeated these experiments on sample A. Figure 1 gives $[\eta]$ and $\overline{M}_w$ changes versus the time of storage of a xanthan-cadoxen solution at 20°C. Contrary to Sato et al., there is a large degradation of the xanthan in cadoxen. A double logarithmic plot of $[\eta]$ against $\overline{M}_w$ is shown in Figure 2. The stars in this figure represent the data of Sato et al. (15) obtained from different $\overline{M}_w$. The agreement with our data points is quite good. So, because the points as a function of storage time and from samples with different $\overline{M}_w$ lie on the same curve, we think that the difference between $\overline{M}_w$ in cadoxen and $\overline{M}_w$ in 0.1M NaCl is due to a degradation process in alkaline conditions. We can notice (Figure 1) that the ratio $\overline{M}_{w\,0.1\,NaCl} / M_{wcadoxen}$ is 2 for a storage time in cadoxen near to 7 h, corresponding nearly to the experimental conditions of $\overline{M}_w$ measurements adopted by Sato et al. (3,15). This degradation in cadoxen may be favoured as the xanthan conformation is disordered in this solvent,

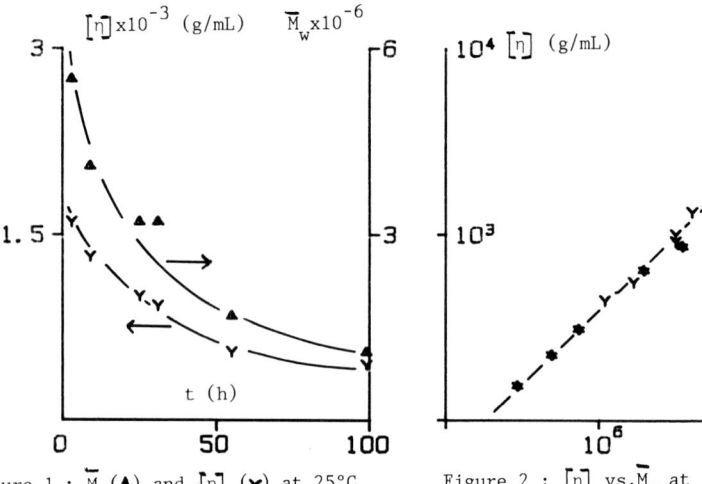

Figure 1 : $\overline{M}_w$(▲) and $[\eta]$ (Y) at 25°C
vs. storage time in cadoxen (sampleA).

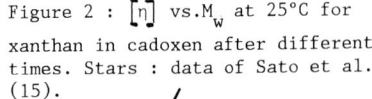

Figure 2 : $[\eta]$ vs.$\overline{M}_w$ at 25°C for
xanthan in cadoxen after different
times. Stars : data of Sato et al.
(15).

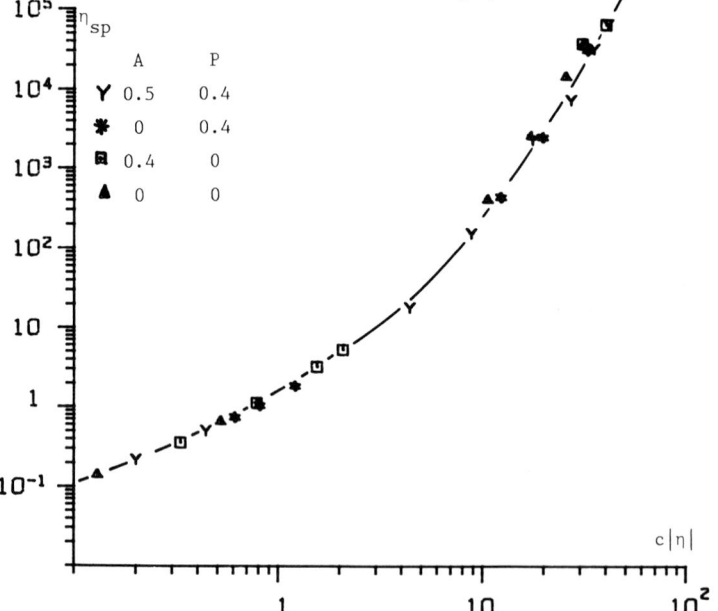

Figure 3 : Specific viscosities vs.overlap parameter,$c[\eta]$ for xanthans
with different acetyl (A) and pyruvate (P) contents in 0.01M NaCl
(Sample A) t = 25°C.

increasing the sensitivity to degradation (16,17).

Stokke et al. (5) show from the molecular size and shape of xanthan revealed by electron microscopy that the xanthan behaves as a double stranded molecule. Depending on the conditions, they found partial dissociation of this double strand. From our experimental results with De Murcia (18), we conclude that electron microscopy is not valid to follow changes in conformation ; the results point out mainly aggregates, like side by side intrachain associations, whatever   the experimental conditions.

So from our work as a wole, we conclude that xanthan conformations are single chains differing by side-chain arrangements (1).

Table I. CHARACTERIZATION AND PROPERTIES OF XANTHAN IN THE NATIVE (N) AND RENATURED (R) CONFORMATIONS.

Sample	$\eta_{1\ g/L}$ (cp)		$[\eta]$ (mL/g)		$\bar{M}_w \times 10^{-6}$		Acetyl	Pyruvate
	N	R	N	R	N	R		
A	75	350	7500	12000	9.2	9.2	0.75	0.4
B	23	190	4700	9300	6.6	6.7	0.64	0.38
C	12	412	5000	12500	7.0	7.0	0.6	0.7

RHEOLOGICAL PROPERTIES

Recently we have shown (13) that acetyl and pyruvate content had no influence on the xanthan solution viscosities in dilute solutions. In semi-dilute solutions, where molecule entanglements exist this conclusion remains valid. We find a unique curve independent   of the substituent content for the double logarithmic plot η_{sp} versus the overlap parameter $c\,[\eta]$ up to $c\,[\eta]$ values equal to 40 (Figure 3). By this representation we take into account the dependence of $\bar{M}_w$ on the viscosity, so, we were able

F.Callet et al.

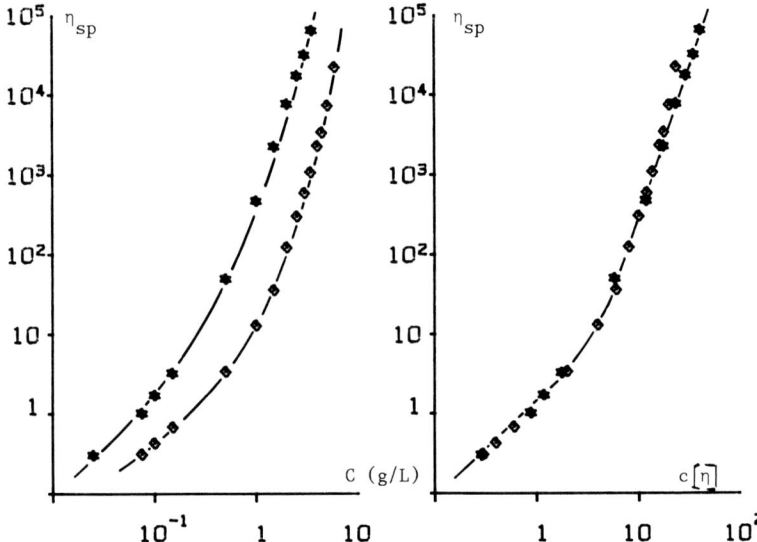

Figure 4 : η_{sp} vs. xanthan concentration C and overlap parameter C|η|
         in 0.01M NaCl (Sample A) ◈ native and ✻ renatured forms.
         t = 25°C

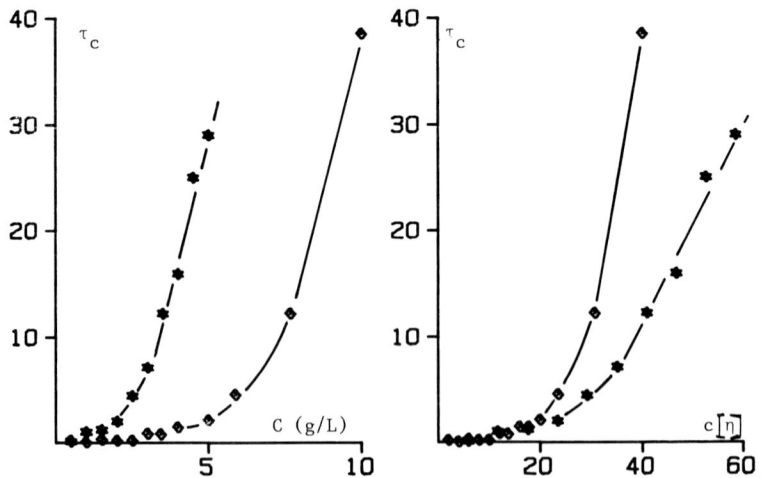

Figure 5 : Yield value vs.xanthan concentration C and overlap parameter c|η|
         in 0.01M NaCl (Sample A) ◈ native and ✻ renatured forms.t=25°C.

to compare samples with different $\bar{M}_w$ (19). In Figures 4 and 5, the role of the conformation on η_{sp} and on yield value τc are given. As for different substituent contents, a unique curve is found when η_{sp} is plotted versus c $[\eta]$ for the native and renatured structures. For τc the conclusions are quite similar.

Therefore, the intrinsic viscosity of xanthan seems to be the parameter which most contributes to the rheological properties of xanthan solutions in dilute and semi-dilute solutions as described by the curves η_{sp} and τc versus the overlap parameter c $[\eta]$.

ACKNOWLEDGEMENTS

The authors thank Professor M. RINAUDO for valuable suggestions and helpful discussions and RHONE-POULENC and SHELL for providing polymer samples.

REFERENCES

1. Milas, M. and Rinaudo, M. (1986) Carbohyd. Res. 158, 191.

2. Milas, M. and Rinaudo, M. (1984) Polym. Bull. 12, 507.

3. Sato, T., Norisuye, T. and Fujita, H. (1984) Polym. J. 16, 341.

4. Lecourtier, J., Chauveteau, G. and Muller, G. (1986) Int. J. Biol. Macromol. 8, 306.

5. Stokke, B.T., Elgsaeter, A., Skjak-Braek, G. and Smidsrod, O. (1987) Carbohyd. Res. 160, 13.

6. Morris, E.R., Rees, D.A., Young, G., Walkinshaw, M.D. and Darke, A. (1977) J. Mol. Biol. 110, 1.

7. Norton, I.T., Goodall, D.M., Morris, E.R. and Rees, D.A. (1980). J. Chem. Soc. Chem. Commun. 545.

8. Frangou, S.A., Morris, E.R., Rees, D.A., Richardson, R.K. and Ross-Murphy, S.B.(1982). J. Polym. Sci. Polym. Lett. Ed. 20, 531.

9. Sandford, P.A, Pittsley, J.E., Knutson, C.A., Watson, P.R., Cadmus, M.C.and Jeanes, A. (1977). Microbial extracellular polysaccharides, ACS Symposium 45, 81.

10. Smith, I.H., Symes, K.C., Lawson, C.J. and Morris, E.R. (1981). Int. J. Biol. Macromol. 3, 129.

11.Bradshaw, I.S., Nisbet, B.A., Kerr, M.M. and Sutherland, I.W. (1983) Carbohydr. Polym. $\underline{3}$, 23.

12.Tako, M. and Nakamura, S. (1984) Agric. Biol. Chem. $\underline{48}$, 2987.

13.Callet, F., Milas, M. and Rinaudo, M. (1987). Int. J. Biol. Macromol.(in press).

14.Rinaudo, M. and Milas, M. (1980) Int. J. Biol. Macromol. $\underline{2}$, 45.

15.Sato, T. Kojima, S. Norisuye, T. and Fujita, H. (1984) Polym. J. $\underline{16}$, 423.

16.Chen, C.S.H. and Sweepard, E.W. (1979). J. Macromol. Sci., Chem. Ed. $\underline{A}$ 13, 239.

17.Lambert F. and Rinaudo, M. (1985)Polymer $\underline{26}$, 1549.

18.De Murcia, G. I.B.M.C. Strasbourg, Unpublished results.

19.Milas, M., Rinaudo, M. and Tinland, B. (1985). Polym. Bull. $\underline{14}$, 157.

Mechanical properties of whey protein gels in relation to composition and microstructure

K.R.Langley, M.L.Green, B.E.Brooker and A.C.Smith[+]

AFRC Institute of Food Research, Reading Laboratory, Shinfield, Reading, Berkshire RG2 9AT, UK
[+]*AFRC Institute of Food Research, Norwich Laboratory, Colney Lane, Norwich NR4 7UA, UK*

SUMMARY
   When heated to denaturation temperatures, solutions of whey proteins form random network gels. The influence of protein composition on micro-structure and compression, tension and impact properties, and the effect of incorporating glass spheres was investigated.
   Compressive strengths and elastic modulii were maximal when the α-lactal-bumin/β-lactoglobulin (mole/mole) ratio was 0.64. The strongest gels had integrated and finer network structures whereas the weaker gels were more globular.
   The strength of gels, with added glass spheres, were dependent on both the glass volume fraction and sphere size.

INTRODUCTION

   In general terms food texture describes the consumer's mouthfeel. Mouthfeel may be considered in terms of e.g. hard as in boiled sweets and soft as in marshmallows. These properties, hard and soft, can be measured instrumentally. However, the underlying factors influencing the mechanical properties of the food, and hence the texture are the composition and microstructure.
   This study considers the rheological properties of denatured whey protein gels and how these relate to the microstructure. Three types of protein gel systems are considered including those derived from protein mixtures and those with an added particulate phase.
   There are many instruments available to assess the mechanical properties of foods. These range from special purpose machines e.g. the Farinograph, to universal ones e.g. the Instron. Nearly all of the machines available operate at drive rates of 5-500mm min^{-1}, which is below the human mastication rate, typically 1,200mm min^{-1} (Bourne)(1), and in this study some measurements were made with an Impact tester which operates at rates in excess of that of the human mastication rate.

MATERIALS AND METHODS

Whey Protein Concentrates (W.P.C.)
    Powders rich in whey proteins, prepared by the method of
Skudder (2) were stored in air-tight glass jars in a
refrigerator. The powders contained 0-12% w/w α-lactalbumin
(α-la), 44-94% w/w β-lactoglobulin (β-lg), with the rest as
non-gelling proteins and peptides derived from casein. Further
ion-exchange chromatography was used on these powders to produce
powders rich in either α-la or β-lg; the purity of the powders
being >98% for α-la and >99% for β-lg.

Glass beads
    Ballotini solid glass spheres of defined size ranges (Jencons
Scientific Ltd., Leighton Buzzard, Bedfordshire, U.K), were
cleaned by soaking in excess conc HNO_3 for a minimum of 3 days,
followed by several washings in distilled water until the pH of
the wash water was neutral. The glass spheres were filtered,
oven dried overnight and stored at ambient temperature in sealed
glass containers prior to use.

Gelation
    Solutions containing 15% w/w whey protein powder were
prepared by dissolving the powders in distilled water using
gentle stirring. The pH of the solutions was adjusted to 7.0 by
dropwise addition of conc. NaOH or conc HCl. The solutions were
placed in glass tubes, 10cm x 1.5cm diameter which were sealed
and immersed for 1h in a water bath at 90 ± 1°C. The tubes were
held vertically in the water bath.
    To ensure adequate mixing and to eliminate settling of the
glass spheres prior to gelation of the protein solution a simple
mechanical rotor was built. This supported the gelation tubes
at their centre, while they were totally immersed in the water
bath. By suitable gearing the filled tubes previously freed of
air bubbles could be rotated at 12 to 30 r.p.m., the latter rate
being used with the largest glass spheres.

Microscopy
    Samples of gel were fixed with glutaraldehyde, stained with
osmium tetroxide, dehydrated and either critical point dried
(scanning E.M), or embedded in araldite (transmission EM).
Scanning photographs of the gels were taken, using an Hitachi
S-570 microscope, at magnifications of x60 and x6000 and
transmission photographs were taken at x6000 with an Hitachi
H-600 microscope.

Mechanical Tests

Compression. Gels were removed from the glass tubes and cut
into 1.5cm lengths using a thin wire guillotine. Gel samples
were then loaded to failure in an Instron Universal Food Testing
Machine, model 1140, at a cross head speed of 50mm min^{-1}. The
sample cross section, at fracture, assuming cylindrical shape,
was used to determine the strength. A minimum of 5 samples were
tested for each gel.

<u>Tension</u>. Tensile samples were prepared using the test mould described earlier (Langley et al) (3). Gels were loaded to ultimate failure in the Instron at a cross head speed of 50mm min^{-1}.

<u>Impact</u>. Gels were removed from the tubes and cut into 5cm lengths. Measurements were made using an instrumented pendulum constructed at IFR,NL and a Zwick (Leominster, Herts, UK) Pendulum Impact Tester, Model 5101, operating in the Charpy mode with a 30° wedge hammer. The former enabled the penetration of the hammer into the sample to be measured and the impact strength calculated. The potential energy of the hammer was adjusted to give an impact speed of 2.9 ms^{-1}.

RESULTS AND DISCUSSION

<u>Gels made from mixtures of α-la and β-lg.</u>
The strength of the gels was found to be maximal when the α-la/β-lg ratio was 0.55:0.45 (concentration); this being equivalent to a mole ratio of 0.64. A similar synergistic effect with α-la and β-lg mixtures has been observed in a study of heat denaturation rates (Baer et al) (4). The gels fractured suddenly in compression by bursting at the surface parallel to the axis of the cylinder. This was especially evident for gels high in β-lg. The fracture surfaces of gels made at high α-la content had a tightly bound network (Figure 1). This network got tighter as the β-lg content increased, until at a β-lg/α-la ratio of 1:1, corresponding to the maximum in strength, the fractured surface was found to be very smooth. As the β-lg content increased the structure was found to be more open and indicative of cleavage planes. This pattern of structures was observed at both low (Figure 1) and high magnifications. This may be related to the involvement of disulphide bonding in gel formation, the β-lg molecule having 1 sulphydryl and 2 disulphide bonds, with the α-la molecule having 4 disulphide bonds.
    The surfaces of gels cut perpendicular to the fracture plane, across the axis of the cylinder were examined. Those containing high amounts of α-la were similar to that found for the fracture surfaces. However as the β-lg content increased these surfaces became increasingly smoother, contrary to that observed for the fractures surfaces. The different types of structure observed for the gels containing high level of β-lg are indicative of an anisotropic structure with weak cleavage planes running perpendicular from the centre of the gel towards the outer diameter.
    This anisotropy is almost certainly due to the method of gel preparation i.e. in a water bath, where there will be a temperature gradient from the outside towards the inside, so that setting occurred first on the edge and the non-gelling material was concentrated in the centre (Green et al) (5).

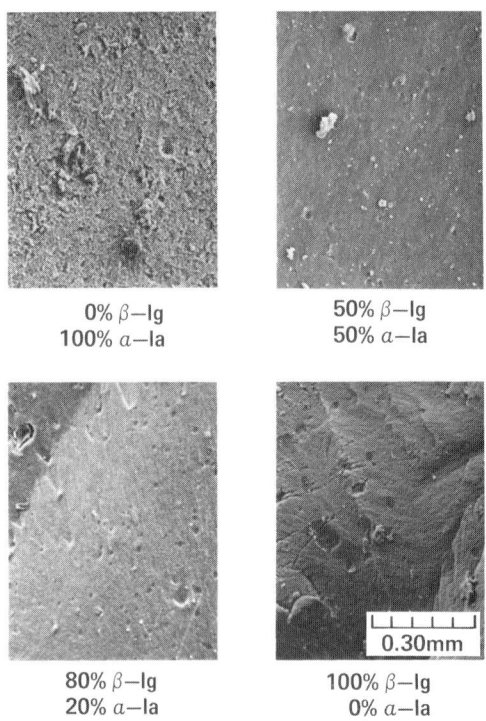

0% β—lg   50% β—lg
100% α—la   50% α—la

80% β—lg   100% β—lg
20% α—la   0% α—la

Fig 1.   Scanning electron micrographs of compression fractured surfaces of gels made from
mixtures of α—lactalbumin and β—lactoglobulin.

Gels made from W.P.C.

The strength of gels made from powders containing high
concentrations of β-lg and casein components and relatively low
concentrations (10% w/w or less) of α-la were found to be very
dependent on the β-lg content, with the strength increasing with
increase in β-lg content (Figure 2).

The fracture surfaces observed in the 3 test methods were
quite different. In compression the gels failed by shear along
a cleavage plane, as observed for the gels made from mixtures of
α-la and β-lg containing high concentrations of β-lg (section
above); in tension the failure occurred across the sample
forming a convex and concave surface (Green et al) (5).   In
impact, both surfaces were irregular, without evidence of
tearing, indicating that they were formed by cracking.

At low magnification (Figure 3) the damaged surfaces produced
by impact of strong gels had an open honey-comb structure

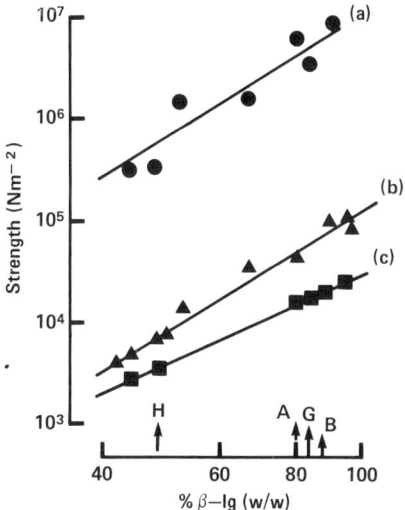

Fig 2. Strength of gels made from solutions of 15% W.P.C., pH 7; tested in
(a) Impact, (b) Compression and (c) Tension.

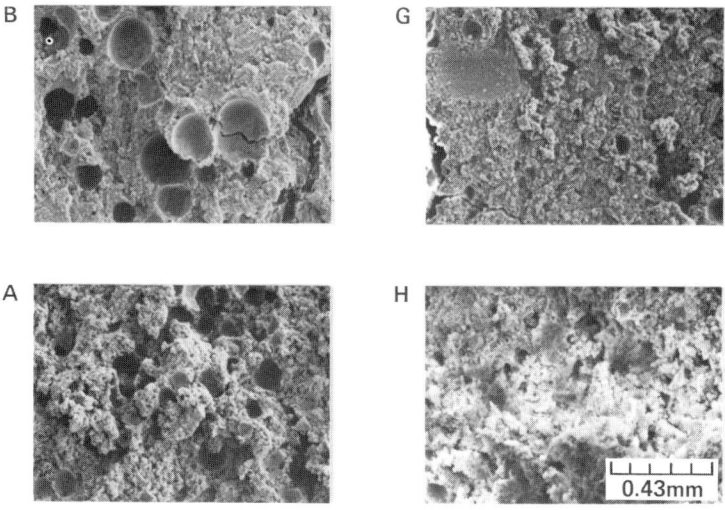

Fig 3. Scanning electron micrographs of impact fracture surfaces of gels made from W.P.C. The
strength of the gels decreases in the order: B, G, A, H, as shown in Fig 2.

whereas the weaker gels had a powdery/flakey appearance. At
higher magnification the strong gels had a closely bound
structure. This structure became more open as the strength of
gel decreased, until finally the weak gels were micellar in
appearance.

W.P.C. gels containing glass spheres
    The strength of W.P.C. made with large diameter glass spheres
increased with increase in sphere concentration (Figure 4).
These results are consistent with a model where there is little
or no interaction between the glass and the protein matrix (Ross
Murphy and Todd) (6). For smaller diameter glass beads the
strength of the composite was minimal at glass sphere volume
fractions of about 0.06, consistent with a model where there is
high interaction between the glass and the matrix. Gels made
with glass spheres with diameters between 80 and 600μm were

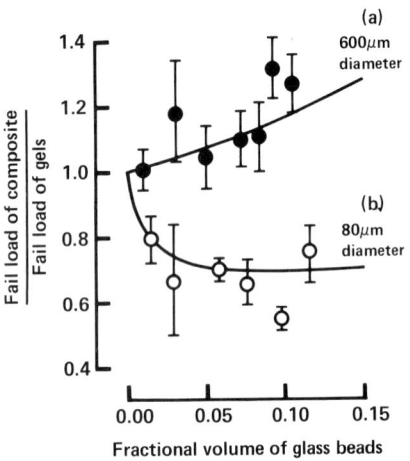

Fig 4. Compression strength of whey protein gels containing glass beads. Lines (a) and (b) are
        fitted to theory for low adhesion between glass and protein matrix (a) and
        high adhesion (b).

found to have compression strengths lying between the upper and
lower limits. These results would be consistent with a model
where there was interaction/adsorption of protein gel onto the
surface of the glass, the effect being small for large diameter
spheres, and considerable for the small diameter spheres, which
have a much larger surface area for a given total sphere volume.
    Gels containing small spheres were ductile under impact
force, with gels containing large concentrations of spheres
showing little or no sign of fracture at or near the time of
impact. Gels containing the large spheres were brittle;
becoming more brittle as the sphere concentration increased.
This is expected from the previous result. When the particles
were bound to the matrix, the composite was more ductile than

when they were not.

CONCLUSION
    In this paper we have shown qualitatively that mechanical
properties of heat denatured whey protein gels were related to
their microstructures, for instance, the strongest gels were
those with a very tight microstructure network and the weakest
those where the network was open or loose. This is expected
from the relative number of cross links observed in the network.
    By the addition of glass spheres to the gel, it has been
possible to model the whole range from complete bonding to the
matrix to no bonding. This is enabling the effect of
particle-matrix interactions or mechanical properties to be
studied.

ACKNOWLEDGEMENTS
    We would like to thank Ruth Oakman, Anne Roberts (E.M.), J.
Taylor (Mechanical Properties) and R. Young (High Speed
Photography) for their technical expertise.

REFERENCES
1) Bourne, M.D. (1976) In 'Rheology and Texture in Food Quality'
    (Eds. J.M. De Man, P.V. Voisey, V.F. Rasper and D.W.
    Stanley), Air Publishing Co., Westport, C.T.
2) Skudder, P.J. (1985) J. Dairy Res. 52, 167-181.
3) Langley, K.R., Millard, D. and, Evans, E.W. (1986) J. Dairy
    Res. 53, 285-292.
4) Baer, A., Oroz, M., and Blanc, B. (1975) J. Dairy Res. 43,
    419-432.
5) Green, M.L., Langley, K.R., Marshall, R.J., Brooker, B.E.,
    Willis, A., and Vincent, J.F.V. (1986) Food Microstructure
    5, 169-180.
6) Ross-Murphy, S.B., and Todd, S. (1983) Polymer 24, 481-486.

The texture of gellan gum gels

G.R.Sanderson, V.L.Bell, R.C.Clark and D.Ortega

Kelco Division of Merck and Co. Inc., San Diego, CA, USA

ABSTRACT

The texture of gels can be described in terms of the parameters, hardness, modulus, brittleness, elasticity and cohesiveness, by analysis of the force/deformation profiles obtained by
compressing gel samples on an INSTRON. This technique, texture
profile analysis, has been applied to gels from gellan gum, agar,
κ-carrageenan and gelatin. Low acyl gellan gum produces gels
which have a high hardness and modulus but are brittle and non-
elastic. It is possible to reduce the brittleness of these gels
to produce a wide range of desirable gel textures. Agar and κ-
carrageenan are texturally similar to low acyl gellan gum and,
hence, this form of gellan gum is a suitable replacer for these
gelling agents. The replacement of agar by low acyl gellan gum
in selected Japanese food products is illustrated. There are
advantages in using gellan gum to formulate products traditionally prepared with gelatin.

KEYWORDS

Gellan gum; low acyl; high acyl; texture profile analysis;
hardness; modulus; brittleness; elasticity; agar; κ-carrageenan;
gelatin; Japanese foods.

In production situations, the quality of gels is often assessed simply by shaking, squeezing or prodding with the finger.
These subjective techniques usually provide the required information, especially when used by experienced personnel. The most
common objective measure of gel quality is gel strength. Gel
strength measurements can be obtained using a variety of instruments but different instruments measure different textural parameters (1). For example, the Bloom Gelometer measures gel firm-
nesss at small strain levels, while the Marine Colloids Gel
Tester determines rupture strength, the force required to break
the gel. A much more comprehensive understanding of gel texture

is obtained by analysis of the force/deformation curve generated
by compressing a gel sample using an INSTRON Universal Testing
Machine. This technique, known as texture profile analysis, was
originally developed by General Foods in the '60's during the
course of their studies on correlation between sensory evaluation and objective measurements of food texture (2, 3, 4).

In our laboratory, texture profile analysis of gels involves
compression of free-standing gel discs using an INSTRON 4201
interfaced to a HEWLETT PACKARD 86B computer. The gels are
prepared by allowing hot solutions to gel on cooling in ring
molds. The INSTRON is programmed to compress the sample twice
in succession to 30% of its original height, using a compression
rate of 2 inches/minute. An idealized texture profile for a gel
and the parameters calculated by the computer are shown in Figure 1.

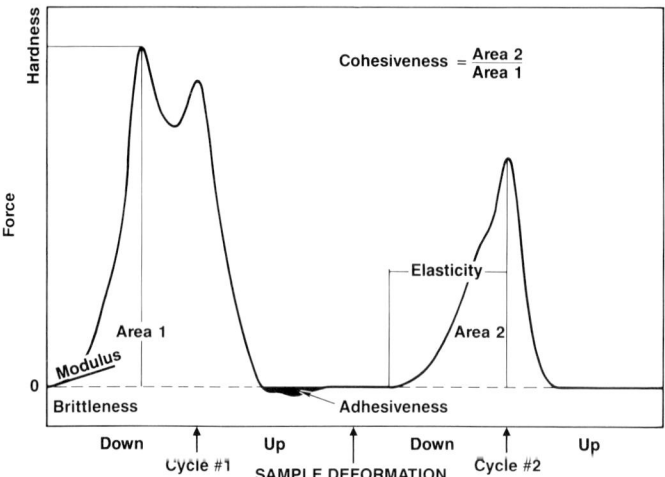

Figure 1.   Idealized texture profile for a gel.

The modulus is the initial slope of the force-deformation
curve. This is a measure of how the sample behaves when compressed a small amount. The modulus usually correlates very
closely with a sensory perception of the sample's firmness and
is expressed in units of force per unit area (usually in pounds
per square inch or Newtons per square centimeter).

Hardness is defined as the maximum force that occurs at any
time during the first cycle compression. It may occur when the
gel initially breaks (as shown in the "idealized" plot) or it
may occur later in the test as the sample is flattened and de-
formed. In most cases, the hardness is correlated to the rupture strength of the material. It is similar to gel strength
measured on a Marine Colloids Gel Tester and is expressed in
units of force (pounds or Newtons).

The first significant drop in the force/deformation curve during the first compression cycle is defined as the brittleness. It is the point of first fracture or cracking of the sample. A gel that fractures very early in the compression cycle is considered to be more brittle or fragile than one that breaks later. Brittleness is measured as the % strain required to break the gel. As the number becomes smaller, it indicates a more brittle gel because the gel breaks at a lower strain level. Since the gels are normally compressed to a 70% strain level, the maximum value for brittleness is 70%.

Following the first compression cycle, the force is removed from the sample as the INSTRON crosshead moves back to its original position. If the material is at all sticky or adhesive, the force becomes negative. The area of this negative peak is taken as a measure of the adhesiveness of the sample. There are no real units for this parameter which is expressed in the internal integrator units of the computer.

As the second compressison cycle is begun, elasticity of the sample is determined. By noting where the force begins to increase during this second compression cycle, a measure of the sample height may be obtained. If the sample returned to its original height, the elasticity would be 100%. The elasticity is a measure of how much the original structure of the sample was broken down by the initial compression. In sensory terms, it can be thought of as how "rubbery" the sample will feel in the mouth. The units are dimensionless (a length divided by another length) and are usually expressed as a percentage.

Cohesiveness is measured by taking the total work done on the sample during the second cycle and dividing it by the work done during the first cycle. Work is measured as the area under the respective curves. This ratio is usually expressed as a percentage. Samples that are very cohesive will have high values and will be perceived as tough and difficult to break up in the mouth. As with elasticity, there are no units since the number is dimensionless.

The influence of calcium ion concentration on the hardness and modulus of 0.25% GELRITE* gellan gum is shown in Figure 2. The gels were prepared by adding calcium chloride solutions of the appropriate concentrations to gellan gum presolubilized in deionized water by heating to 70-75°C. Between 0.016% and 0.05% added calcium (400 ppm and 1250 ppm $CaCO_3$ hardness), the modulus remains relatively constant, between 5.5 and 6.5N/cm^2. Hardness decreases slightly with increasing calcium level within this range from around 4.0 to 3.0 lbs. Thus gel strength in terms of rupture strength or initial firmness is fairly constant over this range of calcium ion concentrations. The brittleness and elasticity of GELRITE at 0.25% under the same ionic strength conditions are shown in Figure 3. Between 0.016% and 0.05% calcium ion concentration, the conditions under which the gel has a maximum modulus, the gel has a high brittleness of around 30 to 35%. Elasticity is low and constant at around 10% under these same conditions.

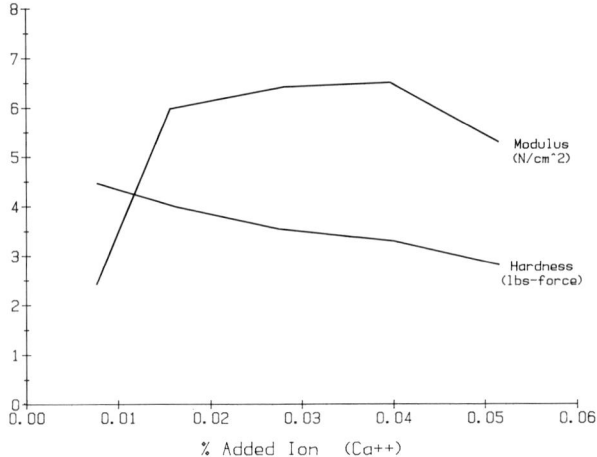

Figure 2.   GELRITE (0.25%) - Hardness and modulus vs. Ca++ concentration ('as is' pH).

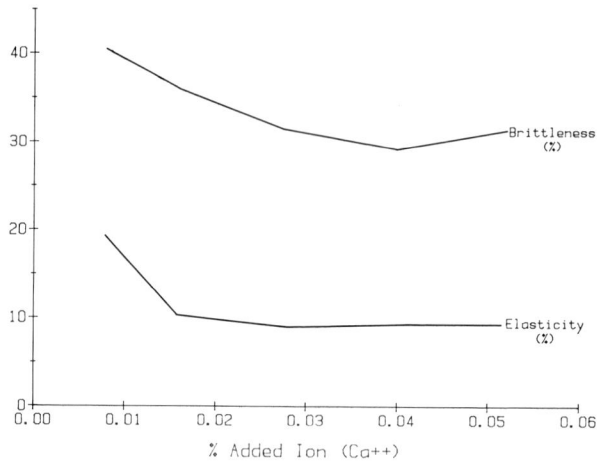

Figure 3.   GELRITE (0.25%) - Brittleness and elasticity vs. Ca++ concentration ('as is' pH).

   Hardness, modulus, brittleness and elasticity as a function
of GELRITE concentration in 0.028% calcium ions are shown in
Figures 4 and 5.   The hardness and modulus increase almost linearly with increasing gum concentration.   At higher gum levels,
the gels are more brittle and more elastic.   There is generally
little change in all four parameters in going from as is   pH
to pH 4.0.

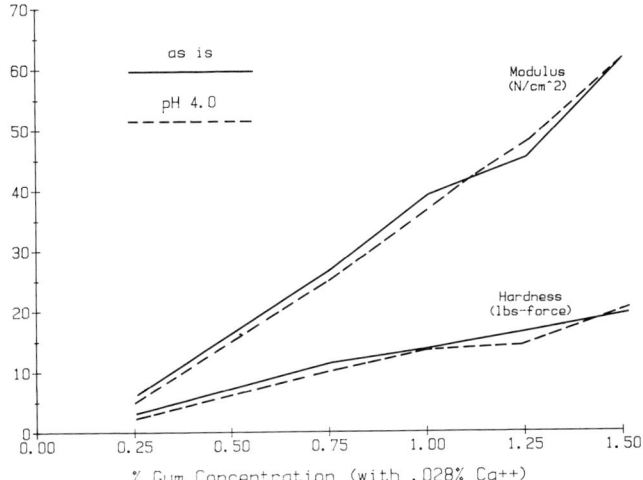

Figure 4. GELRITE - Hardness and modulus vs. gum concentration.

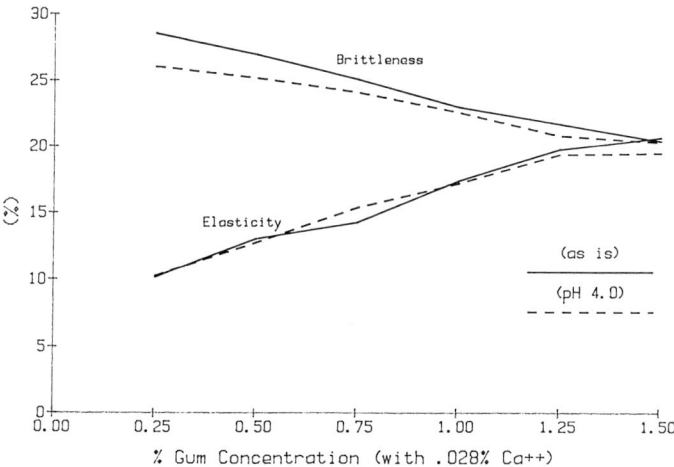

Figure 5. GELRITE - Brittleness and elasticity vs. gum concentration.

    As part of this study, the influence of mono- and divalent
ions on the texture of gels from κ-carrageenan, agar and gelatin
were also investigated. Not surprisingly, potassium ion concentration had a marked effect on the texture of κ-carrageenan.
Potassium ion concentration had a minor influence on the texture of gelatin gels but essentially no effect on the texture of
agar gels. The potassium ion concentration required to produce

the maximum modulus for each of these gelling agents, namely
0.63% for κ-carrageenan, 0.35% for gelatin and 0.70% for agar,
was used to prepare gels from them for determination of textural
characteristics as a function of hydrocolloid concentration.
The values for hardness, modulus, brittleness and elasticity are
respectively shown in Figures 6, 7, 8 and 9. At equal gum concentrations, κ-carrageenan gave the hardest (Figure 6) and most
firm (Figure 7) gel at both neutral and acid pH. However, κ-carrageenan showed the biggest reduction in hardness as the pH was
lowered. The gelatin gels were substantially less hard and firm
than those from agar and κ-carrageenan. As shown in Figure 8,
the gels from agar and κ-carrageenan were both brittle, with
brittleness levelling off above around 0.5% gum concentration.
At and above this concentration, the carrageenan gels increased
in brittleness from 45% to 33% in going from neutral to acid pH.
The low pH carrageenan gels had a brittleness similar to the low
pH agar gels; the neutral pH agar gels were slightly less brittle
than these two. The gelatin gels, in contrast, were not brittle
and, in fact, did not rupture when compressed during the first
compression cycle to 30% of their original height, the limit of
compression used in these studies. The difference between gelatin, on the one hand, and agar and κ-carrageenan, on the other,
was also observed in terms of elasticity (Figure 9). The gelatin gels were considerably more elastic and gave elasticity
values in excess of 90% at higher concentrations. Agar above
0.5% generally had an elasticity of around 20% while κ-carrageenan had an elasticity of approximately 15% at all concentrations
tested.

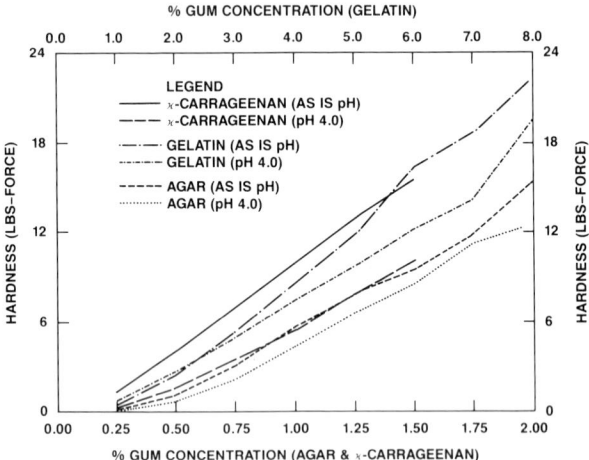

Figure 6. Hardness vs. gum concentration for κ-carrageenan in
0.63% K⁺, agar in 0.70% K⁺, and gelatin in 0.35% K⁺.
(κ-carrageenan- Gelcarin GP 812 (Marine Colloids); gelatin - 250
Bloom (Atlantic Gelatin); agar - USP Medium Gel (Colony Imports))

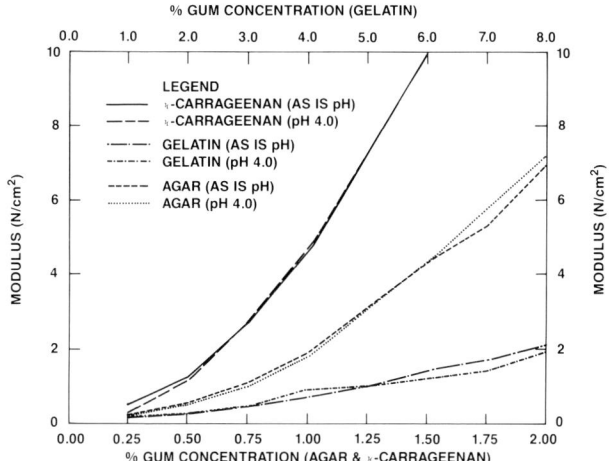

Figure 7. Modulus vs. gum concentration for κ-carrageenan in 0.63% K⁺, agar in 0.70% K⁺, and gelatin in 0.35% K⁺.

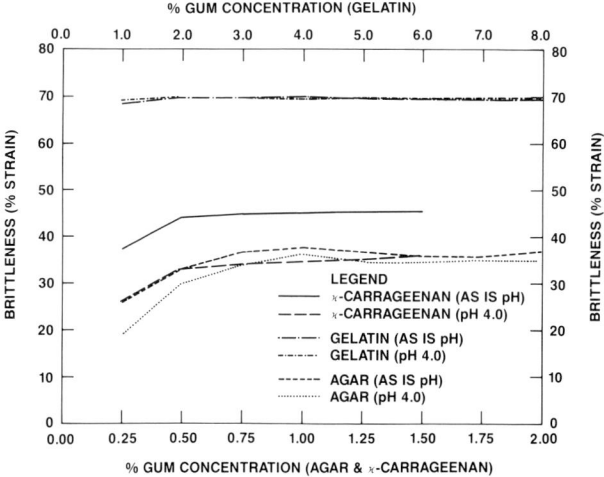

Figure 8. Brittleness vs. gum concentration for κ-carrageenan in 0.63% K⁺, agar in 0.70% K⁺, and gelatin in 0.35% K⁺.

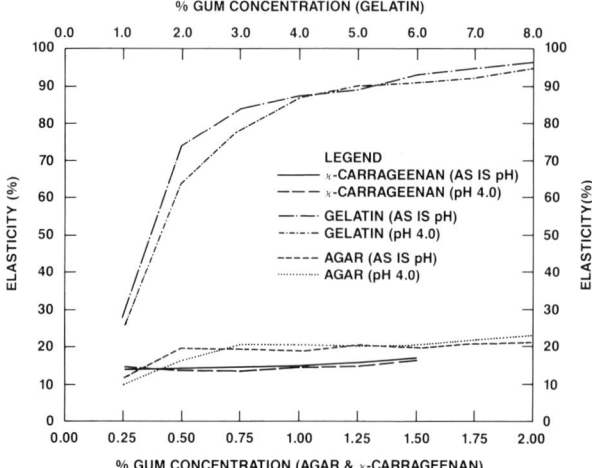

Figure 9.   Elasticity vs. gum concentration for κ-carrageenan in 0.63% K+, agar in 0.70% K+, and gelatin in 0.35% K+.

The textural parameters for 0.5% GELRITE gellan gum gels are compared to those for gels made from typical in-use concentrations of agar (1.0%), κ-carrageenan (0.4%) and gelatin (1.5%) in Figure 10.   Even at this relatively low use level, GELRITE, which

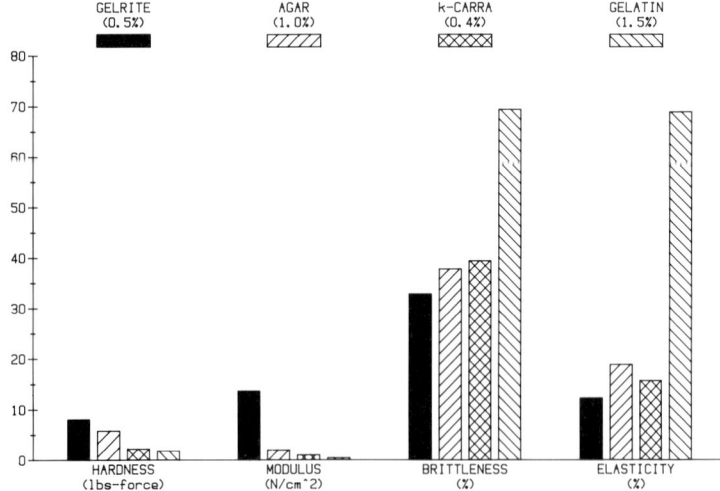

Figure 10.   Textural characteristics of GELRITE compared to agar, κ-carrageenan and gelatin.

is texturally similar to KELCOGEL* food grade gellan gum, provides higher hardness and significantly higher modulus than the other three gelling agents. In other words, at these relative use levels, KELCOGEL and GELRITE would be perceived as considerably more efficient gelling agents when tested for gel strength on a Marine Colloids Gel Tester or a Bloom Gelometer. Figure 10 also shows that GELRITE is similar to agar and κ-carrageenan in terms of brittleness and elasticity. The distinction between gelatin and the other three hydrocolloids is evident. The gelatin gels are not at all firm to the touch (modulus), low in hardness, non-brittle and highly elastic.

From these studies, gellan gum would appear to be a suitable and more efficient replacement for agar and κ-carrageenan. Evaluation of gellan gum in a number of food products as an alternative to these other two hydrocolloids has shown that this is indeed the case. The use levels of gellan gum required are typically one half to one third of those required with agar and κ-carrageenan. The textural similarity that can be achieved is apparent by examining the textural parameters of four Japanese foods prepared with KELCOGEL gellan gum and with agar (Table 1). In each product, the modulus or perceived firmness is, however, higher with gellan gum. This is considered to be preferred organoleptically. In the case of the tokoroten noodles, a desirable resilience is imparted by combining the gellan gum with xanthan gum and locust bean gum. The effects of combining gellan gum with other hydrocolloids are discussed in another paper (5). It should be noted that not only hydrocolloids but also other food ingredients such as fats, sugars and ions are capable of modifying the properties of gellan gum gels.

PRODUCT	KELCOGEL USE LEVEL (%)	AGAR USE LEVEL (%)	TEXTURE (KELCOGEL) h m b e c	TEXTURE (AGAR) h m b e c
MITSUMAME JELLY CUBES	0.4	1.4	4, 6, 27, 17, 5	3, 2, 31, 15, 7
HARD RED BEAN JELLY	0.5	1.5	6, 13, 27, 25, 10	8, 6, 37, 42, 10
SOFT RED BEAN JELLY	0.1	0.35	1, 2, 24, 43, 12	1, 1, 34, 47, 14
TOKOROTEN NOODLES	0.4	0.8	3, 2, 35, 23, 8	3, 1, 36, 25, 10

Table 1   KELCOGEL vs. agar in Japanese foods.

(Texture determined by the texture profile analysis procedure described in this paper.
h = hardness (lbs), m = modulus (N/m^2), b = brittleness (%),
e = elasticity (%) c = cohesiveness (%))

There are three basic forms of gellan gum, native or high acyl, low acyl, clarified and low acyl, unclarified (6). The studies discussed above relate to low acyl gellan gum. Native or high acyl gellan gum has also been evaluated by texture profile analysis. A typical texture profile for a gel prepared from this material by heating and cooling in the presence of ions is shown in Figure 11. The gel is low in firmness, non-brittle and elastic and is, in fact, similar texturally to gels prepared by heating and cooling mixtures of xanthan gum and locust bean gum. Figure 11, which includes a texture profile of low acyl gellan gum for comparison, also shows that by combining low acyl gellan gum with high acyl gellan gum, gels with textures intermediate between those obtained with either form alone can be obtained. The progressive reduction in brittleness in going from low acyl to high acyl gellan gum is particularly noticeable.

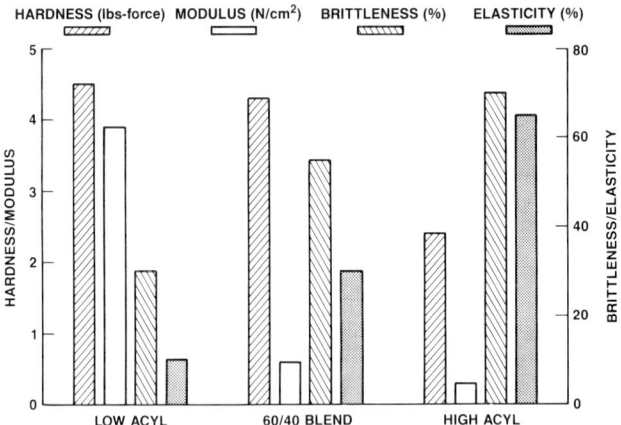

Figure 11. Comparison of the textures of low acyl, high acyl and low acyl/high acyl gellan gum gels.

The difference between the texture of gelatin gels and the textures of low acyl gellan gum, agar and κ-carrageenan gels has already been discussed. Texture, although of paramount importance, is not the only factor determining the organoleptic acceptability of food gels. In the case of gelatin, one of its most valuable features is its so-called 'melt-in'the-mouth' characteristic. Low acyl gellan gum gels do not melt in the mouth but the flavor release is excellent. This is because the gels contain an extremely low level of gum and the water structured within the gel is instantaneously released upon mastication. Gellan gum thus offers the food industry a new tool for the development of novel, appealing products.

REFERENCES

1. Mitchell, J.R. (1976) J. Texture Studies, 7, 313-339.
2. Szczesniak, A.S. (1963) J. Food Science, 28, 385-389.
3. Friedman, H.H., Whitney, J.E. and Szczesniak, A.S. (1963)
   J. Food Science, 28, 390-396.
4. Szczesniak, A.S., Brandt, M.A. and Friedman, H.H. (1963)
   J. Food Science, 28, 397-403.

5. Sanderson, G.R., Bell, V.L., Burgum, D.R., Clark, R.C. and Ortega, D. (1988) ibid.
6. Sanderson, G.R. and Clark, R.C. (1984) in Gums and Stabilisers for the Food Industry 2, (eds. Phillips, G.O. et al.) pp 201-210. Pergamon Press, Oxford and New York.

GELRITE*, Trademark of Merck & Co., Inc., Kelco Div., U.S.A.
KELCOGEL*, Trademark of Merck & Co., Inc., Kelco Div., U.S.A.
INSTRON is a trademark of Instron Corp., Canton MA
HEWLETT PACKARD is a trademark of Hewlett Packard Co., Palo Alto, CA

A method for determining the absolute shear modulus of a gel from a compression test

D.G.Oakenfull, N.S.Parker and R.I.Tanner[+]

CSIRO Division of Food Research, PO Box 52, North Ryde, NSW 2113, Australia
[+]Department of Mechanical Engineering, Sydney University, NSW 2006, Australia

ABSTRACT

The standard method for measuring the shear modulus of a gel is the U-tube method of Saunders and Ward. This method is slow, requires special glassware and presents problems for gels which synerese and do not adhere to glass. The method described overcomes these problems and is particularly suited to soft, weak gels. The gel is formed in a cylindrical dish and subjected to a very small compression by a flat probe. Any apparatus which measures applied force as a function of penetration can be used to obtain an apparent Young's modulus. This is converted into the absolute shear modulus by multiplying by a factor which depends only on the geometry of the measuring system. Values for this factor were calculated numerically using a boundary element simulation procedure. There was excellent agreement with values obtained experimentally for gelatin gels for which shear modulus had been measured by the method of Saunders and Ward.

INTRODUCTION

A compression test is one of the easier rheological measurements to make on a gel (1,2). There are many instruments available which make it possible to subject a firm gel to a compressive force and measure the subsequent deformation as shown in Fig.1(a). From such measurements it is possible to calculate the apparent Young's modulus of the gel – i.e. the ratio of the change of tensile stress to the relative change of length (strain). However it is the shear modulus of the gel which is of more fundamental interest. The shear modulus is the ratio of shear stress to the relative displacement sideways of parallel surfaces (2), as shown in

Fig. 1(b). This is much more difficult to measure.

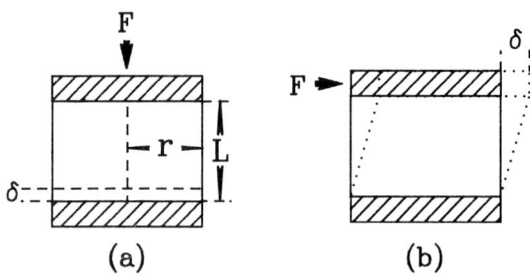

Figure 1.   Deformation of an elastic solid under (a) compression and (b) pure shear. F is the applied force and δ the resulting displacement.

The shear modulus is independent of the geometry of the measuring device and the size of the gel sample. Consequently it is a particularly appropriate measure of elasticity for comparing one gelatinous material with another – for quality control or product development. It is also a particularly useful quantity because it can be used to derive information about the size and stability of junction zones in gel networks (3).

The standard method for measuring shear modulus has long been the U-tube method of Saunders and Ward (4). This method is slow, requires special glassware and presents problems for gels which synerese and consequently do not adhere to glass (5). We present here a method which overcomes these problems – converting experimental data from a simple compression test into an absolute shear modulus. The method was specifically designed for soft gels too weak to be self supporting.

EXPERIMENTAL
All measurements were made with gels prepared from gelatin at 15° and at a concentration of 50 g/kg. The procedure was to form the gel in a cylindrical dish and subject it to compression by slowly inserting a flat-ended cylindrical probe as shown in Fig. 2. The extent of penetration was always

small (< 1% of the depth of the gel). We measured the applied
force vs the extent of penetration in two ways.

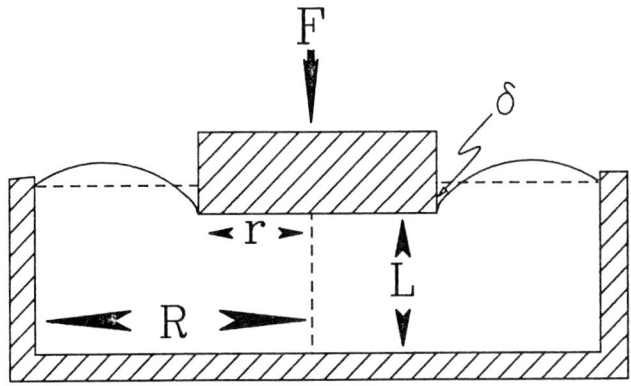

Figure 2. Geometry of the measuring device. Apparent Young's
modulus is defined by $Y = FL/\pi r^2 \delta$.

(1) Large Dishes (Radius = 67.5 mm)

      An Instron Universal Testing Machine was used to lower
the probe (radius 27.5 mm) into the gel at a speed of 5 mm per
minute. Because we were looking at very small deformations,
the net applied force was measured with an electronic
top-loading balance ($\pm 10^{-2}$g). In electronic balances there is
no movement of the pan and this arrangement gives a more
sensitive means of measuring the force applied to the gel than
the Instron's own load cell.

(2) Small Dishes (Radii 14-28 mm)

      This variation of the method was developed for
measurements on small samples. An electronic analytical
balance was used to measure the applied force ($\pm 10^{-4}$g). The
probe was inserted through an opening in the top of the balance
and penetrated the sample with a velocity of 30 mm per minute.
Because of the greater probe velocity, the procedure adopted
was to give the probe a 5 s pulse and record the balance
reading after 55 s for equilibration.

      Both methods gave linear plots of load vs penetration.

The apparent Young's modulus of the sample was calculated from
the slope after correcting for buoyancy.

THEORY

    We regard the gel as an incompressible elastic solid
undergoing small deformations as in the classical theory of
elasticity (6). Let the shear modulus be μ. Then stress (σ_{ij})
is related to strain (ε_{ij}) by

$$\sigma_{ij} = -p\delta_{ij} + 2\mu\varepsilon_{ij} \tag{1}$$

where p is the pressure (to be found), δ_{ij} is the unit tensor
and the components ε_{ij} of the strain tensor are given in terms
of displacements.   Because we are concerned only with
cylindrical samples, it was convenient to define the problem in
terms of cylindrical polar coordinates. We used two numerical
methods to solve these equations – a finite element code based
on Herrmann's algorithm for incompressible materials (7) and a
boundary element code (8). Both methods gave similar results
(agreement well within 1%) and we shall refer only to results
obtained by the boundary element method;   they appeared to be
slightly more accurate for our specific problem.
    Consider firstly a simple case in which a cylindrical
sample of radius r and height L is subjected to a small
compression as shown in Fig. 1.   If the platens are
frictionless, allowing free slippage of the gel against the
platen surface, the equations can be solved analytically giving
the well-known relationship for incompressible linear- elastic
materials, E = 3μ. Our boundary element simulation procedure
gave results agreeing with this simple theory to within ± 0.1%.
    Using the same definitions we can now analyse for the
test geometry shown in Fig. 2.   Here the depth of the gel
sample is L and the platen (probe) radius is r with a dish of
radius R. Defining now an 'effective Young's modulus' (Y)
operationally in terms of the total load (F) on the platens and
the axial displacement of one platen towards the other (δ)

$$Y = (F/\pi r^2) / (\delta/L) \tag{2}$$

The parameters r/R and R/L define the geometry of the system completely and we find that

$$\mu = f.Y \tag{3}$$

where f is a function of r/R and R/L. The numerical factor f can therefore be calculated for any geometry and we give in Table I a selection of values covering a practical range of values of r, R and L. As an example, when r = 39.9 mm, R = 67.5 mm and L = 27.9 mm, we find that the factor f = 0.0647, which is expected to be accurate to about ± 0.0005.

Thus the sample confined in the dish appears more than five time stiffer than the same sample under simple compression. This stiffening effect may also be described by computing the force F for a given μ and any given δ; we then find that

$$F = 3\mu \ (\delta/L) \ \pi r^2 f^* \tag{4}$$

where f^* is the stiffening factor which is related to our conversion factor f by, $f = 1/(3f^*)$. For simple compression $f^* = 1.0$.

Table I.   FACTORS (f) FOR CONVERTING APPARENT YOUNG'S MODULUS INTO ABSOLUTE SHEAR MODULUS FOR DIFFERENT SIZES OF DISH AND PROBE

r/R	L/R= 0.1	L/R= 0.2	L/R= 0.5	L/R= 1.0
0.05	0.157	0.0752	0.0381	0.0186
0.10	0.208	0.104	0.0634	0.0331
0.20	0.185	0.102	0.0912	0.0522
0.40	0.0549	0.0632	0.0930	0.0574
0.50	0.0264	0.0490	0.0803	0.0485
0.60	0.0138	0.0384	0.0622	0.0359

We have also investigated the possibility of error due to distortion of the surface with penetration of the plunger. Our calculations indicated that according to linear theory the net effect would be zero. It is, though, necessary to apply a correction for the buoyancy of the plunger as indicated in the Experimental Section. The simplest correction is to assume that a movement of the plunger of δ downwards gives an average upward movement of the surface of the sample of $\delta r^2/(R^2 - r^2)$. The force on the plunger due to hydrostatic pressure is then given by

$$F_h = \pi\sigma g\delta r^2 \cdot R^2/(R^2 - r^2) \qquad (5)$$

This must be subtracted from the measured force to get F.

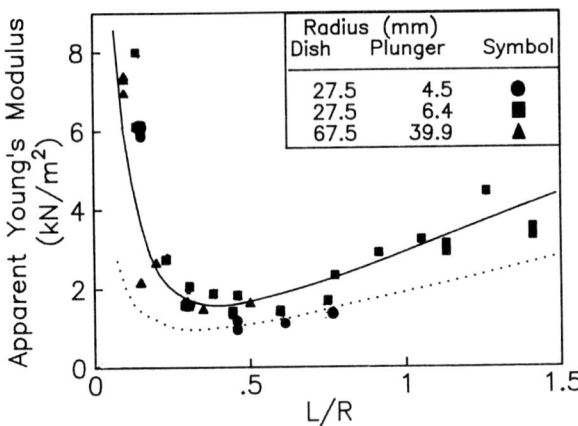

Figure 3. *Comparison of computed and experimental results for a single gel (gelatin at 20 g/kg). Apparent Young's modulus is plotted for different combinations of dish and plunger radius. Solid curve, r/R = 0.593; dotted curve r/R = 0.164.*

COMPARISON WITH EXPERIMENT

(1) Comparison with the Method of Saunders and Ward

   Using the method of Saunders and Ward (4) we found that
   $$\mu = 111.6 \pm 2.6 \; N/m^2$$
Forming the gel in dishes of radius 67.5 mm with a depth of 27.9 mm and subjecting it to compression with a plunger of radius 39.9 mm gave
   $$Y = 1560 \pm 50 \; N/m^2$$
converting this into a shear modulus by our computed factor f of 0.0647 then gave
   $$\mu = 101 \pm 4 \; N/m^2$$

There is thus a discrepancy of about 10% between the methods or, allowing for error ranges, about 4%.

(2) Effect on f of the Geometry of the Measuring System

In Fig. 3 we show how Y varies with the aspect ratio of the sample (L/R) for two very different values of R (67.5 and 27.5 mm). The data points are experimental and the curves were calculated. These results show that the theory adequately accounts for the effect on f of the geometry of the measuring device over a broad range of dimensions. In particular they confirm the minima in Y with increasing L/R predicted by the theory (Table I).

DISCUSSION

(1) The Effect of the Geometry of the Measuring System on the Conversion Factor

The calculations show that the conversion factor f has a maximum as a function of r/R with L/R fixed (Table I) and also as a function of L/R with r/R fixed. This complicated dependence on the geometry of the measuring system appears to result from three factors operating to generate the total force.

(a) When the plunger is small compared with the dish (small r/R) the influence of the walls of the dish must also be small. It is known (9) for incompressible materials generally that when L/r becomes very large

$$F = (8/3\pi) \cdot L/r \qquad (6)$$

This factor would then explain the sharp downturn in f when r/R is small.

(b) The sharp upturn of the curves in Fig. 3 for shallow gel samples (small L/R) is explained by a squeeze-flow action. When the wall effects are small we find, by analogy with lubrication theory, that the stiffness factor

$$f^* = 0.5 \ (r/L)^2 \qquad (7)$$

When r/L >> 1 this becomes an important contribution.

(c) When r approaches R, wall effects become important. The material being squeezed out from under the plunger has to 'escape' though a small gap of width (R – r), causing extra apparent stiffness. An approximate analysis suggests that f^* must have the form

$$f^* = K(L/R)/(R/r - 1)^2 \qquad (8)$$

where K is a constant. This factor explains the sharp downturn in the conversion factor f as the radius of the plunger approaches that of the dish.

It is reasonable to assume that the total response from confining the sample in the dish is the sum of these three effects so that

$$f^* = 8/3\pi.(L/r) + 0.5(r/L)^2 + 3.38(L/R)/(R/r - 1)^2$$

The constant 3.38 has been fitted from the results in Table I by using the single computed point for L/R = 1.0, r/R = 0.6 so we are not advocating using this equation to calculate the conversion factor for practical purposes. The point is to illustrate the form of the relationship between the apparent stiffness of the sample and the geometry of the dish and plunger. The equation very adequately describes the experimental results as shown in Fig. 3, showing the appropriate minima. If required, an equation of this form with empirically fitted coefficients would be suitable for interpolation from Table I.

CONCLUSIONS

We have shown by computer simulation that the shear modulus of a sample of gel formed in a cylindrical dish can be obtained from a compression test using, for example, an Instron Testing Machine. The apparent Young's modulus can be converted into an absolute shear modulus by multiplying by a factor which depends only on the geometry of the measuring system. We have tabulated these factors making it possible to use the method with any combination of sizes of dish and plunger within the normal practical range. The method is much quicker and easier than the Saunders and Ward U-tube method. It can be used for gels which do not adhere to glass and it is particularly suitable for soft, weak gels.

ACKNOWLEDGEMENTS

We thank Dr F. Sugeng for doing the calculations leading to Table I and Mr A.G. Scott who carried out the experimental measurements. This project forms part of the CSIRO/University of Sydney Cooperative Research Programme. This support is greatfully acknowledged.

REFERENCES

1. Mitchell, J.R. (1980) J. Texture Stud., 11, 315-337.
2. Sherman, P. (1970) Industrial Rheology, 423 pages, Academic Press,
   London, UK.
3. Oakenfull, D.G. (1984) J. Food Sci, 49, 1103-1104 & 1110.
4. Saunders, P.R. and Ward, A.G. (1953) Proc. Int. Cong. Rheol. 2nd, Oxford, UK, 294-290.
5. Stainsby, G., Ring, S.G. and Chilvers, G.R. (1984) J. Texture Stud., 15, 23-31
6. Jaeger, J.C. (1964) Elasticity, Fracture and Flow, 323 pages, Methuen, London, UK.
7. Nickell, R.E., Tanner, R.I. and Caswell, B. (1974) J. Fluid Mech., 65, 189-206.
8. Bush, M.B. and Tanner, R.I. (1983) Int. J. Numer. Methods Fluids, 3, 71-72.
9. Timoshenko, S. P. and Goodier, J.N. (1954) Theory of Elasticity, 2nd edn, 298 pages, McGraw-Hill, New York, U.S.A.

Relationship of electromyographic evaluation of semi-fluid model food systems with dynamic shear viscosity

I.C.M.Dea, A.Eves, D.Kilcast and E.R.Morris[+]

Leatherhead Food R.A., Randalls Road, Leatherhead, Surrey KT22 7RY, UK
[+]Cranfield Institute of Technology, Silsoe College, Silsoe, Bedford MK45 4DT, UK

ABSTRACT

Subjective responses to perceived thickness of semi-fluid model systems used in this study correlate closely with the activity of the muscles that control tongue movement, as measured by electromyography. In addition, the muscular activity involved in eating the semi-liquid model systems shows a power law relationship to the objective measurement of dynamic viscosity, which has the same gradient as the power law relationship reported earlier between perceived thickness and objective measurement of viscosity for a range of liquid model systems.

INTRODUCTION

A novel method of food texture measurement known as electromyography (EMG) has recently been developed.[1,2] The technique involves the measurement of electrical potentials in the facial muscles involved in the manipulation and breakdown of food during eating, as they relax and contract. Initial studies concentrated on the muscles involved in the mastication or chewing of solid foods, and consequently were concerned with the collection and examination of EMG data obtained from the masseter muscle, which controls the opening and closing of the jaw.[3] Such studies have demonstrated that EMG data give true force-time curves for the breakdown of solid foods in the mouth. An extension of this approach in the evaluation of the texture of confectionery gum products appears elsewhere in this volume.[4] We have now applied the EMG technique for the texture evaluation of semi-fluid model food systems by measuring the electrical potentials produced in the muscles, located under the chin, that control tongue movement. This EMG information was collected throughout the period when the subject was assessing the perceived thickness of the model food. The range of derived EMG data was then compared with measurements of both perceived thickness and dynamic shear viscosity made using a Sangamo viscoelastic analyser.

It has previously been demonstrated that rotational viscosity measurements at ~ 50 rad/s correlate well with sensory evaluation of thickness for disordered ('random coil') polysaccharide systems, but substantially underestimate the perceived thickness of samples that have a tenuous 'weak gel' structure, which is broken down by shear.[5] Non-destructive small-deformation oscillatory measurements, however, yield data (dynamic shear viscosity at 50 rad/s) that correlate directly with perceived thickness of both Newtonian and non-Newtonian systems.[6]

MATERIALS AND METHODS

Electromyography

Electromyographic patterns were recorded using a Grass Polygraph (Model 7D). The system has a regulated power supply consisting of two dc driver amplifiers. One of these is connected to a pre-amplifier and displays the raw signal. The other is connected to the pre-amplifier and an integrator module, and displays the integrated data. Both driver amplifiers include a 50-Hz filter rejecting interference from ac sources. All recordings were made with the amplifiers' frequency bands as wide as possible (10 Hz–40 kHz). This ensures a flat frequency response up to 200 Hz, the maximum response rate of the pens. The time base was set at 0.2 s.

Surface electrodes were used to detect the electrical signals. The skin under the chin and the earlobe of the subject were cleansed with 95% (V/V) ethanol to remove traces of dust and perspiration, which may interefere with the signal. The EMG data were collected from a group of muscles under the chin that control the movement of the tongue. Two electrodes were placed under the chin, approximately 5 cm apart, on a line connecting the cleft of the chin to the centre of the clavicle. The exact position of these electrodes was selected to achieve a maximum electrical inflection when the tongue was moved during assessment of the thickness of semi-fluid model systems by the subject. A third electrode was placed on the earlobe, a point of no muscular activity, which acted as an earth. Electrode cream, a conductive paste, was applied to the electrode surface in all cases.

After the terminal ends of the electrodes had been located in the Polygraph, the subject was presented with a sample to assess. All samples were assessed on four separate occasions. While EMG measurements were being carried out, the subject manipulated the samples in the mouth in order to assess the perceived thickness. This ensured that a reasonable sequence of EMG data was obtained and analysed, in addition to permitting a comparison of EMG data with perceived thickness to be carried out. A typical example of the raw signal and integrated data from a semi-fluid model system is shown in Fig. 1, and is of a similar form to the EMG data collected during the chewing of solid food, except that there was no trend to smaller peaks with time in the mouth. The derived data from the EMG measurements were obtained from the integrated data. These were collected, via an a/d converter, on to a computer disc at a rate of fifty points per second. The stored data were analysed by a computer program calculating various trace parameters — peak height (PH), adjusted height (AH), pre-maximum gradient (PRG), post-maximum gradient (POG), and area under peak (A) (Fig. 2). Gradients and area were calculated from minima rather than baseline, and all parameters were adjusted in relation to a pre-recorded baseline. Small inflections on the integrated trace were eliminated by means of a rejection factor. Mean values for EMG parameters were calculated from all integrated peaks over the entire time for which the sample remained in the mouth, and from the repeat assessments. These values were then compared with perceived thickness and measurements of dynamic shear viscosity.

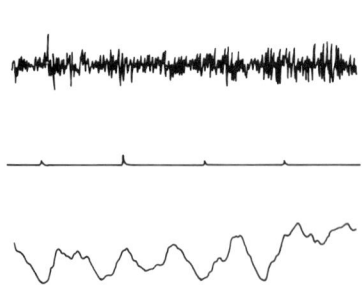

Figure 1. EMG output.

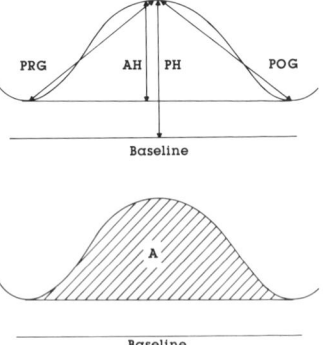

Figure 2. Integrated EMG trace parameters.
PH = peak height; AH = adjusted height;
PRG = pre-maximum gradient; POG = post-maximum
gradient; A = area

Model liquid foods

Two model systems were used. Golden Syrup was used as representative of a Newtonian system, and was tested at four different dilutions (undiluted, solution 1; 8 parts Golden Syrup to 1 part water, solution 2; 6 parts Golden Syrup to 1 part water, solution 3; 4 parts Golden Syrup to 1 part water, solution 4). Aqueous solutions of xanthan gum were used as representative of a non-Newtonian system. These were prepared from 3% (m/V) xanthan gum by making four dilutions with full-cream homogenised milk to make the solutions more palatable. The concentration of these xanthan gum preparations were 2% (solution 1), 1.5% (solution 2), 1% (solution 3) and 0.5% (solution 4).

Sensory assessment

Assessment of perceived thickness was made using the ratio scaling technique. No formal training was given in either ratio scaling techniques or in thickness assessment. For each of the two model systems, solution 3 was used as the reference and assigned an arbitrary score of 10. The other solutions in each set were scored relative to this fixed control, so that, for example, a solution considered to be twice as thick as the reference would be given a score of 20, a solution considered to be half as thick a score of 5, and so on. The reference was presented first, with the other three samples in each set presented in a random order. Each sample was assessed on four separate occasions.

Rheological measurement

Dynamic shear viscosity (η^*) was measured at 25°C on a Sangamo viscoelastic analyser using a cone and plate configuration of diameter 5 cm and cone angle 2 degrees.

RESULTS AND DISCUSSION

As expected from previous work,[5,6] the subjective evaluation of thickness for both sets of samples correlated closely with the objective measurement of dynamic shear viscosity at 50 rad/s.

For the xanthan gum set of samples, of the five EMG parameters defined in Fig. 2, peak height showed the best correlation with subjective evaluation, with, on a log-log scale, peak height increasing linearly with perceived thickness (Fig. 3). This is a relationship that would be expected if this new EMG technique truly monitored the perceived thickness of semi-liquid fluids in the mouth. It might be expected that the use of adjusted peak height or peak area (Fig. 2) would further improve this relationship. However, although the trend of increase in magnitude of EMG parameter with increase in perceived thickness still occurred, the correlations were poor. In earlier EMG studies on the chewing of firm foods it has been found that pre-maximum gradient and post-maximum gradient can correlate with certain perceived textural parameters.[4] For the xanthan gum series of samples no correlation was found between pre-maximum gradient and perceived thickness (Fig. 4). However, post-maximum gradient decreased linearly with log (perceived thickness), and showed a good correlation (Fig. 4).

Similar correlations between the EMG parameters and perceived thickness were also found for the Golden Syrup set of samples. Thus again the best correlation was obtained between log (peak height) and log (perceived thickness), as indicated in Fig. 5. In the case of these samples the relationships between the gradient measurements and perceived thickness were better. Thus pre-maximum gradient increased linearly with log (perceived thickness), while post-maximum gradient decreased linearly with log (perceived thickness), as indicated in Fig. 6.

It is therefore clear that subjective response to perceived thickness of the semi-fluid model systems used in this study correlates closely with the activity of the muscles that control tongue movement, as measured by electromyography. The correlation appears to be best for the Newtonian system studied, although this could have resulted from the greater palatability of this set of samples.

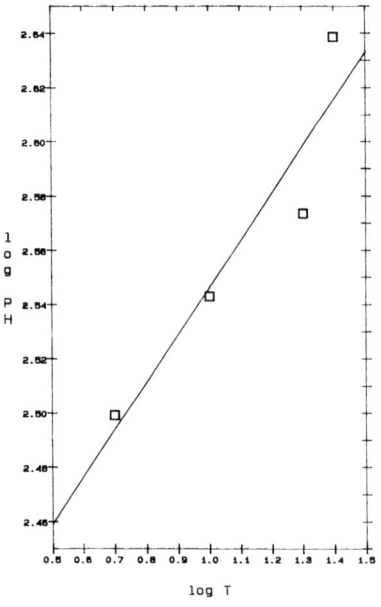

Left:
Figure 3. Relationship of log perceived thickness
(log T) with log peak height (log PH) for xanthan
gum samples.

Below left and right:
Figure 4. Relationship of log perceived thickness
(log T) with pre-maximum gradient (PRG) and post-
maximum gradient (POG) for xanthan gum samples.

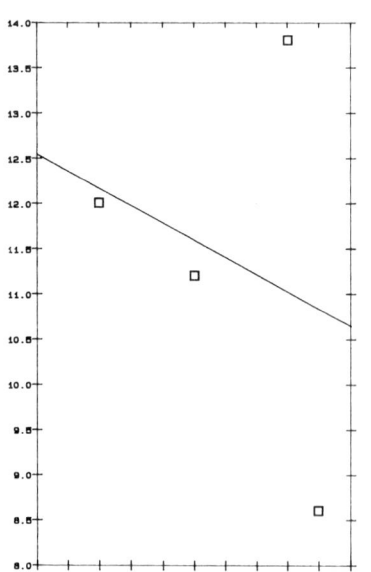

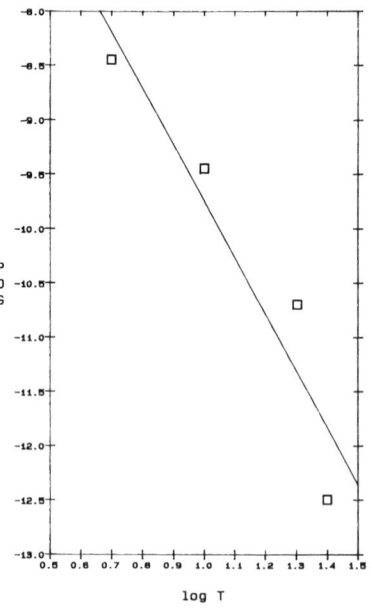

Right:
Figure 5. Relationship of log perceived thickness
(log T) with log peak height (log PH) for Golden
Syrup samples.

Below left and right:
Figure 6. Relationship of log perceived thickness
(log T) with pre-maximum gradient (PRG) and post-
maximum gradient (POG) for Golden Syrup samples.

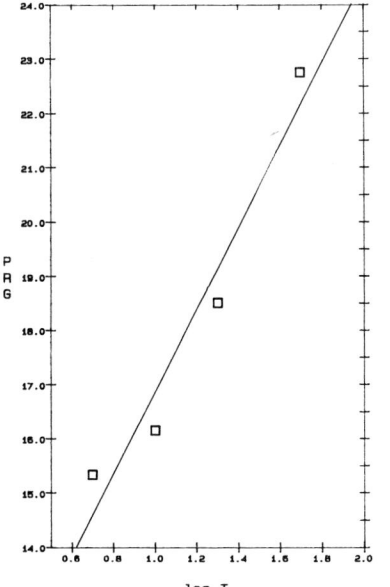

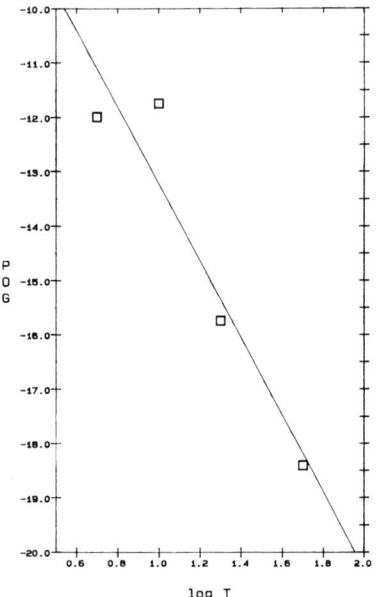

Because of the success of these correlations, the relationship between EMG peak height and dynamic shear viscosity (η^*) was examined. A good correlation was found between log (peak height) and log η^* (Fig. 7). It is therefore clear that muscular activity on eating semi-liquid foods has a power law relationship to the objective measurement of dynamic shear viscosity. In previous studies a power law relationship has been observed between perceived thickness of a range of model systems and objective measurement of viscosity,[7] and it is interesting to note that the gradient of the power law relationship in this study (0.25) is, within experimental error, the same as that of the earlier study (0.22). This is compelling evidence that the EMG measurements truly monitor the subjective response to thickness of semi-liquid model food systems.

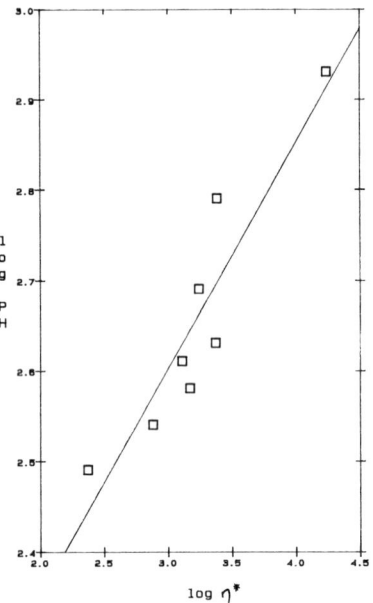

Figure 7. Relationship of log dynamic viscosity
(log η^*) measured at 50 rad/s with log peak height
(log PH) for xanthan gum and Golden Syrup samples.

REFERENCES

[1]   Boyar, M.M. and Kilcast, D. (1986) J. Text. Stud., **17**, 221–252.
[2]   Boyar, M.M. and Kilcast, D. (1986) J. Food Sci., **51** (3), 859–860.
[3]   Morrey, C.W. and Nelson, R.J. (eds) (1970) Dental Science Handbook, DHEW Publication No. 72-336, U.S.A.
[4]   Eves, A., Boyar, M.M. and Kilcast, D. This volume.
[5]   Morris, E.R. (1984) in Gums and Stabilisers for the Food Industry 2: Application of Hydrocolloids (eds Phillips, G.O., Wedlock, D.J. and Williams, P.A.) pp 57–78. Pergamon, Oxford.
[6]   Morris, E.R. (———) in Recent Developments in Industrial Polysaccharides — The Impact of Biotechnology and Advanced Methodologies (eds Crescenzi, V., Dea, I.C.M. and Striala, S.S.). Gordon & Breach Science Publishers, New York (in press).
[7]   Cutler, A.N., Morris, E.R. and Taylor, L.J. (1983) J. Text. Stud., **14**, 377–395.

Part 3

APPLICATIONS

The role of hydrocolloids in stabilising particulate dispersions and emulsions

Eric Dickinson

Procter Department of Food Science, University of Leeds, Leeds LS2 9JT, UK

ABSTRACT

The underlying principles governing the stabilization of particulate dispersions and emulsions by hydrocolloids are reviewed from a predominantly thermodynamic viewpoint. The main topics discussed are the role of hydrocolloid thickeners in inhibiting sedimentation or creaming, the phase behaviour of mixtures of protein + polysaccharide, the influence of polymers on the flocculation of particles and droplets, and the formation of adsorbed layers by polysaccharide hydrocolloids.

INTRODUCTION

Addition of high-molecular-weight water-soluble biopolymers—so-called 'hydrocolloids'—makes an important contribution to the control of texture and stability in many food products. The hydrocolloid stabilizer may be involved in preventing or at least retarding any one or more of a number of physical phenomena: sedimentation of suspended solid particles (chocolate, fruit, etc.); creaming of oil droplets or gas bubbles; crystallization of water or sugar; aggregation (or disaggregation) of dispersed particles; syneresis of gel networks. In technological terms, food hydrocolloids are associated (1) with an extensive range of functional properties, including such varied roles as binding agent (sausages), clarifying agent (beer, wine), clouding agent (fruit juice), emulsifier (salad cream), gelling agent (desserts, puddings), thickening agent (gravies, sauces), and so on. With regard to particulate dispersions and emulsions, what we usually mean by stability is, however, rather more limited: the maintenance of an apparently homogeneous structure and texture throughout the system through the prevention of any discernible sedimentation or creaming, or any separation of free fat or aqueous serum.

Chemically speaking, most food hydrocolloids are complex high-molecular-weight polysaccharides. The conventionally perceived role whereby they stabilize food colloids against sedimentation or creaming is by modifying the rheological properties of the aqueous continuous phase, i.e., by acting as a thickening or

gelling agent. There is, therefore, a strong connection between
the control of stability and the conferring of a particular
structure and texture arising from a delicate interplay of the
large number of weak interactions between macromolecules in the
aqueous medium. What all hydrocolloids have in common, it seems,
despite obvious chemical differences, is an ability to interact
with water, small ions, other polymers, as well as groups
residing at interfaces (protein—water, oil—water or air—water),
with the formation of an aqueous structured material with useful
viscoelastic mechanical properties under conditions of low shear
stress. Understanding the detailed chemistry of these polymer—
polymer interactions, and relating them to the observed rheology,
is a significant aspect of the study of hydrocolloids, and,
indeed, such considerations are the concern of many of the papers
at this meeting. In this article, however, the focus is not on
specific aspects of macromolecular structure or biochemical
function, but rather on the relevant underlying principles of
colloid and polymer science (2-6).

Important though it undoubtedly is, bulk-phase viscosity
modification is by no means the only way in which polysaccharides
can affect the stability of a particulate dispersion or an oil-
in-water emulsion. We also need to consider the extent to which
a hydrocolloid polymer might be adsorbed at solid—liquid or
liquid—liquid interfaces, and the extent to which this polymer
adsorption, or lack of it, has an effect on the degree of
flocculation of the dispersed particles. In the food context, we
must consider not only the surface activity of the hydrocolloid
itself, but also its adsorption and thermodynamic phase behaviour
in the presence of proteinaceous emulsifier. A hydrocolloid
that is thermodynamically incompatible with adsorbed protein will
be distributed differently within the system from a hydrocolloid
that forms a chemical complex with the adsorbed protein. This
thermodynamic aspect is crucial in determining the degree of
particle flocculation and the tendency of the system to phase
separate. An important factor here is the solvent quality of the
aqueous medium for the dissolved macromolecules, since this
influences the partition of hydrocolloid polymer between bulk
phase and surface, or between two bulk phases. The variables
commonly affecting solvent quality in food colloids are pH, ionic
strength, sugar content, alcohol content, and temperature.

THICKENING OF HYDROCOLLOID SOLUTIONS

For a typical hydrocolloid solution, the apparent viscosity η
depends on the shear rate D as illustrated in figure 1a. The
value of D below which a limiting zero-shear viscosity η_0 is
reached depends on the solvent conditions, the molecular weight
of the polymer, and the polymer concentration c. It is found
experimentally (7) that the relationship between the limiting
viscosity η_0 and the concentration c is as shown in figure 1b.
The slope in the plot of $\log \eta_0$ against $\log c$ changes abruptly at
a certain concentration $c = c^*$. The presence of the hydrocolloid
makes little impact on the aqueous viscosity in dilute solution
($c < c^*$), but in the semi-dilute régime ($c \geqslant c^*$) the viscosity
increases strongly with the polymer concentration:

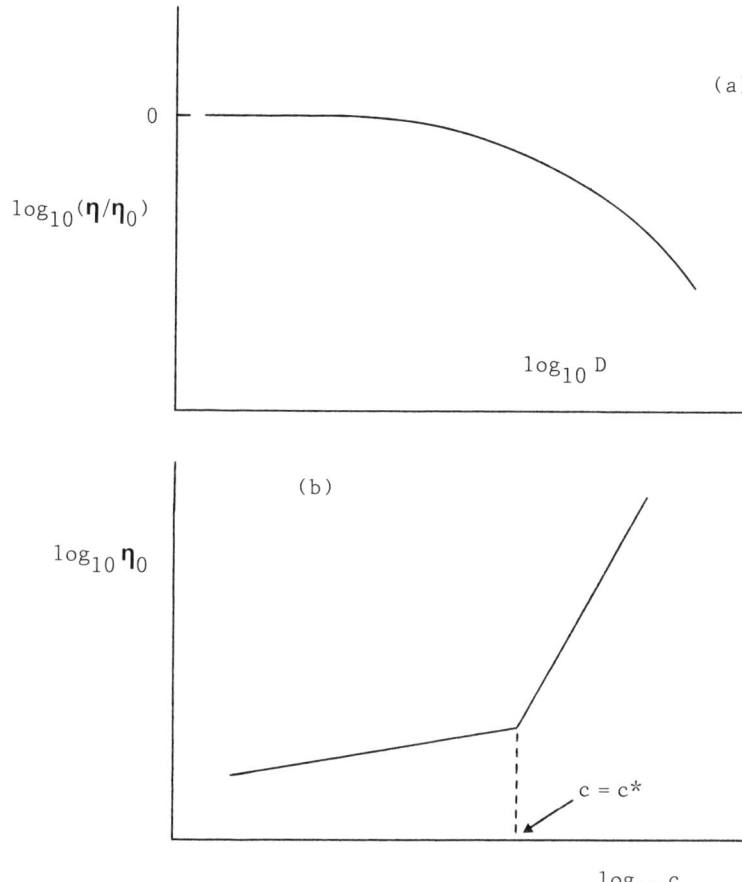

Figure 1. Dependence of apparent viscosity η of a random coil polymer solution on the rate of shear and the polymer concentration. (a) Logarithm of viscosity ratio η/η_0 is plotted against logarithm of shear rate D. (b) Logarithm of limiting zero-shear viscosity is plotted against logarithm of concentration c. The transition from dilute to semi-dilute behaviour occurs at c = c*.

$$\eta_0(c) \sim \left\{ \begin{array}{ll} c^{1.3} & (c \lesssim c^*) \\ c^{3.3} & (c \gtrsim c^*) \end{array} \right. \tag{1}$$

The concentration c* is interpreted (2) as corresponding to the overlap between individual hydrocolloid molecules perceived as

'random coil' polymers. There is an abrupt change in the flow
properties at c = c* because, while in dilute solution the solute
molecules are free to move individually, in semi-dilute solution
they can only move by wriggling past their neighbours in the
entangled network. For a wide range of random coil carbohydrate
polymers (e.g., dextran or alginate, but excluding those with
gel-like specific interactions), Morris and coworkers have found
(8) that the onset of overlap corresponds to

$$\eta_0(c^*) \approx 10 \text{ mPa s} \quad , \qquad c^* \approx 4/[\eta] \quad . \tag{2}$$

The quantity $[\eta]$ in equation (2) is the intrinsic viscosity
defined by

$$[\eta] = \lim_{c \to 0} \left\{ [(\eta/\eta_s) - 1]/c \right\} \quad , \tag{3}$$

where η_s is the viscosity of the aqueous solvent; it is directly
proportional to the actual volume occupied by individual polymer
molecules in dilute solution. The low flexibility of the
carbohydrate polymers as compared with proteins leads to $[\eta]$
values that are relatively large. For the same molecular mass,
$[\eta]$ is much smaller for a compact branched hydrocolloid (gum
arabic or amylopectin) than for a highly disordered hydrocolloid
(alginate or carrageenan). So, a 10 wt% solution of gum arabic
is dilute (c < c*) and of low relative viscosity, but a 1 wt%
solution of alginate is semi-dilute (c > c*) and viscoelastic.

The above discussion refers to semi-dilute solutions whose
rheological behaviour is determined by non-specific chain—chain
entanglements. Slightly different behaviour is observed with
some hydrocolloids (e.g., locust bean gum) which have gel-like
specific interactions between regular sequences on different
chains: η_0 increases with a power of c significantly greater than
3.3, and the onset of the semi-dilute régime occurs at a lower
value of c. There is an advantage in using hydrocolloids with
specific interactions for stabilizing particulate dispersions and
emulsions since, under the low shear-rate conditions involved
(see below), they will give higher viscosities at lower polymer
concentrations than hydrocolloid thickeners relying on the effect
of non-specific chain entanglements.

The shear rate at which the limiting zero-shear viscosity is
reached decreases with the polymer concentration. This is a
reflection of the fact that the time-scale for macromolecular
disentanglement increases with concentration. For a wide range
of hydrocolloids, even including those showing gel-like junction
behaviour, Morris and coworkers (7,8) were able to represent the
shear-thinning behaviour in terms of just two parameters: the
limiting zero-shear viscosity, and the shear rate required to
reduce the apparent viscosity to 10% of η_0.

KINETICS OF SEDIMENTATION OR CREAMING

Sedimentation (creaming) is a potential problem in aqueous
systems containing solid particles (oil droplets) of diameter
greater than approximately 0.5 µm. The settling speed v(d) of an
isolated sphere of diameter d is given by the Stokes equation,

$$v(d) = |\rho - \rho_0| g d^2 / 18\eta \quad , \tag{4}$$

where g is the acceleration due to gravity, and ρ and ρ_0 are the densities of particle and continuous phase, respectively. When the particle-size distribution is not too broad, it is convenient to think in terms of a mean settling speed, which is given by equation (4) with d^2 replaced by

$$\langle d^2 \rangle = \sum_i n_i d_i^5 \Big/ \sum_i n_i d_i^3 \quad , \tag{5}$$

where n_i is the number of particles of diameter d_i. A rough rule of thumb is that a system has good stability with respect to sedimentation (creaming) if the mean settling speed is less than 1 mm per day.

The most direct way by which a hydrocolloid stabilizer may retard sedimentation (creaming) is by increasing the viscosity η in equation (4). At the surface of a particle settling under gravity, the stress is of magnitude

$$S \sim d|\rho - \rho_0|g \quad , \tag{6}$$

which means a value in the range 10^{-3} to 10^{-1} N m^{-2} for particles of size 1 to 10 µm in a food colloid. The local shear rate at the particle surface is of magnitude

$$D \sim v(d)/d \quad , \tag{7}$$

corresponding to a numerical value in the range 10^{-4} to 10^{-1} s^{-1}. The values of S and D are of little consequence when the fluid continuous phase is water, but they become significant when the medium is a non-Newtonian hydrocolloid solution. The appropriate viscosity coefficient is the limiting zero-shear viscosity η_0 if the medium behaves according to figure 1a. The restriction, based on equation (6), that this viscosity should be measured under conditions of very low applied stress means that, if the hydrocolloid forms a weak gel, a yield stress of 10^{-2} Pa may be enough to prevent the particles from moving at all. This type of stabilization by a weak gel network may be reinforced if the hydrocolloid polymer interacts chemically with groups at the particle surface; conversely, it may be less effective if the hydrocolloid polymer is thermodynamically incompatible with the polymer molecules adsorbed directly at the interface.

Hydrocolloids may affect sedimentation (creaming) indirectly through their influence on particle (droplet) flocculation. In disperse systems of low volume fraction, flocculation is usually undesirable as flocs settle more rapidly than individual single particles owing to their larger hydrodynamic size. In colloids of high volume fraction, however, flocculation may be a positive factor, since sedimentation (creaming) may be severely restricted if particles (droplets) are flocculated into an open, but highly connected, semi-continuous gel-like structure permeating the entire system volume. Then, large flocs are geometrically unable to pass by one another without extensive rearrangement, and so settling is effectively halted by the formation of a weak gel network of colloidal particles (cf. stabilization by a weak gel network of polymer molecules). Such flocculated networks are, however, prone to serum separation by a process of gel syneresis. Whether there is serum separation in any particular instance will depend on the water-holding capacity of the hydrocolloid, and its interactions at the surface of the flocculated particles. For

discussion of the relationship between interparticle interactions
and the structure of colloidal aggregates, the reader is referred
elsewhere (9,10).

   The Stokes settling speed is proportional to the difference in
density between particle and medium. When small colloidal
particles are coated with a thick stabilizing layer, the density
of the macromolecular layer makes an important contribution to
the net particle density. With a stabilizing layer composed of
surface-active hydrocolloid material which is denser than the
medium (e.g., gelatin), and a dispersed phase which is less dense
(e.g., vegetable oil), the effect of the adsorbed layer may be to
reduce substantially, or even reverse, the particle motion. In
such circumstances, the role of the hydrocolloid would be to some
extent a 'weighting agent'.

THERMODYNAMIC PHASE BEHAVIOUR

   The thermodynamic properties of hydrocolloid solutions are of
particular relevance to their stabilizing action. Under the
heading of thermodynamic properties, we mean the solubility in
the aqueous dispersion medium, the state of aggregation (gelation)
in the aqueous medium, the tendency to adsorb at interfaces, and
the compatibility with other macromolecular components. What all
these properties have in common is a dependence on the solvent
quality of the medium with respect to dissolved polymer. Good
solvent quality is usually a prerequisite for satisfactory
stabilization.

   In polymer science, a widely used measure of solvent quality
is the Flory—Huggins χ-parameter (11), as defined by

$$\mu_s - \mu_s^\circ = RT \left[\ln(1 - \emptyset) - (1 - m^{-1})\emptyset + \chi\emptyset^2 \right] \ , \qquad (8)$$

where μ_s is the solvent chemical potential in a solution of
polymer volume fraction $\emptyset$, μ_s° is the chemical potential of pure
solvent at the same temperature T, m is the polymer chain length
(m = 1 for the solvent), and R is the gas constant. For the case
of an athermal solution ($\chi = 0$), the only contribution to the free
energy of mixing of polymer + solvent is the entropy of mixing
associated with the configurational chain statistics (11). The
resulting negative deviation from thermodynamic ideality is due
simply to the large difference in molecular size between polymer
and solvent. In real systems, however, χ normally deviates from
zero due to the uneven balance between the various types of
intermolecular interaction: solvent—solvent, solvent—polymer
and polymer—polymer. Polymer molecules in a good solvent ($\chi \approx 0$)
are distributed fairly uniformly throughout the solution, but as
solvent quality gets poorer (χ increasing) the polymer molecules
become, on average, more associated or aggregated. Moreover,
when χ exceeds a certain critical value χ_c, the system separates
spontaneously into two coexisting phases, one rich in solvent and
the other rich in polymer. The value of χ_c depends only on the
polymer chain length:

$$\chi_c = (1 + m^{\frac{1}{2}})^2 / 2m \ . \qquad (9)$$

In the limit of infinite chain length, χ_c is equal to 0.5. The
value of χ for a particular hydrocolloid in water will depend on

the temperature and the aqueous phase composition. Reduction in solvent quality by addition of alcohol, sugar, salt, etc., may lead to precipitation of previously soluble polymer. In a system containing dispersed particles, this may take the form of polymer accumulation at the particle—solvent interface, as the polymer—surface interaction becomes energetically preferred over the solvent—polymer interaction.

Accumulation of thick layers of polymer at a surface in direct contact with a polymer solution is known as <u>prewetting</u>. The phenomenon is maximized in a relatively poor solvent with χ just slightly less than χ_c, or in homogeneous systems with $\chi > \chi_c$ and where the amount of polymer does not exceed the solubility limit. The effect of the interface in prewetting is to perturb the local thermodynamic state of the system into the two-phase region by acting as a template for polymer precipitation when the bulk composition is in the one-phase region but is close to the phase coexistence curve. The polymer concentration in the prewetting layer is generally high; so, if the polymer forms a gel phase at high concentrations, the surface will be covered with a thick gel-like film whose thickness depends on bulk polymer volume fraction (12). Many food hydrocolloids do indeed gel above a certain concentration threshold, and are therefore susceptible to surface gelation under prewetting thermodynamic conditions. We note, however, that the hydrocolloid polymer will not in general be accumulating at a bare particle surface, but at one already covered with more surface-active polymeric material, i.e., the proteinaceous emulsifier. That is, the role of the hydrocolloid in this instance is to form a secondary stabilizing layer around the dispersed particle or droplet.

The complex nature of food colloids means that we are dealing invariably with systems containing more than a single polymer component. In systems containing two species of polymer, we can distinguish three possible situations as shown in figure 2:

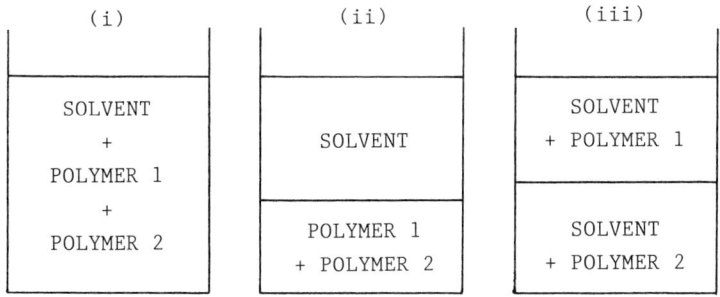

Figure 2. Phase behaviour for mixture of solvent + polymer (1) + polymer (2): (i) one phase [e.g., water + gelatin + casein]; (ii) coacervate [e.g., acid + gelatin + gum arabic]; (iii) incompatibility [e.g., water + gelatin + starch].

(i) <u>complete miscibility</u>—a homogeneous solution is obtained;

(ii) <u>complex coacervation</u>—phase separation occurs, with the two polymer components being mainly together in one phase, and the other phase consisting almost entirely of solvent;

(iii) <u>incompatibility</u>—phase separation occurs, but the two polymer components are mainly in different phases.

The ternary mixture of solvent (s) + polymer (1) + polymer (2) is described by not one but three Flory—Huggins parameters: χ_{s1}, χ_{s2} and χ_{12}. There is complete miscibility when the solvent is of good quality with respect to both polymers ($\chi_{s1} \approx 0$, $\chi_{s2} \approx 0$) and when the polymer—polymer interactions are roughly in balance ($\chi_{12} \approx 0$). When the solvent quality is poor with respect to just one of the polymer components, partial immiscibility occurs as for the binary system. There is complex coacervation when the solvent is poor with respect to both polymers, or when there is a relatively strong intermolecular attractive interaction between the different polymer species ($\chi_{12} < 0$). Incompatibility is the result when the balance of the various interactions is such that there is an effective repulsion between the two polymers ($\chi_{12} > 0$). Coacervation normally implies some sort of specific interaction between two macromolecular components (e.g., electrostatic interaction between positively charged gelatin molecules and negatively charged gum arabic molecules at low pH). On the other hand, incompatibility is a more universal phenomenon, since it is necessary only for the system polymer (1) + polymer (2) to be very slightly endothermic for the effect to occur above a certain polymer concentration. In practice, the liquid mixing of non-electrolytes is inherently endothermic (13), and the same is also true for many electrolytes. So, for a concentrated solution of two high polymers ($m \rightarrow \infty$), where entropy of mixing effects are negligible (11), incompatibility is the rule, and complete miscibility the exception.

The conditions associated with the thermodynamic compatibility of various food macromolecules in aqueous solution have been reported by Tolstoguzov and coworkers (14,15). Phase behaviour in mixtures of solvent + protein + polysaccharide has been studied as a function of pH, ionic strength, polymer concentration and temperature. Figure 3 shows a typical phase diagram, in this case for casein + gum arabic under alkaline conditions. In food colloids containing proteinaceous emulsifier, the most likely consequence of incompatibility is separation into a serum phase containing mainly the carbohydrate polymer solution and a cream layer containing most of the emulsion droplets. In some systems (e.g., 0.15 M NaCl + casein + dextran), the transition from one-phase to two-phase equilibrium is observed (14) near to ambient temperature (20 °C). Such a transition would be very deleterious to the stability of a colloidal system subjected to changes in temperature during long-term storage.

Complex formation between proteins and polysaccharides in biocolloidal systems has been known for many years (16). It is recognized that the formation of a phase-separated protein—polysaccharide complex around the surface of emulsion droplets will generally improve the steric stabilization (see below), and may also be beneficial in controlling the emulsion rheology by flocculating droplets into a gel-like structure which is held

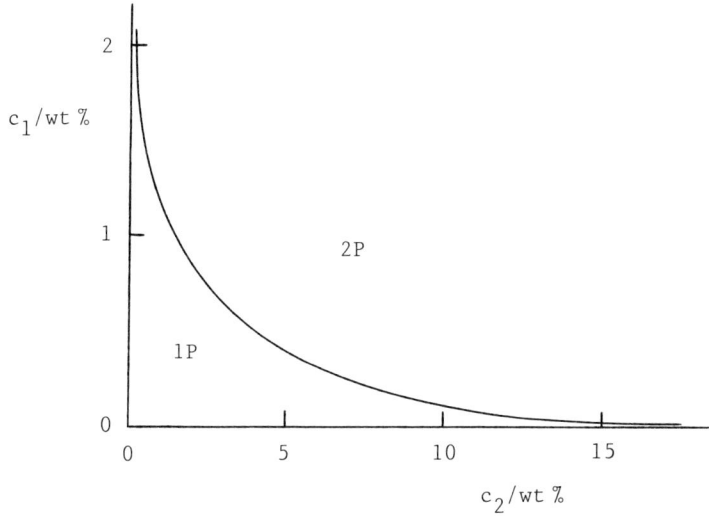

Figure 3. Phase diagram for gum arabic + casein + 0.1 M NaOH at 20 °C showing one-phase (1P) and two-phase (2P) regions. Gum arabic concentration c_1 is plotted against casein concentration c_2. [After ref. 14.]

together, at least in part, by protein—polysaccharide linkages. The use of complexes of proteins and anionic polysaccharides, held together by electrostatic interactions, to stabilize oil-in-water emulsions has been discussed by Tolstoguzov and Braudo (17). The authors describe how the addition of calcium ions promotes the controlled complex coacervation of casein with alginate or pectin at the droplet surface. The good long-term stability of the emulsion is the result of optimizing the combined functional properties of protein and polysaccharide.

STABILIZATION VERSUS FLOCCULATION

In considering stability with respect to flocculation, it is important to distinguish between adsorbing polymers and non-adsorbing polymers.

Covering particle surfaces with a thick protective layer of adsorbing polymer molecules inhibits flocculation. The coated particles strongly repel one another at close range through a combination of polymeric elastic and osmotic effects (3-5), and the colloid is said to be 'sterically stabilized'. This type of stabilization is most effectively achieved with copolymers having a combination of anchoring groups and dangling chains (figure 4a). The anchoring groups should have a strong affinity for the

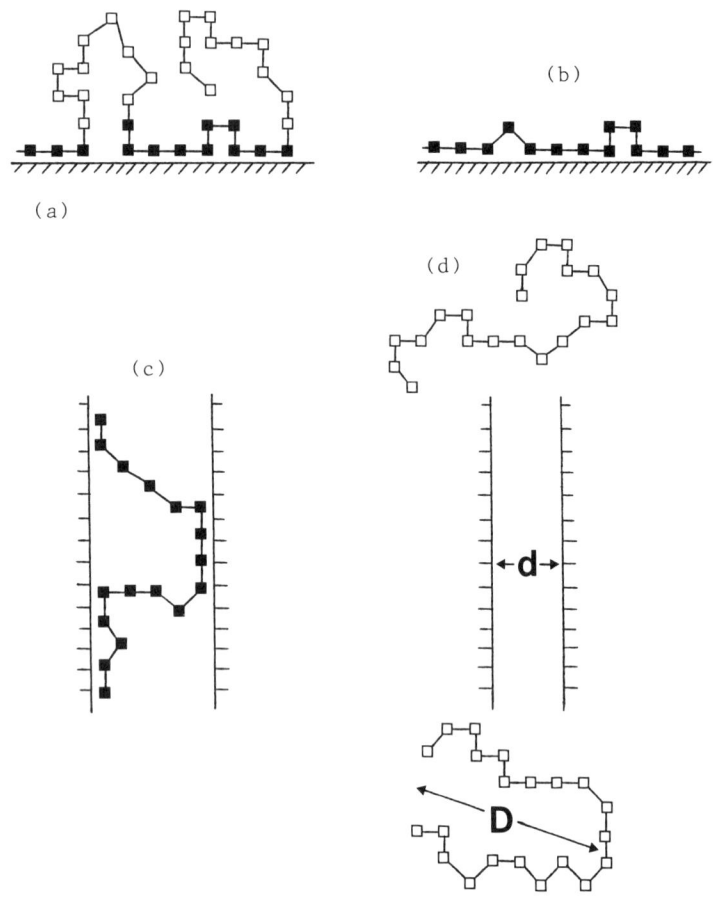

Figure 4.   Sketches of interfacial configurations of chains
with adsorbing (■) and non-adsorbing (□) groups: (a)
copolymer; (b) homopolymer; (c) bridging; (d) depletion.

surface and a low affinity for the continuous phase, and vice
versa for the dangling chains. In addition, there should be
enough adsorbing copolymer to cover the particle surfaces quite
completely, and the layer thickness should be sufficient for two
particles to be prevented from approaching to a pair separation
at which the interparticle van der Waals attraction is of an
appreciable magnitude to the thermal energy kT, as illustrated in

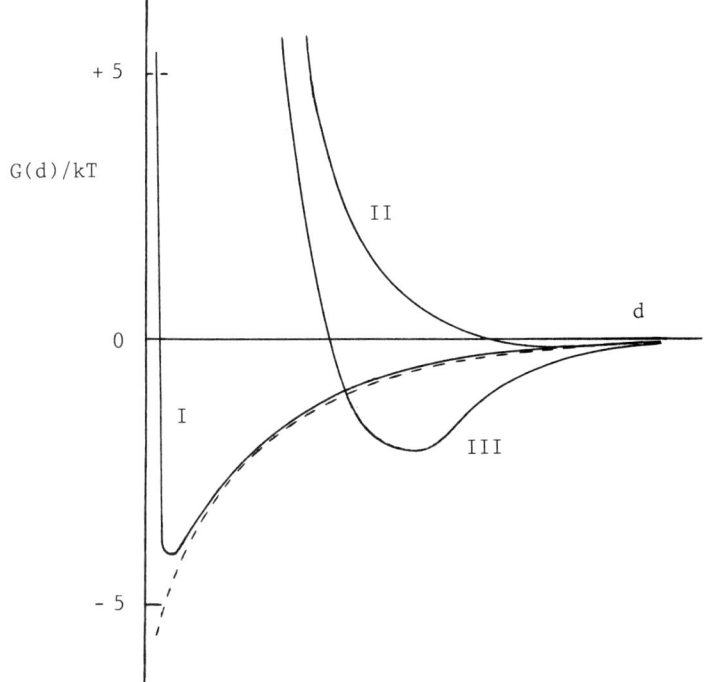

Figure 5. Interaction free energy G(d), in units of kT, as a function of surface-to-surface separation d for pair of polymer-coated particles: (I) thin polymer layer; (II) thick polymer layer (good solvent); (III) thick polymer layer (poor solvent); — — —, van der Waals interaction.

figure 5. A high molecular mass favours thick film formation and thus good steric stabilization. A homopolymer, composed of many identical adsorbing segments (figure 4b), is not a good steric stabilizer since it produces too thin a layer, leading to a free energy of interaction with an attractive minimum whose depth may be several kT (see figure 5).

Where surface coverage is incomplete, a single polymer chain may become attached to more than one particle as illustrated schematically in figure 4c. Such bridging flocculation is most likely when the polymer concentration is approximately half that required for saturation coverage.

Flocculation may occur in a previously sterically stabilized system if the solvent quality of the continuous phase is reduced to such a level that the Flory-Huggins χ-parameter exceeds χ_c for the dangling chains. We call this solvent-induced flocculation

because the solvent conditions required to produce it are similar
to those required for phase separation in the absence of any
particles. Reducing the solvent quality leads to a substantial
minimum in the free energy curve G(d) in figure 5. Such
flocculation can be readily reversed by subsequently improving
the solvent quality.

Non-adsorbing polymers may induce particle flocculation by the
mechanism of depletion. When two particle surfaces approach
within a distance d less than the diameter D of a free polymer
molecule (see figure 4d), the exclusion of polymer from the
intervening gap is associated with an attractive force between
the particles due to the tendency of solvent to flow out from the
gap under the influence of the osmotic pressure gradient (18).
Depletion flocculation occurs at polymer volume fractions $\emptyset > \emptyset^\circ$,
where the value of $\overline{\emptyset^\circ}$ decreases with increasing particle volume
fraction and increasing molecular mass of non-adsorbing polymer.
The free energy of transfer of solvent molecules from the
depletion zone to the bulk solution is (19)

$$\Delta G_d = 2\pi a \delta^2 (\mu_s - \mu_s^\circ)[1 + (2\delta/3a)]/v_s \, , \qquad (10)$$

where a is the particle radius, δ is the thickness of the polymer
depletion layer, and v_s is the solvent molecular volume. The
size of the free energy change depends on δ and $\mu_s - \mu_s^\circ$, the
latter quantity being proportional to the osmotic pressure. In
dilute solution, δ is approximately equal to the radius of
gyration of the polymer molecules, and $\mu_s - \mu_s^\circ$ is proportional to
$\emptyset$; but, in semi-dilute solution, δ is less than the radius of
gyration, and $\mu_s - \mu_s^\circ$ increases more strongly [see equation (8)].
The system separates into particle-rich and solvent-rich phases
at a polymer volume fraction $\emptyset^\circ$ at which the magnitude of the
free energy change associated with equation (10) exceeds that
arising from the entropy of mixing of particles and solvent.

Proteinaceous or protein-containing hydrocolloids (e.g.,
gelatin, various gums) are good steric stabilizers since they
contain a sufficient proportion of flexible hydrophobic groups to
acts as anchoring points and many hydrophilic groups to form the
stabilizing chains. However, most polysaccharide polymers are
sufficiently hydrophilic to be regarded as non-adsorbing (3).
Bridging flocculation by hydrocolloids may be induced by the
precipitation of polymer on or complexation with the surface of
dispersed particles. The flocculation of fat globules in cold
fresh milk is normally attributed to bridging by agglutinin (3),
and bridging flocculation by proteinaceous emulsifier occurs
during homogenization at low protein load. Depletion flocculation
undoubtedly occurs with many hydrocolloid polymers, although its
effect may be masked in many cases by changes in rheological
behaviour. So far, few well-documented studies of colloid phase
separation or depletion flocculation have been reported for
systems containing food hydrocolloids, although Napper reports
(5) that phase separation occurs when a concentrated dispersion
of silica particles is mixed with a solution of Xanthan gum, and
recent observations (20) on casein-stabilized emulsions with
added dextran are consistent with enhanced instability induced by
non-adsorbing polymer. More quantitatively, Robins and coworkers
(21,22) have recently attributed the enhanced creaming kinetics
in hydrocolloid-thickened emulsions to depletion flocculation.

SURFACE ACTIVITY OF HYDROCOLLOIDS

Being predominantly hydrophilic, carbohydrate polymers have a low surface activity at oil—water or air—water interfaces. This means that they are not expected to form primary adsorbed layers in systems which also contain protein or a low-molecular-weight surfactant. This may not be the case, however, if the particular polysaccharide hydrocolloid contains auxiliary hydrophobic groups (methyl, acetyl, etc.), if it is contaminated with protein or polypeptide, or if it is present in overwhelming excess over any other possible surface-active components. Chemically modified derivatives, like highly-substituted methyl cellulose, give a surface activity at the oil—water interface which is similar to that found with the best food emulsifiers (23), and sugar beet acetylated pectin shows good surface activity at the air—water interface (24).

An extracellular polysaccharide having an especially large proportion of hydrophobic side-chains attached to the sugar backbone is produced by the oil degrading bacterium *Acinetobacter calcoaceticus* RAG-1 (25). This anionic biopolymer, called simply 'emulsan' in recognition of its excellent emulsifying properties, has amphiphilic character arising from the presence of fatty acids linked to the amino-sugar backbone. Emulsan is a good steric stabilizer because it combines the hydrophobicity of a protein with the hydrophilicity and high molecular weight of a carbohydrate polymer.

Table I   SURFACE PRESSURE Π AND SURFACE VISCOSITY η_i FOR ADSORBED FILMS OF GUM ARABIC OF VARIABLE NITROGEN CONTENT AFTER 24 HOURS AT THE n-HEXADECANE—WATER INTERFACE (pH 7, 0.005 M, 25 °C, 10^{-3} WT %)

Sample	% N[a]	Π/mN m^{-1}	η_i/mN m^{-1} s
A. eriopoda	5.27	19	220
Commercial gum	0.36	4	10
A. ampliceps	0.10	< 1	< 1

[a]Composition determined by Dr D. M. W. Anderson of the University of Edinburgh

The work of Anderson and colleagues has established (26) that gums such as guar, Xanthan, acacia, etc., contain low levels of nitrogen which may be associated with bound polypeptide or protein. Table I shows a strong correlation between the nitrogen content of samples of gum arabic and the surface properties of films adsorbed at the oil—water interface. The measurements were made in our laboratory (27) using samples kindly supplied by Dr Anderson. Following adsorption for 24 hours at the interface between n-hexadecane and a 10^{-3} wt % aqueous solution, the gum sample with the highest nitrogen content (5.3 %) gives a surface pressure and surface viscosity larger than for gelatin under

similar experimental conditions (28). In marked contrast, the
sample with the lowest nitrogen content (0.1%) has negligible
surface activity. These results would support the view that the
variable surface activity and emulsifying properties of samples
of commercial gum arabic is due to differing amounts of protein
in the samples. It should be noted, however, that there is also
evidence elsewhere (29) to suggest a correlation between the film
properties of gum arabic and its average molecular weight; but
whether nitrogen content and molecular weight are themselves
correlated is not known.

Supporting evidence for a link between protein content and
hydrocolloid surface activity and emulsifying capacity comes from
recent work (30) with the biopolymer emulsan, which is known to
exist as a complex of lipoheteropolysaccharide (apoemulsan) and
protein. Apoemulsan, or emulsan treated with proteolytic enzymes,
has little ability to form emulsions due to its relatively low
surface activity. Optimum emulsification and good surface
activity are found (30) when the emulsan contains ca. 10% protein.

With food colloids, of course, it is not enough just to
consider the surface activities of components in isolation. There
will inevitably be competition between macromolecules and low-
molecular-weight amphiphiles, between polysaccharides and proteins
(glycoproteins), and between hydrocolloid polymers of differing
chemical composition and molecular weight. Other things being
equal, the smaller polymers will diffuse faster from bulk aqueous
solution to the interface, but in time they will be displaced by
the larger polymers, since the latter are preferentially adsorbed
in the thermodynamic limit (31). Solvent conditions that cause a
hydrocolloid polymer to complex with protein or surfactants will
tend to enhance its surface activity and its ability to form
thick stabilizing layers. A full understanding of these relevant
competitive and co-operative effects is required before we shall
properly be able to unravel the various roles of hydrocolloids in
stabilizing these systems.

REFERENCES

1 Glicksman, M. (1969) Gum Technology in the Food Industry,
   Academic Press, New York.
2 de Gennes, P.-G. (1979) Scaling Concepts in Polymer Physics,
   Cornell University Press, Ithaca, NY.
3 Dickinson, E. and Stainsby, G. (1982) Colloids in Food,
   Applied Science, London.
4 Tadros, Th. F. and Vincent, B. (1983) in Encyclopedia of
   Emulsion Technology (ed. Becher, P.), vol. 1, pp. 129-285.
   Marcel Dekker, New York.
5 Napper, D. H. (1983) Polymeric Stabilization of Colloidal
   Dispersions, Academic Press, London.
6 Mitchell, J. R. and Ledward, D. A. (eds) (1986) Functional
   Properties of Food Macromolecules, Elsevier Applied Science,
   London.
7 Morris, E. R. (1984) in Gums and Stabilisers for the Food
   Industry (eds Phillips, G. O., Wedlock, D. J. and Williams,

P. A.), vol. 2, pp. 57-78. Pergamon Press, Oxford.

8 Morris, E. R., Cutler, A. N., Ross-Murphy, S. B., Rees, D. A. and Rice, J. (1981) Carbohydrate Polymers, 1, 5-21.

9 Dickinson, E. (1987) in Food Structure: Its Creation and Evaluation (eds Mitchell, J. R. and Blanshard, J. M. V.), in press. Butterworths, London.

10 Dickinson, E. (1987) J. Colloid Interface Sci., 118, 286-289.

11 Flory, P. J. (1953) Principles of Polymer Chemistry, Cornell University Press, Ithaca, NY.

12 Kim, M. W., Peiffer, D. G. and Pincus, P. (1984) J. Physique Lett. (Paris), 45, L953-L959.

13 Hildebrand, J. H. and Scott, R. L. (1962) Regular Solutions, Prentice-Hall, Englewood Cliffs, NJ.

14 Tolstoguzov, V. B., Grinberg, V. Ya. and Gurov, A. N. (1985) J. Agric. Food Chem., 33, 151-159.

15 Tolstoguzov, V. B. (1986) in Functional Properties of Food Macromolecules (eds Mitchell, J. R. and Ledward, D. A.), pp. 385-415. Elsevier Applied Science, London.

16 Boojy, H. L. and Bungenberg de Jong, H. G. (1956) Biocolloids and their Interactions, Springer-Verlag, Vienna.

17 Tolstoguzov, V. B. and Braudo, E. E. (1985) J. Dispersion Sci., 6, 575-603.

18 Asakura, S. and Oosawa, F. (1954) J. Chem. Phys., 22, 1255-1256.

19 Fleer, G. J., Scheutjens, J. H. M. H. and Vincent, B. (1984) ACS Symp. Ser., 240, 245-263.

20 Bullin, S. R., Dickinson, E., Impey, S. J., Narhan, S. K. and Stainsby, G. (1988) This volume, pp.

21 Gunning, P. A., Hennock, M. S. R., Howe, A. M., Mackie, A. R., Richmond, P. and Robins, M. M. (1986) Colloids Surf., 20, 65-80.

22 Hibberd, D. J., Howe, A. M., Mackie, A. R., Purdy, P. W. and Robins, M. M. (1987) in Food Emulsions and Foams (ed. Dickinson, E.), pp. 219-229. Royal Society of Chemistry, London.

23 Darling, D. F. and Birkett, R. J. (1987) in Food Emulsions and Foams (ed. Dickinson, E.), pp. 1-29. Royal Society of Chemistry, London.

24 Dea, I. C. M. and Madden, J. K. (1986) Food Hydrocolloids, 1, 71-88.

25 Zosim, Z., Gutnick, D. and Rosenberg, E. (1982) Biotechnol. Bioeng., 24, 281-292.

26 Anderson, D. M. W. (1986) in Gums and Stabilisers for the Food Industry (eds Phillips, G. O., Wedlock, D. J. and Williams, P. A.), vol. 3, pp. 79-86. Elsevier Applied Science, London.

27 Busby, R. (1986) M. Sc. Thesis, University of Leeds.

28 Dickinson, E., Murray, A., Murray, B. S. and Stainsby, G. (1987) in Food Emulsions and Foams (ed. Dickinson, E.), pp. 86-99. Royal Society of Chemistry, London.

29 Nakamura, M. (1986) Yukagaku, 35, 554-560.

30 Zosim, Z., Gutnick, D. L. and Rosenberg, E. (1987) Colloid Polym. Sci., 265, 442-447.

31 Dickinson, E. (1986) Food Hydrocolloids, 1, 3-23.

Cellulose ethers—the role of thermal gelation

Alan Henderson

Dow Chemical Europe, CH-8810 Horgen, Switzerland

Introduction

Cellulose ethers are used widely within the Food and other industries. Often, these products are narrowly thought of as "just another thickener" - although they are efficient viscosity modifiers their other physical properties fundamentally influence their use in foodstuffs. A unique property of methylated cellulose derivatives - particularly methylcellulose (MC) and hydroxypropyl methylcellulose (HPMC), is the ability of aqueous solutions containing them to reversibly thermally gel. That is heating a solution of MC or HPMC causes a gel to form - the gel reverts back to being a solution (at the original viscosity) on cooling.

Today I will cover the principles behind this gelation phenomenon and how this effect can be used to advantage by the food industry.

We will cover viscosity control at elevated temperature, oil absorption reduction and some useful effects found during baking.

Thermal Gelation

Heating of the three solutions of MC (tube A) and HPMC (tubes F and K) shown in Figure 1 cause a gel to be formed, the consistency of the gel formed depends upon which type of ether is involved.

Thermal gelation is affected by a number of polymer dependent character-istics. The most important determinant of gel strength is the concentration of methyl groups and the methyl: hydroxypropyl ratio. As the methyl concentra-tion increases, the gel formed on heating becomes firmer. Conversely, as the hydroxypropyl substitution increases, the gel becomes softer. The most probable explanation involves dehydration followed by hydrophobic association of the chains. At temperatures greater than the thermal gel point (TGP), the vibrational and rotational energy of the water molecules increases, exceeding the ability of hydrogen bonding to order the dipolar water molecules around the chain. The energized water molecules then tend to disengage from the fragile envelope of ordered water surrounding the chain. The dewatered polymer chains begin to associate with each other via hydrophobic interac-tions between the methyl groups.

As the temperature or the time at high temperature increases, the methyl: methyl hydrophobic interactions increase in number, producing an ever firmer gel.

Further support of the methyl: methyl interaction theory comes from data on the impact of concentration on gel formation. As the polymer concentration increases, the gel temperature decreases. From the standpoint of equilibrium, this makes sense because as the polymer concentration increases, there are more methyl groups available for potential interactions, which would tend to favor gel formation at lower temperatures.

The addition of hydroxypropyl groups to methyl cellulose always tends to diminish the rigidity of the gel and increase its critical thermal gelation temperature. The probable explanation for this observation fits well with the theory as advanced above. Hydroxypropyl substituents are more hydrophilic than methyl groups, and hence are better able to retain water of hydration when exposed to heat. Because they hold on to their water more tightly the temperature needed to drive them apart is correspondingly greater than with methyl cellulose alone. Also, the equilibrium association between water and hydroxy propyl substituents will tend to cause a more hydrophilic gel than is possible with methyl cellulose.

Thus for example, for the three tubes in Figure 1

A   = methyl cellulose

F and K = hydroxypropyl methylcellulose

A   gives a firm sliceable gel characteristic of MC,

F   a semi-firm gel and

K   with a greater hydroxypropyl to methyl ratio, a soft gel

The gelation temperatures also rise in this order. (Figure 2)

For a given type of methyl cellulose ether the strength of the gel increases with increasing molecular weight until a maximum strength is reached at about Mn 40,000. This corresponds to a 2% solution viscosity of about 400 mPa.s as determined by the ASTM standards. Further increase in molecular weight will not increase the gel strength. Additionally, the molecular weight has no effect on the thermal gel temperature. This suggests that gel formation is essentially dependent on chemical and thermal kinetics and not on inherent viscosity contributions due to molecular weight.

Figure 3 shows graphically the effect of thermal gelation on solution viscosity. Initially, just like the other food grade cellulose ethers, viscosity decreases with temperature. However, as the gel point is reached viscosity increases, through the onset of the gelation process. On cooling, the original solution state and viscosity are regained.

THERMAL GELATION OF MC AND HPMC

FIGURE 1

ETHER TYPE	MeO %	HPO %	GEL TYPE	GELATION TEMP. °C (2% SOLUTION)
A	30	--	FIRM	50 - 55
F	28	5.0	SEMI-FIRM	62 - 68
K	22	8.1	SOFT	70 - 90

THERMAL GELATION DATA FOR MC AND HPMC[+]

+ DATA REFERS TO METHOCEL* A, F AND K CELLULOSE ETHER

FIGURE 2

CHARTING THERMAL GELATION

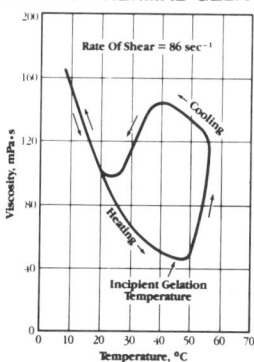

FIGURE 3

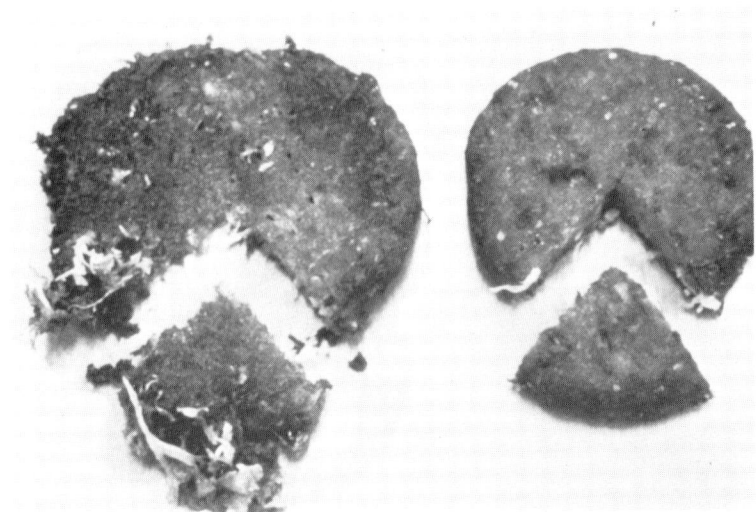

FIGURE 4

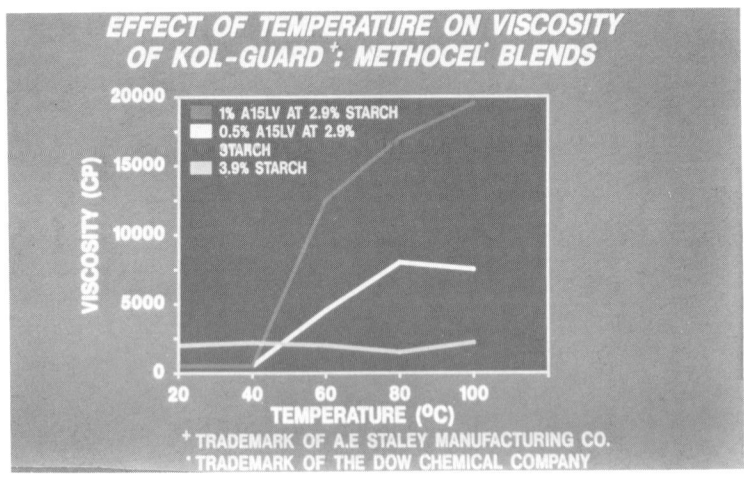

FIGURE 5

The effect of gelation can be dramatic as Figure 4 shows. This is a Reuben Patty - corned beef plus Swiss cheese and sauerkraut after deep-frying. Gelation of the MC (Methocel A4M Premium Gum 0.7%) has protected the products during frying. Without MC the result is disasterous.

Interaction of Methyl Cellulose with Starch

Loss of viscosity of starch solutions usually occurs at elevated temperatures - an effect which can be undesirable in gravies, soups etc. Blends of starch with methyl cellulose can be formulated that boost viscosity with increasing temperature.

Figure 5 shows the behaviour of a low viscosity methyl cellulose (METHOCEL* A15 LV Premium) when mixed with Kol-Guard. The viscosity increases steadily from 40-95°C. Furthermore, the final viscosity or the viscosity at an intermediate temperature can be maintained by varying the concentration of METHOCEL A15LV and the starch.

Using a higher viscosity methylcellulose (METHOCEL A4C Premium) a slight to substantial viscosity increase is seen between 40-60°C, the viscosity then remains essentially stable between 60-95 °C. (Figure 6) The above examples show the interactive effect that thermal gelation can exact on normally heat-thinning starch solutions.

Fat Barrier Properties of HPMC and MC

A gel formed from either MC or HPMC is both water and oil insoluble. As a consequence in fried foods, for example, a coating containing methylcellulose can both retain moisture in the product and reduce the amount of oil penetration. (Figure 7)

We have carried out two studies on this phenomena, the first in a simple way using a model system, the second using a sophisticated measurement technique in a food experiment.

In the model system, a brass tube holds securely a stack of 30 filter papers - of which only the lower one is exposed to the frying oil (Figure 8). By the simple method of measuring the oil penetration, either as a function of the weight of oil through the stack or by counting the number of filter discs that had become translucent, a quick measure of the barrier properties of a coating could be made. The device was jointly developed by Dow Chemical Europe and Leatherhead Food RA. All the data shown in Figure 9 was generated at the FRA. As can be readily seen from the data on this slide MC shows superior properties as a fat barrier to other food gums. An almost 60% reduction versus the blank is obtained with MC.

Model systems are extremely useful for demonstrating a general point. However, we were very interested in understanding how real food behaves when coated with MC containing batters. To this end we made use of the fact that bromine can be made radioactive by neutron bombardment. Chicken pieces (coated and uncoated) were fried in brominated soy oil. Samples of flesh-skin and/or coating were removed, irradiated and the oil uptake measured by gamma counting. Blanks were of course run to measure background bromine levels and the stability of the brominated oil.

A. Henderson

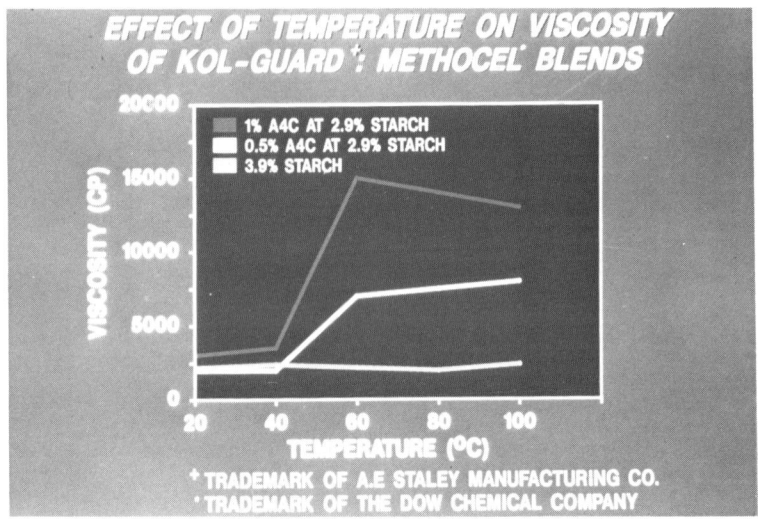

FIGURE 6

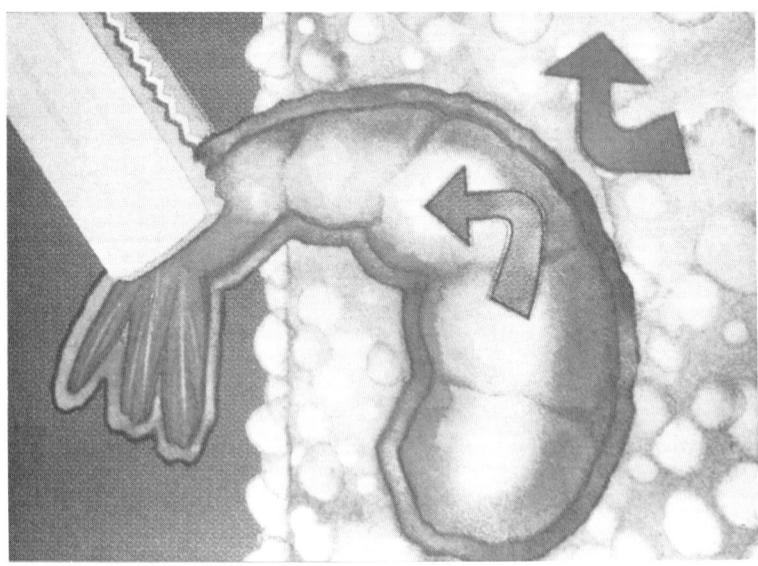

FIGURE 7

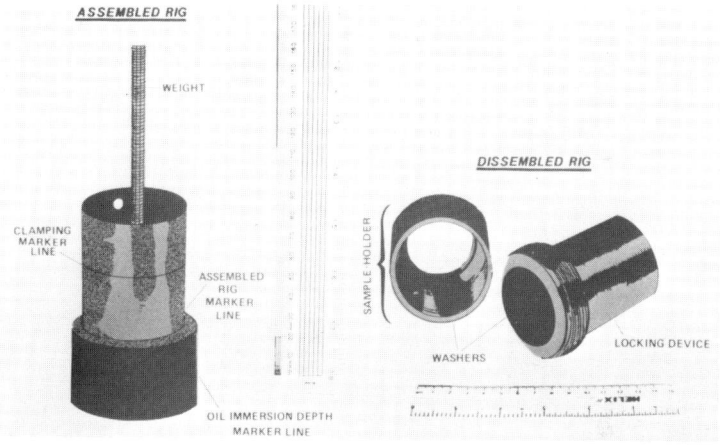

Figure 8

FAT BARRIER PROPERTIES OF METHOCEL* A4M AND K4M PREMIUM CELLULOSE ETHERS AND OTHER FOOD GUMS

TREATMENT SOLUTION (2% AQUEOUS)	OIL %/MM (MEAN)	% REDUCTION IN OIL PENETRATION
BLANK	26.1	--
HYDROXYPROPYLCELLULOSE	22.9	12.3
NA CARBOXYMETHYLCELLULOSE	18.7	28.4
STARCH**	22.0	15.7
METHOCEL* K4M PREMIUM (HPMC)	17.9	31.4
METHOCEL* A4M PREMIUM (MC)	10.5	59.8

** WAXY MAIZE, RECOMMENDED FOR FRIED FOOD COATINGS
* TRADEMARK OF THE DOW CHEMICAL COMPANY

Figure 9

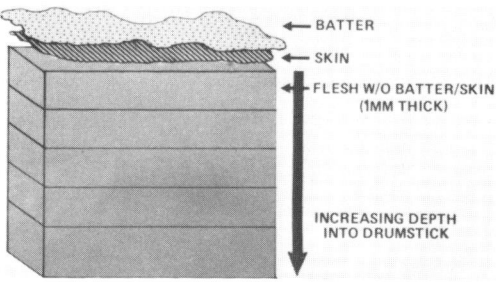

Figure 10

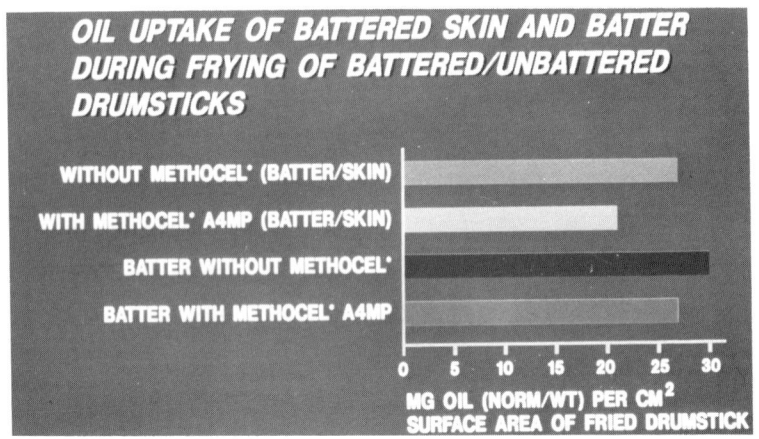

FIGURE 11

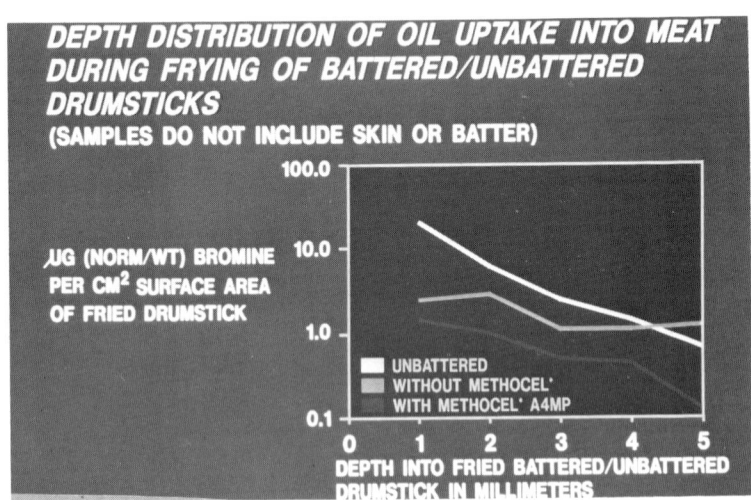

FIGURE 12

To obtain figures for "batter alone" samples, an egg was first blown then dipped in batter and fried. The batter was then easily removed for irradiation and counting.

Batter, skin and a portion of flesh 4 x 2 x 5 cm were removed as shown in Figure 10.

Dealing with the coating first Figure 11 shows the weight of oil per square cm of chicken surface.

The methylcellulose used in all cases was METHOCEL A4M Premium.

Two key points can be made.

1. The methylcellulose reduced oil absorption in the batter/skin system compared to the control by 22%.

2. In "pure" batter systems (i.e. the egg) METHOCEL A4M Premium batter showed a 10% decrease in oil absorbed.

Figure 12 shows in a semi-log plot the depth distribution of oil within the flesh sample. Oil concentration decreases with increasing depth.

Oil concentration of the battered (without MC) section are 65% on average less than those of the unbattered meat.

Oil concentrations of flesh coated with a METHOCEL containing batter are 78% lower on average than that of the control - a considerable improvement.

It must be stressed that this data is preliminary in nature. Further work is needed to fully prove the method, duplicate results and expand and improve the technique.

Baked goods

Both HPMC and MC are mildly surface active, reducing surface tension in dilute solutions to about 50 and 60 dynes/cm respectively. Normally surfactants, as well as most cellulosics, have lower surface tensions at elevated temperatures. - The MC and the HPMC in the table apparently behave in the opposite way. (Figure 13)

Figure 14 shows an explanation of this contradictory behaviour, the surface tension measurements are dynamic measurements. As a surfactant the methylcellulose concentrate at the interface and although surface tension does continue to drop, the occurrence of interfacial gelation (driven by increasing temperature as well as increasing concentration) takes over. Thus it appears that the surface tension rises with temperature in De Nouy measurements.

How can all the above be relevant to baking?

In baked goods, thermal gelation aids in the formation of more uniform gas cells during baking. The result is lighter, more evenly textured products with increased baked volume. The gel also provides added strength to gas cell

SURFACE TENSION OF CELLULOSIC THICKENERS (.05% THICKENER CONCENTRATION AT 10°C AND 40°C)		
THICKENER	10°C (DYNES/CM)	40°C (DYNES/CM)
METHOCEL* A4M PREMIUM	61.0	64.0 (APPARENT)
METHOCEL* E4M PREMIUM	50.1	56.0 (APPARENT)
HPC	53.9	41.1
HEC	66.6	61.3
CMC	72.8	65.8
* TRADEMARK OF THE DOW CHEMICAL COMPANY		

FIGURE 13

SURFACE FORCES OF A DILUTE SOLUTION
OF METHOL CELLULOSE ETHER

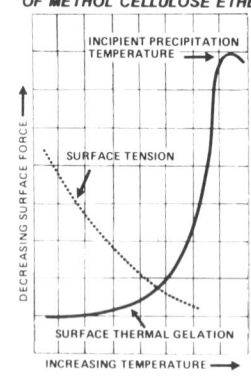

FIGURE 14

Comparing baked volume and texture

Without METHOCEL   *With METHOCEL*

FIGURE 15

walls, so delicate baked goods are less likely to fall apart, even when handled by high speed, automatic equipment. During cooling, the gel reverses and does not interfere with texture or taste. (Figure 15).

Also, due to their ability to increase baked volumes, METHOCEL Premium Food Gums have proved useful in compensating for fluctuations in flour protein levels. In low gluten wheat and rice flour breads, METHOCEL gums can contribute to the body and texture normally associated with full gluten flours.

It was my aim to show that HPMC and MC are "more than thickeners". Hopefully you now agree with me that these molecules are much more interesting and useful than at first meets the eye.

Evaluation of the texture of confectionery gum products using electromyography

Anita Eves, Michael Boyar and David Kilcast

Leatherhead Food R.A., Randalls Road, Leatherhead, Surrey KT22 7RY, UK

ABSTRACT

Various aspects of the electromyography (EMG) technique for assessment of texture have been studied using confectionery gums amongst the samples tested. EMG data were found to be reproducible between occasions but to show greater variability between subjects. EMG was able to differentiate between three types of confectionery gums.

INTRODUCTION

Texture plays a vital role in our enjoyment of food, and gums and stabilisers help to produce some of the wide range of textures found in processed foods. For example, adding gelatin to a syrup and whipping produces a mallow. The degree of aeration of such a product is dependent on the syrup composition and the type and quantity of gelling agent used. Starch-deposited mallows, for instance, are more chewy. Addition of gelling agents to syrups also leads to the production of gelled products — jellies. Such products range from hard gums through to soft jellies. The gelling agent used profoundly affects the texture of the product. Gelatin generally gives tough, chewy jellies, while pectin tends to produce softer, shorter jellies.[1]

For product development purposes there is a need to be able to evaluate the textural characteristics of foods, for instance if a need arises to change the gelling agent in a product. Texture is a subjective response to a tactile stimulus, usually arising in the mouth, although visual clues may also affect the perception of texture.

A number of methods exist for the textural evaluation of food products. The most comprehensive of these is the sensory texture profile. Typically, a panel of assessors develops a series of terms to describe the texture of products similar to those being studied. They are trained in the use of these terms and then score them in relation to the products of interest. Results are often presented in the form of a star diagram. Although this method is the most complete method of texture evaluation, it is a very time-consuming and so a costly procedure. Typically, a Quantitative Descriptive Analysis method requires approximately 20 h of each panellist's time for development of terms and training in their use.[2]

Because of the time and cost constraints of profiling, instrumental tests are often used. Such tests include penetration tests and compression tests using such instruments as the Instron Universal Testing Machine or the Stevens CR Analyser. For example, compression and shear properties of solid chocolate and various properties — melting, adhesion and particle size distribution — of dispersed components of molten chocolate were evaluated.[3] Results were considered to be reproducible and to give an accurate evaluation of the textural properties of the samples investigated; results were not, however, correlated with data from a formal sensory panel.

Other researchers have investigated the forces involved in penetration tests on chewing gum[4] and methods by which to evaluate the eating quality of toffee.[5] Although valuable results may be obtained through such investigations, it is frequently difficult to gain meaningful correlations between instrumental data and subjective data.

A novel method of food texture measurement is being developed at the Leatherhead Food Research Association, known as electromyography (EMG). The technique involves the measurement of electrical potentials in muscles as they relax and contract, in this case the muscles involved in mastication. The present paper considers solid foods (confectionery gums), which are chewed, and consequently examines data gathered from the masseter muscle, which controls the opening and closing of the jaw.[6] Specific aspects considered are as follows.

1. Reproducibility of the technique between occasions with the same subject and between subjects.

2. Textural differences between three confectionery gum samples, as assessed firstly using the Stevens CR Analyser and secondly using the EMG technique.

METHODS

Electromyography

Electromyographic patterns were recorded using a Grass Polygraph (Model 7D). The system has a regulated power supply, consisting of two dc driver amplifiers. One of these is connected to a pre-amplifier and displays the raw signal. The other is connected to the pre-amplifier and an integrator module and displays the integrated data (Fig. 1). Both driver amplifiers include a 50-Hz filter rejecting interference from AC sources. All recordings were made with the amplifiers' frequency bands as wide as possible (10 Hz–40 kHz). This prevents the chart from undulating with the movement of the jaw and ensures a flat frequency response up to 200 Hz, this being the maximum response rate of the pens. The time base was set at 0.2 s at the recommendation of the manufacturer.

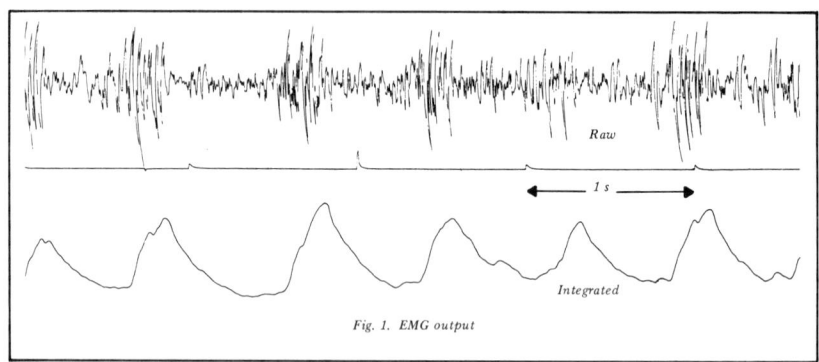

Figure 1. EMG output.

Surface electrodes were used to detect the electrical signals. The skin on the cheek and earlobe were cleansed with 95% (V/V) ethanol to remove traces of dust and perspiration, which may interfere with the signal. A position was located on the cheek at the maximum point of inflection of the masseter muscle and two electrodes were located above and below this point in line with the muscle and approximately 0.5 cm apart. A third electrode was placed on the earlobe, a point of no muscular activity, which acted as an earth. Electrode cream, a conductive paste, was applied to the electrode surface in all cases.

After the terminal ends of the electrodes had been located in the Polygraph, the subject was presented with a sample to chew. All samples were assessed in triplicate and all samples were of the same size and geometry. Data were recorded from the time the sample was put into the mouth to the time of swallowing. The results given below were all measured from the integrated data. These were collected, via an AC/DC convertor, on to a computer disc at a rate of fifty points per second. The stored data were analysed by a computer program calculating various trace parameters — peak height (PH), adjusted height (AH), pre-maximum gradient (PRG), post-maximum gradient (POG), and area under peak (A) (Fig. 2).

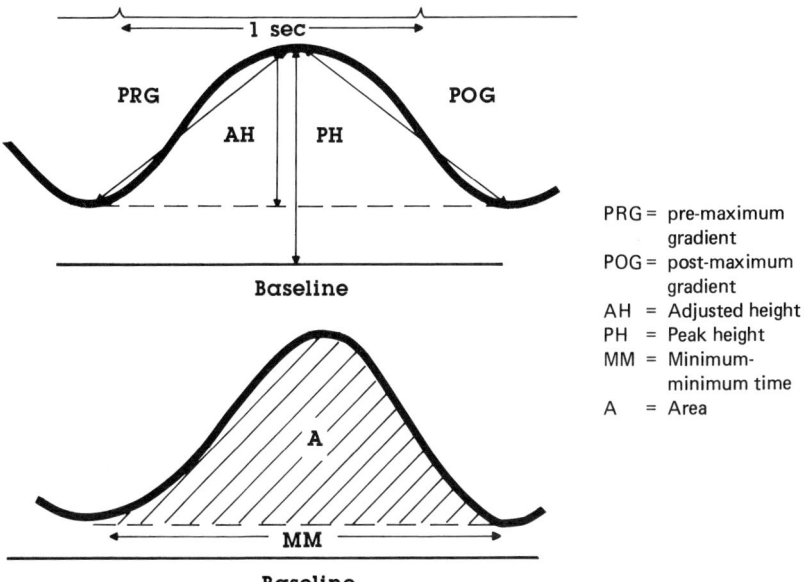

PRG = pre-maximum
gradient
POG = post-maximum
gradient
AH = Adjusted height
PH = Peak height
MM = Minimum-
minimum time
A = Area

Figure 2. Integrated EMG trace parameters.

Gradients and area were calculated from minima rather than baseline, and all parameters were adjusted in relation to a pre-recorded baseline. Small inflections on the integrated trace were eliminated by means of a rejection factor. The rejection factor was based on the ratio of the highest and smallest peaks required; in this case the factor used was 30%, so that any peak less than 30% of the height of the largest peak was rejected. Any of the trace parameters can be plotted against time, a line or curve being fitted to the data using the Genstat statistical package.

Integrated peak height has previously been shown to be very highly correlated with biting forces; hence integrated peak height versus time plots can be considered as true force-time relationships during chewing.

Reproducibility

Between Occasions. A subject linked to the Polygraph was presented with 4-g samples of toffee. The procedure was repeated on a further two occasions and by a further four subjects, also over three days. Integrated peak height was evaluated for each subject on each day, a broken stick model being fitted to the data:

Calcium and water-binding activity during alginate gelation

N.M.Barfod

Department of Basic Application Research, Grindsted Products A/S, 8220 Brabrand, Denmark

ABSTRACT

Reactions of Ca^{2+} with alginates in water solutions have been studied by low resolution pulsed nuclear magnetic resonance spectroscopy. Multiexponential T2 relaxation analyses show that gelation induces an immobilization or hydrogen bonding of free water, whereas the bound fraction of water remains constant. There is no exchange between the free and bound water fractions in alginate gels which may partially explain their good diffusion properties. Experiments using D_2O instead of H_2O indicate that the alginate polymer-Ca^{2+} complex influences water mobility some distance from the original complex thus inducing long-range structuring effects on the water present.

The Ca^{2+}-binding of alginate has been analysed using differential scanning calorimetry by measuring the effect of alginate on the dissolution rate of calcium lactate in water. Different types of alginates with varying mannuronic/guluronic ratios and molecular weights have been studied and compared to gel breaking strength measurements. Ca^{2+}- and water-binding seem to interplay in systems with low gel strength whereas the Ca^{2+}-binding seems to be the most important for systems with high gel strength.

1. INTRODUCTION

The functions of food hydrocolloids are generally explained by the conformational arrangement of the macromolecular chains (thickening effect) and the more specific interchain associations in junction zones (structuring effect). Despite numerous studies conducted of the water-binding properties, the exact nature of the forces holding water in gels is still largely

unknown. It is generally believed that the water in a gel is
very similar to free water and largely unrestricted [1].

However, there are measurable differences between pure water
and that held in gels. One of the characteristics of gel-bound
water is a reduced mobility as measured by nuclear magnetic
resonance (NMR).

The polymer-water interaction during gelation of sodium alginate has been investigated by low resolution pulsed NMR. The
Ca^{2+}-binding of alginate has been analysed by measuring the
dissolution rate of calcium lactate in water in the presence of
alginate using differential scanning calorimetry (DSC). The
contribution of water- and Ca^{2+}-binding to gel breaking
strength has been investigated.

2. MATERIALS AND METHODS

Sodium alginate with a ratio of mannuronic/guluronic residues of 1.7 was extracted from Laminaria digitata at Sobalg,
France.

Calcium alginate gels were prepared 1) by simple blending
of a sodium alginate solution with a $CaCl_2$ solution or 2) by
using an internal setting method where insoluble $CaHPO_4$ is
slowly dissolved due to the spontaneous hydrolysis of gluconodeltalactone in water resulting in a controlled gelation of alginate. The weight proportion in a standard gel was 0.7% sodium
alginate, 0.5% gluconodeltalactone, 0.15% $CaHPO_4 \cdot 2H_2O$ and
98.65% water. The prepared gels were allowed to set in NMR tubes or plastic beakers (for gel strength analysis).

The mobility of water was assessed by T_2 proton relaxation time analysis with the Carr-Purcell- Meiboom-Gill method
using a pulsed NMR spectrometer (model Minispec PC/20B from
Bruker Spectrospin, West Germany). The fractions of free and
bound water in gels were estimated by computerized multiexponential regression analysis on the T_2 relaxation data obtained
using the spin-echo technique [2].

The firmness of the gels was measured with a Voland
Stevens Texture Analyzer using a cylindric probe 12.7 mm in diameter, adjusted to a penetration depth of 10 mm at a speed of
0.5 mm/sec. The gels were allowed to set in 150 ml plastic
beakers with a diameter of 63 mm filled up to 5 mm from the
edge.

The calcium-binding of alginate was evaluated by a
Perkin Elmer DC-4 DSC using 40% easily soluble calcium lactate
encapsulated in hardened soy fat (in the proportion 1:1) dispersed in 10% sodium alginate in water. The mixture was placed in
50 µl aluminium pans and heated from 50 to 80°C at a rate of
2.5°C/minute.

3. RESULTS AND DISCUSSION

Figure 1a shows the T_2 relaxation time in msec. of alginate gels with varying concentrations of Ca^{2+}. Before adding
Ca^{2+} the T_2 time is of the same length as that of free water ($T_2 \sim 2000$ msec.). The gelation is accompanied by a
strong immobilization of free water. Increased water immobilization is also obtained at increased concentration of sodium alginate (fig. 1b).

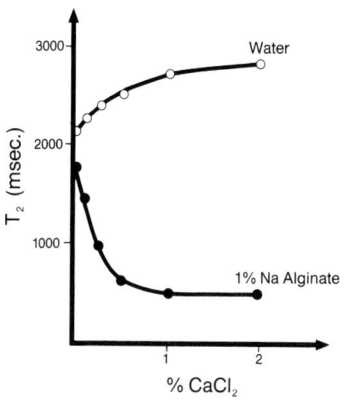

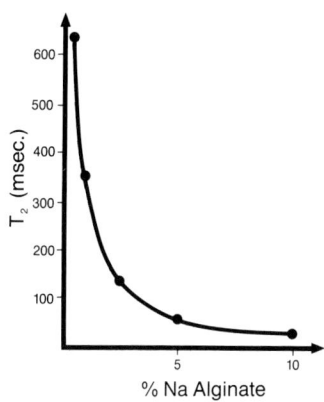

Fig. 1a.: Effect of Ca^{2+} on
T_2 relaxation of water in
alginate gels

Fig. 1b.: Effect of Na alginate
concentration on T_2-time in
alginate gels

Water-binding can be studied in more detail by further calculations of the T_2 relaxation data shown in figure 2. Two types of water represented by two distinct relaxation decay
curves seem to be present in an alginate gel, that is, strongly
bound water with short T_2 value and more loosely bound (free)
water with longer T_2. Ca^{2+} gelation seems to affect only
the free type of water.

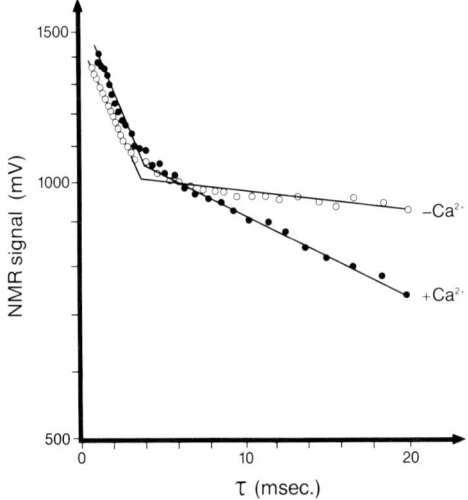

Fig. 2.: Water proton spin-echo relaxation in a 5% alginate gel

    The molecular dynamics of water in the gel can be studied
by analyzing the relaxation times and their relative abundance
at varying ratios of water and polymer [3]. If exchange processes take place between the free and bound water compartment,
the T_2-times and their relative abundance would be expected
to vary with alginate concentrations. It is very clear from figure 3 that the two parameters are remarkably constant over the
concentration interval studied. The only variation actually
observed is that the T_2-time of free water decreases with increasing alginate concentration.

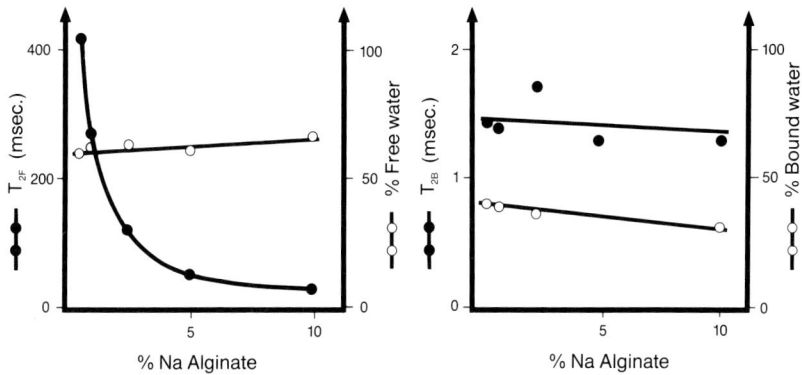

Fig. 3.: Changes in the T_2-time of free and bound water and their relative abundance in gels at varying alginate concentration.

Since the T_2-time of bound water in a non-gelled system (T_2 = 1.40 msec.) is similar to that in a gelled system (T_2 = 1.30 msec.) as seen in table 1, it is assumed that the bound type of water is directly associated with the alginate molecule and independent of gelation. However, the free type of water represents more loosely bound water very sensitive to orientation into a more ordered state induced by the Ca^{2+} alginate complex.

More than 90% of the protons present in an alginate gel are found in water molecules which means that it is largely the behaviour of water which is investigated. By substituting water with heavy water, D_2O, which does not produce a NMR signal, the contribution of dissolved alginate protons to the relaxation of the whole system may be explored. Table 1 shows that the T_2 time of both bound and free types of dissolved alginate protons are remarkably similar in H_2O and D_2O.

TABLE 1: T_2 RELAXATION TIME IN MSEC. OF WATER AND ALGINATE PROTONS IN A 10% GELLED AND NON-GELLED SYSTEM

Solvent	Gelled ($+Ca^{2+}$)		Non-gelled ($-Ca^{2+}$)	
	T_{2bound}	T_{2free}	T_{2bound}	T_{2free}
D_2O	1.43	27.36	2.06	245.66
H_2O	1.30	25.73	1.40	268.56

The results indicate that the protons in the alginate molecule may determine the mobility of water protons associated
with the polymer both in solution and in the gelled state. In
other words, alginate has a long-range structuring effect on
water in the gel. Since there are no indications of exchange
between free and bound types of water in alginate they must
exist as individual compartments in the gels. The behaviour of
water in such gels may explain the excellent results obtained
with alginate-gel-immobilized enzymes and microorganisms. In
these systems the gel behaves as a semipermeable membrane allowing rather unrestricted diffusion of low molecular weight,
water soluble molecules in the "free" water compartment[4]. If
there were a measurable exchange between the free and bound
water compartment in the gel, the water diffusion would be
expected to be more restricted. The interplay between relaxing
water and alginate protons is shown in figure 4.

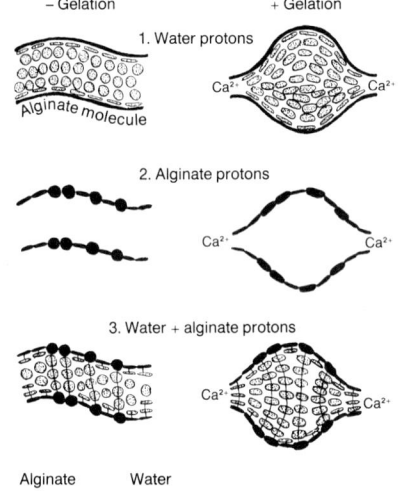

Fig. 4.: Organization of
water in alginate gels.
Upper part: Gelation
affects only the
free type of water.
Middle part: The
relaxation of dissolved
alginate protons is
similar to that of water
both in a gelled and
non-gelled system. Lower
part: The alginate pro-
tons determine the
relaxation behaviour and
thus induce long-range
structuring effects on
the water present

The Ca^{2+}-binding of sodium alginate was evaluated by DSC
using calcium lactate encapsulated in hardened soy fat with a
melting point of about 60°C. Upon heating a sodium alginate solution in the presence of calcium lactate beads the salt will
dissolve after the fat core has melted. The endotherm observed

at about 62°C in figure 5 is due to the melting of hardened soy fat. The second endotherm is supposed to be due to the dissolution of calcium lactate. In the presence of sodium alginate an increased second endotherm is observed probably due to an increased dissolution rate of calcium lactate caused by the binding of Ca^{2+} to alginate.

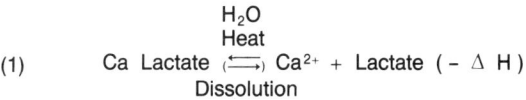

$$\text{(1)} \quad \text{Ca Lactate} \underset{\text{Dissolution}}{\overset{\overset{\textstyle H_2O}{\overset{\textstyle Heat}{}}}{\rightleftharpoons}} Ca^{2+} + \text{Lactate} \ (- \Delta H)$$

$$\text{(2)} \quad Ca^{2+} + \text{Alginate}^- \underset{Ca^{2+} \ \text{binding}}{\rightleftharpoons} \text{Ca Alginate}$$

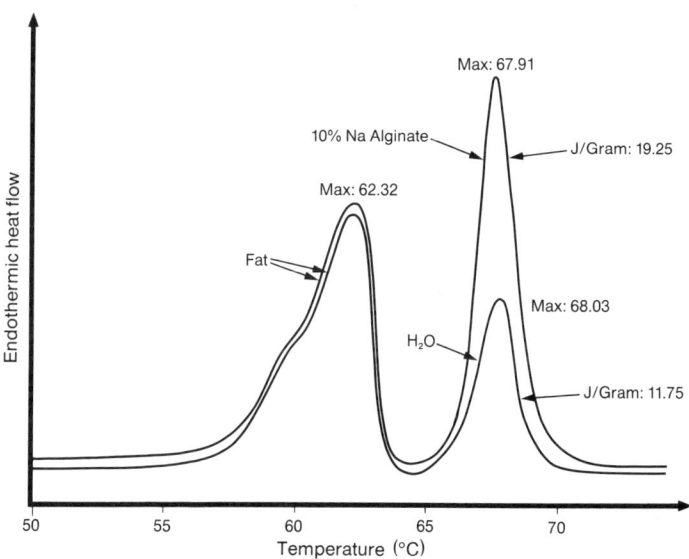

Fig. 5.: Increase of calcium lactate dissolution endotherm in the presence of alginate

The Ca^{2+}-binding is modified using different types of alginates. Alginates with high guluronic/mannuronic ratios bind Ca^{2+} a little stronger than low G/M types like the Sobalg alginates. However, a more decisive factor for Ca^{2+}-binding is the viscosity or molecular weight of a given alginate. Figure 6 shows that the increase in Ca^{2+}-binding is strongly influenced up to a molecular weight about 100.000 D, after which the effect is somewhat less. The effect of molecular weight on water-binding (T_2) is strong up to about 100.000 D after which there is no further contribution to water-binding. The results indicate that Ca^{2+}-and water-binding interplay in systems with low gel strength whereas Ca^{2+} binding is most important for systems with high gel strength.

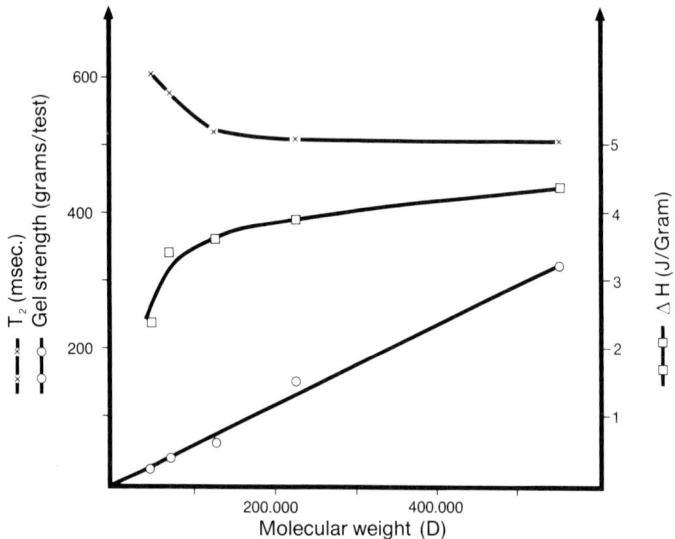

Fig. 6.: Water binding (pNMR), Ca^{2+}-binding (DSC) and gel breaking strength of sodium alginates with varying molecular weights (estimated from viscosity measurements)

4. CONCLUSION

    Investigations as those outlined above may lead to a better understanding of the water- and Ca^{2+}-binding properties of alginates and hopefully, to better product concepts.

REFERENCES

1. Labuza, T. P. & Busk, G.C. J. Food Sci. 44, 1380-85 (1979).
2. Brosio, E., Altobelli, G., Shi Yun Yu & DiNola, A. J. Food Tech. 18, 229-236 (1983).
3. Brosio, E., Altobelli, G. & DiNola, A. J. Food Tech. 19, 103-108 (1984).
4. King, A. H. Food Hydrocolloids 2, 115-188 (1983) CRC Press.

Interactions between carrageenan and polyols and their application

Richard J.Tye

FMC Corporation, Marine Colloids Division, Box 308, Rockland, ME 04841, USA

ABSTRACT

Carrageenans interact strongly with polyols to give systems with unique rheological properties. The rheology resulting from these interactions are examined.

Carrageenan can usefully control the flow and texture properties of any system containing a polyol. It is shown how the thixotropy of carrageenan/polyol systems can be used to improve the stability and appearance of cosmetic and pharmaceutical preparations containing polyol humectants (e.g., hand lotion and dentifrice).

Glycol/water/carrageenan systems exhibit a definite yield point followed by marked shear thinning which can be utilized to give very effective industrial (spray-applied) de-icers ranging from aircraft to car windshields to protecting exposed machinery.

Carrageenan readily entraps a wide variety of oils from heavy hydrocarbons to light and volatile flavor systems. The presence of small amounts of polyol can significantly change the elastic properties of the carrageenan matrix.

INTRODUCTION

Carrageenan, a linear-sulfated polysaccharide consisting of D-galactose and 3,6-anhydro-D-galactose(2) interacts very strongly with the water/polyol systems currently used in foods and industry to give gelled or highly thixotropic systems with well-defined yield points. The rheology of these systems cannot be matched by a water/gum only system or explained in terms of changes to the water activity coefficients. Indeed, iota carrageenan (predominantly Na$^+$/K$^+$) will dissolve in hot, anhydrous glycerine or ethylene glycol to give clear solutions. For example, Na$^+$/K$^+$ iota carrageenan begins to dissolve in anhydrous glycerine around 60°C as shown in the Brabender heating and cooling cycles of Figure 1.

Iota Concentration: 2%

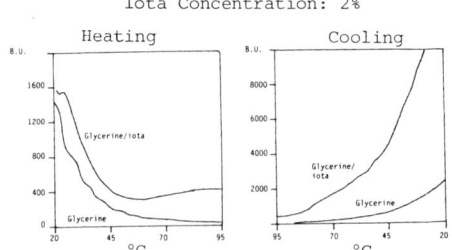

Figure 1. Solvation of an iota carrageenan in glycerine assayed at 99.6%.

Solubility of Carrageenan in Dry Polyols
    Na^+/K^+ iota carrageenan is soluble in hot glycerine or ethylene glycol but not in propylene glycol, 1,3 propandiol or any other polyol with a hydroxyl content (OH /Mol. Wt.) less than around 0.53.

Polyol	(OH ./Mol.Wt.)	Solubility in Anhydrous Polyol
CH_3OH	0.53	Sodium iota solutions will not coagulate
CH_2OHCH_2OH	0.55	Soluble hot
$CH_2OHCHOHCH_2OH$	0.55	Soluble hot
$CH_2OHCH_2CH_2OH$	0.45	Insoluble up to 95°C
$CH_2OHCHOHCH_3$	0.45	Insoluble up to 95°C

    The cooled ethylene glycol or glycerine solutions exhibit very well-defined yield points followed by thixotropy. Figure 2 shows a typical stress/strain profile generated with a rotating cylinder viscometer.

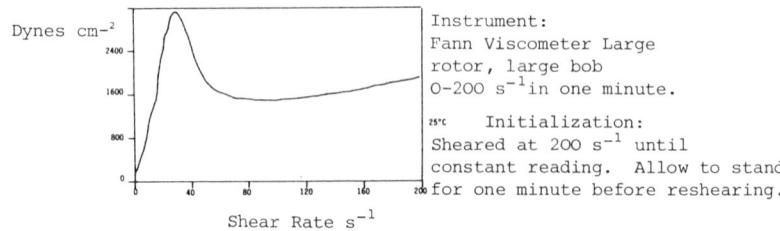

Instrument:
Fann Viscometer Large rotor, large bob
0-200 s^{-1} in one minute.

25°C   Initialization:
Sheared at 200 s^{-1} until constant reading. Allow to stand for one minute before reshearing.

Figure 2. Stress strain profile for a carrageenan solvated in 99.6% glycerine.

    Although both yield point and sheared viscosity increase rapidly as the iota concentration increases, the systems maintain their thixotropy with very marked shear thinning. For example, a solution of 0.2% sodium iota in glycerine will be gelled when quiescent and yet will flow more readily than glycerine alone when subjected to low shear (Figure 3.).

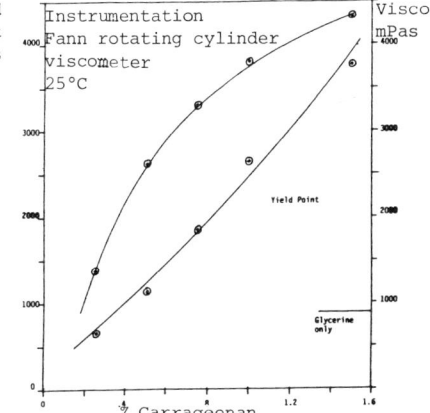

Figure 3.   Yield points and apparent viscosity for Gelcarin SA 911 carrageenan in glycerine, no added water.

Solubility of Carrageenan in Water/Polyol Mixes.

The solvation characteristics of an iota carrageenan are critically dependent on the cation balance.   An iota carrageenan which is predominantly Na^+/Ca^{++} will not dissolve in any dry polyol up to 95°C; kappa carrageenans are also insoluble in dry polyols.

However, all carrageenans can be dissolved in water/polyol mixes.   The water/polyol ratio required for coagulation of the carrageenan depends on the carrageenan, polyol and ionic environment.

Typical Coagulation Profiles (up to 2% carrageenan).

	100/0 Water/Polyol		0/100 Water/Polyol
Kappa type Glyerine	40/60		
Na^+/Ca^{++} Iota Glycerine		30/70	
Na^+/K^+ Iota Glycerine			0/100
Na^+/Ca^{++} Iota PEG-600	40/60		
Na^+/K^+ Iota PEG-600		20/80	
Na^+/Ca^{++} Iota Ethylene glycol		10/90	
Na^+/K^+ Iota Ethylene glycol			0/100

Rheology of Carrageenan/Water/Polyol Systems.
Kappa carrageenans in water/polyol systems form gels which
break with a well-defined fracture. Figure 4 shows a typical
Instron™ gel breakforce profile for water/glycerine systems at
constant gum concentration. The gel breakforce passes through
a maximum which represents the water/polyol ratio at which
coagulation of the carrageenan just begins.

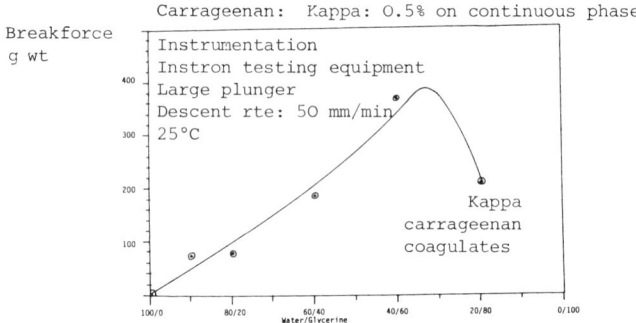

Figure 4. Gel strength of Gelcarin™ GP 812 kappa carrageenan
in water/glycerine.

Iota carrageenans in water/polyol form true thixotropes
with well-defined yield points. Although the appearance of a
yield point after shearing is very rapid, these systems do not
regain maximum gel structure for at least half an hour (Figure
5). Once the yield point has been exceeded, the systems flow
readily with an apparent viscosity which can be below that of
the pure polyol (e.g., Figure 6, water/glycerine systems).

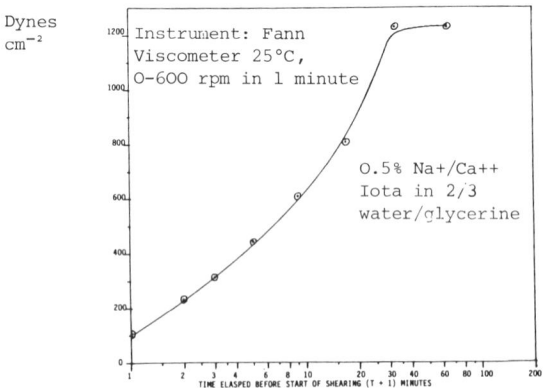

Figure 5. Yield point kinetics for an iota carrageenan in
water/glycerine.

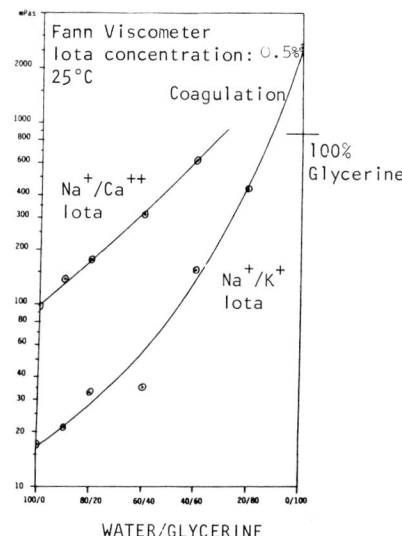

WATER/GLYCERINE

Figure 6. Apparent viscosity of iota carrageenans in water/glycerine.

Texture Profile for Carrageenan in Water Polyols.

Brittle Gels		Flaccid Gels		Viscous Solutions	

Rheological
characteristics:

Well-defined gel breakforce		Well-defined yield points		No observable yield points	

High Break	Low Break	High Yield	Low Yield	High Viscosity	Low Viscosity

_____INCREASING WATER/POLYOL RATIO_____ →

←_____INCREASING CARRAGEENAN CONCENTRATION_____

←_(e.g.,Kappa)__→

    ←_(e.g., Na$^+$/Ca^{++} Iota__→

        ←_(e.g., Na$^+$/K$^+$Iota_→

Typical Applications.

The unique combination of high yield points and extreme shear thinning makes carrageenan particularly useful in controlling the rheology and organoleptic properties of systems formulated with polyols. The range of potential applications is vast, from cosmetics to de-icers. This discussion will concentrate on three very different uses: cosmetics, entrapping technology and de-icers.

Cosmetic Applications.

The most obvious use of carrageenan is to control the texture of a cosmetic preparation, how it flows and its organoleptic properties. A small amount of iota carrageenan can turn a thin and runny product into one with good handfeel and retention properties. Figure 8 illustrates the viscosity of a calamine preparation with and without iota carrageenan.

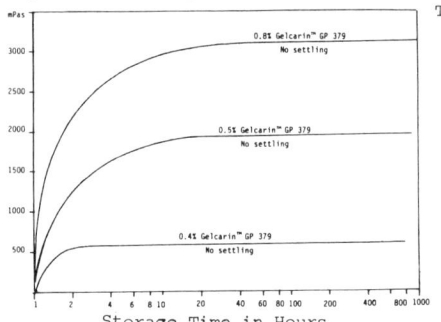

Figure 7. Viscosity of calamine lotion as a function of time.

Entrapping Technology.

Gelcarin™ GP 911 carrageenan is a kappa type, which, in distilled water (no added salts), gels at low temperatures (less than 20°C for a 1% solution). When such a solution is dropped into aqueous potassium chloride, gel beads are formed immediately. The gel phase will trap any dispersed phase, emulsion or even soluble compounds which were present in the carrageenan solution at the time of bead formation. On drying, the carrageenan forms a network which traps the dispersed phase or soluble compounds to form a dry bead with a coating of carrageenan. The bead size can be readily controlled by adjusting the initial drop size.

The rapid transition from solution to gel phase which occurs when a carrageenan solution encounters potassium ions can thus be exploited as an economical method of trapping, stabilizing or even encapsulating expensive flavor oils.

Few methods are available for encapsulating liquids, restricting the choice of wall material. Coating solids with boundary films is a much easier proposition. Carrageenan can be used to immobilize a liquid in a dry bead (with very high liquid/carrageenan ratios) which can then be coated with another polymer.

Gelcarin™ GP 812 kappa-type carrageenan gels at much higher temperatures than Gelcarin™ GP 911. Hot solutions of this carrageenan will also form gel beads instantaneously when dropped into potassium chloride solution. The resultant beads will have different elastic properties from those produced with a carrageenan of lower gelation potential.

The mechanical properties of the bead material can be modified with the addition of polyols to give beads which are more elastic and capable of deformation. The presence of a polyol such as glycerine can also increase the oil-carrying capacity of the bead matrix.

De-Icers.

Heavy Machinery:
The unique gel/yield point and thixotropy of carrageenan-glycol systems is ideally suited to de-icer applications which range from thickening aircraft fluids to those intended for heavy machinery exposed to the most severe conditions as with the North Atlantic shipping routes during winter months.

Heavy machinery requires a thick coat capable of giving long-term protection even during severe precipitation. The gel coat must not clog or interfere with moving parts and should be easy to apply and remove. Kappa carrageenan glycol sols can be sprayed hot and will set to a firm but soft gel (Figure 4) which will remain attached to the substrate during the most arduous conditions. The visco-elasticity of the gels ensures that machinery will not clog; the crushed gels will in fact act as a lubricant.

Protective oils can be emulsified with the carrageenan/glycol anti-freeze to give a corrosion-resistant coating which can be readily removed by washing with hot water.

Aircraft:
There are currently two types of de-icer used by aircraft operators, unthickened type I fluids and the thickened type II systems. Type II (thickened) fluids protect the aircraft during ground operations and are used primarily in Europe. The problem with current fluids is that they do not flow completely from the wings during the run-up to take-off, prior to rotation.

A thickened fluid should ideally have a yield point high enough to prevent viscous flow during ground operations but shear thin to allow the de-icer to flow readily from the wings during the take-off run.

Carrageenan-thickened systems do just that. Their extreme shear thinning not only ensures clean wings but also makes them readily pumpable. They do not permanently lose viscosity during normal pumping. The shear thinning profile is compared to that of a commercially-available thickened de-icer in Figure 8.

Carrageenan-thickened systems present no environmental hazard beyond that of the glycol, and readily wash from the runway.

The non-gelling hydrocolloids or thickeners evaluated were
xanthan gum, guar gum, locust bean gum, carboxymethylcellulose
and tamarind gum, a commonly used thickener in the Far East.
All behaved similarly when combined with low acyl gellan gum
and their influence on the texture of gellan gum is illustrated
in the case of xanthan gum in Figure 1. Hardness and modulus
decrease, elasticity increases slightly and brittleness remains
approximately constant. In other words, these non-gelling hydrocolloids do not alter the brittleness of low acyl gellan gum
gels. They act as diluents and the texture profiles of the
blends are similar to those for low acyl gellan gum alone at
the concentrations in the blends. It is frequently desirable,
however, to use thickeners in combination with gellan gum in
the preparation of food products to satisfy the necessary processing, stability and organoleptic requirements. They can
be used, for example, to improve freeze/thaw stability, reduce
syneresis, prevent undesirable interactions between ingredients,
and modify mouthfeel. In some products, use of a thickener
with gellan gum is essential. Thus, in pet foods consisting
of comminuted meat in a gelled gravy, guar gum is required to
provide 'fill viscosity' and prevent settling out of the meat
prior to formation of the gel network. In low pH dairy products, carboxymethylcellulose, pectin or guar gum are usually
required to control gellan gum/milk protein reactivity (2).

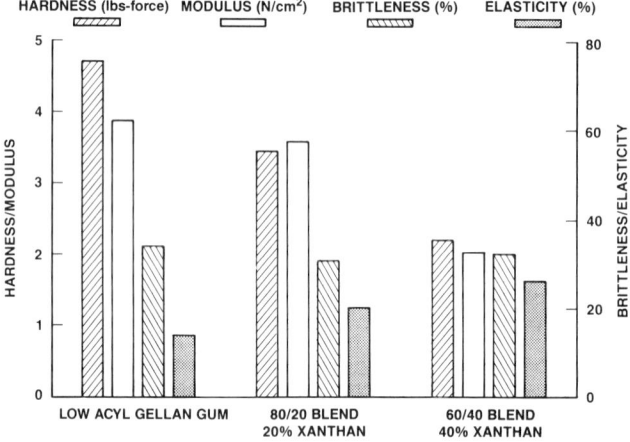

Figure 1.   Textural changes on blending low acyl gellan gum with
xanthan gum.

When low acyl gellan gum is combined with other brittle gelling agents, namely agar and κ-carrageenan, the brittleness of
the low acyl gellan gum gels is not reduced. Although this, at
first sight, may appear to be a statement of the obvious, this
is not the case since, as pointed out by Morris and co-workers
(3), combination of more than one gelling agent can result in
several options for network formation. In the case of agar,

using 4mMCa^{2+} as the gelling ion, a progressive reduction in hardness and modulus was observed as the blends became richer in the agar component. For κ-carrageenan in 0.16MK$^+$, hardness and modulus were found to drop sharply when low levels of carrageenan were substituted for part of the low acyl gellan gum.

As indicated in our previous paper (1), combinations of high acyl or native gellan gum and low acyl gellan gum are capable of providing a range of gel textures. In particular, replacement of low acyl gellan gum with progressively increasing amounts of native gellan gum results in a progressive decrease in brittleness. The xanthan gum/locust bean gum gelling system produces similar textural changes to native gellan gum when combined with the low acyl gum. The gradual reduction in brittleness is shown in Figure 2 while the other textural changes that accompany the change in brittleness are shown in Figure 3. The texture profile of xanthan gum/locust bean gum in Figure 3 is remarkably similar to the texture profile of high acyl gellan gum in Figure 11 of our previous paper (1). The triple combination of low acyl gellan gum, xanthan gum and locust bean gum has been successfully used in a number of products to provide optimal texture and stability. Room deodorant gels and certain varieties of petfoods are specific examples. The stabilizer currently used in these products is usually κ-carrageenan/locust bean gum, the locust bean gum being included to reduce the brittleness of the gel obtained using carrageenan alone. Using the gellan gum blend, the stabilizer concentration can be reduced to between one half and one third the level required with carrageenan/locust bean gum.

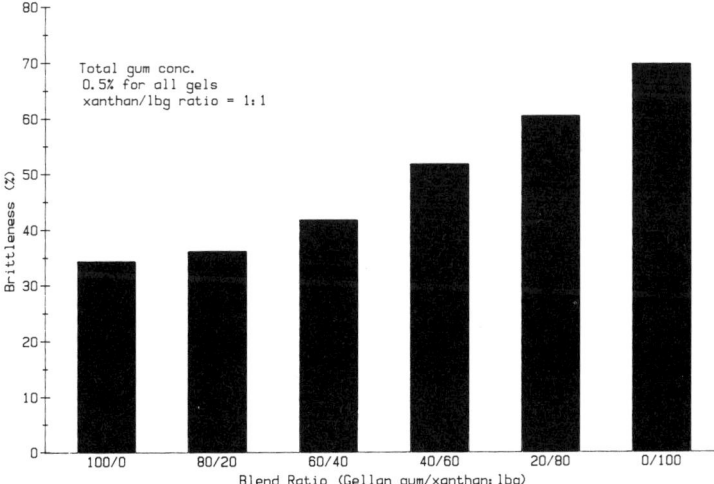

Figure 2. Brittleness values for low acyl gellan gum/xanthan gum/locust bean gum gels in 0.004MCa^{2+}.

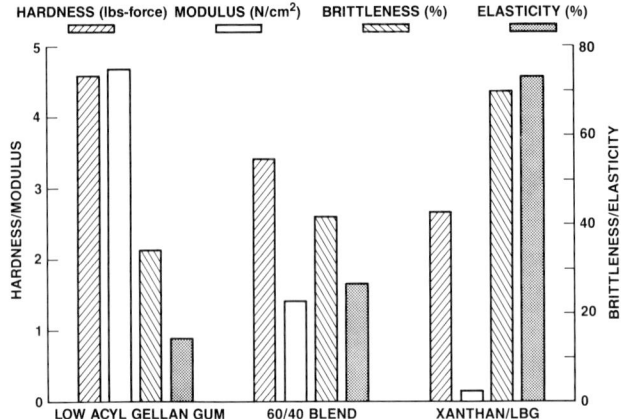

Figure 3. Comparison of the textures of low acyl gellan gum,
xanthan gum/locust bean gum and low acyl gellan gum/xanthan gum/
locust bean gum gels.

Gelatin is the most widely used gelling hydrocolloid. As
discussed (1), although gelatin and low acyl gellan gum have
vastly different textures, novel, appealing variants of food
products currently containing gelatin can be prepared using low
acyl gellan gum instead of the gelatin. Dessert gels are good
examples. The effects of combining low acyl gellan gum and gelatin are, as for all polysaccharide/protein combinations, dependent on a number of factors including pH, temperature, ionic
strength, time, total and relative hydrocolloid concentrations,
and hydrocolloid type. Studies in our laboratory have shown
that manipulation of these factors can result in precipitation
of the two hydrocolloids, or, alternatively, allow low concentrations of low acyl gellan gum to be added to gelatin without
substantial change in the characteristic gelatin texture. Under
conditions where low acyl gellan gum and gelatin precipitate, it
is possible to induce coacervation and the formation of micro-
capsules (4). Independent studies have also demonstrated this
effect (5). Some early work by Kelco showed that low acyl gellan gum and gelatin may have utility in the preparation of soft
gelatin capsules and in photographic film. Further work is required to substantiate these findings.
    Starch is by far the most widely used food hydrocolloid and
is available in a wide variety of forms, both natural and modified. In some cases, the starches are pre-cooked or pregelati-
nized to allow hydration in cold water. In most cases, however,
the starch has to be solubilized by cooking. Although much
attention has been focused on the performance of starches during
the cooking cycle, the final determinant of performance is the
quality of the final paste, notably in terms of viscosity, texture and stability. For many products, raw starches do not provide the desired properties and modified starches are required.

In general terms, these modifications consist of crosslinking, to modify the swelling and stability characteristics of the starch granules, and/or substitution, to reduce retrogradation of the amylose chains and alter the water binding properties of the cooked starch. Attempts have been made to achieve the benefits obtained from modified starches by using raw starches in combination with other hydrocolloids (6). We have recently investigated the effects of inclusion of low levels of low acyl gellan gum on both natural and modified starches. The specific starches tested were native cornstarch (5%), native waxy cornstarch (5%), cross-linked waxy cornstarch (2 types, 4.5 and 7% respectively), acetylated, crosslinked waxy cornstarch (5%) and hydroxypropylated, cross-linked waxy cornstarch (6%). Evaluations, made in both water and water containing 2% 100 grain vinegar, consisted of determination of standard Amylograph curves, evaluation of viscosities (after cooking and cooling overnight, immediately after subsequent shearing, and twenty four hours after shearing), and estimation of syneresis from the cooked starch pastes over a ten day storage period. GELRITE* low acyl gellan gum was used at 0.1%, except in some of the Amylograph studies when, as indicated below, 0.5% was used.

Several major effects were observed. The inclusion of gellan gum did not markedly alter the initial peak viscosities obtained with all of the starches on cooking. This is in contrast to the behavior of other hydrocolloids, which cause a significant increase in peak viscosities. The effects of adding GELRITE, xanthan gum and carboxymethylcellulose to cross-linked, waxy cornstarch are shown in Figure 4. As expected, in view of the sensitivity of starch swelling and gellan gum to the presence of ions, changes in ionic concentration brought about some minor changes in peak viscosities. After cooking and cooling, the gellan gum imparted a short, gel-like texture to all of the starch pastes. A significant viscosity loss was observed immediately after shearing the cooked and cooled pastes containing gellan gum. However, these viscosities after shearing were always higher than the corresponding viscosities of the pastes with starch alone. The viscosities of the gellan gum containing starch samples after shearing and recovery were higher than the viscosities of the pastes with starch alone both before shearing and after recovery after shearing. These effects are shown for acetylated, cross-linked waxy cornstarch in Figure 5. The high viscosities of the starch/gellan pastes are undoubtedly due to the gel-like structure imparted by the gellan gum. The reset properties of gellan gum gels after shearing and time to recover are shown in Figure 6. The above effects, taken together, suggest that addition of low acyl gellan gum to starch could be useful in the preparation of a variety of starch based fillings, which are generally prepared by cooking, stored in containers, sheared upon depositing and allowed to recover viscosity on standing.

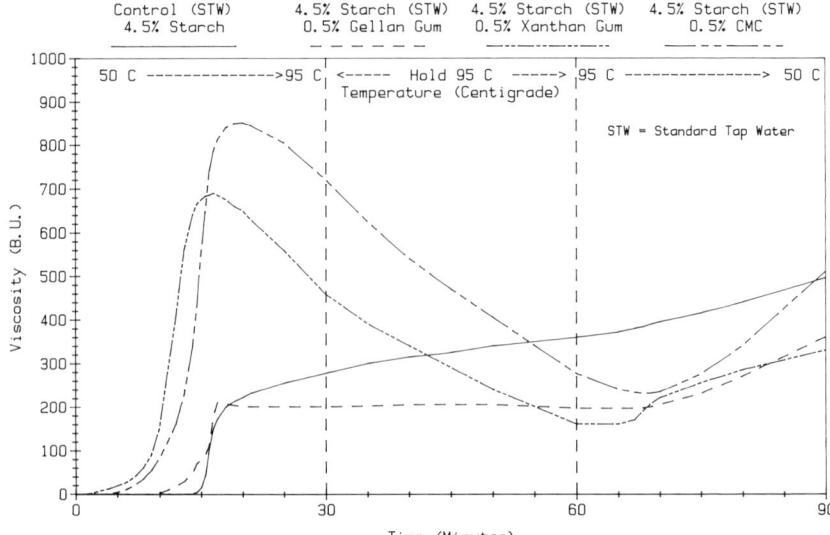

Figure 4. BRABENDER Amylographs of 4.5% cross-linked waxy corn-starch with 0.5% added gum.

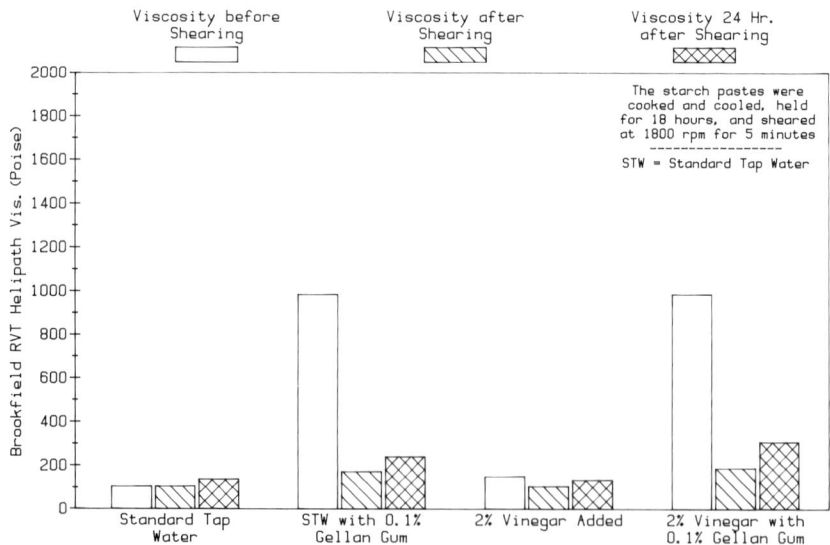

Figure 5. Shear stability of 5% acetylated, cross-linked waxy cornstarch with and without GELRITE gellan gum.

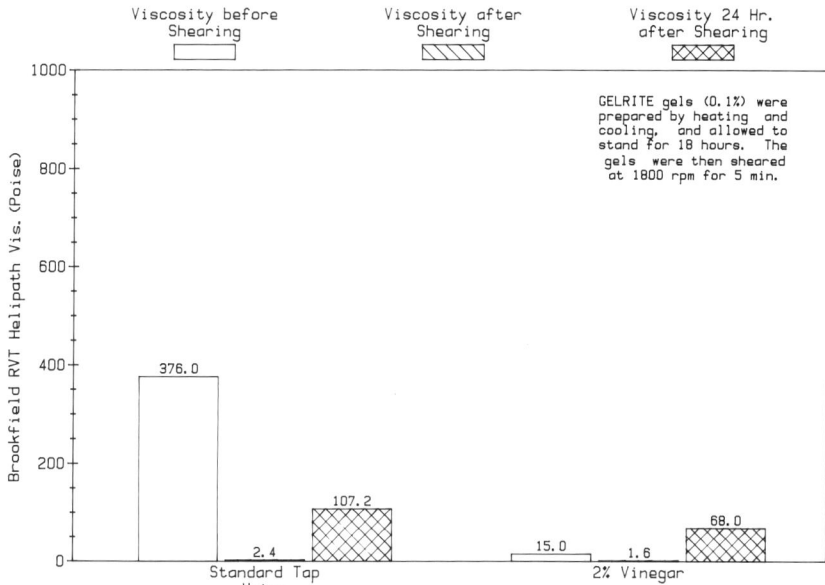

Figure 6.   Shear stability of GELRITE gellan gum.

Results of evaluation of the stability of the starch pastes
with and without gellan gum in terms of syneresis upon storage
indicated that gellan gum does reduce but does not totally eliminate syneresis. Reduced syneresis or improved water holding
capacity of the pastes was more noticeable when the gellan gum
was added to the unmodified or slightly modified starches than
with the highly modified starches. Results for waxy cornstarch
are shown in Figure 7. "Percent area covered on filter paper"
refers to the area of a filter paper wetted by water released
from the starch paste, which was used as a measure of the water
binding capacity of the paste. Work of several years ago indicates that use of low levels of gellan gum can permit formulation
of a number of food products with a reduced level of starch (7).
Pastiness is reduced and flavor release improved relative to the
products made with starch alone.

Algin, another important gelling agent has, like low acyl
gellan gum, a strong affinity for calcium ions. However, algin
gels are normally prepared by reaction with calcium ions in the
cold while those from low acyl gellan gum are usually formed
by heating and cooling in the presence of calcium ions. These
differences in functionality can lead to problems when preparing
mixed gels from gellan gum and algin and careful control of the
reaction conditions is required. In contrast to calcium ions,
magnesium ions do not form gels with algin. They do, however,
react similarly to calcium ions with gellan gum (2, 7). Use of
this ion should therefore increase the number of options available in formulating mixed gels from algin and low acyl gellan
gum.

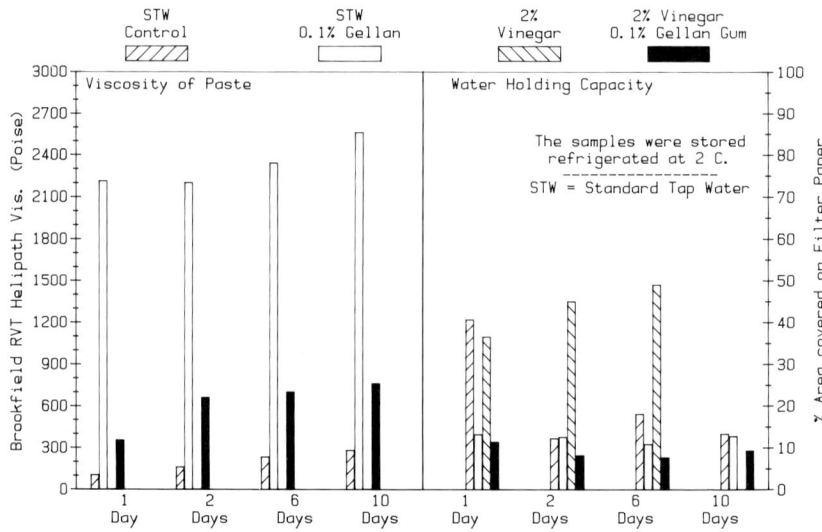

Figure 7.   Storage stability of 5% waxy cornstarch pastes with and without 0.1% GELRITE gellan gum.

ACKNOWLEDGEMENTS

    The authors wish to thank American Maize Products Co., Chicago, for provision of samples of native and modified starches.

REFERENCES

1. Sanderson, G.R., Bell, V.L., Clark, R.C. and Ortega, D. (1988) ibid.
2. Sanderson, G.R. and Clark, R.C. (1984) in Gums and Stabilizers for the Food Industry 2, (eds. Phillips, G.O. et al.) pp 201-210. Pergamon Press, Oxford and New York.
3. Brownsey, G.J. and Morris, V.J. (1986) presented at Food Structure - Its Creation and Evaluation, University of Nottingham, Sutton Bonington, U.K.
4. Lindroth, T.A. (1983) Commercial Development Bulletin CD-31, Kelco Division of Merck and Co., Inc., San Diego, California.
5. Chilvers, G.R. and Morris, V.J. (in press) Carbohydrate Polymers.
6. Christianson, D.D. (1981) presented at Institute of Food Technologists Annual Meeting, Atlanta, Georgia.
7. Sanderson, G.R. and Clark, R.C. (1983) Food Technology, 37, 63-70.

GELRITE*, Trademark of Merck & Co., Inc., Kelco Div., U.S.A.
BRABENDER is a trademark of C.W. Brabender, Inc., South Hackensack, NJ

The role of hydrocolloids in food preparations based on 'Tonyu' (soya milk) and/or 'Tofu' (soyabean curd)

B.Villaudy, G.Tilly and R.Rizzotti

Mero-Rousselot-Satia, D.A.M., Food Application, Baupte, 50500 Carentan, France

ABSTRACT

Amongst the traditional foodstuffs based on soya, "Tonyu" (soya milk) and "Tofu" (soyabean curd) can be used as basic ingredients for many food preparations. The soya nutritional properties are of great interest in dietary foods. Free of lactose and cholesterol, soya is rich in polyunsaturated fatty acids as well as in proteins containing the essential amino-acids. Its vegetable origin is also of significance in vegetarian or kosher diets.

Hydrocolloids : originating from protein macromolecules (gelatine) or polysaccharide molecules (carrageenans, alginates, xanthan gum, galactomannans) hydrocolloids are effectively used as textural agents in food applications and are of great advantage in the foodstuffs based on soya.

The rheological behaviours of these hydrocolloids in "Tonyu" either singly or in combination with "Tofu" have been tested in various food preparations : beverages, desserts, ice-creams and related products, fermented products, sauces, dietary meat products...

KEYWORDS

Soya, Soya milk (Tonyu)
Polysaccharides, Carrageenan, Guar gum, Locust bean gum, Xanthan gum, Alginates, Agar-agar, Starch, CMC
Protein, Gelatine.

INTRODUCTION

Whereas the use of soya *(Glycine max)* is traditional in the far-east, it represents a new development in the west where it is of great interest for industrial food processing.

Its dietary benefit in combination with its vegetable origin make soya of particular value for processing new trend foodstuffs.

In Europe, food products based on soya is now an expanding market and provide new opportunities for the production of new food developments. In conjunction with the lack of flavour, Tonyu and Tofu are very similar in appearance to milk and fresh curd respectively. These factors help promote new food products by partial or complete substitution of cow's milk by vegetable milk.

Preparations such as beverages, fresh desserts, frozen desserts, comminuted products require the present day technology in combination with optimizing agents. Polysaccharides (carrageenans, alginates, xanthan gum, galactomannans...) and proteins (gelatine...) are particularly suitable in modern food manufacture.
Hydrocolloids, as water binding agents, with their functional properties, i.e. the thickening and gelling abilities, modifiy the rheological behaviour of the systems in which they are used.

The rheological behaviours of carrageenans, guar gum, locust bean gum have been compared in water, in milk and in Tonyu. These behaviours have been analyzed on a Brabender viscograph and by measuring the gel strength (penetration test method) or determining viscosity.

HYDROCOLLOID MAIN FEATURES

Among the ingredients which are necessary for food preparations, the textural agents (thickeners, gelling agents, stabilizers) are water-binding macromolecules, of the utmost importance in determining the final physical state :

■ **They thicken :** *Increased viscosity as a result of reduced mobility within the milk or water system.*
The presence of macromolecules causes a decrease in the mobility within the system, a phenomenon which is generally enhanced by the various interactions that may occur.

■ **They gel :** *Gel formation and final physical form.*
The interaction between the macromolecules results in the formation of a three-dimensional network and a gel structure. This structure can be formed with macromolecules of similar or of a different nature (synergy).

■ **They stabilize :** *Maintaining suspension equilibrium of the components which have a natural tendency to separate.*
They maintain in suspension the components by either increasing the viscosity whenever the stabilizer properties are compatible with the profile of the final product, or by entrapping particles within a three-dimensional gelled network.

These textural agents are hydrocolloids and are generally classified as polysaccharides such as carrageenans, alginates, xanthan gum, pectins, galactomannans,... or as protein such as gelatine.

CARRAGEENAN

Carrageenans are cellular components of some seaweeds of the Rhodophyceae class.
The denomination "carrageenan" refers in fact to a product with a high chemical heterogeneity. The macromolecule is made up of a chain of more or less sulphated galactose units alternately linked in 1,3 and 1,4 glycosidic linkages. The classification of carrageenans (E 407) into "kappa", "iota", "lambda", "mu" and "nu" originates from these differences.
The three main types of carrageenans : kappa, iota and lambda (mu and nu carrageenans being the chemical precursors of kappa and iota respectively) are not found in isolation. Kappa carrageenan is soluble after heating and gives a rigid and brittle gel on cooling. Iota carrageenan is entirely soluble after heating, and on cooling produces an elastic and cohesive syneresis-free gel. Lambda carrageenan is cold soluble.

ALGINATE

Alginates are extracted from brown seaweeds of the Phaeophyceae class.
The sodium alginate (E 401) is the salt of the alginic acid. It is a linear polymer of β - $(1 \rightarrow 4)$ - D - mannuronic acid residues and α - $(1 \rightarrow 4)$ - L - guluronic acid units. The relative proportion of the ratio mannuronic acid/guluronic acid units varies with biological source, season and harvesting area.
The carboxyl group prevents the chains from associating by electrostatic repulsion. Sodium alginate is a cold water soluble thickener.
A decrease in ionic forces is required to form a network. It is effective by the acidification of the medium or by the addition of a divalent or a trivalent cation, i.e. the cation of the calcium. Gelation of sodium alginate solutions is also possible in milk.

AGAR-AGAR

Agar-Agar (E 406) is extracted from red seaweeds. The main chain consists of 3,6 anhydro-galactopyranose residues and D - galactopyranose units. This little sulphated polysaccharide is soluble in hot systems and sets on cooling to form a firm and brittle gel. It displays neither thickening capacity nor reactivity with milk proteins.

GELATINE

Gelatine is from animal origin. It is obtained from collagen contained in hides from tanneries or meat packing plants, or bones from slaughter houses and butcher-shops.
As with other proteins, gelatine exhibits an amphoteric character. The pI is dependent on the treatment used
■ *Acid process :*   Pig skin and ossein   pHi between 6.5 and 9.5
■ *Alkaline process :* Ovine skin and ossein   pHi between 4.7 and 5.0
Gelatine dissolves in hot system and sets after cooling to give a resilient and thermoreversible gel.

GALACTOMANNANS

Guar gum

Guar is obtained from the guar plant (*Cyamopsis tetragonolobus*) which belongs to the family of Leguminosae. Guar gum, as mentioned in the EEC list under the number (E 412) is a galactomannan which consists of a linear chain of (1 → 4) - D - mannose residues with D - galactose units attached by (1 → 6) linkages. The ratio of D - galactose to D - mannose in guar gum is 1 : 2. As guar gum is highly substituted, it is a thickener which dissolves easily in cold water solutions.

Locust bean gum (carob)

Carob is the fruit of the carob tree (*Ceratonia siliqua*). Locust bean gum (E 410) is a galactomannan which consists of a linear chain of (1 → 4) D - mannose units with D - galactose residues attached by (1 → 6) linkages. The ratio of D - galactose to D - mannose in locust bean gum is 1 : 4 . The locust bean gum chain is thus very often interspersed by longer regions of essentially unsubstituted mannan backbone. This is the reason why it is only soluble in hot water. When used alone it is a thickener but by the addition of kappa carrageenan or xanthan gum it will form gels.

XANTHAN GUM

Xanthan gum is an heteropolysaccharide produced by a culture fermentation of a carbohydrate and growth factors with *Xanthomonas campestris*.
The main chain of the xanthan gum (E 415) contains β - D - glucose units linked through the 1 - and 4 - positions. A trisaccharide side chain is linked to the 3 - position of every alternating glucose unit in the main chain. It contains two mannose units and a glucuronic acid unit. The D - mannose unit adjacent to the main chain contains an acetyl group and the terminal D - mannose unit is linked to a pyruvate group. The presence of substituted backbones and repulsive charges explain that xanthan gum will dissolve easily in cold water and produces thickened solutions .

STARCH

Starch is the major nutritional reserve food supply of the higher plants. The starch molecule is constituted of :
- *amylose :* linear chain made up of a - (1 → 4) anhydrogalactose units.
- *amylopectin :* linear chain made up of (1 → 4) anhydrogalactose residues and branched in a - (1 → 6).

Starch dissolves in hot conditions and forms thickened solution or a gel on cooling. The final texture depends on the rate of use and on the systems used. Amylose is responsible for the gelling properties of starch and consequently of the retrogradation which may occur. Amylopectin is the thickening component of starch.

CARBOXYMETHYLCELLULOSE

Carboxymethylcellulose (E 466) is a water-soluble cellulose ether, processed by reacting monochloroacetate or monochloroacetic acid with native cellulose. The quantity of the sodium monochloroacetate is dependant on the degree of substitution (DS) required. For food purposes, the crude product is purified by alcohol washing to extract the salts.
The CMC chain consists of a linear backbone more or less polymerized with β - D - glucose. The polysaccharide is characterized by its degree of substitution : hydroxyl groups substituted by carboxymethyl groups. CMC is a cold soluble thickener.

TONYU AND TOFU - PRESENTATION

Under the term soya, a differentiation should be made between :
- **Green type soya**
or mungo bean (*Phaseolus mungo*), rich in proteins, glucids, and fairly free of lipids.
- **Yellow type soya**
(*soya hispana*) rich in proteins and lipids. Yellow type soya enables the production of a "filtrate" called tonyu or soyamilk because of its white colour and a composition closely ressembling that of cow's milk.
Tofu is obtained by coagulating tonyu with coagulating agents (nigari or minerals).

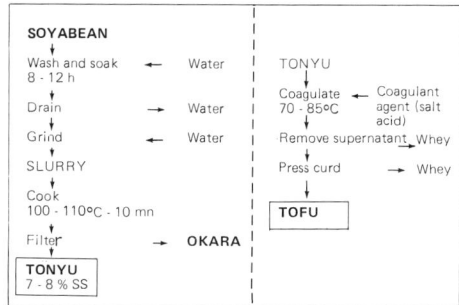

EQUIPMENT AND METHODS

TEST SAMPLES

SPECIFICATIONS

Carrageenan A (Kappa)	MR : 320 REK : 330
Carrageenan C (Kappa)	MR : 310 REK : 780
Carrageenan G (Lambda)	Viscosity in a 1 % aqueous solution : 300 cP
Carrageenan H (Iota)	LD : 32 RES : 211
Guar Gum Re 3	Viscosity in a 1 % aqueous solution : 460 cP
Guar Gum Re 30	Viscosity in a 1 % aqueous solution : 3025 cP
Locust Bean Gum	Viscosity in a 1 % aqueous solution : 1800 cP
Xanthan Gum	Viscosity in a 0.2 % aqueous solution + 1 % KCl : 157 cP
	Viscosity in a 1 % aqueous solution + 1 % KCl : 1680 cP

Remarks on the specifications :

MR : Measured on a Stevens Texture Analyser.
Determine the resistance of a milk-gel to deformation.
Penetration distance : 4 mm.

REK : Measured on a Stevens Texture Analyser.
Determine the resistance of a water-gel in presence of KCl, to deformation.

RES : Measured on a "Sommer & Runge" penetrometer.
Determine the stress required to perform deflection in a resilient gel in presence of NaCl.

LD : Measured on a Brookfield RVT viscometer, with the Helipath stand, equipped with the TC spindle and capable of rotating at 2.5 rpm.
Determine thixotropy after the breaking of the gelled structure upon stirring.

Viscosity : Measured on a Brookfield RVT viscometer, equipped with the N°1 or N°2 spindle and capable of rotating at 20 rpm.

PREPARATION AND CONTROL TESTS FOR SOL OR GEL

☐ Disperse while stirring from 0.5, 1, 1.5, or 2 % hydrocolloid into the system : cold water, cold water + 8 % of sugar, cold half skimmed milk or cold Tonyu.
☐ Heat to boiling point and fill into beakers (carrageenan G, galactomannans and xanthan gum) or in crystallizing dishes (carrageenans A, C, H, I).
☐ Cool at room temperature and control the viscosity or the gel strength, 24 hours after processing.

☐ Viscosity control test : It is carried out with a Brookfield RVT viscometer, equipped with the appropriate spindle ranging from N°1 to N°6 and rotating at 20 rpm.

☐ Gel strength   It is measured on an Instron Universal Testing Instrument, (Re : 1122) equipped
   Penetration test :   with a plunger 25.4 mm in diameter.

BRABENDER BEHAVIOUR

☐ Disperse the hydrocolloid in the cold system.
☐ Heat and cool the solution through the Brabender viscograph.
☐ Rotating speed : 80 rpm - Scale : 250 to 350 cmg, except gelatine : Rotating speed : 50 rpm - Scale : 1000 cmg.

RESULTS · DISCUSSION

Hydrocolloids are macromolecules which modify the rheological behaviours of the system in which they are used. They exhibit the functional properties, i.e. thickening and gelling abilities, which makes them useful in foodstuffs.

THICKENING AGENTS

The thickeners should be readily soluble and little liable to create junction zones between the macromolecules. Among the existing thickening systems, galactomannans (guar gum and locust bean gum), xanthan gum, lambda carrageenan, starch, and CMC have been compared in water, milk and Tonyu.

The rheological behaviour of an hydrocolloid depends on total solids content. Each comparative test with water, Tonyu (8 % TS) and milk (8 % TS) has been carried out under the same conditions : the total solids of the water was increased by the addition of 8 % sugar.

Tab. 1

**VISCOSITY COMPARISON BETWEEN THICKENING AGENTS
USED IN WATER, IN WATER WITH SUGAR, IN MILK AND IN TONYU**

HYDROCOLLOIDS	GUAR GUM 3	GUAR GUM 30	LOCUST BEAN GUM	XANTHAN GUM	LAMBDA CARRAGEENAN (G)
RATE OF USE	1 %	0.5 %	1 %	1 %	1.5 %
WATER*	860 cP	205 cP	800 cP	3150 cP	860 cP
WATER + 8 % SUGAR*	1140 cP	380 cP	2750 cP	3200 cP	1600 cP
MILK*	2250 cP	550 cP	3200 cP	6000 cP	30000 cP
TONYU*	3300 cP	860 cP	9700 cP	9800 cP	25750 cP

* Viscosities measured on a Brookfield RVT viscometer, equipped with the appropriate disk spindle, and rotating at 20 rpm (expressed in centipoises).

Fig. 1 - VISCOSITY MEASUREMENTS / CONCENTRATION RELATIONSHIP

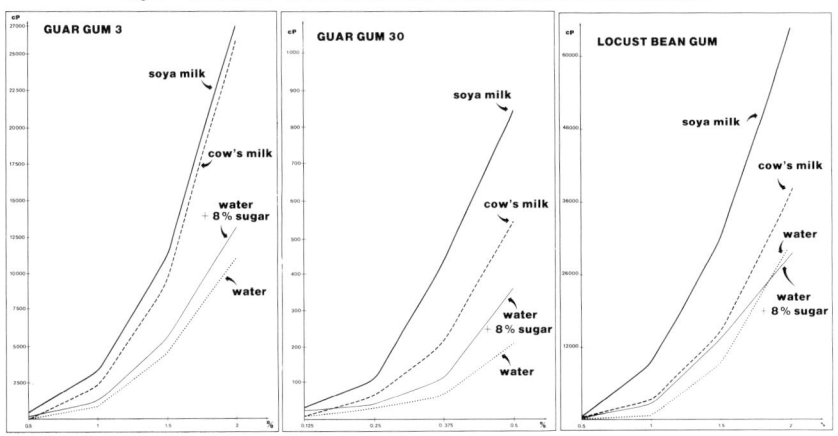

Fig. 1 (continued)

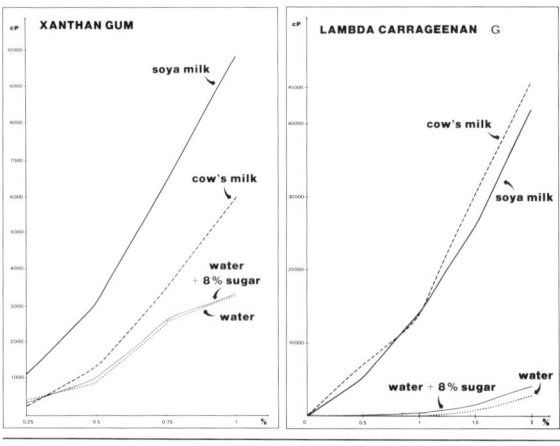

Viscosity measurement of an heterogeneous solution (presence of protein particles) has been carried out on a Brookfield viscometer and has been found unusually high. These high figures observed for galactomannans and xanthan gum in Tonyu are not significant.

Lambda carrageenan in milk or Tonyu gives an homogeneous solution, and is not dependent on the level of use. Viscosities obtained with lambda carrageenan in milk, however, are higher than those recorded in Tonyu.

Fig. 2

BRABENDER CURVES
FOR GALACTOMANNANS AND XANTHAN GUM

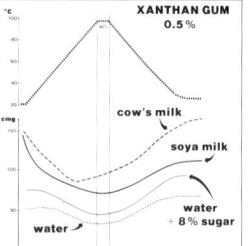

As in cow's milk, the addition of galactomannans or xanthan gum leads to the flocculation of Tonyu proteins. Up to a level of 1 % of hydrocolloid, the effect on viscosity is insignificant and separation of the protein with setting is observed within the system.

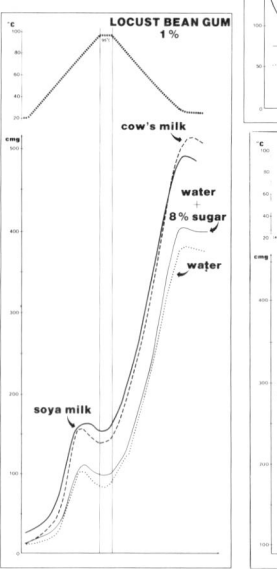

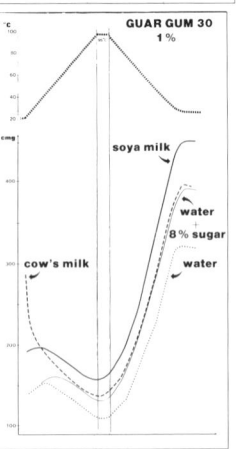

Fig. 3 - BRABENDER CURVES FOR STARCH AND CMC

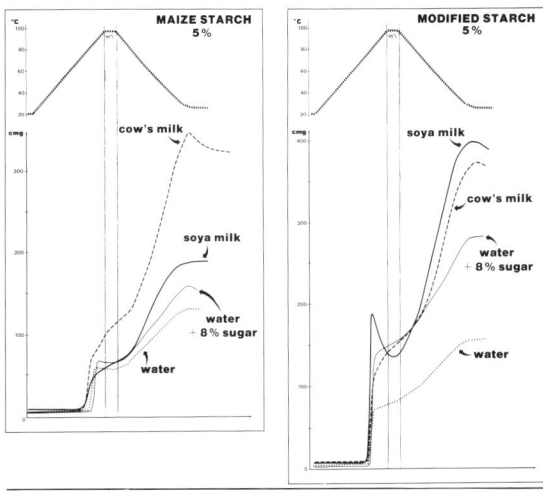

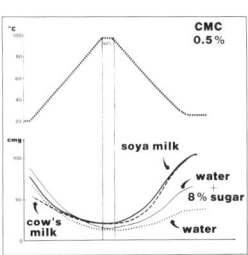

The reaction to heat treatments has been investigated on a Brabender viscograph for starch and CMC. Any significant differences have been observed in the discussed systems milk/Tonyu.

No significant differences in the behaviour of the thickening hydrocolloids in these systems (milk and Tonyu) have been observed, except lambda carrageenan which seems to have a different behaviour from the other hydrocolloids in milk and Tonyu.

GELLING AGENTS

A gelling agent must be :
- hot water soluble and gel forming on cooling (thermo-reversible gels)
- cold water soluble and gel forming in the presence of reactive salts,
- hot water soluble and gel forming in the presence of acid (non-thermo-reversible gels).

Among the various existing gelling systems, the behaviour of carrageenan (kappa and iota fractions), gelatine, alginate and agar-agar has been compared.

Fig. 4 - BRABENDER CURVES FOR GELATINE, AGAR-AGAR AND ALGINATE

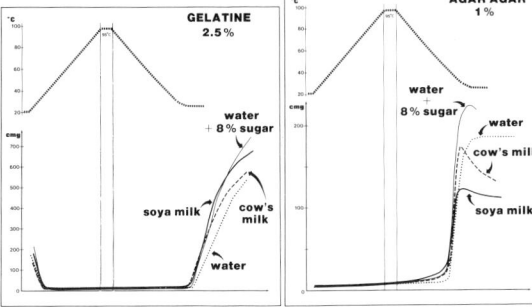

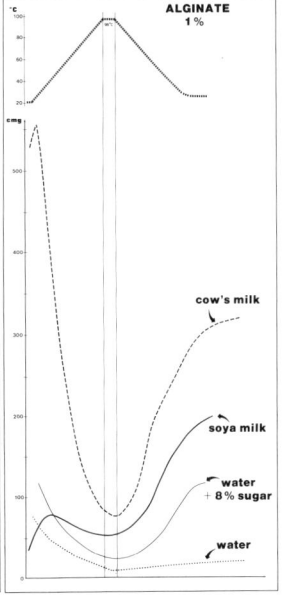

The reactions to heat treatments, measured on a Brabender viscograph, do not display any significant differences between the systems when gelatine and agar-agar are used. However, alginate seems to be slighly more viscous in milk than in Tonyu (Fig. 4).

Tab. 2

GEL STRENGTH COMPARISON BETWEEN CARRAGEENANS USED IN WATER, IN WATER WITH SUGAR, IN MILK, AND IN TONYU

| HYDROCOLLOID | KAPPA CARRAGEENAN | | IOTA CARRAGEENAN | | LAMBDA CARRAGEENAN |
	A	C	H	I	G
RATE OF USE (%)	1 %	1 %	1 %	1 %	1.5 %
WATER * Penetration test method ** Viscosity	2.5	8.4	.***	0.6	860
WATER + 8 % SUGAR * Penetration test method ** Viscosity	3.6	8.2	.***	0.6	1600
MILK * Penetration test method ** Viscosity	22.0	21.5	2.3	2.3	30000
TONYU * Penetration test method * Viscosity	11.5	14.2	1.6	1.9	25750

* Gel strength measured by penetration test method on an Instron Universal Testing Instrument equipped with a 25.4 mm in diameter plunger (expressed in Newton).
** Viscosities measured on a Brookfield RVT viscometer, equipped with the appropriate disk spindle and rotating at 20 rpm (expressed in Centipoises).
*** Non measurable.

Fig. 5 - INSTRON PENETRATION CURVES FOR KAPPA AND IOTA CARRAGEENANS

A comparison between the behaviours obtained with carrageenan in different systems and in relation to those obtained in water or milk, shows that this hydrocolloid displays an intermediate behaviour in Tonyu.

Carrageenans, as sulphated polygalactoses, have a strong anionic character and, as a result, will react with cationic polyelectrolytes. For example, proteins with an amphoteric characteristic precipitate with carrageenans when the pH of the systems is below the isoelectric point. At higher pH values, the protein/carrageenan interaction can take place.

In milk, micelles result from the internal associations of the various fractions of casein and are mainly due to the presence of calcium ions.

However large the carrageenan fraction may be, the graph recorded for Tonyu is always between those of milk and water. Since the Tonyu composition is quantitatively similar to that of the milk, the difference in behaviour observed with carrageenan in these two systems is probably due to a variation in reactivity to mineral salts and/or proteins.

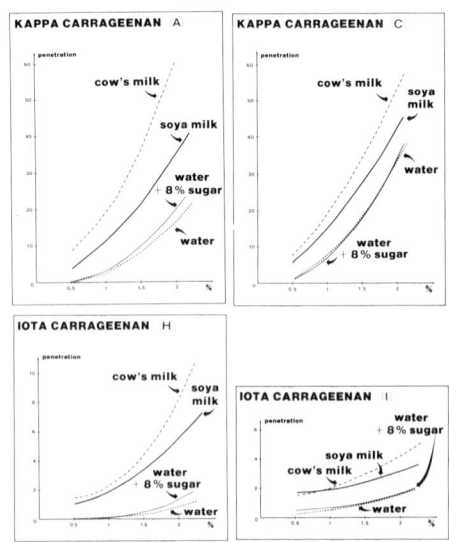

Fig. 6 - BRABENDER CURVES FOR KAPPA AND IOTA CARRAGEENANS

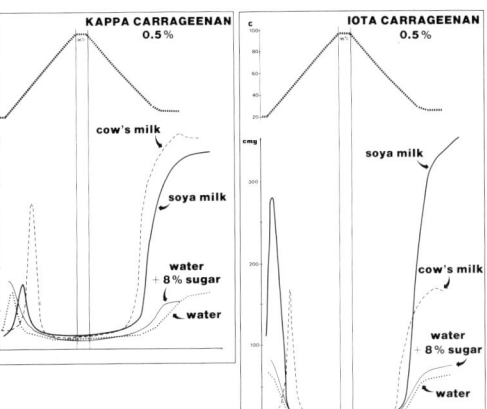

KAPPA CARRAGEENAN 0.5%
cow's milk
soya milk
water + 8% sugar
water

IOTA CARRAGEENAN 0.5%
soya milk
cow's milk
water + 8% sugar
water

Tab. 3

SUCCINCT COMPOSITION OF TONYU AND MILK

	TONYU*	MILK**
MOISTURE CONTENT	90.8 %	87.7 %
PROTEINS	3.6 %	3.4 %
LIPIDES	2.0 %	3.3 %
GLUCIDES	2.9 %	4.7 %
MINERAL SALTS	0.7 %	0.9 %
pH	6.8 - 7	6.6 - 6.8

Source : * The book of Tofu   W. Shurtleff - A. Aoyagi
Ed. Ballantine Books
** Le lait   C. Alais
4è édition Sepaic

MINERAL SALTS

Tab. 4

SALT INFLUENCE

The addition of 0.2 % KCl to kappa carrageenan allows the gel strength to increase in water and to a lesser extent in Tonyu. In milk, carrageenan/casein synergy displays such a large effect that the electrolyte addition, in practice, has no influence in the gel strength.

KAPPA CARRAGEENAN	A + 0.2 % Sugar	A + 0.2 % KCl	C + 0.2 % Sugar	C + 0.2 % KCl
RATE OF USE	1 %	1 %	1 %	1 %
WATER *Penetration test method	2.5	13.8	8.4	18.1
WATER + 8 % SUGAR *Penetration test method	3.6	13.3	8.2	13.8
MILK *Penetration test method	22.0	20.0	21.5	23.3
TONYU *Penetration test method	11.5	16.5	14.2	15.5

* Gel strength expressed in Newton

Tab. 5

MINERAL COMPOSITION MILK/TONYU

MINERAL SALTS	TONYU* (mg/100 g)	MILK** (mg/100 g)
CALCIUM	15	125
SODIUM	2	40
PHOSPHORUS	49	90
POTASSIUM	?	150
IRON	1.2	0.7

Source : * The book of Tofu   W. Shurtleff - A. Aoyagi
Ed. Ballantine Books
** Le lait   C. Alais
4è édition Sepaic.

The potassium cation (K^+) shows a significant effect on kappa carrageenan. Because of its small size, in the hydrated state, it can fit into the coil and partially neutralize the sulphate groups. Thus, the double helices can cluster together : that will promote the aggregation of K^+ salt, which will cause skrinkage and hardening of the gel structure.

PROTEINS

Tab. 6

INFLUENCE OF THE RATE OF PROTEINS

HYDROCOLLOID	KAPPA CARRAGEENAN (A)
RATE OF USE	1 %
WATER + 8 % SUGAR * Penetration test method	3.6
MILK * Penetration test method	22.0
TONYU * Penetration test method	11.5
MILK/WATER (50/50) * Penetration test method	10.1
MILK/WATER (30/70) * Penetration test method	7.3

* Expressed in Newton (N).

When used in milk, the carrageenan gel strength diminishes as the protein content of the milk decreases, because, in the gel system, the reaction carrageenan/protein is modified.

For the kappa carrageenan gel strength to be equivalent in Tonyu and in milk, the latter must be diluted by half.

The protein content of milk is identical to that of Tonyu (as per table of composition) but the nature of the proteins is different.

The main proteins in soyabeans are glycinine (or globuline 11S) and β conglycinine (or globuline 7S) ; the proportion of the two globulines varies with the cultivar of soya.

□ The glycinine, composed of 12 sub-units (6 acid and 6 alcaline) linked together by di-sulfur links, has a globular formation.
□ The β conglycinine is a glycoprotein composed of 3 acid sub-units.

The glycinine of soya and the milk casein, which have different primary structures (amino-acid chain) and an identical isoelectric pH (4.6) will therefore have the same pH and a different distribution of charges along the macromolecules.

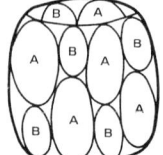

FIG. 8 - Glycinine molecule
A : acid sub-unit
B : alkaline sub-unit

Since the protein/polysaccharide interaction is of an electrostatic or ionic nature (carrageenan/milk casein) it seems that the difference in distribution of electric charges on the protein macromolecule is responsible for the reaction carrageenan/soya protein.

The reaction carrageenan/protein will be more significant as the number of sulphate groupings increases (fractions kappa, iota, lambda). Note that from kappa to lambda carrageenan, the difference in behaviour increases when observed in water and in Tonyu.

In solution, alginate is an anionic polysaccharide, due to the dissociation of the acid functions. These negative charges can react with the positive charges of the protein macromolecules which could explain the difference of behaviour of alginate with milk and with Tonyu.

In the tests carried out, the rheological behaviour in milk of various thickeners (guar, locust bean gum, xanthan gum, starch, CMC) and gelling agents (gelatine, agar-agar) is identical to the behaviour observed in Tonyu.

For the various fractions of carrageenans (kappa, iota, lambda), and for alginates, a polysaccharide/soya protein synergy seems to occur. The latter is less significant that those observed with milk casein.

In practice, it will be possible to increase the rheological characteristics of carrageenans in Tonyu by addition of electrolytes.

FOOD APPLICATIONS

Tonyu (soya milk) can be used as a basic ingredient in the manufacture of beverages, fresh or frozen desserts, sauces, soups and low-calorie comminuted meat products.

Carrageenan is well-suited to neutral dairy applications for the following reasons :
- its low dosage, due to its reactivity to milk proteins,
- the wide range of textures it produces, either singly or in association with other hydro-colloids (starch, locust bean gum, guar),
- its ease of use,
- its organoleptic qualities : lack of flavour and sticky feel, and excellent release of flavours.

With tonyu, these properties will remain, but the synergistic effect with soya proteins will be less significant.

- In chocolate drinks, cocoa suspension will be obtained with a higher percentage of carrageenan (500 ppm as compared to 200 ppm in milk),
- In fresh, gelled or creamy desserts, a range of texture will be available by capitalising on carrageenan extracts and their salt contents. The association of carrageenans/thickeners makes it possible to obtain smoother, more cohesive, syneresis free textures.
- Since carrageenan is little suited to applications in acid system, it will be replaced by textural agents traditionally used in acid dairy systems (gelatine, agar-agar, galactomannans, xanthan gum, starch...),
- In tonyu, gelatine and agar-agar display a texture similar to that obtained in milk.

For food applications, such as frozen desserts, sauces, soups, low calorie meat products, which require the use of thickeners (guar gum, locust bean gum, xanthan gum, starch...), the behaviour of hydrocolloids will be very similar to that observed in milk systems.

REFERENCES

Alais C., Science du lait, ed. Sepaic, Paris, 4 eme Edition, 1984.

Shurtleff W. and Aoyagi, The Book of Tofu (Vol.1), Ed.Ballentine Books, New York, 1983.

Shurtleff W. and Aoyagi, Tofu and Soymilk Production - The Book of Tofu (Vol.2), Ed.Soyfoods Centre, Lafayette (California), 2nd Edition, 1984.

Part 4

EMULSION STABILISATION

The role of proteins in the stabilisation of emulsions

Pieter Walstra

Department of Food Science, Wageningen Agricultural University, The Netherlands

ABSTRACT

Proteins are very suitable in stabilizing edible oil-in-water emulsions, though they vary greatly in suitability. To help understand this, the various processes occurring during emulsification are briefly discussed first. The nature and the concentration of protein affect the thickness of the layer of protein around the droplets (protein load) and the droplet size obtained, and these two variables are interrelated. Moreover, energy density during emulsification and volume fraction of oil have considerable influence. The stability to coalescence of the emulsion droplets depends on protein type, surface load and droplet size, but if droplet size is small and surface load not too low, most proteins give very stable emulsions. The same variables may affect the stability to aggregation of the droplets, but aggregation greatly depends on other conditions as well. Because of the intricacy of emulsion stability, relations between protein characteristics and suitability for emulsion stabilization are difficult to give, but one simple aspect is that the protein should be well soluble, i.e. be present as separate, not very large molecules in the solvent.

INTRODUCTION

Proteins consist of large amphiphilic macro-ions and therefore tend to adsorb strongly at air-water and oil-water interfaces. They then impart considerable repulsion (mostly of a steric nature) between such interfaces and they are thus eminently suitable to stabilize emulsions. Since they are water soluble, this only applies to oil-in-water emulsions. Since proteins are natural components of foods, they are commonly applied in edible emulsions. Different proteins give, however, considerable differences in emulsion properties. This paper does not attempt to review the work done on the role of proteins in emulsions (and foams) or on proteins at interfaces; there are several reviews (e.g. 1-4), as well as books (5-7) and review articles (8-11) covering several aspects of food emulsions. Instead, some general considerations and selected results, which the author considers to be particularly relevant, will be given. Some work on synthetic polymers in emulsions will be taken into consideration.

Before coming to relevant aspects of proteins, it may be useful to have a brief look at the emulsification process.

EMULSIFICATION

See ref. (12) for an extensive review. In an emulsifying machine, the oil is dispersed into coarse droplets and these are successively broken up into smaller ones. The latter process is crucial, because the smaller the droplet, the more difficult its disruption; its Laplace pressure $\Delta p = 4\gamma/d$ (where d is diameter and γ interfacial tension), which counteracts the disrupting forces, increases with decreasing d. In practice, oil-in-water emulsions are made in high-speed stirrers or high-pressure homogenizers, which cause a very intense turbulence in the liquid. Disruption is for the most part by inertial forces (due to pressure fluctuations caused by turbulent eddies of a size comparable to that of the droplets), though some shearing forces (due to elongational flow between comparatively large eddies) may play a part. If inertial forces are determinant, the theory of Kolmogorov for isotropic turbulence predicts that

$$\text{average } d = K \, \epsilon^{-2/5} \, \gamma^{3/5} \, \rho_c^{-1/5} \qquad [1]$$

where K is a constant (of the order of 1), ϵ the effective energy density (or rather power density: energy dissipated per unit volume and per unit of time) and ρ_c the mass density of the continuous phase. Equation [1] is often obeyed very well, at least for dilute emulsions, though K may vary significantly according to conditions. Incidentally, many literature results were obtained by using very small laboratory homogenizers, which produce a low Reynolds number, hence no turbulence, hence conditions very different from those in an industrial machine.

Equation [1] is deficient in that it fails to explain the differences due to variation in oil viscosity, but this can be amended (12, 13). Of far greater significance is that, besides droplet disruption, several other processes occur during emulsification. This is illustrated in Figure 1. Adsorption of surface-active material (i.e. protein) onto the droplet and recoalescence of newly formed droplets (line 4) are especially important. These processes each have their own time scales, which depend in different manners on such variables as ϵ, volume fraction of oil (ϕ), protein concentration (c) and viscosity of the continuous phase (η_c). Some of these aspects are discussed below. It should be realized that the processes shown in Figure 1 can mostly happen numerous times (for each droplet) during the process.

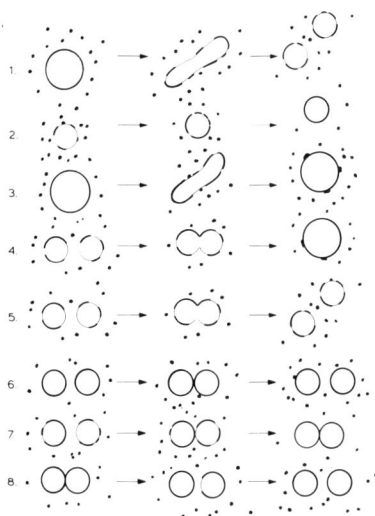

Figure 1. Processes taking place during (the latter stages of) emulsification. Protein material is depicted by heavy lines and dots. Not to scale and highly schematic. From ref. 12, by courtesy of Marcel Dekker Inc.

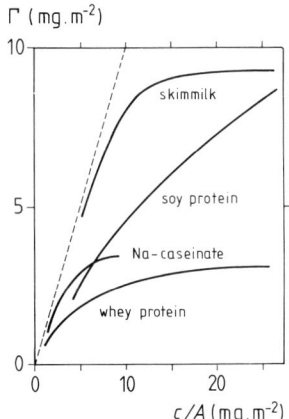

Figure 2. Surface load (Γ) as a function of the concentration of protein (c) per unit oil surface area (A) created by emulsification for some proteins. Approximate results from various sources, mainly (14, 19). However, results depend on conditions like pH and ionic strength. The dotted line corresponds to complete adsorption of all protein.

POINTS OF INTEREST

In order to assess the suitability of a protein to stabilize emulsions, the following points should be considered.
1. How thick a layer is formed at the o/w interface? This determines how much protein is needed to produce an emulsion of a desired specific surface area of oil, and it may also be relevant for emulsion stability (points 3 and 4). Thickness is best expressed as surface excess or protein load Γ (e.g. in $mg \cdot m^{-2}$). It is difficult and expensive to determine Γ with accuracy (from depletion and specific surface area (e.g. (14)).
2. What is the effect on the droplet size distribution? Droplet size is of paramount importance since it strongly affects coalescence stability, creaming rate and tendency to flocculate, smaller droplets commonly giving greater stability. Droplet size also determines specific surface area (A), hence amount of protein needed ($A = 6 \, \phi/d_{vs}$, where ϕ is volume fraction of oil and d_{vs} is the volume-surface average diameter). Size distributions can conveniently be determined by spectroturbidimetry (e.g. 15), Coulter counting (e.g. 16) or other methods, but it is by no means easy to obtain reliable results.
3. What is the coalescence stability of the emulsion droplets in conditions as desired (pH, ionic strength, temperature, etc.)? Even for the same d and Γ there may be differences between proteins. Coalescence is mostly undesirable since it leads to rapid creaming and phase separation. It is a rate process, mostly comparatively slow and of first order. It can be assessed by determining $-dlnN/dt$, or more conveniently $dlnd/dt$ (where N = number of droplets per unit volume and t = time).
4. Do the droplets tend to aggregate in conditions as desired? Aggregation may happen during emulsification or afterwards, e.g. when conditions change. It leads to rapid creaming and to altered rheological properties. If it happens, it is usually a fast process, and it is fairly easy to establish (e.g. by microscopy), although confusion with partial coalescence may occur.

These aspects will be discussed below and it will be seen that they are, to a considerable extent, interrelated. Needless to say that the proteins should also fulfill all other conditions for a food ingredient.

SURFACE LOAD

Protein layers at o/w interfaces may show a fairly wide range of thickness, from about 1 to several times 10 $mg \cdot m^{-2}$. If $\Gamma \approx 1$ $mg \cdot m^{-2}$, it concerns extended polypeptide chains, i.e., the protein molecules are (almost) fully unfolded. If $\Gamma \approx 3 \, mg \cdot m^{-2}$, we may have a monolayer of globular proteins, or (partly) unfolded molecules that are adsorbed with trains, loops and tails. If $\Gamma > 5 \, mg \cdot m^{-2}$, probably either aggregates of proteins or multilayers have adsorbed, although some proteins of very high molecular weight may also give high Γ.

Γ depends on protein concentration during emulsification: see Figure 2 (which compiles almost all results available). Are these curves to be compared with adsorption isotherms as determined by adsorption from a bulk solution to a plane interface or even from spread layers in a Langmuir trough? Although there must be some similarities, the curves are not equal.

a. The preparations are mixtures of proteins (and impurities) and in a bulk experiment we have a very low surface to volume ratio, enabling the components that give the lowest surface tension to predominate in the interface, even if present in low concentrations. In an emulsion the surface to volume ratio is much higher (e.g. by 3 orders of magnitude); hence, the composition of the adsorbed layer may be greatly different.

b. Although proteins are not adsorbed completely irreversibly, and some desorption and exchange between the bulk and interface is possible (22), such processes are often slow compared to the observation time (4). Adsorption during emulsification may be extremely fast: to a first approximation (12) the time needed for the protein to cover a droplet is given by

$$t_{ads} \approx 10 \; \Gamma \; \eta_c^{\frac{1}{2}}/dc\epsilon^{\frac{1}{2}} \qquad\qquad [2]$$

Assuming $\Gamma = 3 \times 10^{-6} kg \cdot m^{-2}$, $\eta_c = 2 \times 10^{-3} Pa \cdot s$, $d = 2 \times 10^{-6}$ m, $c = 20 \; kg \cdot m^{-3}$ and $\epsilon = 10^{11} W \cdot m^{-3}$, we find $t_{ads} = 10^{-7}$ s. Due to several uncertainties this may in fact be an order of magnitude longer and the gradual depletion of protein during emulsification may cause another increase by a factor 10, but then still the adsorption times are shorter than those in bulk experiments by several orders of magnitude. This may affect possibilities for unfolding and also relative abundances of proteins in mixtures (also see point 3, below).

c. During emulsification, compression of the o/w interface occurs (see lines 3 and 4 in Figure 1) and this may lead to a higher Γ.

It is therefore not surprising that plots of Γ versus bulk concentration after emulsification (as are commonly made for adsorption isotherms), often do not coincide if the emulsions were made at varying volume fractions of oil, as was found for synthetic polymers (17,18). But if plotted as in Figure 2, i.e. Γ against total concentration over the surface area produced, the curves do coincide (12,19). Often an approximately linear relation is found when plotting Γ against log (c/A) (11). But, as shown below, A itself will depend on the protein-to-oil ratio, and is thus not known beforehand. It is no surprise that Γ may also depend on ϵ, hence droplet size, for the same c/A. The considerations given imply that it is very difficult to draw conclusions about emulsification properties from adsorption isotherms (Γ against bulk concentrations, determined by diffusion to a plane interface). It is certainly not warranted to draw conclusions from the much easier to obtain plots of γ against concentration or γ against time, obtained by diffusion. This is because high surface loads (in terms of mass per unit area) are

(7) for a full discussion. Any flocculation is mostly in the secondary minimum, which implies that small droplets are more stable than big ones. However, adsorbed proteins usually provide sufficient repulsion to prevent flocculation. But if conditions change in such a way that the protein itself becomes poorly soluble, the droplets will probably flocculate. For instance, near the isoelectric pH of the protein and/or at high ionic strength electrostatic repulsion will be very slight and steric repulsion may also be less, because the protein tends to obtain a compact conformation. Likewise, lowering solvent quality (adding ethanol or specific salts) may diminish steric repulsion, hence cause flocculation.

3. Cross-linking reactions between proteins adsorbed at different but closely approaching droplets, such as S-S-bridges formed during heating. Such reactions may occur between droplets pressed together in a cream layer, but also in fleeting encounters, since steric repulsion between droplets as caused by protruding peptide chains, although it prevents flocculation of the droplets, does not prevent contact between the protruding chains. For droplets in a cream layer, even very slow cross-linking reactions may cause aggregation, otherwise the reaction must be very fast to achieve this.

CONCLUDING REMARKS

It will be clear that a simple answer to the question about the suitability of various proteins to help form and keep stable emulsions cannot be given: there are too many variables during and after emulsification. Moreover, we have no real understanding of the relation between the primary, secondary and tertiary structure of a protein and its emulsifying properties. Further studies of the interaction between proteins and oil droplets during emulsification would be most welcome, and the considerations given above may provide some clues.

Evaluation of the suitability of a certain protein by one simple test is hardly feasible. Emulsifying capacity, for instance, gives very little useful information (1). The emulsifying activity index (27) makes more sense, but a far better way is to determine what droplet size is obtained for various protein-to-oil ratios; some information about the surface load (from depletion studies) gives useful additional information. Of course, such tests should be done in emulsifying conditions (energy density, type and size of machine) comparable to those in industrial practice.

To evaluate 'stability', creaming tests are often applied, either gravity or centrifugal creaming. Such a test only gives an indication about droplet size and is only reliable if it is done at the right conditions (meaning that on average roughly half of the droplet volume reaches the cream layer), if the droplets are not aggregated and if the various proteins used do not give a markedly different viscosity (or even a yield stress) to the aqueous phase. To determine the effect of the protein on stability, the emulsion should be observed at conditions as encountered in practice. Rapid tests for coalescence stability are never reliable; for example, the coalescence behaviour during ultracentrifugation mostly does not correlate well with that under quiescent conditions. But slow coalescence can be detected in an early stage by following average droplet size (or just turbidity at a suitable wavelength) with time; changes of a few percent can be detected and extrapolated to longer times with reasonable success.

Finally: in most instances, a more stable emulsion can be obtained by making smaller droplets and/or using more protein.

REFERENCES

1. Halling, P.J. (1981). Crit. Rev. Food Sci. Nutr. **15**, 155-203.
2. Phillips, M.C. (1981). Food Technol. **35**(1), 50-51, 54-57.
3. Dickinson, E. (1986). Food Hydrocolloids **1**, 3-23.
4. Mitchell, J.R. (1986). In: Developments in Food Proteins, Vol. 4. (Ed. Hudson, B.J.F.) pp. 290-338, Elsevier Applied Sci., London.
5. Mulder, H. and Walstra, P. (1974). The Milk Fat Globule: Emulsion Science as applied to Milk Products and Comparable Foods, 296 pages, CAB, Farnham Royal, Pudoc, Wageningen.
6. Friberg, S. (ed.) (1976). Food Emulsions, 480 pages. Dekker, New York.
7. Dickinson, E. and Stainsby, G.E. (1982). Colloids in Foods, 533 pages, Applied Science Publ., London.
8. Stainsby, G.E. (1986). In: Functional Properties of Food Macromolecules. (Eds. Mitchell, J.R. and Ledward, D.A.). pp. 315-353, Elsevier Applied Sci., London.
9. Darling, D.F. and Birkett, R.J. (1987). In: Food Emulsions and Foams. (Ed. Dickinson, E.). pp. 1-29. Royal Soc. Chem., London.
10. Walstra, P. (1987). In: Food Emulsions and Foams. (Ed. Dickinson, E.). pp. 242-257. Royal Soc. Chem., London.
11. Walstra, P. (1987). In: Food Structure and Behaviour, Vol. 1. (Eds. Blanshard, J.M.V. and Lillford, P.J.). pp. 87-106. Academic Press, London.
12. Walstra, P. (1983). In: Encyclopedia of Emulsion Technology, Vol. 1. Basic Theory. (Ed. Becker, P.). pp. 57-127. Dekker, New York.
13. Davies, J.T. (1985). Chem. Engin. Sci. **40**, 839-842.
14. Oortwijn, H. and Walstra, P. (1979). Neth. Milk Dairy J. **33**, 134-154.
15. Walstra, P. (1968). J. Colloid Interface Sci. 27, 493-500.
16. Walstra, P. and Oortwijn, H. (1969). J. Colloid Interface Sci. 29, 424-431.
17. Lankveld, J.M.G. (1970). Meded. Landbouwhogeschool Wageningen, **70-21**, 1-114.
18. Böhm, J.T.C. (1974). Meded. Landbouwhogeschool, Wageningen, **74-75**, 1-109.
19. Tornberg, E. (1978). J. Sci. Food Agriculture **29**, 867-879.

20. De Feijter, J.A. and Benjamins, J. (1982). J. Colloid Inter-
    face Sci. 90, 289-292.
21. Walstra, P. and Oortwijn, H. (1982). Neth. Milk Dairy J. 36,
    103-113.
22. MacRitchie, F. (1985). J. Colloid Interface Sci. 105,
    119-123.
23. Musselwhite, P.R. (1964). Proc. 4th Int. Congress Surface
    Active Substances, p. 947.
24. Walstra, P. and Jenness, R. (1984). Dairy Chemistry and Phy-
    sics, 467 pages, Wiley, New York.
25. Lankveld, J.M.G. and Lyklema, J. (1972). J. Colloid Inter-
    face Sci. 41, 475-483.
26. Böhm, J.T.C. and Lyklema, J. (1976). In: Theory and Practice
    of Emulsion Technology. (Ed. Smith, A.L.) pp. 11-27. Acade-
    mic Press, London.
27. Pearce, K.N. and Kinsella, J.E. (1978). J. Agric. Food
    Chem. 26, 716-723.
28. Muschiolik, G., Dickinson, E., Murray, B.S. and Stainsby, G.
    (1987). Food Hydrocolloids 1, 191-196.
29. Fisher, L.R., Mitchell, E.E. and Parker, N.S. (1987). In:
    Food Emulsions and Foams. (Ed. Dickinson, E.). pp. 230-241.
    Royal Soc. Chem., London.
30. Kato, A. and Nakai, S. (1980). Bioch. Biophys. Acta 624,
    13-20.
31. Van den Tempel, M. (1960). Proc. 3rd Int. Congr. Surface
    Activity, 2, 573-579.
32. Lucassen, J. (1979). In: Physical Chemistry of Surfactant
    Action. (Ed. Lucassen-Reynders, E.H.), pp. 217-265, Dekker,
    New York.
33. Oortwijn, H. and Walstra, P. (1982). Neth. Milk Dairy J. 36,
    279-290.
34. De Wit, J.N., Klarenbeek, G. and Swinkels, G.A.M. (1976).
    Voedingsmiddelentechnologie 9 (17) 14-17.
35. Van Boekel, M.A.J.S. and Walstra, P. (1981). Colloids Surf.
    3, 109-118.
36. Ogden, L.V., Walstra, P. and Morris, H.A. (1976). J. Dairy
    Sci. 59, 1727-1737.

Stability aspects of casein-containing emulsions: effect of added alcohol or dextran

Stephen Bullin, Eric Dickinson, Sarah J.Impey, Sunit K.Narhan and George Stainsby

Procter Department of Food Science, University of Leeds, Leeds LS2 9JT, UK

ABSTRACT

The effect of (i) ethanol and (ii) dextran on the stability of casein-containing oil-in-water emulsions has been investigated. A moderate level of alcohol (10−20 wt%) reduces stability with respect to flocculation by added electrolyte, but improves stability with respect to creaming and serum separation, through a reduction in droplet size, whereas a high level of alcohol ($\geqslant$ 30 wt%) markedly impairs stability due to protein insolubility. A low level of dextran ($\geqslant$ 1 wt%) reduces stability with respect to creaming, but only if added electrolyte ($\geqslant$ 0.05 M) is present also. This last result is discussed in relation to the phase behaviour of mixtures of dextran and caseinate.

INTRODUCTION

Food emulsions are inherently unstable. The technological art is to delay the perception of instability through control of the main processes that promote it—namely creaming and flocculation. Some creaming may be acceptable, provided there is no obvious separation of serum, and some flocculation may also be acceptable, provided the accompanying textural changes are not too great. Extensive creaming and flocculation are generally undesirable.

An important factor affecting the formation and stability of food emulsions is solvent quality. Most studies so far on the effect of the aqueous phase composition on the properties of protein-stabilized emulsions have been concerned with changes in pH. The main conclusion (1,2) is that the interfacial film is thickest and strongest and the emulsion most stable at the isoelectric pH where protein solubility is at a minimum. In the experiments reported here, we keep the pH constant; and instead we change the quality of the aqueous phase solvent by addition of either ethanol or dextran.

While ethanol is a potent precipitant for proteins, stable cream liqueurs containing, say, 15 wt% alcohol can be made with a shelf-life of several years. Addition of ethanol certainly causes considerable change in the physical properties of aqueous systems: there is a decrease in surface (or interfacial) tension, a decrease in dielectric constant, and an increase in viscosity.

At concentrations below that which causes precipitation, the
presence of alcohol favours adoption of the α-helix conformation
where the amino-acid sequence of the protein makes this possible.
Overall, though, it operates as a denaturant through hydrophobic
effects (3).

High-molecular-weight polysaccharides are added to many food
emulsions (e.g., salad dressing) in order to thicken the aqueous
phase and so reduce the rates of creaming and flocculation (4).
Nevertheless, such non-adsorbing polymers may also reduce colloid
stability by inducing depletion flocculation (4), and recently
Hibberd et al. (5) have interpreted the more rapid than expected
rate of creaming for emulsions containing Xanthan gum as being
due to this depletion effect. Here, we choose the neutral linear
polymer, dextran, as our high-molecular-weight polysaccharide,
so as to avoid any complications due to polysaccharide gelation
or charge—charge interactions between the polysaccharide and the
protein emulsifier.

There is an analogy between colloid instability induced by
non-adsorbing macromolecules and the phase separation that can
occur when different polymers share a common solvent (4). Except
for the work of Tolstoguzov and colleagues (6,7), the importance
of such macromolecular incompatibility to foods has been largely
unrecognized, though the effect is well known in the synthetic
polymer field. In this paper we are concerned with the rate at
which an emulsion separates into a cream layer, rich in protein-
coated droplets, and a serum layer, rich in polysaccharide
molecules. We are particularly interested in the effect of ionic
strength on stability, since it is known (6) that the phase
behaviour of mixed solutions of dextran + protein is sensitive to
this variable.

In this study, n-tetradecane is chosen as the oil phase so as
to avoid any possible complications arising from the presence of
surface-active lipids in food oils. Commercial sodium caseinate
is the sole emulsifier.

MATERIALS AND METHODS

Materials
Spray-dried sodium caseinate ('Scottish Pride') was obtained
from the Scottish Milk Marketing Board. 'Absolute' ethyl alcohol
was obtained from Hays Chemicals. The dextran T2000, obtained
from Pharmacia Fine Chemicals (lot IK32207), was a fraction with
a narrow distribution of molecular weights (average value 2×10^6
daltons). AnalaR grade n-tetradecane (> 99 wt %) was obtained
from Sigma Chemicals, and sodium azide from BDH Chemicals. The
phosphate buffer (pH 7.0, 0.005 M) was prepared from AnalaR grade
reagents and double-distilled water.

Emulsion Preparation
Oil-in-water emulsions containing alcohol were prepared with
1 or 3 wt % sodium caseinate as emulsifier and 15 wt % tetradecane
as the oil phase. The aqueous protein solution (pH 7.0, 0.005 M
phosphate buffer), containing alcohol as required, was blended
with the hydrocarbon in a small-scale Silverson-type homogenizer,
and the resulting coarse emulsion was then passed at 300 bar

through the valve mini-homogenizer described previously (8).
    Oil-in-water emulsions containing no alcohol were prepared
with 1.5 wt% sodium caseinate and 30 wt% tetradecane as the oil
phase. The aqueous protein solution (pH 6.2, distilled water)
was blended with the hydrocarbon and homogenized as above.
Aqueous dextran solutions were separately prepared, and aliquots
were thoroughly mixed with equal weights of freshly made emulsion,
to give final emulsions of 15 wt% oil phase.
    To retard microbial attack, sodium azide (0.01 or 0.1 wt%) was
present in all emulsions containing less than 14 wt% alcohol.

Emulsion Stability
    Samples were stored in vials (height 75 mm, diameter 10 mm)
with tightly fitting lids at a constant temperature (5 °C for the

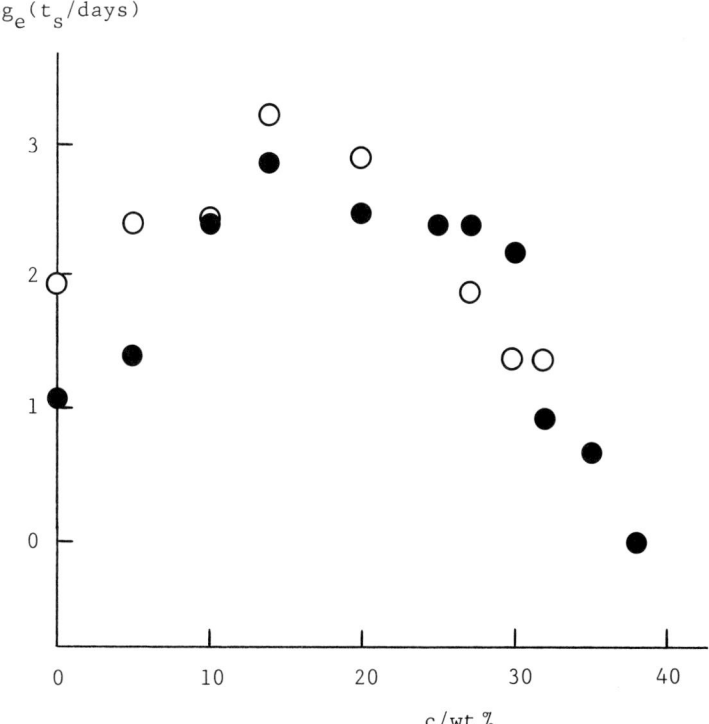

Figure 1. Effect of alcohol on emulsion storage stability
at 5 °C as measured by first appearance of discernible serum
separation. Logarithm of serum separation time t_s is plotted
against ethanol concentration c in mix prior to blending and
homogenization: ●, 1 wt% caseinate; o, 3 wt% caseinate.

alcohol systems, 8 °C for the dextran systems). Samples were
inspected daily for the onset of serum separation or creaming or
both. Droplet-size distributions were determined from counts of
10^5 droplets using a Coulter counter model TAII with a 30 μm
orifice tube and 0.18 M sodium chloride as suspending electrolyte.
There was no evidence for any microbiological deterioration of
the samples over the observational time-scale reported below.

RESULTS AND DISCUSSION

Emulsions Containing Alcohol
    Figure 1 shows the effect of ethanol concentration on the
storage stability, as judged by the onset of serum separation,
for two sets of emulsions with different protein loads. We see
that the presence of a moderate level of alcohol (10—20 wt%), in
the mix before homogenization, improves the stability. But, the
effect is rather less marked when there is an excess of protein
present (3 wt% caseinate is typical for a commercial dairy
emulsion containing alcohol).
    The improvement in stability recorded in figure 1 is probably
a consequence of the reduction in droplet size: the most-probable
droplet diameter decreases from ca. 1 μm in the absence of any
alcohol to < 0.6 μm in the presence of alcohol (0.6 μm is the
lower limit of particle size on the Coulter counter). When, in
separate experiments, we added alcohol after homogenization, we
found that there was no change in stability up to an ethanol
concentration of ca. 20 wt%, which shows that it is the presence
of alcohol during homogenization which is crucial to the increase
in stability. The better emulsification in the presence of the
alcohol can be explained in terms of the substantial lowering by
alcohol of the interfacial tension between a hydrocarbon and a
phosphate buffer containing various amounts of sodium caseinate
(9). Our results on these model n-tetradecane-in-water emulsions
are in keeping with studies of cream liqueurs by Banks et al. (10)
who found that less severe conditions of homogenization were
needed to produce a fine emulsion when 14 wt% alcohol was
present. Banks et al. (10) ascribed the improvement to the
change in mix viscosity.
    The favourable effect of alcohol on emulsion stability is lost
at ethanol concentrations $\geqslant$ 30 wt%. The serum separation time
becomes lower than that for comparable systems without alcohol,
and the most-probable droplet size becomes sharply dependent on
alcohol content (see figure 2). The poor stability at ethanol
concentrations above 30 wt% is probably due to caseinate
aggregation and insolubility in the poor quality solvent. Legal
requirements for cream liqueurs in the U. K. demand that the
alcohol content exceeds 14 wt%. It is fortuitous, and fortunate,
that this value corresponds approximately to optimal stability,
based on the observations recorded in figure 1.
    The stability of a casein-stabilized oil-in-water emulsion
towards added electrolyte is very sensitive to the composition of
the interfacial layer, especially when there is only just enough
protein present to confer stability in the absence of added salt
(11,12). Hence, it is interesting to examine how alcohol affects
salt stability. For the 1 wt% caseinate systems, it was found

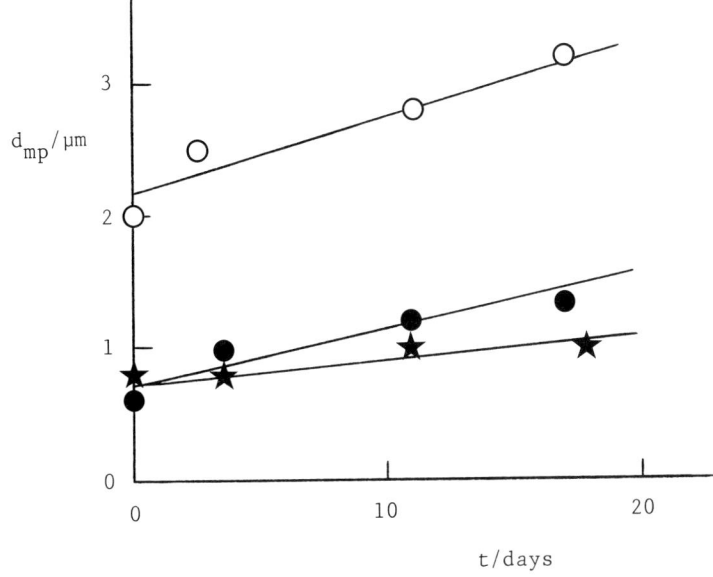

Figure 2. Effect of alcohol content on storage stability at 5 °C for 1 wt % caseinate emulsions as measured by Coulter counter. Most-probable droplet size d_{mp} is plotted against storage time t: ⋆, 27 wt % ethanol; ●, 30 wt % ethanol; o, 32 wt % ethanol.

that, whereas 1:1 addition of 2 M sodium chloride had no effect on the measured droplet-size distribution in the absence of alcohol, the presence of just 5—10 wt % alcohol in the freshly prepared emulsion led to flocculation on addition of the same electrolyte. Figure 3 shows the changes in apparent droplet-size distribution for ethanol concentrations of 10 and 14 wt %. There seems to be no simple explanation for this result, though one can speculate on possible contributory factors. In the first place, alcohol lowers the dielectric constant of the aqueous phase, and thus increases the magnitude of repulsive electrostatic interactions between charged droplets at low ionic strength. Secondly, it is known that alcohol, as an unfavourable solvent, will induce conformational changes in proteins, and can decrease steric stabilization in dispersions of casein micelles (13,14). It is possible that ethanol may cause a similar shrinkage of the casein stabilizing films in our emulsions. In addition, quite apart from inducing conformational change, alcohol may act as a partial displacer of caseinate from the droplet interface, and so sensitize the system to salts. This explanation is given some

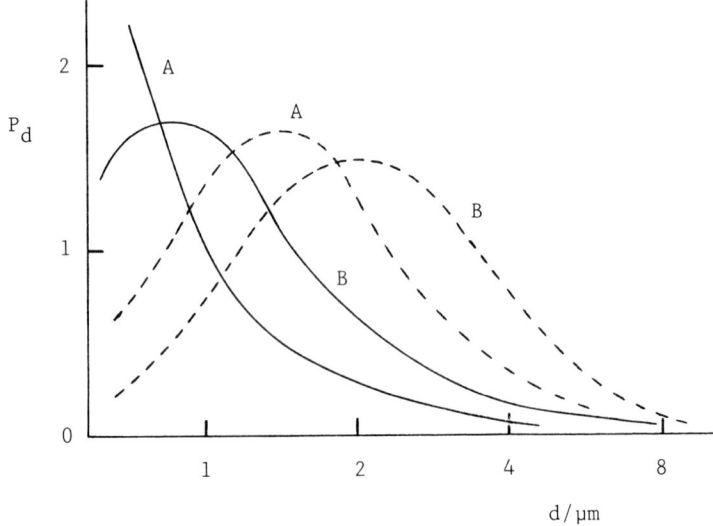

Figure 3. Salt stability of 1 wt% caseinate emulsions. The volume-weighted droplet-size distribution function P_d is plotted against droplet diameter d: —, emulsion diluted 1:1 with phosphate buffer (pH 7, 0.005 M); ---, emulsion diluted 1:1 with 2 M NaCl; A, 10 wt% EtOH; B, 14 wt% EtOH.

support by recent measurements in this laboratory (9) which show that the surface viscosity of a caseinate film adsorbed at the oil—water interface is substantially changed by the presence of alcohol even at the 1 wt% level.

Emulsions Containing Dextran
    We first present results illustrating the limited miscibility of casein + dextran in aqueous solution (no n-tetradecane, no homogenization). Table I shows how the phase behaviour of one

Table I   EFFECT OF TEMPERATURE AND IONIC STRENGTH ON PHASE
           BEHAVIOUR OF A MIXTURE OF EQUAL VOLUMES OF 10 WT%
           DEXTRAN + 5 WT% CASEINATE

Ionic Strength/M	Number of Phases		
	0 °C	18 °C	45 °C
0.015	1	1	1
0.05	1	1	2
0.15	1	2	2

Table II EFFECT OF MACROMOLECULAR CONCENTRATIONS (BEFORE
MIXING) ON PHASE BEHAVIOUR OF MIXTURE OF EQUAL
VOLUMES OF DEXTRAN (D) AND CASEINATE (C) SOLUTIONS
AT 18 °C AND IONIC STRENGTH 0.15 M

Wt % D	Number of Phases		
	0.75 wt % C	1.5 wt % C	3 wt % C
2.5	1	1	1
5.0	1	1	1
10.0	1	2	2
20.0	1	2	2

particular mixture changes with temperature and sodium chloride
concentration. Table II shows how the phase behaviour depends on
the concentrations of dextran and caseinate. The behaviour is in
good general agreement with that reported by Tolstoguzov et al.
(6). We can understand the ionic strength dependence of phase
separation in a mixture of protein and neutral polysaccharide in
the following way. At high salt concentrations, electrostatic
interactions are screened, and the system behaves like a mixture
of uncharged polymers, which typically exhibits incompatibility
at high polymer concentrations. At low ionic strengths, however,
there is strong electrostatic repulsion between the protein
molecules, and the electrostatic contribution to the free energy
of the system is minimized if these molecules are distributed as
uniformly as possible throughout the whole system. So, phase
separation is inhibited because it would require the protein
molecules, on average, to get closer together by being largely
concentrated in one region of the system.

Moving now to the emulsions, figure 4 shows how the creaming
stability depends on dextran concentration and ionic strength for
a set of emulsions containing 0.75 wt % caseinate and 15 wt %
n-tetradecane. Various known levels of sodium chloride were
present in the oil/water mix before homogenization, and sodium
azide (final concentration 0.01 wt %) was present to retard
microbial attack. Each point in figure 4 corresponds to a sample
prepared by mixing equal volumes of freshly made emulsion and an
aqueous dextran solution. Points are distinguished according to
whether or not a cream layer had visibly separated after 3 days
of storage. We stress that the behaviour represented in figure
4 describes a kinetic phenomenon and not a thermodynamic one:
after 1—2 weeks all the samples had separated into two layers,
even those devoid of dextran.

The results in figure 4 clearly show that stability in oil-in-
water emulsions containing caseinate and dextran is reduced with
increasing ionic strength and increasing dextran concentration.
These trends are qualitatively the same as those found with
solutions of caseinate + dextran, though it is clear that the
concentration of dextran required to destabilize the emulsion is
much lower than that required to cause thermodynamic phase
separation in the absence of oil at the same ionic strength and
caseinate concentration (see tables I and II). All the emulsions

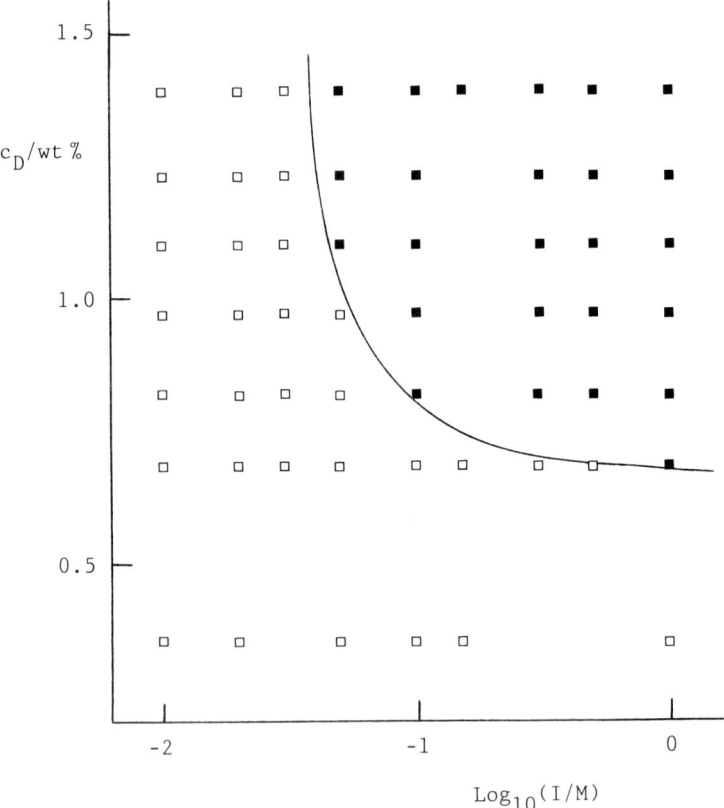

Figure 4.   Creaming stability at 8 °C of emulsions containing
0.75 wt % caseinate and 15 wt % tetradecane.   The aqueous
phase dextran concentration c_D is plotted against logarithm
of ionic strength I: □, one phase visible after 3 days; ■,
two phases visible after 3 days.

containing dextran were of similar low viscosity, and could be
poured readily when first prepared.   Following separation in the
presence of dextran, the cream layer exhibited extensive droplet
flocculation, and it could not be poured.   These results are
consistent with dextran reducing stability by a depletion
flocculation mechanism, but only if added electrolyte is present
at concentrations above ca. 0.05 M.
      Whether thermodynamic incompatibility between caseinate and
dextran is the primary factor relevant to the effect of added

dextran on emulsion stability is still uncertain. Analytical data on compositions of separated phases would be an invaluable aid to resolving the issue.

ACKNOWLEDGEMENTS

We thank Mr Alan Laycock and Mrs Linda Witkowski for their experimental assistance.

REFERENCES

1 Phillips, M.C. (1981) Food Technol., 35, 50-57.
2 Waniska, R.D., Shetty, J.K. and Kinsella, J.E. (1981) J. Agric. Food Chem., 29, 826-831.
3 Herskovits, T.T., Gadegbeku, B. and Jaillet, H. (1970) J. Biol. Chem., 245, 2588-2598.
4 Dickinson, E. (1988) This volume, pp. 249 - 264.
5 Hibberd, D.J., Howe, A.M., Mackie, A.R., Purdy, P.W. and Robins, M.M. (1987) in Food Emulsions and Foams (ed. Dickinson, E.), pp. 219-229. Royal Society of Chemistry, London.
6 Tolstoguzov, V.B., Grinberg, V.Ya. and Gurov, A.N. (1985) J. Agric. Food Chem., 33, 151-159.
7 Tolstoguzov, V.B. (1986) in Functional Properties of Food Macromolecules (eds Mitchell, J.R. and Ledward, D.A.), pp. 385-415. Elsevier Applied Science, London.
8 Dickinson, E., Murray, A., Murray, B.S. and Stainsby, G. (1987) in Food Emulsions and Foams (ed. Dickinson, E.), pp. 86-99. Royal Society of Chemistry, London.
9 Woskett, C.M. (1987) Unpublished results.
10 Banks, W., Muir, D.D. and Wilson, A.G. (1982) J. Sci. Dairy Technol., 35, 41-44.
11 Chesworth, S.M., Dickinson, E., Searle, A. and Stainsby, G. (1985) Lebensmittel Wiss. Technol., 18, 230-232.
12 Dickinson, E., Whyman, R.H. and Dalgleish, D.G. (1987) in Food Emulsions and Foams (ed. Dickinson, E.), pp. 40-51. Royal Society of Chemistry, London.
13 Horne, D.S. (1984) Biopolymers, 23, 989-993.
14 Griffin, M.C.A., Price, J.C. and Martin, S.R. (1986) Int. J. Biol. Macromol., 8, 367-371.

Influence of vegetable protein on freeze−thaw-stability of O/W emulsions

G.Muschiolik, A.Dahme, G.Schmidt and H.Schmandke

Academy of Sciences of the GDR, Central Institute of Nutrition, Arthur-Scheunert-Allee 114−116, Bergholz-Rehbrucke, DDR-1505, GDR

ABSTRACT

The effect of acetylated faba bean protein isolate on the freeze-thaw-stability and rheology of acidified 36/64 oil/water systems was studied. Variables evaluated include pH, protein concentration, and pre-heating of emulsions. The influence of potato starch syrup, gelatin and potato starch on the stability of emulsions to centrifugation is also shown. In the case of acetylated faba bean protein isolate, emulsions produced show good freeze-thaw-stability. The protein acts as an emulsifier and stabilizer. Apparent viscosity of emulsions was enhanced by preheating and freezing.

INTRODUCTION

Acetylated faba bean protein isolate (AFBPI) is particularly suitable for the stabilisation of O/W emulsions in acid and neutral pH regions. An essential feature of the consistency of O/W emulsions with AFBPI is an attractive creamy texture and non-slimy mouthfeel (1). AFBPI has two of the most desirable physico-chemical characteristics for effective emulsion formation and stabilisation: low interfacial tension and high surface visco-sity, both at the oil-water interface (2).

The aim of the present paper is to test AFBPI as an emul-sifier and stabilizer for dressings and to investigate the influence of AFBPI on the freeze-thaw-stability of O/W emulsions, especially dressing type, which is of interest for use with frozen salad.

MATERIALS AND METHODS

Materials

Neutralized and spray-dried acetylated faba bean protein isolate (AFBPI) with a degree of 97 % acetylation was prepared from field beans (Vicia faba L. minor) as described previously (3). Sucrose and rapeseed oil was obtained from a local super-market. Potato starch syrup powder (DE 30) and potato starch were obtained from VEB Stärkefabrik Kyritz and gelatin (Gelatine 170) was obtained from VEB Gelatinewerk Calbe. Acetic acid solution was prepared from reagent A grade glacial

Table II. SHEAR STRESS OPTIMUM (Pa) OF EMULSIONS
          (FORMULATION 3)

Preheating	Freezing	pH 3.5		pH 4.5	
		Non-sheared	Presheared	Non-sheared	Pre-sheared
−	−	20.5	8.1	23.0	8.0
+	−	31.6	13.0	57.0	17.4
−	+	11.2	6.2	13.0	6.2
+	+	32.2	11.2	53.3	8.7

Shear rate $\dot{\gamma}$ = 4.45 s^{-1}

This could be explained by the high quantity of adsorbed
acetylated protein at the interface(4) and by the most compact
configuration of the adsorbed protein molecules at the
isoelectric point pI, where less water is bound by proteins.
The surprisingly low ES at pH 2.5 was found by Aoki (4) also.
    At the pI of AFBPI (pI = 3.8 - 4.3) the apparent viscosity
(beginning of the second linear part of the flow curve) of
acidified emulsions is increased (Fig. 1). Results of the
rheological evaluation show the shear stress optimum at pH 4.5

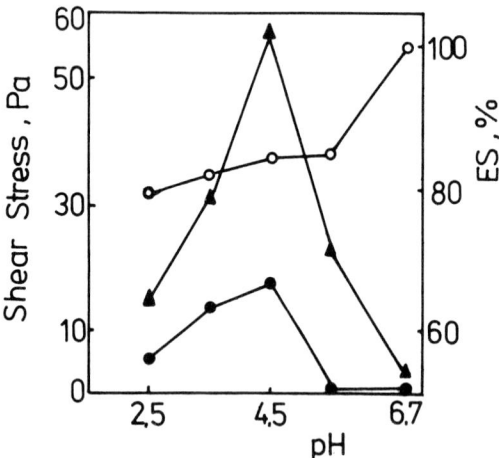

Figure 2.   Influence of pH on emulsion stability (ES,○—○)
and shear stress optimum of preheated and non-frozen emulsions
(formulation 3, shear rate $\dot{\gamma}$ = 4.45 s^{-1}).
▲, before shearing; ● , after measurement of flow curve, cycle 18 min
(see Figure 4).

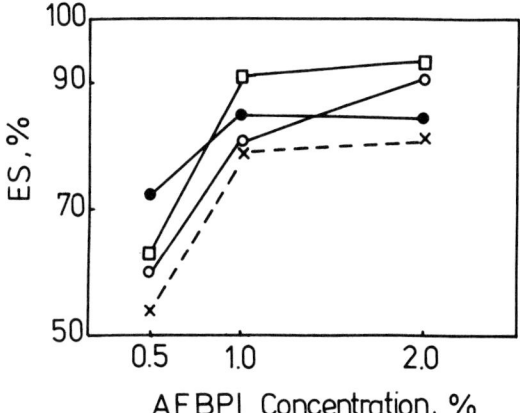

Figure 3. Influence of acetylated faba bean protein isolate
(AFBPI) on emulsion stability (ES) to centrifugation.
Formulations 1 to 3; pH 3.8; 0.7 %vol/wt acetic acid.
□ , preheated, frozen; O , preheated, non-frozen
X , non-heated,frozen; ● , non-heated, non-frozen.

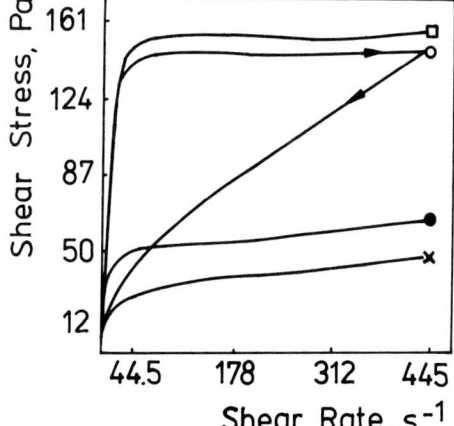

Figure 4. Flow curves of emulsions with AFBPI depending on
thermal treatment and freezing.
Formulation 3; pH 4.5; pH adjusted with 3M HCl.
□ , preheated, frozen; O , preheated, non-frozen
X , non-heated,frozen; ● , non-heated,non-frozen.

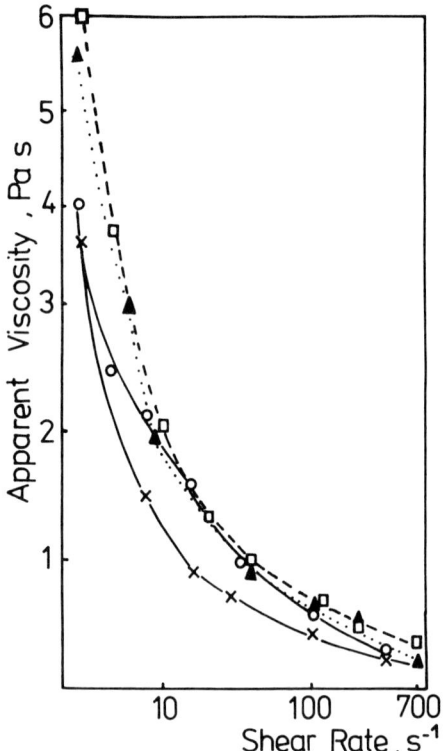

Figure 5.   Flow curve of preheated emulsions with 2 %wt/wt
AFBPI depending on hydrocolloid additives.
Formulations 4 to 7; pH 3.8; 0.7 %vol/wt acetic acid.
▲ , 4 % starch syrup; □ , 1.5 % potato starch
○ , 1.5 % gelatin; ✗ , without additive.

was increased and with the further decreased pH,the shear stress
optimum was decreased (Fig. 2, Fig. 4, Tab. II). Shear stress
optimum means here the beginning of a breakdown of the internal
structure which was found to be maximal at $\dot{\gamma}$ = 4.45 s^{-1}. The
dependence of the shear stress optimum on the shear rate is a
result of the shear induced particle interaction of the
undestroyed structure in the low shear region. This interaction
may be caused by chemical interaction and entanglement of

polymer chains. After measuring the flow curve (as a preshearing,(cycle 18 min) much of the internal structure of the emulsions is reduced,and the value of shear stress optimum at $\dot{\gamma} = 4.45$ s^{-1} is reduced (Fig. 2).
Here unpublished results show that depending on pH, the linkages restore differently at storage.
At pI, the adsorbed AFBPI molecules at the O/W interface are in a more compact configuration than at higher pH values. These conditions influence here the rheological behaviour of emulsions also (increased apparent viscosity at lower pH range, Fig. 1).

Preheated and frozen emulsions are more stable to centrifugation compared with non-heated and frozen emulsions (Fig. 1). This means in our case less centrifuged water.   Aoki (4) found the same decrease of ES (emulsions were not centrifuged) with a reduction   of pH for acetylated soybean protein.
Increasing AFBPI concentration of emulsions from 0.5 to 1.0 % enhances ES. Between 1.0 and 2.0 % AFBPI is not as big a difference in ES (Fig. 3). These emulsions did not show an oil volume after centrifugation.

Table III. INFLUENCE OF DIFFERENT HYDROCOLLOIDS ON STABILITY OF EMULSIONS WITH 2 % AFBPI

Hydrocolloid	%	Emulsion stability to centrifugation, %			
		non-frozen		frozen	
		non-heated	heated	non-heated	heated
Control		81	81	79	77
Gelatin	0.5	78	75	74	79
"	1.0	89	86	84	87
"	2.0	98	100	99	99
Potato	1.0	82	96	70	86
starch	1.5	86	99	77	98
"	3.0	91	100	79	100
Starch	2.0	85	82	79	84
syrup	4.0	86	92	92	94
"	6.0	84	91	82	93
"	8.0	95	96	94	95
"	10.0	96	96	92	98

pH of emulsions 3.8; 0.7 %vol/wt acetic acid; emulsions centrifugated at 1000 x g, 10 min.

Acidified emulsions stabilized by 7S soybean protein showed a continuous increase in mean drop size during storage, due to coalescence (5). Similar investigations on the influence of AFBPI on storage stability and drop coalescence are now in progress.

Increased shear stress optimum at pI (Fig. 2) could be
indicative of enhanced emulsions stability in low stress
situations e.g. transportation. Under centrifugation
conditions used no correlation was observed between shear stress
optimum and ES (Fig. 2). Further investigations are necessary
to find the optimal rheological behaviour for high emulsion
stability.

Addition of other hydrocolloids increases ES (or water
binding) of acidified emulsions. The volume of centrifuged
water is reduced. Potato starch shows the highest effect on
ES (Tab. III) and on viscosity at low shear range (Fig. 5).

CONCLUSIONS

AFBPI improves freeze-thaw-stability of acidified O/W
emulsions. After freezing emulsions are creamy and do not
break up.

Flow behaviour of AFBPI containing emulsions depends
on pH.

Centrifuged emulsions show a decrease in water binding
on lowering pH.

Preheating and freezing improves water binding of
acidified emulsions.

Water binding of acidified emulsions can be improved by
adding other hydrocolloids, e.g. potato starch, starch syrup
or gelatin.

ACKNOWLEDGEMENTS

The authors are indebted to Prof. P. Sherman, King's
College London, for critical hints on this subject and to
Mrs K. Ackermann, Mrs G. Aust and Mrs I. Masch for their
skillful assistance.

REFERENCES

1. Muschiolik, G., Schmidt, G., Andersson, O., Schneider, Ch.,
   and Schmandke, H. (1986) in Gums and Stabilisers for the
   Food Industry 3, (ed. Phillips, G.O., Wedlock, D. J. and
   Williams, P. A.) pp. 419-428. Elsevier Applied Science
   Publishers, London and New York.
2. Muschiolik, G., Dickinson, E., Murray, B. S. and
   Stainsby, G. (1987) Food Hydrocolloids, 1, 191.
3. Schneider, Ch., Schultz, M. and Schmandke, H. (1985)
   Nahrung, 29, 785-791.
4. Aoki, H. (1981) New Food Industry, 23, 57-71.
5. Reeve, M. and Sherman, P. (1986) Food Microstructure,
   5, 163-168.
6. Watanabe, M., Tsuji, R. F., Hirao, N. and Arai, S. (1985)
   Agric. Biol. Chem., 49, 3291-3299.
7. Berger, K. G. (1976) in Food Emulsions, (ed. Friberg, S.)
   pp. 141-210. Marcel Dekker, New York, Basel.

Applications of isolated soy proteins as functional ingredients in the food industry

D.Welsby

Purina Protein Europe, Excelsiorlaan 13, B-1930 Zaventem, Belgium

ABSTRACT

Soy products are well known and have been widely used in the Food Industry for a number of years, in a variety of different forms, and for a wide range of applications. Isolated Soy Protein is the purest form of soy protein commercially available, and as such, exhibits the widest range of useful functional properties. Where an understanding of the physical properties needed of the isolated soy protein in a particular system exists, these features can be engineered into the protein during its manufacture in order that it may perform optimally. Among those properties whose control is well understood are solubility, gelation properties, and emulsifying capability.

INTRODUCTION

The soybean is one of mankind's oldest crops, first cultivated probably in Northern China or Mongolia perhaps as many as 3000 years ago (Shurtleff and Aoyagi (1)). Although its modern cultivation is principally as a source of excellent edible oil, it first found widespread use in the Far East as a highly productive food crop in its own right, providing large populations with a significant proportion of their protein and energy requirements.

The soybean plant itself, Glycine max, is a member of the family leguminosae, a family which also includes peas, beans, lupins and peanuts. The beans themselves are produced in pods, each containing two or three seeds. Each seed has an approximate composition as shown in Table 2, and as can be seen in Table 1, soybean protein is one of the world's most important protein resources.

COMMERCIAL SOYBEAN PRODUCTS

Soybeans are commercially processed into crude oil and defatted fractions. Cleaned, dehulled soybeans are crushed between smooth rolls to facilitate oil extraction, and the defatted, desolventised flakes are then available for soy protein production. Three general categories of food ingredients are produced from defatted flakes - soy flours and grits, soy concentrates and isolated soy protein. Typical compositions are given in Table 2.

Table 1 - WORLD PLANT PROTEIN SUPPLY - SOME IMPORTANT SOURCES

Crop	Production (1) 10^6 tonnes	Protein Content %	Protein Product 10^6 tonnes
Cereals			
Wheat	458	12	55
Rice	414	8	33
Maize	451	9	41
Barley	158	12	19
Oilseeds			
Soybeans	88	42	37
Ground Nuts	19	26	4.9

(1) 1981 FAO Production Yearbook.

ISOLATED SOY PROTEIN

Isolated soy protein can be defined as the major proteinaceous fraction of the soybean, prepared by removing most of the non-protein components, and having a protein content not less than 90% (Nx6.25) on a moisture-free basis (Waggle and Kolar (2)). Figure 1 illustrates the processing procedure for isolated soy protein, which takes advantage of the fact that the major proteinaceous components have very low solubility at or near their isoelectric point (Figure 2), and are readily precipitated, washed, further processed and dried.

Table 2 - TYPICAL COMPOSITION OF SOYBEANS AND SOYBEAN PRODUCTS (1)

	Protein (Nx6.25) %	Oil %	Total Carbo- hydrates(2) %	Ash %	Crude Fibre %
Whole Soybean	42	20	35	5.0	5.5
Defatted Soy Flour	54	1.0	38	6.0	3.5
Soy Protein Concentrate	70	1.0	24	5.0	3.5
Isolated Soy Protein	92	0.5	2.5	4.5	0.5

(1) Moisture-free basis
(2) Includes crude fibre

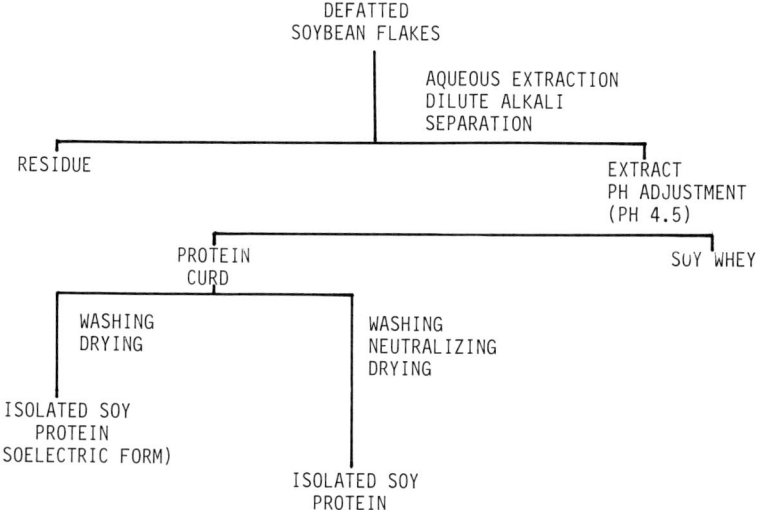

Figure 1. Production of Isolated Soy Protein from Defatted Flakes.

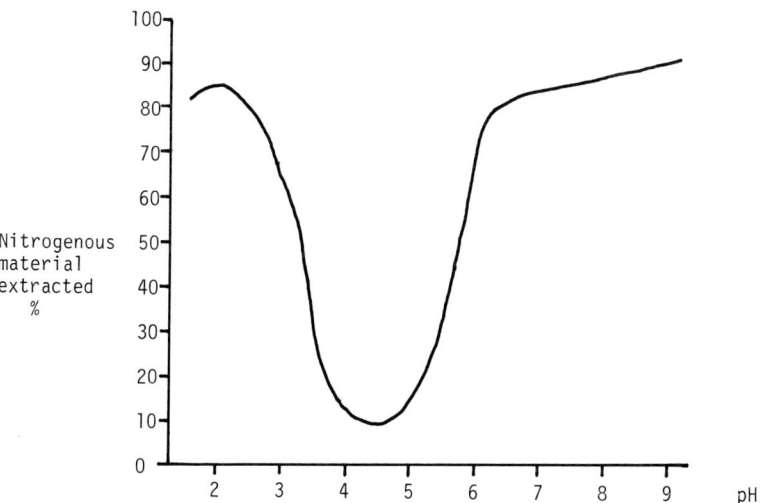

Figure 2. Effect of pH on the extractability of soy proteins.

NATURE OF SOYBEAN PROTEINS

The bulk of the proteins in soybean seeds is found in the protein bodies of the cotyledons and it is composed almost exclusively of the two fractions characterised by their sedimentation coefficients, the 7S and 11S proteins (Wolf (3)). The remainder exists as 2S and 15S fractions. Isoelectric precipitation separates most of the 7S and 11S fractions, together with small amounts of the 15S fraction. 7S fractions of soybean proteins are glycoproteins and are extremely complicated, consisting of several subunits, which associate and dissociate under different conditions of pH, ionic strength, temperature treatments etc... (Thanh, et al. (4)). Similarly, 11S particles are extraordinarily complex, consisting as illustrated in Fig. 3, of two hexagonal rings of six subunits each stacked one on top of the other, rather like two doughnuts (Badley et al (5)). As with the 7S particle, the 11S subunits can remain associated, or can dissociate according to conditions.

Although the quaternary structures of these two main storage protein forms are reasonably well understood, information concerning the more basic levels of organisation and their interactions is not available at a molecular level. However, over the past twenty years, a great deal of understanding has been reached on how to change conditions during the final stages of manufacturing in order to engineer into isolated soy protein products the necessary characteristics to perform optimally in particular food industry applications.

When we are dealing with molecular structures of such size and complexity, clearly the number of possibilities to specify particular properties is very large, and it is therefore more proper to consider isolated soy protein not as one single product, but rather as a large and continuously expanding "family" of products, as new applications are demanded.

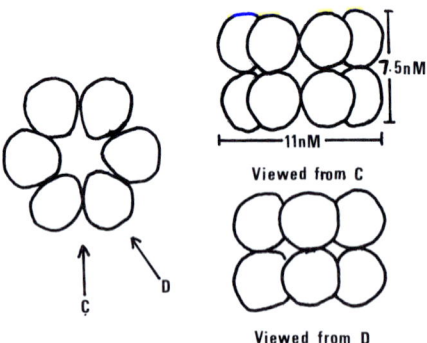

Figure 3. Model of subunit arrangement in soybean 11S protein (from Badley et al (5)).

NUTRITIONAL PROPERTIES OF ISOLATED SOY PROTEIN

Since the basic material for all the isolated soy proteins in the "family" of products is the same acid precipitated, washed and neutralised protein fractions, their amino acid composition remains constant, and is given in Table 3.

The recent FAO/WHO report on human protein requirements (WHO (6)) following many years of intensive research on the subject (Ralston Purina Publications, (7)) indicates that isolated soy proteins when properly prepared and processed, have the same high protein quality as animal proteins long recognised as of high nutritional value, such as proteins from meat, milk or eggs.

Table 3 - AMINO ACID COMPOSITION OF ISOLATED SOY PROTEIN

Amino Acid	g AA/ 100 g Product	g AA/ 100 g Protein
Alanine	3.8	4.3
Arginine	6.7	7.6
Aspartic Acid	10.2	11.6
Cysteine	1.1	1.3
Glutamic Acid	16.8	19.1
Glycine	3.7	4.2
* Histidine	2.3	2.6
* Isoleucine	4.3	4.9
* Leucine	7.2	8.2
* Lysine	5.5	6.3
* Methionine	1.2	1.3
* Phenylalanine	4.6	5.2
Proline	4.6	5.1
Serine	4.6	5.2
* Threonine	3.3	3.8
* Tryptophan	1.2	1.4
Tyrosine	3.3	3.8
* Valine	4.4	5.0
* Total Sulphur A.A.	2.3	2.6
* Total Aromatic A.A.	7.9	9.0

* = Essential Amino Acid

FUNCTIONAL PROPERTIES OF ISOLATED SOY PROTEIN IN FOOD SYSTEMS

Different foods are recognised for what they are by their consumers because of a whole range of uniquely combined physical, chemical, textural, and sensorial properties. The physical and textural properties of many meat, poultry, fish, dairy, egg and bakery products are generally related to the types and levels of proteins contained in them. Indeed, it may be said that in many such products, the proteins are the characterising components, and in acting in this way, they are manifesting one or several of the functional properties of proteins, lists of which are to be found in almost every review on the subject, like for example, that by Kinsella in 1976 (8).

As well as being found in foods fulfilling this rôle of characterising the
product, proteins from various sources are used in the modern food industry
as technical ingredients, where again one or more of the functional
properties listed by Kinsella may be important.

Dispersibility

Very few, if any, protein containing foods are consumed dry.   A food
processor, therefore, when faced with dry protein ingredients, is faced
with the task of getting that protein into an aqueous system, and the first
step is the product's dispersal in water. With modern, high energy mixing
equipment, this seldom presents any problem, and where it does, products
treated by using common powder instantisation techniques resolve it without
difficulty.

Solubility

Once dispersed in water, the nature of the protein-water and
protein-protein interactions is of critical importance in determining how
the protein will function in a food system. Proper hydration of the protein
is of vital importance, and the minimum water to protein ratio for each
product has to be respected. Care must be taken to control water
availability in formulations, and to ensure that the critical hydration
step occurs in the absence of certain salt ions.   Practical experience
indicates that solubility is readily controllable by relatively minor
process adjustments to produce NSI values   (Nitrogen Solubility Index)
ranging from just over zero to over 90 (Fig. 4) depending on the cation
used for neutralisation, and other processing factors. At the same time,
even insoluble isolated soy proteins may affect water absorption, for
example in doughs to produce protein rich baked goods.

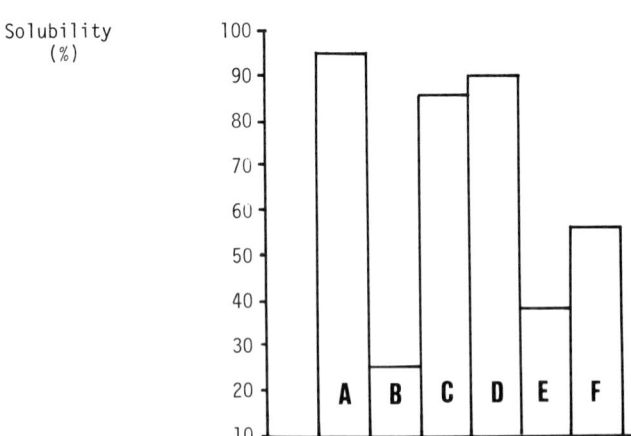

Solubility
  (%)

Isolated soy protein type

Figure 4. Solubilities, for some isolated soy proteins.

Viscosity

By choosing different processing conditions, the extent to which interactions between the various molecular species in solution occur can be controlled, and different isolated soy proteins can exhibit markedly different viscosity characteristics in solution, thus directing them at particular applications (Fig. 5).

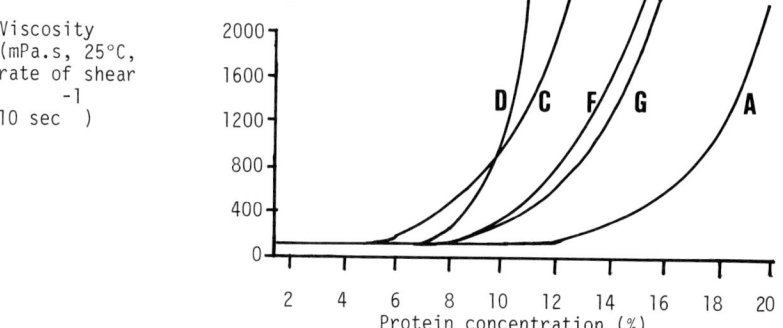

Viscosity
(mPa.s, 25°C,
rate of shear
10 sec^{-1})

Protein concentration (%)

Figure 5. Viscosity/concentration relationships for different isolated soy protein products.

For example, low viscosity isolated soy protein would be important in special beverages which have to remain liquid even at the high protein concentrations demanded for nutritional reasons. Protein gels are required in many applications in the meat industry in order to control final product resilience and bite.

Emulsifying capacity

Process control plays an important part in determining the emulsifying capacity of isolated soy proteins, since the structural stability of the protein units in solution determines how easily the molecule can be denatured in the oil-water interface, orientating its hydrophobic groups into the oil and its hydrophilic ones into the water. Emulsion capacity is therefore engineered into products whose principal applications render this a critical factor (Fig. 6).

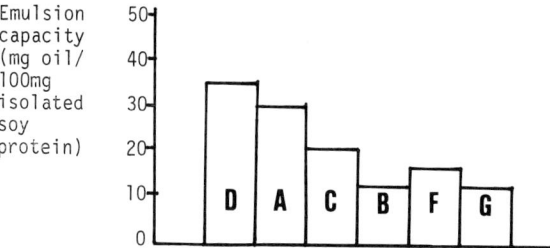

Emulsion
capacity
(mg oil/
100mg
isolated
soy
protein)

Figure 6. Emulsion capacity for different isolated soy proteins

Table 4 - APPLICATIONS OF ISOLATED SOY PROTEINS

Application	Properties	Products
Instant, blended products	Dispersibility, solubility, mineral compatibility, bland flavour	PP660 PP530
Liquid, ready-to-serve drinks - UHT drinks - Hospital feeding	Solubility, heat stability, emulsification, bland flavour, mineral composition	PP760 PP1500 PP1611
Wet-blended, dried products - infant formula - adult nutritionals	Solubility, low viscosity, low allergenicity, nutritional availability, bland flavour	PP1751
Meat products	Gel formation, emulsification, texture, water binding, dispersibility, compatibility with meat proteins	PP500E PP530 PP660
Frozen desserts	Freeze-thaw stability, foam stabilisation, solubility, bland flavour	PP760 PP710
Fermented products	Solubility, smooth gelation, bland flavour	PP760
Baked products	Water absorption (high or low)	PP810 PP860

CONCLUSIONS

Table 4 lists some common applications of isolated soy protein and some of the properties required for their optimal performance in these applications. Through the judicious evaluation of the essential properties required for particular applications, new products are regularly being developed as food manufacturers seek to exploit the interesting marketing and technical benefits brought by these versatile food ingredients.

REFERENCES

1.   Shurtleff, W. and Aoyagi, A. (1983) The book of Tofu, 3rd edn., 335 pages, Ten Speed Press, Berkeley, U.S.A.
2.   Waggle, D.H. and Kolar, C.W. (1979) in Soy Protein and Human Nutrition (eds. Wilcke, H.L., Hopkins, D.T. and Waggle, D.H.) pp 19-51. Academic Press, London, U.K.
3.   Wolf, W.J. (1970) J. Am. Oil Chem. Soc., 47, 107-108.
4.   Thanh, V.H., Okubo, K., and Shibasaki, K. (1975) Plant Physiol. 56, 19-22.
5.   Badley, R.A., Atkinson, D., Hauser, H., Oldani, D., Green, J.P., and Stubbs, J.M. (1975) Biochim. Biophys. Acta 412, 214-228.
6.   World Health Organisation Technical Report Series 724 (1985) Energy and Protein Requirements. Report of a joint FAO/WHO/UNU consultation.
7.   Ralston Purina Company (1986) Nutritional Aspects of Ralston Purina Isolated Soy Protein. Clinical Studies Summary. St Louis, U.S.A.
8.   Kinsella, J.E. (1976) Critical Reviews in Food Science and Nutrition, April, 219-280.

Gums as stabilisers for emulsifier covered emulsion droplets

Björn Bergenståhl

Institute for Surface Chemistry, Box 5607, S-114 86 Stockholm, Sweden

ABSTRACT

The ability of a number of industrial gums to act as steric stabilisers of emulsions, in the presence of different low molecular weight emulsifiers, has been compared. The flocculation rate of dilute oil in water emulsions has been measured by using a turbidimetric method. It is found that several gums show pronounced effects on the stability already at the ppm level of concentration. To explain these results, we propose that the gums adsorb onto the surfactant layer, forming a combined structure of a primary surfactant layer covered by an adsorbed polymer layer.

INTRODUCTION

The use of polymer thickeners or stabilisers together with regular low molecular emulsifiers is very common in technical emulsion systems. Examples are mayonnaise emulsions where gums interact with emulsion droplets covered by egg lecithin as well as protein- lipoprotein mixtures or ice cream emulsions where added gum stabilisers interact with mono/diglyceride surfaces and milk proteins. Despite the common use of gums as stabilisers in emulsion systems there are very few investigations that have been aimed at clarifying the basic mechanisms by which gums stabilise emulsions. Most textbooks suggest that the stabilisation mainly is due to the increased viscosity (1-3) and that steric stabilisation due to an adsorbed layer at the droplet surface (in the gum literature usually termed "protective colloid properties") only applies to a few cases such as gum arabic (2).

In our earlier investigation we studied the stabilising effect of gums on a latex dispersion (4). It was found that most gums acted as steric stabilisers. It was also clear that this affinity of the gums for the latex particle surface, and hence also their stabilising capacity, was very strongly dependent on the type and origin of the gum.

However, latex surfaces are quite different compared to the oil/water interface in true food emulsions. We have therefore

363

extended the previous study on the model latex to a true emulsion
system. This extension has also allowed us to study the influence
of low molecular weight emulsifiers on the adsorption of gums at
the oil/water interface and how the adsorption depends on the type
of gum.

MATERIALS

The origin and a very brief description of the different gums
and emulsifiers used in this investigation is given in table 1.
All polymers were used without further purification. The polymers
were dissolved in cold water and left stirring over night.
All water was doubly distilled. The oil phase was a soy bean oil
of technical grade (Karlshamns AB, Sweden). All other chemicals
were of analytical grade.

Table 1   Origin of gums and emulsifiers used.

Gum/Emulsifier.	Trade name	Description	Manufacturer
Pectin	XSS 100	DM 58-62	Grindsted
	RS 400	DM 71-75	Products
Dextran	T-500	MW 500 000	Pharmacia
Xanthan gum	Keltrol T		Kelco
Methylcellulose	MB 3000 P		Carl Roth
Gum Tragacanth	G1		Gumix Handels
			GmbH Hamburg
Ethylhydroxyethylcellulose	E 320 G		Berol
	E 150 G		Kemi
Soybean lecithin	Epikuron 200	98 % PC	Lucas Meyer
Monoolein		98 % pure	Grindsted
			Products
Sorbitan monooleate	Span 80		Kao-Atlas
Saccharose ester	F-160		DKS
			international

METHODS

The flocculation rates were estimated from turbidimetric
measurements. The theory of light scattering from spherical
particles predicts that when the particle size increases the
turbidity increases, as long as the particles are small enough in
comparison with the wavelength of the scattered light (5). Droplets of
vegetable oil dispersed in water should be smaller than 1.5 μ.

The increase in turbidity will also depend on the mechanism
associated with the removal of stability itself. The turbidity
increases roughly as the square of the particle volume and
linearly with the particle   number, so that the change in
turbidity will depend on whether the particles coalesce or not (6).
The theory has mainly been applied to monodisperse systems. Egusa
(7) has checked its applicability on polydisperse systems empirically.

He measured the turbidity increase in bimodal systems and found that it was similar to the increase in the corresponding monodisperse system with the same weight average diameter. Melik and Fogler (9) have also adopted the turbidimetric method to estimate flocculation rates in bidisperse systems.

Here the initial linear turbidity increase will be used as a direct measure of the instability, and all graphs show the turbidity increase normalized against the turbidity increase with no gum added
($[d\tau/dt]_{t=0\,added\,gum}$ /$[d\tau/dt]_{t=0\,no\,gum\,added}$)

Highly dispersed emulsions suitable for the turbidometric measurement, were obtained by using a special high pressure homogenisation technique (800-1000 bar pressure drop), termed "microfluidization" (8) (Microfluidics inc., Boston, Mass. USA), which yields droplets with average sizes around 0.3 to 0.8 μ.

The emulsion samples were prepared with the emulsifier dissolved in the oil phase, usually about 3% of the oil phase to allow for a complete coverage of the oil droplets. The stock emulsions were diluted 100-200 times to give a low particle concentration and a final turbidity of approximately 0.3 absorbance units. This reduces the multiple scattering, increases the sensitivity of the spectrophotometer and gives the process a suitable flocculation rate. The diluted emuision was mixed with the polymer solution and left to equilibrate for about 4 hours. Then flocculation was induced by the addition of $MgCl_2$ which eliminates the electrostatic contribution to stabilisation.

RESULTS.

The flocculation rate as a function of added gum, with sodium octanoate as emulsifier, is given in figure 1.

For several polymers (such as methylcellulose, ethylhydroxy-ethylcellulose, Gum tragacanth, pectin etc) the flocculation rate curves decline sharply already at the ppm level of added gum. At these very low concentrations any effect of change in solvent viscosity on the rate can be safely neglected and the obvious explanation for the decreasing rate is steric hindrance due to adsorption of the polymers onto the droplet surface. However, some polymers (dextran and starch) influence the flocculation rate only at much higher concentrations. A similar observation was made in our earlier experiments on polystyrene latex particles (4). We estimated how much the viscosity increase reduced the diffusion rate of the particles in solution of very weakly adsorbed polysaccharides. It was quite clear that in these cases a reduced diffusion rate, or the increased viscosity, only explains a minor part of the observed stability increase. Hence, it seems to be doubtful if there are any polysaccharides at all that mainly act by increasing the viscosity.

B.Bergenstahl

The *flocculation rate of emulsion* droplets covered with:

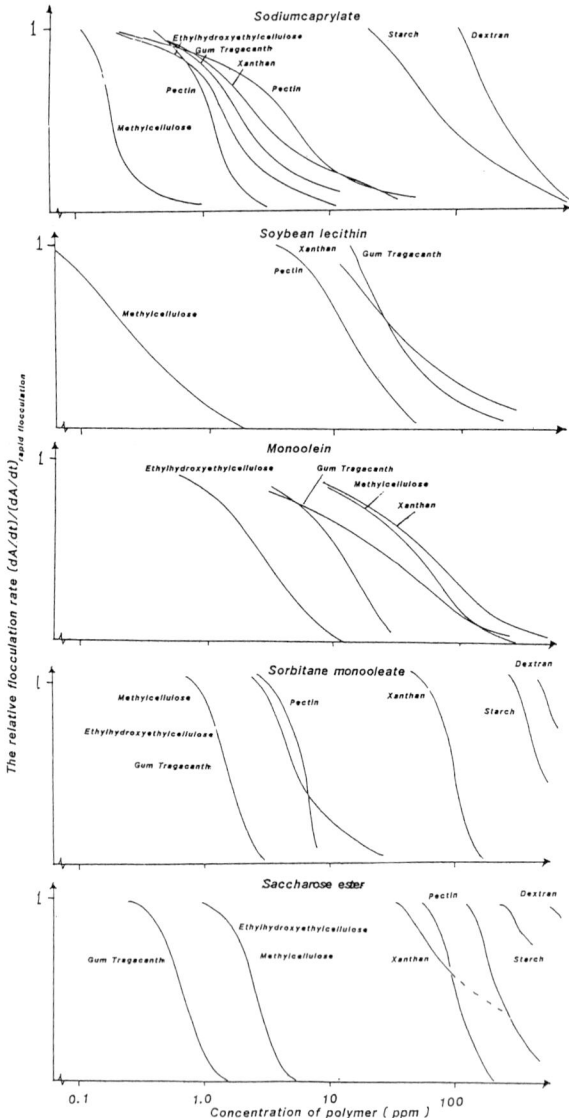

Figure 1. The flocculation rate as a function of the concentration of the added gum. a/ sodium octanoate b/ soybean lecithin c/ monoolein d/ sorbitanmonooleate and e/ a saccharose ester .

The experiments were repeated with several different emulsifiers: soybean lecithin (pure phosphatidyl choline), monoolein, sorbitan oleate and a saccharose ester. The results are shown in figure 1. Basically the results are similar to those obtained for sodium octanoate. Most gums show a pronounced ability to stabilize the emulsions already at the ppm level. Hence, it is obvious that most gums are able to act as steric stabilisers due to the adsorption of the gums at the oil/water interface even in emulsions including surfactants.

The critical concentration necessary to achieve the stability depends strongly on how the type of gum and the type of emulsifier are combined.

An anionic emulsifier tends to weaken the adsorption of anionic polymers. Emulsifiers that form surfaces densely covered with hydroxy groups (the monoolein surface) tend to reduce the adsorption of most gums. The surface formed by the disaccharide ester tends to promote the adsorption of Gum tragacanth.

DISCUSSION

It can be conceived that the adsorption of hydrocolloids onto a surface covered by low molecular surfactants could result in three different type of layers:

i/ The polymer might replace the emulsifier in the oil water interface (competitive adsorption (10,11))

ii/ The polymer might adsorb outside the intact emulsifier layer or

iii/ a mixed new layer might be formed through the penetration of a second component into the primary layer.

These surfactants are all much more surface active, than the gums, and reduce the interfacial tension between water and oil to around 1-10 mN/m (originally 25 mN/m, table 2). It can therefore, be directly concluded that it is impossible for the much less surface active gum to squeeze the emulsifier out of the oil water interface (10). We therefore propose that the gums are adsorbed on the top of the emulsifier layer and a combined emulsifier polymer structure, as schematically presented in figure 2, is developed. Some penetration of the polymer through the surfactant layer, as proposed for instance for β-casein interacting with a lecithin surface (12), cannot be excluded, but the hydrophilicity and the weak surface activity of   most of the polysaccharides suggests that it should be of minor importance. The observed interaction effects corroborate this model. Thus, for instance, a negatively charged surface reduces the adsorption of a negatively charged polymer and a hydrophilic surface reduces the adsorption of most polymers.

Table 2. The soybean oil/water interfacial tension (the pendant
        drop method) in the presence of surfactants or
        polysaccharides.

Substance	Interfacial tension (mN/m).
Pure water	25.2
1% Monoolein disolved in the oil	8.3
1% Soybean lecithin dissolved in the oil	1.2
0.2% Methylcellulose dissolved in the water.	13.6
0.2% Ethylhydroxyethyl cellulose dissolved in the water.	16.4
0.2% Pectin (DM 71-75) dissolved in the water.	24.5

   The experiments described were all made on quite dilute
systems. The stabilising concentrations of the polymers therefore
cannot be directly applied to technical dispersions. However, the
relative effectivness of the different polymers as   stabilisers in
dilute systems gives indications of the capabilities of the
different gums to function even in concentrated systems.

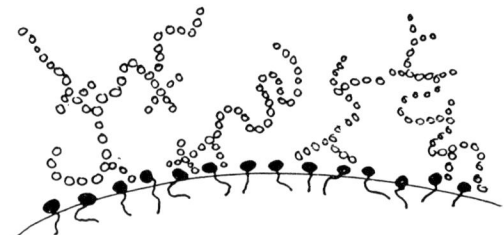

Figure 2.   The proposed structure at the emulsion droplet surface
when the emulsifier and the gum is present.

   The structure proposed in figure 2 should exist even in
concentrated systems as it is caused by direct interactions
between the surfactant covered surfaces and the dissolved
polymers.

   The adsorption of a hydrocolloid onto the droplet surface
will completely change the surface properties. It is well known
from different measurements of the forces between surfaces that
hydrophilic emulsifiers, such as lecithin, are able
to create strong shortrange   repulsive surface forces (13-17). Such
forces will together with the electrostatic repulsion contribute
to the repulsive particle-particle interactions that stabilize
emulsions, or the repulsive surface-surface interaction that
stabilize the thin liquid films in concentrated emulsions. In
non-ionic and zwitterionic systems these forces are generally
rather short range. In experiments on a number of pure

phospholipids Lis and coworkers (17) found that the range of the force usually is limited to about 30 Å. This distance is close to theoretical estimations of the minimum film thickness criteria for stability in emulsions (18,19). In agreement with this it is also observed that.most low molecular surfactants are fairly weak stabilisers in food emulsions (18).

If we accept the structure in figure 2 , we can get an idea of the change that will be created by the adsorbed polymer layer. A much more long range steric repulsion will be added to the short range repulsion due to hydration. The upper limit of the total surface repulsion will expand from 30 to about 200 Å. This film thickness will be large compared to the critical values and will therefore increase the stability.

In experiments under progress we have been able to show even the formation of an emulsifier-protein-gum surface, with properties (flocculation rates, surface potential etc) that are determined mainly by the outermost layer. We hope that this knowledge will give a new light to the understanding of many very complex food emulsions.

REFERENCES
1. Sanderson G. R., Food Technology 50 (1981)
2. Glicksman, M. "Gum technology in the food industry", Academic Press New York (1969)
3. Sandford, P. A. and Baird J. in "The polysaccharides " Vol 2 Ed. Aspinall G. O. Academic Press 1983
4. Bergenståhl, B., Fogler S. and Stenius P. in Phillips G. O., Wedlock D. J. and Williams P. A. (eds) "Gums and Stabilisers for the food industry" Elsevier Applied Science, London. Vol 3 p 286
5. Heller, W., McCarty, H. J., J. Chem. Phys., 29, 78 (1958).
6. Lichtenbelt, J. W. T., Ros, J. M. C. and Wiersema P. H., J. Colloid Interface Sci., 46, 522 (1974)
7. Egusa S., J. Colloid Interface Sci., 86 ,135 (1982)
8. Cook, E. J. and Lagace, A. P., US Pat. 4533254 (1985)
9. Melik, D. H. and Fogler, H. S., J. Coll. Interface Sci., 108, 503 (1985)
10. Kronberg, B. ,Kuortti, J. and Stenius, P., Colloid and Surfaces, 18, 411 (1986)
11. Dickinson,E. Food Hydrocolloids, 1, 3 (1986)
12. Phillips, M. C., Evans, M. T. A. and Mauser, H. ACS Adv Chem Ser 144,217 (1978)
13. Le Neveu, D.M., Rand, R.P., Parsegian, V.A. and Ginzell, D., Biophys. J. 18, 209 (1977).
14. Parsegian, V.A., Fuller, N. and Rand, R.P., Proc. Natl. Acad. Sci. USA, 76,2750 (1979).
15. Marra, J. and Israelachvili, J., Biochemistry 24, 4600 (1985).
16. Marra, J., J. Colloid Interface Sci. 109, 11 (1986).
17. Lis, L.J., McAlister, M., Fuller, N., Rand, PR.P. and Parsegian, V.A., Biophys. J. 37, 657 (1982).
18. Walstra, P in McLoughlin J. V. and McKenna, B. M. (eds) "Research in food science and nutrition" Vol 5 Boole Press Dublin (1984).
19. Ivanov, I. B., Jain, R. K. and Somasundaran, P. in Mittal K. L. (ed) "Solution chemistry of surfactants" Vol 2 Plenum, New York (1979)

Part 5

CURRENT DEVELOPMENTS

Food hydrocolloids in Japan

K.Nishinari

National Food Research Institute, Tsukuba 305, Japan

ABSTRACT
   Present aspects of Japanese dietary pattern and of research on
food hydrocolloids in Japan are described. Physico-chemical
studies on some examples of food hydrocolloids used in the food
industry are described, in the domain of soft drinks, fish paste,
polysaccharide gels, soybean proteins and some other colloids.

I. INTRODUCTION
Longest Life Expectancy in the World
   According to statistics, life expectancy is the longest in
Japan. The reason is attributed to dietary pattern. What are
the characteristics in the Japanese diet? Many kinds of foods,
but the quantity of each food is very small. Although Japanese
people are now changing to take much more animal protein than
before World War II, a considerable amount of it is still taken
from fish. One of the peculiar characteristics of Japanese
dishes is to eat fish or vegetables as they are, e.g. sasimi,
sliced raw fish. In addition to these, Japanese people take much
fibre from vegetables, and also take many seaweeds.
   There have been many traditional dishes which contain fibres
and seaweeds, and recently some new type foods which consist of
seaweeds or hydrocolloids appeared on the market. These are
considered not only delicious but also very healthy.

Management of Academic - Industry Interaction
   As is almost always the case with Japan such as in
electronics, automobile and camera industries, Japanese food
research concentrates on applications. Of course, food research
itself is an application of science, but basic research which
elucidates the functional characteristics of food hydrocolloids
must be quite important for further development.
   Biotechnology such as genetic engineering, cell fusion etc.
has attracted much attention recently, and in this situation many
petroleum or iron or machine companies are now making new
divisions concerning biotechnology. They are now trying to
approach food processing. Since the research on physico-chemical
properties and utilization of these substances are done
separately in many laboratories and factories in Japan, it must
be very useful to discuss the problems and exchange ideas in
order to understand the properties and develop the utilization.

Thus, symposia were organized in 1985 and 1986 in Tokyo (1,2). Two symposia entitled "Food and Gels", "Water Soluble Polymers in Food Systems" are being held also in Tokyo and Kyoto respectively in the autumn of 1987.

II. PHYSICO-CHEMICAL STUDY ON FOOD HYDROCOLLOIDS

Some examples of basic research and application of food hydrocolloids which are used frequently in Japanese food industry are described below.

Agar

Agar is extracted from red seaweeds, especially from Gelidium and Gracilaria. Commercially available agar is often a blend of several polysaccharides of different origins.

Agar is composed of two main polysaccharides agarose and agaropectin (3). Gel forming ability is mainly governed by agarose which has been shown to consist of D-galactose and 3,6-anhydro-L-galactose (4). A copolymer of D-galactose and 3,6-anhydro-L-galactose is, however, an idealised chemical structure for agarose. The content of 3,6-anhydro-L-galactose is less than 50% in real agarose molecules.

The alkaline pretreatment enhances remarkably the gel forming ability (5-7) and is an important procedure in manufacturing (8). Density of sulfate esters decreases and 3,6-anhydro-L-galactose content increases with increasing concentration of alkaline pretreatment. Gel forming ability increases discontinuously just above the 3,6-anhydro-L-galactose content 30% which corresponds to the concentration of alkaline pretreatment 3%. This is shown by relaxation spectra in Fig. 1 (5). Since the change in intrinsic viscosity and saccharide content was not so great, the possibility of degradation by alkaline pretreatment was ruled out, and it was confirmed that the desulfation and the increase

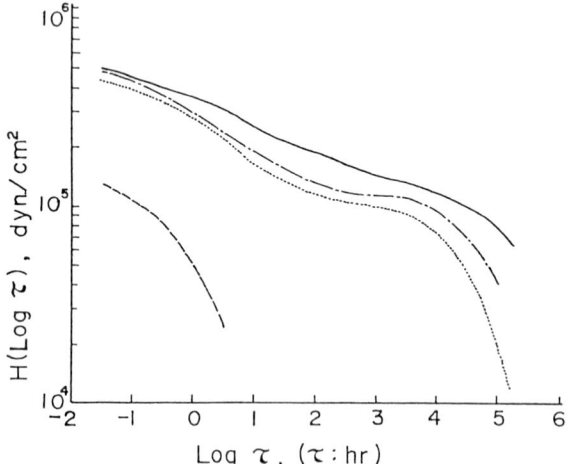

Fig. 1   Relaxation spectra of 3 wt% agar gels at 25°C,
         pretreated by 2% NaOH (-----), 3% NaOH (⋯⋯⋯⋯),
         7% NaOH (—·—·—) and 10% NaOH (———) (5).

of 3,6-anhydro-L-galactose must be the main factor in gel formation.

Like many synthetic polymer gels, the gel forming ability of agarose is influenced by its molecular weight. The rheological properties of gels prepared from four fractions of agarose of different molecular weight but with almost the same content of sulfate esters and 3,6-anhydro-L-galactose were examined (9). The square root of the elastic modulus of agarose gels divided by the concentration has been shown to increase linearly with the intrinsic viscosity of aqueous agarose solutions in the case of dilute gels less than 2.5% w/w, just as in the case of gelatin gels (10). Both the breaking stress and the breaking strain of the agarose gels increased with increasing molecular weight. It was suggested that only cross-links connected by long, flexible chains will remain at large deformations as in the case of carrageenan and alginate (11). The number of these long chains will increase with the molecular weight of the primary molecules.

It is empirically known by seaweed collectors that agar gels prepared from summer seaweeds are stronger than those prepared from spring seaweeds in Japan. The chemical analysis shows that polysaccharides extracted from summer seaweeds contain more sulfate groups and their intrinsic viscosity is higher than those extracted from spring seaweeds. These two factors are opposing from the view point of gel formation: the sulfate group increases the solubility and hence inhibits the gelation, while the higher intrinsic viscosities lead to larger values of elastic modulus as stated above. As for the different viscoelastic behaviours in agar gels prepared from spring and summer seaweeds, the intrinsic viscosity seems more influential than the sulfate group content (12). On longer time scale, the molecular entanglement plays a dominant role, and so the molecular weight (or more precisely the intrinsic viscosity of the dilute solution) is more important than the content of sulfate esters. On the other hand, the content of sulfate esters seems more important in the shorter time scale.

Using the similarity between the concentration dependence of the gelation temperature, helix forming temperature and melting temperature and the phase diagram for rod-like polymers proposed by Flory(13), Hayashi interpreted the swelling and related properties together with the retrogradation-like phenomenon of aqueous agarose solutions (14, 15).

The temperature dependence of the elastic modulus was reversed in agarose and kappa-carrageenan gels by the addition of glycerin or ethylene glycol (16). This was discussed on the basis of a model consisting of junction zones which are connected by Langevin chains (17).

Tokita and Hikichi (18) observed the dynamic elastic modulus of agarose gels near the gelation point. According to scaling concepts, the elastic modulus G in the immediate vicinity of the sol-gel transition is given by $G = (\phi - \phi_g)^t$ as a function of concentration at a constant temperature, where ϕ is the concentration of the gel and ϕ_g is the critical concentration i.e. the concentration beyond which gelation occurs. The elastic modulus G as a function of temperature at a constant concentration can be written by $G = (T - T_g)^t$, where T is the temperature and T_g is the sol-gel transition temperature.

Since the exponent t = 1.9 was found to take almost the same
value for both the concentration dependence and for the
temperature dependence, they supported the notion of
universality; the critical exponents are independent of the path
to the sol-gel phase boundary.

Recently, agarose specimens of very low molecular weight ($[\eta]$
= 1.7 in 0.01 mol/l NaSCN) were prepared, and gels of higher
concentrations up to 35% w/w were examined. The gel showed a
sharp X-ray diffraction peak at 13°, 19° and 26°. Low
temperature heating DSC thermograms showed an endothermic peak
which has been attributed to the melting of frozen bound water in
addition to an endothermic peak for free water (19).

Konjac mannan

Konjac mannan (KM) is composed mainly by β-1,4 linked mannose
and glucose, and is called glucomannan. There are some branches
linked by β-1,3 linkages, and it contains a very small quantity
of acetyl groups. Maekaji (20) observed the change of turbidity
and viscosity of KM sol, and monitored the IR spectra and the
alkaline quantity which was consumed in the gelation of KM. He
found that acidic moieties with C=O groups were eliminated and
that the gelation occurred through the formation of a network
structure supported by hydrogen bonding. He identified the
eliminated acidic moieties to be acetic acid, and the molar ratio
of this component to hexose residues in KM was estimated to be
1:19 if KM is represented by a molecular formula of
$(CH_3CO)_m (C_6H_{10}O_5)_n$ (21). The junction zones in KM gel were not
considered to consist of chemical cross-links because the gel is
solubilised under mild conditions with reagents such as
salicylate or urea which disrupt the hydrogen bonds. Gelation of
KM is promoted by heating in contrast to many other thermo-
reversible gels. The details of the gelation mechanism for KM
are not yet known.

Torigata et al. prepared nitro-KM and studied the molecular
weight and conformation in aqueous solution by light scattering
and viscometry (22). They obtained the root mean square end-to-
end distance as $<R^2>^{1/2}$ = 1380A, the weight average molecular
weight as M_w = 2.7 x 10^5, and the Mark-Houwink-Sakurada relation
between intrinsic viscosity and degree of polymerization as $[\eta]$ =
1.16 x 10^{-3} $P^{0.95}$. Then, they concluded that the nitro-KM molecule
is rather stiff in aqueous solution.

Sugiyama et al. (23) prepared the water soluble KM sample by
centrifugation, dialysis and freeze drying. They studied the
solution properties by light scattering, and obtained

$$M_w = 6.8 \times 10^5 \sim 1.9 \times 10^6, \quad <S^2>^{1/2} = 9.1 \times 10^2 \sim 2.3 \times 10^3 \text{Å}.$$

Both molecular weight and root mean square radius of gyration
$<S^2>^{1/2}$ were found to be quite dependent on strain and   place of
production.

Kishida et al. (24) prepared partially methylated KM samples,
and studied the solution properties by light scattering. Their
results were quite different from those of Sugiyama et al. (23);
molecular weight $M_w \sim 1 \times 10^6$, root mean square radius of
gyration $<S^2>^{1/2} \sim 1.2 \times 10^3$Å, intrinsic viscosity 19 (dl/g), were
not dependent on strain and   place of production. They obtained

$$<S^2>^{1/2} = 4.2 \times 10^{-1} \ M^{1.08} \ , \ [\eta] = 6.37 \times 10^{-4} \ M^{0.74}$$

and α_s (expansion coefficient) = 1.3, and concluded that KM takes on a random coil form in water. The expansion coefficient α_s was found to be about 1.3, and the methyl-KM molecule in water was considered to be in the form of stretched random coils.

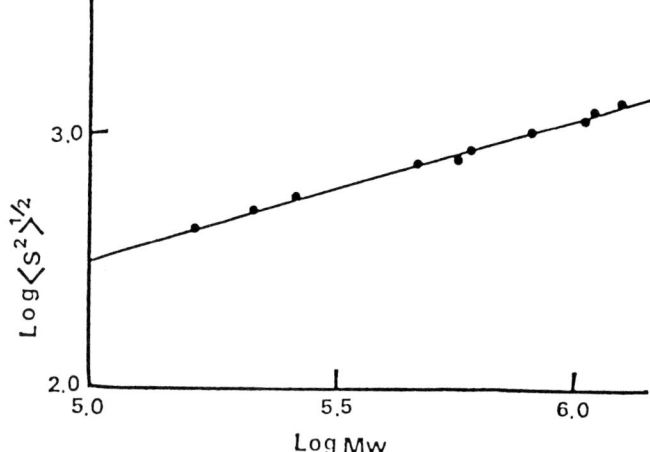

Fig. 2 Double logarithmic plot of $\langle S^2 \rangle^{1/2}$ vs. M_w for methyl-KGM solution (24).

Recently, Nishinari et al. (25) prepared fractionated KM samples, and carried out viscometric measurements. They plotted the zero shear specific viscosity against coil overlap parameter and found similar behaviour as observed for many polysaccharides by Morris et al. (26).

Xanthan
    Solution properties of xanthan were extensively examined by light scattering, sedimentation velocity and equilibrium measurement and viscometry by Sato, Norisue and Fujita (27-29). Comparing the dependence of the radius of gyration on the number of main chain glucose residues for xanthan in cadoxen with that for similar β-1,4-D-glucans, cellulose and the Na salt of carboxymethyl cellulose in 1:1 water-diluted cadoxen, they concluded that the Na salt of xanthan dissolves in cadoxen as single chains. They examined the ratio of the molecular weight of xanthan in different solvents, and found that M_w(in 0.1 M NaCl)/M_w(in cadoxen) is close to 2. Therefore, it is concluded that xanthan molecules exist as dimers in 0.1 M aqueous NaCl. The contour length per main chain glucose residue of the xanthan dimer was found to be 0.47 nm, which agreed fairly well with the pitch of the 5_1 double-stranded helix proposed for the crystalline structure of xanthan. Therefore, they concluded that the xanthan dimer dissolves in 0.1 M aqueous NaCl as 5_1 double-stranded helical molecules.
    From the molecular weight dependence of the root-mean square radius of gyration shown in Fig. 3, it was concluded that the

double helix of xanthan in 0.1 M aqueous NaCl is almost
completely rigid below and semiflexible above $M_w \simeq 3 \times 10^5$. On
the other hand, it was suggested that the single xanthan chain in
cadoxen suffers more excluded volume effect with increasing
molecular weight. Using Yamakawa-Fujii theory (30) for the
root-mean square radius of gyration and S of a long, straight
cylinder, Norisue et al. found that the molar mass per unit
contour length of the cylinder $M_L \simeq 1940$ nm^{-1} and the diameter of
the cylinder d = 2.7 nm.
     In order to estimate the rigidity of the double helix of
xanthan, the persistence length q of Kratky-Porod worm-like chain
(31) was determined as q = 120 nm by curve fitting using the

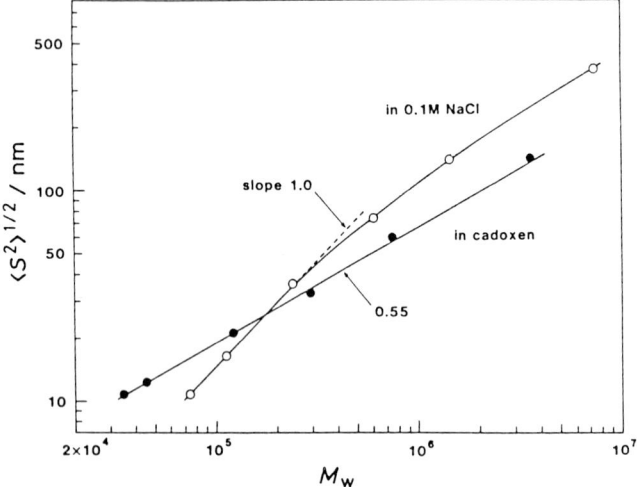

Fig. 3   Double-logarithmic plots of $\langle S^2 \rangle^{1/2}$ vs.M_w for Na
         salt xanthan in 0.1 M aqueous NaCl and cadoxene at
         25°C (27).

Benoit-Doty equation (32) for the root-mean square radius of
gyration. It was found that the rigidity of the xanthan double
helix in 0.1 M aqueous NaCl characterized by q = 120 nm, is
intermediate between that of double stranded DNA (q = 60 nm) (33)
and triple stranded collagen (q = 150 nm) (34) or schizophyllan
(q = 200 nm) (35).
     Viscometric studies also showed that these findings are
consistent; curve fitting of the observed intrinsic viscosity of
xanthan in 0.1 M aqueous NaCl as a function of molecular weight
using Yamakawa-Yoshizaki theory (36) for worm-like chains leads
to almost the same value for the pitch per residue of the 5
double-stranded helix. Double logarithmic plots of the intrinsic
viscosity against molecular weight for the Na salt of xanthan in
0.1 M aqueous NaCl and in cadoxen showed a similar tendency in
the double-logarithmic plot of the radius of gyration against
molecular weight.

Pullulan

Pullulan is produced by Aureobasidium pullulans, and it is a linear polymer in which maltotriose units are linked by α-1,6-glucosidic linkages (37).

Recently, Kawahara et al. (38) emphasized that pullulan can be used as a standard sample for water soluble polymers because pullulan can be easily dissolved and has no branching as in dextran. They set out to fractionate pullulan, and obtained a series with molecular weights in the range 5×10^3 to 8×10^5. They studied the solution properties by viscometry and ultracentrifugation, and obtained

$$[\eta] = (1.91 \pm 0.02) \times 10^{-2} \; M_w^{0.67 \pm 0.01} \qquad (1)$$

for molecular weights higher than 4.8×10^4. For lower molecular weights, the exponent decreased and was about 0.5 for the lowest molecular weight of 3×10^3. The relation between sedimentation coefficient and the molecular weight was also obtained as

$$s = 2.86 \times 10^{-15} \; M_w$$

The Mandelkern and Flory parameter β' (39)

$$\beta' = \frac{N_A \, s_0 \, [\eta]^{1/3} \, \eta_0}{100^{1/3} \, (1 - \bar{v}\rho) M^{2/3}}$$

was estimated as 2.32×10^6, which was regarded as reasonable for a random coil polymer in a good solvent (40).

Almost the same relation to eq. (1) was obtained by Kato et al. (41) using light scattering. They obtained the relation between the Z-average radius of gyration and the molecular weight

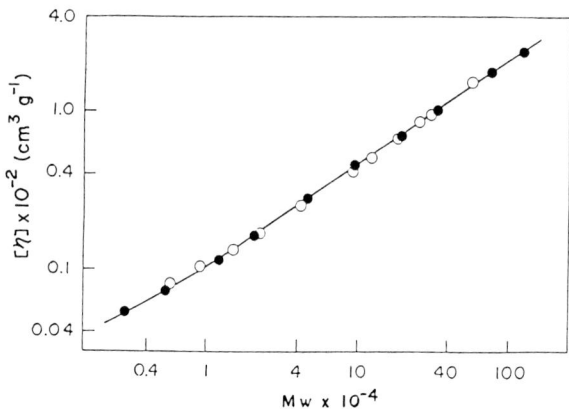

Fig. 4 Double logarithmic plot of $[\eta]$ vs. M_w for pullulan fractions (●) and for unfractionated samples (○) (38).

as

$$\langle S^2 \rangle^{1/2} = 1.47 \times 10^{-2} M_w^{0.58} \; (nm)$$

and also the molecular weight dependence of the second virial coefficient as

$$A_2 = 5.42 \times 10^{-3} \; M_w^{-0.26} \; (cm^3 mol g^{-2})$$

They confirmed that the solution behaviour of aqueous pullulan solutions is very similar to the solution behaviour of polystyrene in good solvents.

It was confirmed by light scattering that pullulan in aqueous solution is degraded quite fast at room temperature if sodium azide was not added (42).

Rheological and dielectric properties of pullulan films have been studied together with amylose and dextran films in order to clarify the relation between structure and functional properties (43-45). The mechanical and dielectric loss at 10 Hz as a function of temperature showed a peak at about -110°C, and it was attributed to the rotational motion of hydroxymethyl groups attached to the C_5 atom in glucose residues. The effect of water content on the molecular motion was also examined. The peak of the mechanical and dielectric loss at -110°C was masked by a steep rise at higher moisture levels, while it became more pronounced with decreasing moisture content at lower moisture levels. A small amount of water molecules play the role of stabiliser for hydrogen bonding at liquid nitrogen temperatures, while they become a plasticizer at higher temperatures.

Curdlan

Curdlan is produced by many strains of Agrobacterium and some strains of Rhizobium, and is composed of β-1,3 glucosidic linkages (46, 47).

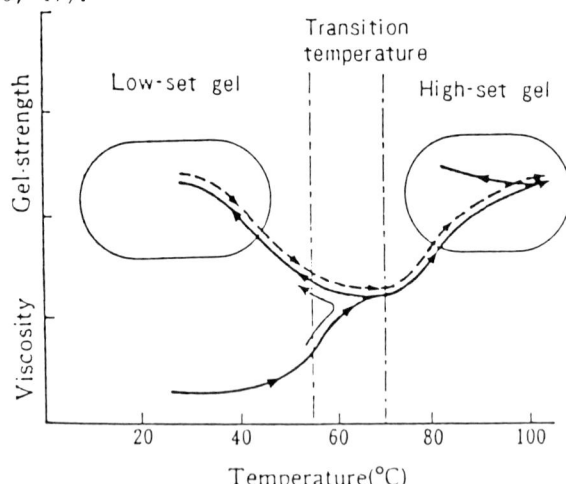

Fig. 5 Gel formation of curdlan. When the suspension is heated to 60°C and then cooled, the viscosity increases. If the suspension is left without stirring, gel is formed (low-set gel). When the low-set gel is heated to 60°C, it becomes soft, and this gel is thermo-reversible. If the suspension is heated above 80°C, the viscosity increases with increasig temperature and the gel is formed (high-set gel). This gel is thermo-irreversible.

Harada et al. examined the gel strength of curdlan as a function of heating temperature, heating time, pH and concentration (48). A gel was formed above 54°C. The gel strength increased remarkably with increasing the heating temperature, in the range from 54°C to 60°C, and from 77°C to 100°C, while the gel strength did not change much between 60°C and 77°C. The gel strength was almost independent of pH between pH=2.5 and pH=10.

Konno et al. (49) found that the specific viscosity of aqueous curdlan suspension as a function of temperature was almost constant up to 50°C with increasing temperature, but it showed a rapid increase at the temperature range from 50°C to 63°C. When the suspension was cooled after that, the specific viscosity showed a rapid increase at 39°C, and the gel was formed. This gel became fluidized when heated to 63°C. When the suspension was heated above 63°C, the specific viscosity increased slightly until 70°C, and then it increased rapidly above 70°C. If the temperature was raised above 70°C, a thermo-irreversible gel was formed.

These experimental facts together with the observation of light transmittance suggest that in thermo-reversible gels, formed on cooling after heating to 63°C, hydrogen bonding creates junction zones. On the other hand, hydrophobic bonding plays a major role in the formation of thermo-irreversible gels which are formed on heating at higher temperatures.

Ogawa et al. (50) examined the conformation of curdlan in alkaline solution, and found that curdlan molecules adopt a helical conformation in neutral or in weak alkaline solution (0.19N NaOH) and a random coil conformation in strong alkaline solution (0.24N NaOH). This conformational change was reversible for the alkaline concentration.

Recently, resistance of curdlan gel against freeze-thawing (51) was reexamined (52). Gel strength and syneresis accompanying the gel formation were measured for curdlan gels heated at 60°C, 85°C and 95°C for 30 min. These quantities were also measured after freeze-thawing. The gel strength increased and the syneresis decreased with increasing concentration of curdlan. All these three kinds of gels were found to be stable even after freeze-thawing, and gel strength and syneresis were increased by freeze-thawing. Since syneresis was repressed by the addition of starch, milk or soy-milk, curdlan is expected to be used as a freeze-thaw resistant hydrocolloid in food processing.

The structure of curdlan, resistant to enzymatic hydrolysis, is formed by heating at the temperatures 95°C or 120°C (53). The gel heated only at 60°C for 30 min did not show enzymatic resistant behaviour. The resistant structure was attributed to hydrophobic interaction.

Soybean protein
Tofu (soybean curd) is one of the most traditional Japanese foods. There have been many investigations on mechanisms of soybean protein gel formation and rheological properties of soybean protein gels. Calcium or magnesium salt or glucono-delta-lactone (GDL) is added as a coagulant to make tofu.

Flow charts of manufacturing soybean milk (tonyu), soybean curd (tofu), soybean sheet (yuba), and frozen-and-dried tofu (koritofu) are shown below:

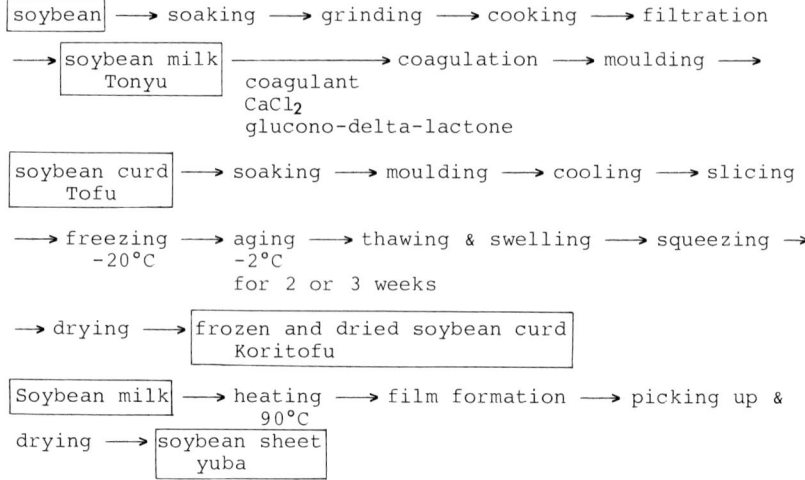

soybean ⟶ soaking ⟶ grinding ⟶ cooking ⟶ filtration

⟶ soybean milk ——————————⟶ coagulation ⟶ moulding ⟶
       Tonyu   coagulant
                   $CaCl_2$
                   glucono-delta-lactone

soybean curd ⟶ soaking ⟶ moulding ⟶ cooling ⟶ slicing
    Tofu

⟶ freezing ⟶ aging ⟶ thawing & swelling ⟶ squeezing ⟶
   -20°C   -2°C
            for 2 or 3 weeks

⟶ drying ⟶ frozen and dried soybean curd
            Koritofu

Soybean milk ⟶ heating ⟶ film formation ⟶ picking up &
            90°C
drying ⟶ soybean sheet
       yuba

Hashizume (54) examined the relation between hardness of gels and the concentration of $CaCl_2$ or GDL. Maximum hardness was 2.5 times larger in GDL gel than in $CaCl_2$ gel. The higher concentration of $CaCl_2$ did not form a gel and induced syneresis, while GDL could make a gel at higher concentrations beyond the isoelectric point.

It was confirmed that gel strength of 11S soy protein was larger than that of 7S soy protein (55, 56). The gel strength of an emulsion of soybean protein increased by the addition of sodium chloride up to 0.05 M and decreased by further addition of sodium chloride (57).

The relation between emulsification characteristics of heat denatured protein and surface hydrophobicity of protein, and the relation between hydrophobicity and foaming ability were studied (58, 59). The emulsification characteristics of soybean protein in the presence of milk casein (60), lecithin (61) were studied.

The effect of alkali salts on flow properties of acid precipitated protein was studied. The solution viscosity was increased by the addition of salt in the order of Hofmeister's lyotropic series (62). The strength of Yuba, a soybean milk sheet was larger in the 11S sheet than in the 7S sheet (63).

Dope, alkaline solubilised concentrated protein, is made into fibre in acidic solution. Rheological properties of dope and the fibre were studied by Hayakawa (64).

Kamaboko

Elongational stress-strain curves of kamaboko were found to obey fairly well the theory of rubber elasticity. Therefore, the texture of kamaboko, similar to rubber, was explained by the idea that the segments of network which take part in the deformation of kamaboko were random coils (65). When starch was added to kamaboko, the elongational stress-strain curves as well as the

texture deviate from rubber-like behaviour. It was thought that
small lumps of starch gels contained in kamaboko play the role of
filler just as the filler contained in rubber. Elongational
stress was separated into an entropic and an energetic part, and
it was found that the entropic part decreased with increasing
starch content (66).

The temperature dependence of storage shear modulus G' and
loss tangent tanδ was observed for kamaboko (67). G' decreased
remarkably over the temperature range from 10°C to 50°C and then
decreased gradually over the range from 50°C to 85°C. The former
and latter ranges have been thought to correspond to transition
regions and rubber regions respectively. The network structure
of kamaboko was thought to consist of a heat stable bond and
various kinds of weak bonds. The temperature dependence of G'
and tanδ of kamaboko containing starch was observed (68). G'
increased with increasing starch content at the range from 10°C
to 85°C. The transition region below 55°C shifted to higher
temperatures with increasing starch content. It was thought that
starch played the role of water absorbing agent and did not
strengthen the network structure very much.

Other Hydrocolloids

Since starch is quite often used as a texture modifier for
sauces, fish pastes, and also used for frozen foods, there have
been many physico-chemical investigations on starch. Annealing
of starch granules, especially warm water treatment and heat-
moisture treatment has been reviewed recently (69). Micro-
Brownian motion and conformation of amylose in solution was
studied by the fluorescence method (70). Structure of starch and
the problems of gelatinization and retrogradation were reviewed
(71). Recent advances in starch science in Japan were reviewed
(72).

There have been many investigations on the physico-chemical
properties of carrageenans because they are used in dessert
jellies and in many other foods, and because they are an
appropriate model substance for studying the gelling mechanism.
I only want to call attention of the readers to papers by Gekko
et al (73,74), Tako et al (75) and Watase et al (76,77).

Generally speaking, gels sometimes contain air bubbles, and so
they are compressible. Therefore, the deviation of Poisson's
ratio from 0.5 must be taken into account. This has been done
numerically using a finite element method, and compared with
experiments for a porous agar-gelatin gel (78,79).

I regret very much that I could not write the details of the
important work on emulsion science (80,81).

The cooking properties of pectin (82,83) and agar (84) or
starch (85,86) have been extensively studied.

Recently, gelation of proteins such as casein and gelatin by
transglutaminase has been reported (87). Since it forms covalent
cross-linkages in protein molecules, the gel is heat-resistant
(88).

Effects of pH and salts on the rigidity of ovalbumin, hen's
egg low density lipoprotein, and bovine serum albumin gels were
studied (89). Usually, heat-induced transparent gels from
ovalbumin solutions can be prepared only at low ionic strength,
which limits their practical uses. A method to prepare such gels

at different concentrations was proposed (90).

It was found that the head and tail of myocin molecules were denaturated at different temperatures, and that they play an important role in the formation of network structures (91).

The functional properties of proteins has been recently reviewed (92).

The rheological properties of spirulinan, a polysaccharide extracted from Spirulina subsulsa, were also studied recently (93).

III. SOME EXAMPLES OF HYDROCOLLOIDS USED IN THE FOOD INDUSTRY

Agar.   A cube of agar gel is often used in mitsumame, Japanese typical sweets, into which red bean and small pieces of rice cake, orange or cherries are added.

Agar jelly containing eggs, milk, fruit juices, plum etc. are cooked at home as sweets.

Agar is often used as a texture modifier for mizuyokan, sweet red bean paste, which people like to eat chilled in summer.

Tokoroten is noodle-shaped agar jelly. People eat it with vinegar and soy sauce with small pieces of laver sheet and mustard.

A review on the functional properties and applications of agar in the food industry will appear (94).

Konjac.   Konjac gel consists mainly of glucomannan, which is extracted from powders of Amorphophallus konjac K. Koch. Since konjac mannan is not hydrolyzed by digestive enzymes in human beings, and is hydrolyzed only by Aerobacter mannanolyticus (95,96), it is regarded as an almost non-calorie food. Konjac gel is used necessarily in oden, Japanese style pot au feu, in which fish paste, fried soybean curd, boiled eggs and vegetables are cooked together in a soup of soy sauce taste. Shirataki, konjac noodle is necessarily used in sukiyaki, which consists of beef, mashrooms, leak, tofu (soybean curd), and other vegetables. Recently, carrageenan was added to konjac mannan so that the latter can form a gel even at an acidic pH. Then, dessert jellies of the blend of carrageenan and konjac mannan were produced. A method of preparing a frozen konjac gel, which recovers its initial state when it is thawed, has been proposed recently. It consists of adding side chains on glucomannan main chains (97).

Xanthan.   A trial of making low calorie mayonnaise using xanthan gum has been made. A mixture of xanthan, locust bean, guar and tamarind gums is used to make a tofu of higher water holding capacity and with a good mouthfeel (98). Xanthan is used to prevent retrogradation of rice-cake (mochi) and also to improve the texture (99). A thickener consisting of xanthan and low methoxylpectin is proposed (100). Japanese noodles containing xanthan is reported to have an elastic texture (101). Xanthan is also used for glaze in order to protect frozen fish and vegetables preventing syneresis (102).

Pullulan.   Edible films of pullulan are now produced commercially on a large scale. They are expected to be eaten as snack foods and to form a composite film containing granular food such as cod roe or powdered cheese in addition to the more conventional use as packaging films for hams (103).

Curdlan.   It is expected to be used as a gelling agent for

dessert starch jelly (uiro, gyuhi), a new type multi-layer
dessert jelly or as a thickener for low calorie salad dressing or
peanut butter (104).
Soy protein.   Soy protein is frequently used in hamburgers,
kamaboko etc to replace animal proteins mainly because of the
healthy intentions of consumers.   By using two axis extrusion
cooking, soy protein is given a special texture resembling beef.
Non-gelling soy protein which is prepared by enzymatic
degradation, is expected to have low viscosity, and to be a
solution even when it is heated.   It is also expected to be a
good emulsifier and to be freeze-resistant.   It is considered to
be used in frozen desserts, soft drinks and soups, and many other
foods (105).   It was found that acid-precipitated soybean protein
and 11S globulin coagulated when they were heated and then
treated with bromelain, and that these treated protein coagula
absorbed twice as much water as native ones (106).   It is
reported that freeze-resistant tofu is developed by immersing
tofu in aqueous solution of citric or lactic acid-containing
saccharides and salts (107).   Yuba, a soybean cream sheet is
manufactured in Kyoto and Nikko.   Soybean milk contains protein,
and no animal fat, so it is considered as a healthy food.   Soy
milk ice cream containing 30% freeze-resistant konjac is also
manufactured.
Kamaboko.   Although the consumption of traditional kamaboko is
decreasing, only the kamaboko with crab taste and texture is sold
massively and about 50,000 tons are exported to the United
States, Europe and Australia per annum.   The main ingredients are
mashed frozen cod with glycine, alanine, crab flavor, crab
extract and egg white.   Kamaboko is cut into strings and then
gathered into a bundle, and this gives a similar texture to crab
legs.   Kamaboko with scallop taste and texture is also made by
gathering strings of kamaboko so that the texture may resemble
scallop.   The main ingredients are mashed frozen cod, scallop
extracts, starch and wheat flour.   Hanpen is a traditional fish
product in Japan.   It contains many air bubbles, and has a
sponge-like texture.   Meat of sharks and cod are mashed with
yams, and then boiled in hot water.   Sharks and yams are
necessary for producing air bubbles.
Imitation ikura (salmon roe).   A drop of salad oil is introduced
into a sol of carrageenan or xanthan gum which does not form a
gel.   This sol is wrapped by a film of alginate or pectin.   The
appearance resembles a natural salmon roe, so people cannot
distinguish between them (108,109).   This is quite often used for
sushi and some other snacks.

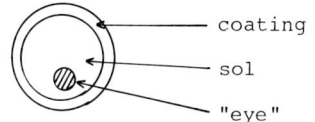

Structure of imitation ikura.

    Cyclodextrin.   A method of high yield production of
cyclodextrin was proposed (110).   Cyclodextrin is used for
wrapping flavors of wasabi, Japanese mustard made from hot
radish, and for masking unfavorable smells of some meat or fish

(111), and for reducing bitterness of naringin and limonin (112).
Many other applications of cyclodextrins for improving food
quality were reviewed by many authors (113-115).
Starch. A trial to use a starch hydrolysate such as maltodextrin
in stead of gum arabic was made to develop a new emulsifier or
stabiliser (116). Kuzu starch has been traditionally used for
Japanese sweets made from red bean paste. Recent advances in
starch fermentation technology and commercial uses of starch in
industrial application were reviewed (117,118).
Other applications. Pectins are used for jam making,
carrageenans are used for jelly desserts.
    Fu is made from wheat gluten and namafu is especially loved in
Kyoto.
    Coffee whitener is made from vegetable oil, casein and
emulsifier. Many tea shops use this kind of coffee whitener.
    The main ingredients of imitation milk are vegetable oil and
solid milk. Since it contains only a little milk fat, the
cholesterol content is very low. Some specially treated milk is
produced especially for people with lactose intolerance (lactase
deficiency). They lack lactase and therefore lactose cannot be
hydrolysed into glucose and galactose in their intestines.
    Coffee jelly is one of the favorite sweets served in Japanese
tea shops. Gelatin is used for this.

REFERENCES
 1. Yano, T. and Nishinari, K. ed. (1985) Proc. of the symposium
    on "Science of Hydrocolloids - Science and Application of
    Texture Modifiers". 29 pages, The Agricultural Chemical
    Society of Japan, Tokyo.
 2. Yano, T. and Nishinari, K. ed. (1986) Proc. of the symposium
    on "Food Hydrocolloids", 34 pages, The Agricultural Chemical
    Society of Japan, Tokyo.
 3. Araki, C. (1956) Bull. Chem. Soc. Japan, 29, 543.
 4. Araki, C. and Arai, K. (1967) Bull. Chem. Soc. Japan, 40,
    1452-1456.
 5. Watase, M. and Nishinari, K. (1981) Rheol. Acta, 20, 155-
    162.
 6. Nishinari, K. and Watase, M. (1983) Carbohydr. Polymers, 3,
    39-57.
 7. Watase, M. and Nishinari, K., (1986) in "Gums and
    Stabilisers for the Food Industry 3" G.O. Phillips, D.J.
    Wedlock and P.A. Williams ed., pp. 185-194, Elsevier,
    Amsterdam.
 8. Glicksman, M. (1969) Gum Technology in the Food Industry,
    Academic Press, New York.
 9. Watase, M. and Nishinari, K. (1983) Rheol. Acta, 22, 580-
    587.
10. Eldridge, J.E. and Ferry, J.D. (1954) J. Phys. Chem., 54,
    992-995.
11. Mitchell, J.R. (1980) J. Texture Stud., 11, 315-337.
12. Watase, M. and Nishinari, K. (1981) Nippon Shokuhin Kogyo
    Gakkaishi, 28, 437-443.
13. Flory, P.J. (1956) Proc. Roy. Soc. London, A234, 73-89.
14. Hayashi, A., Kinoshita, K. and Yasueda, S. (1980) Polymer
    J., 12, 447-453.

15. Hayashi, A. and Kanzaki, T. (1987) Food Hydrocolloids, in press.
16. Nishinari, K. and Watase, M. (1987) Agr. Biol. Chem., in press.
17. Nishinari, K., Koide, S. and Ogino, K. (1985) J. Phys. (Paris), 46, 793-797.
18. Tokita, M. and Hikichi, K. (1987) Phys. Rev. A. 35, 4329-4333.
19. Watase, M. and Nishinari, K. (1987) Presented at 31st IUPAC Macromolecular Symposium, Merseburg, GDR.
20. Maekaji, K. (1974) Agr. Biol. Chem., 38, 315-321.
21. Maekaji, K. (1978) Agr. Biol. Chem., 42, 177-178.
22. Torigata, H., Inagaki, H. and Kitano, H. (1952) Nippon Kagaku Zasshi, 73, 186-188.
23. Sugiyama, N., Shimahara, H., Andoh, T., Takemoto, M. and Kamata, T. (1972) Agr. Biol. Chem., 36, 1381-1387.
24. Kishida, N., Okimasu, S. and Kamata, T. (1978) Agr. Biol. Chem., 42, 1645-1650.
25. Nishinari, K., Kim, K.Y., Kohyama, K. (1987) Presented at the 2nd International Workshop on Plant Polysaccharides, Grenoble.
26. Morris, E.R., Cutler, A.N., Ross-Murphy, S.B., Rees, D.A. and Rice, J. (1981) Carbohydr. Polymers, 1, 5-21.
27. Sato, T., Norisue T. and Fujita, H. (1984) Polym. J., 16, 341-350.
28. Sato, T., Kojima, S., Norisue, T. and Fujita, H. (1984) Polym. J., 16, 423-429.
29. Sato, T., Norisue, T. and Fujita, H. (1984) Macromolecules, 17, 2696-2700.
30. Yamakawa, H. and Fujii, M. (1973) Macromolecules, 6, 407-415.
31. Kratky, O. and Porod, G. (1949) Rec. Trav. Chim. Pays-Bas, 68, 1106-1122.
32. Benoit H. and Doty, P. (1953) J. Phys. Chem., 57, 958-963.
33. Record, M. T. Jr., Woodbury, C. P. and Inman, R. B. (1975) Biopolymers, 14, 393-408.
34. Saito, T., Iso, N., Mizuno, H., Onda, N., Yamato, H. and Odashima, H. (1982) Biopolymers, 21, 715-728.
35. Kashiwagi, Y., Norisue, T. and Fujita, H. (1981) Macromolecules, 14, 1220-1225.
36. Yamakawa, H. and Yoshizaki, T. (1980) Macromolecules, 13, 633-643.
37. Wallenfels, K., Keilich,G., Bechtler, G. and Freudenberger, D. (1965) Biochem. Z., 341, 433-450.
38. Kawahara, K., Ohta, K., Miyamoto, H. and Nakamura, S. (1984) Carbohydr. Polym., 4, 335-356.
39. Mandelkern, L. and Flory, P.J. (1952) J. Chem. Phys., 20, 212-214.
40. Flory, P.J. (1953) in "Principles of Polymer Chemistry", Cornell University, Ithaca, p. 628.
41. Kato, T., Katsuki T. and Takahashi, A. (1984) Macromolecules, 17, 1726-1730.
42. Nishinari, K., Williams, P. A. and Phillips, G.O., Manuscript in preparation.
43. Nishinari, K. Shibuya, N. and Kainuma, K. (1985) Makromol. Chem., 186, 433-438.

44. Nishinari, K. Chatain, D. and Lacabanne, C. (1983) J.
    Macromol. Sci. Phys. Ed., B22, 795-811.
45. Nishinari, K. and Fukada, E. (1980) J. Polym. Sci. Polym.
    Phys. Ed., 18, 1609-1619.
46. Harada T. and Amemura, A. (1981) Mem. Inst. Sci. Ind. Res.,
    Osaka University, p.37-49.
47. Harada, T. (1979) in "Polysaccharides in Food", ed. by J.M.
    Blanshard and J.R. Mitchell, p. 283-300, Butterworths.
48. Maeda, I., Saito, H., Masada, M., Misaki A. and Harada, T.
    (1967) Agr. Biol. Chem. 31, 1184-1188.
49. Konno, A., Kimura, H.,Nakagawa, T. and Harada, T. (1978)
    Nippon Nogei Kagaku Kaishi, 52, 247-250.
50. Ogawa, K. Watanabe, T., Tsurugi J. and Ono, S. (1972)
    Carbohydr. Res., 23, 399-405.
51. Misaki, M., Tsujimoto, Y., Nakagawa, T., Sukenari J. and
    Moritaka S. (Takeda Chem. Ind. Ltd.) (1974): US Pat.,
    3,857,975.
52. Takahashi, F. and Harada, T. (1986) Kaseigaku Zasshi, 37,
    251-256.
53. Takahashi, F., Harada, T., Koreeda A. and Harada, A. (1986)
    Carbohydr. Polymers, 6, 407-421.
54. Hashizume, K. and Ka, G. (1978) Nippon Shokuhin Kogyo
    Gakkaishi, 25, 383-386.
55. Saio, K., Kajikawa, M. and Watanabe, T. (1971) Agr. Biol.
    Chem. 35, 890-898.
56. Yamano, Y., Miki, E. and Fukui, Y. (1981) Nippon Shokuhin
    Kogyo Gakkaishi, 28, 131-136.
57. Aoki, H. and Nagano, H. (1975) Nippon Shokuhin Kogyo
    Gakkaishi, 22, 320-324.
58. Voutsinas, L.P., Cheung, E. and Nakai, S. (1983) J. Food
    Sci., 48, 26-32.
59. Townsend, A.A. and Nakai, S. (1983) J. Food Sci., 48, 588-
    594.
60. Aoki, H., Shirase, Y., Kato, J. and Watanabe, Y. (1984)
    Nippon Shokuhin Kogyo Gakkaishi, 31, 333-338.
61. Miura, M. and Yamauchi, F. (1983) Agr. Biol. Chem., 47,
    2217-2222.
62. Umeya, J., Yamauchi, F. and Shibasaki, K. (1981) Agr. Biol.
    Chem., 45, 233-237.
63. Watanabe, K. and Okamoto, S. (1975) Nippon Shokuhin Kogyo
    Gakkaishi, 22, 325-330.
64. Hayakawa, I. (1983) New Food Industry, 25, No.4, 63-79.
65. Takagi, I. and Simidu, W. (1972) Bull. Jap. Soc. Sci. Fish.,
    38, 471-474.
66. Takagi, I. and Simidu, W. (1972) Bull. Jap. Soc. Sci. Fish.,
    38, 769-772.
67. Hamada, M. and Inamasu, Y. (1983) Bull. Jap. Soc. Sci.
    Fish., 49, 1897-1902.
68. Hamada, M. and Inamasu, Y. (1984) Bull. Jap. Soc. Sci.
    Fish., 50, 537-540.
69. Kuge, T. and Kitamura, S. (1985) Denpun Kagaku (J. Jpn. Soc.
    Starch Sci.) 32, 65-83.
70. Kitamura, S., Tanahashi, H. and Kuge, T. (1984) Biopolymers,
    23, 1043-1056.
71. Kainuma, K. (1986) New Food Industry, 28, No.7, 49-86.
72. Fuwa, E. (1987) Chori Kagaku, 20, 2-8.

73. Gekko, K., Mugishima, H. and Koga, S. (1985) Int. J. Biol. Macromol., 7, 57-63.
74. Gekko, K. and Kasuya, K. (1985) Int. J. Biol. Macromol., 7, 299-305.
75. Tako, M., Nakamura, S. and Kohda, Y. (1987) Carbohydr. Res., 161, 247-255.
76. Watase, M. and Nishinari, K. (1986) Polymer J., 18, 1017-1025.
77. Watase, M. and Nishinari, K. (1987) Makromol. Chem., In press.
78. Shiinoki, Y. and Yano, T. (1986) J. Texture Stud., 17, 175-188.
79. Yano, T., Shiinoki, Y., Miyawaki, O. and Sakiyama, T. (1987) J. Food Eng., 6, 217-230.
80. Matsumoto, S. (1982) Hyomen (Surface), 20, 80-90.
81. Matsumoto, S. (1987) in "Nonionic Surfactants-Physical Chemistry", Ed. M. J. Schick, Marcel Dekker Inc., New York and Basel, 549-600.
82. Kawabata, A., Sawayama, S., Nakahara, H. and Kamata, T. (1981) Agric. Biol. Chem., 45, 965-973.
83. Kawabata, A., Sawayama, S., Nagashima, N. and Uchimura, Y. (1981) Kaseigaku Zasshi, 32, 739-744.
84. Nagura, H., Akabane, H. and Nakahama, N. (1984) Nippon Shokuhin Kogyo Gakkaishi, 31, 339-345.
85. Hirao, K., Murayama, Y., Akabane, H. and Nakahama, N. (1985) Kaseigaku Zasshi, 36, 10-17.
86. Takahashi, S., Hirao, K., Kobayashi, R., Kawabata, A., and Nakamura, M. (1987) Denpun Kagaku, 34, 21-30.
87. Nio, N., Motoki, M. and Takinami, K. (1986) Agric. Biol. Chem., 50, 1409-1412.
88. Motoki, M. (1987) Nippon Nogeikagaku Kaishi, 61, 486-488.
89. Nakamura, R., Fukano, T. and Taniguchi, M. (1982) J. Food Sci., 47, 1449-1453.
90. Kitabatake, N., Hatta, H. and Doi, E. (1987) Agric. Biol. Chem., 51, 771-778.
91. Yasui, T. and Samejima, K. (1985) New Food Industry, 27, No. 5, 76-82, No. 6, 81-88, No. 7, 82-89, No. 8, 55-65.
92. Yamauchi, F. ed. (1986) "Shokuhin Tanpakushitsu no Kagaku (Science of Food Proteins)", Shokuhin Shizai Kenkyukai, Tokyo, 216 pages.
93. Kubota, M., Koyano, T., Shinohara, K. and Nishinari, K. (1987) Proc. 34th Nippon Shokuhin Kogyo Gakkai, p. 39.
94. Matsuhashi, T. in "Food Gels", Harris, P. ed., To be published by Elsevier Applied Science Publi. Londcn.
95. Inoue, N. (1941) Rep. Imp. Govern. Inst. Nutrition, 10, No. 2, 20-31.
96. Okimasu, T. ed. (1984) "Konjac no Kagaku (Science of konjac)", Keisuisha, Hiroshima, 313 pages.
97. Hara, K. (1987) Shokuhin to Kaihatsu, 22, No. 9, 32-35.
98. Yagi, Y., Sato, M., Fujii, K., Iijima, Y., Sayama, Y. and Moriyama, K. (Sanraku Ocean Co., Ltd.) (1984) Jap. Pat. 59-169460.
99. Uchida, K. (San-ei Kagaku Kogyo Co., Ltd.) (1985) Jap. Pat. 60-30651.
100. Sugisawa, K., Shibuki, M., Yamaguchi, N. and Sakaguchi, T. (House Shokuhin Kogyo Co. Ltd.) (1985) Jap. Pat. 60-87741.

101. Toda, Y. and Imamura, M. (Taiyo Kagaku Kogyo Co., Ltd.) (1981) Jap. Pat. 56-5501.
102. Kurihara, M. and Miyasaka, Y. (Chiba Seifun Co., Ltd.) (1985) Jap. Pat. 60-9769.
103. Oogami, T. (1986) Shokuhin to Kaihatsu, 21, No.3, 16-20.
104. Sato, S., Okumura K. and Harada, T. (1978) New Food Industry, 20, No. 10, 49-57.
105. Sato, T. (1986) Shokuhin to Kaihatsu, 21, No.9, 34-38.
106. Mohri, M. and Matsushita, S. (1987) J. Ag. Fd. Sci., 32, 486-490.
107. Yamamoto, K. (Nippon Shokuhin Co., Ltd.) (1984) Jap. Pat. 59-20345.
108. Ueda, T. (QP Co., Ltd.) (1985) Jap. Pat. 60-83570.
109. Kishi, S. (Nippon Carbide Kogyo Co., Ltd.) (1977) Jap. Pat. 52-59079.
110. Kobayashi, S., Kainuma, K., Watanabe, A., Ohtani, T. and Umeda, K. (1982) Jap. Pat. 57-202298.
111. Sato, S. and Saito, Y. (1987) Chori Kagaku, 20, 159-163.
112. Konno, A., Misaki, M., Toda, J., Wada, T. and Yasumatsu, K. (1982) Agric. Biol. Chem., 46, 2203-2208.
113. Szejtli, J. (1982) Starke, 34, 379-385.
114. Yagi, Y. (1987) Food Chemicals, No.4, 59-63.
115. Hara, K. (1987) Shokuhin to Kaihatsu, 22, No.2, 32-37.
116. Hanno, Y. (1985) Food Chemicals, No.11, 78-84.
117. Kainuma, K. (1984) in "Starch, Chemistry and Technology" 2nd Ed., R. L. Whistler, J. N. BeMiller and F. Paschall ed., Academic Press, New York, pp. 125-152.
118. Kainuma, K. (1985) Proc. IVth FAOB Symp. and IUB Symp. (Manila), pp. 81-90.

Enzymatic modification of natural seed gums

J.S.Grant Reid, Mary Edwards and Iain C.M.Dea[+]

Department of Biological Science, University of Stirling, Stirling FK9 4LA, UK
[+]Leatherhead Food R.A., Randalls Road, Leatherhead, Surrey KT22 7RY, UK

ABSTRACT

A series of enzymes capable of hydrolysing specific linkages in
xyloglucan gums have been identified in germinated xyloglucan
containing seeds. Two of these enzymes - an endo-$(1 \to 4)$-β-$\underline{D}$-glucanase
and a β-$\underline{D}$-galactosidase - have been purified to homogeneity. The
endo-β-$\underline{D}$-glucanase is novel in that it is totally specific towards
xyloglucans. The β-$\underline{D}$-galactosidase is capable of cleaving
specifically the terminal $(1 \to 2)$-β-$\underline{D}$-galactosyl residues from tamarind
gum. Progressive removal of galactosyl residues leads to a decrease
in solution viscosity followed by an increase culminating in gel
formation. Gelation occurs on removal of about 50% of the terminal
galactosyl residues. The role of galactose in controlling the
solution properties of xyloglucans is discussed.

INTRODUCTION

The "seed gums" of commerce are plant cell-wall polysaccharides.
During seed development they are deposited as massive thickenings of
endosperm or cotyledonary cell-walls. After germination they are
broken down and used as food reserves for the developing seedling.
Biologically they are termed "cell-wall storage polysaccharides"
(1, 2).
In the seed (as in commerce) the bulk properties of the cell-wall
storage polysaccharides are important. They may confer hardness on
the seed, protecting it from physical damage. Alternatively their
interaction (or non-interaction) with water may be important in the
adaption of the seed to germinate successfully in its natural
environment (1). This has been demonstrated experimentally in the
case of the galactomannan in a leguminous seed. In fenugreek
(Trigonella foenum-graecum L.), a leguminous species adapted to
semi-arid conditions, galactomannan has been shown to be the molecular
basis of a drought-avoidance mechanism (3). The polysaccharide is in
the seed endosperm which is physically interposed between the embryo
and the environment. Most of the water taken up by the seed during

THE ENDO-(1→4)-β-D-GLUCANASE

The endo-(1→4)-β-D-glucanase of the germinated nasturium seed has been purified to homogeneity and some of its properties have been described (8). It has an endo mode of action on nasturium and tamarind seed xyloglucans, reducing solution viscosities rapidly in the initial stages of hydrolysis with a slow, linear release of reducing-power. Even after prolonged digestion, hydrolysates of tamarind xyloglucan contained only oligosaccharides with DP ≥ 6. The pH optimum is 4.5 to 5.0, the isoelectric point is pH 5.0, and the molecular weight is 29,000 (no subunits) (8). This enzyme exhibits a unique type of substrate-specificity, in that is appears to be totally specific towards xyloglucans. It does not hydrolyse cellulose, synthetic cellulose derivatives, cello-oligosaccharides or other polysaccharides containing the (1→4)-β-D-glucosidic link (Table 1).

The molecular basis of the substrate-specificity of this enzyme is under investigation. It may be a true xyloglucanase, recognising structural features of the side-chains whilst hydrolysing the backbone. Alternatively it may be a novel type of endo-(1→4)-β-D glucanase with a subsite binding requirement which is satisfied by xyloglucans but not by any of the other substrates tested (8).

THE β-D-GALACTOSIDASE

The germinated nasturtium seed contains at least two different β-D-galactosidase activities, and the major enzyme has been purified to homogeneity. The details of the purification will be published (9). On SDS-polyacrylamide gel electrophoresis the enzyme behaves as a single polypeptide chain (Mr = 97000). On isoelectric focusing it is resolved into a number of closely related molecular species exhibiting β-D-galactosidase activity and spanning the range pI = 6.6 to pI = 7.1. It is possible to obtain a single isoenzyme by chromatofocusing (9), but the following data were obtained using the mixture of isoenzymes.

It is clear (Table 2) that the enzyme is a β-D-galactosidase, hydrolysing substrates with the β-D-, but not the α-D-galactosyl linkage. The ability of the enzyme to release D-galactose from xyloglucans is of particular interest. Other β-D-galactosidases tested (from E. coli, yeast and Jack bean) were unable to do this. The Km and Vmax values with respect to xyloglucans were not accessible. Using xyloglucan solutions, plots of substrate concentration [S] versus reaction velocity V were linear and double-reciprocal plots (1/V versus 1/[S]) tended to pass through the origin. Consequently substrate-saturation of the enzyme, if it occurs, must be at substrate "concentrations" corresponding to hydrated solids rather than solutions. Biologically this is reasonable, since the enzyme is secreted into hydrated xyloglucan-rich cell walls.

Fig. 3 shows changes in solution viscosity accompanying the removal of terminal β-D-galactopyranosyl residues from tamarind xyloglucan (Fig. 1) by the enzyme. As the reaction proceeds the viscosity of the solution gradually decreases and then increases until gel-formation occurs. Gel formation occurs when about 50% of the galactose residues have been removed. The gel synereses when heated to 100°, but does

Table I. SUBSTRATE SPECIFICITY OF THE endo- β -D-GLUCANASE FROM NASTURIUM COTYLEDONS

Substrate	Increase in reducing power over 6 h (mM glucose equiv.)	Decrease in specific viscosity over 6 h (%)
Nasturtium seed xyloglucan (gal:xyl:glc = 4.4:6.5:10)[1]	1.6	99
Tamarind seed xyloglucan (gal:xyl:glc = 3.1:6.9:10)[1]	1.3	97
Mung bean primary cell-wall xyloglucan (fuc:gal:xyl:glc = 0.8:1.7:11.3:10)[1,2]	0.10	23
Sodium carboxymethylcellulose (DS = 0.7, DP = 1100)	0.01	0
Hydroxyethylcellulose, molar substitution = 2.5	0.01	4.3
Microcrystalline cellulose (Avicel)	0	
Phosphoric acid reprecipitated cellulose	0	
Barley β-glucan, 26% (1→3) 74% (1→4) linkages[3]	0	0
Konjac glucomannan (man:glc = 1.6:1)	0	3.1
Spruce wood galactoglucomannan (gal:glc:man = 1.0:1.9:5.5)[4]	0	
Asparagus seed galactoglucomannan (gal:glc:man = 1.0:6.0:7.1)[4]	0	
cello-oligosaccharides	no hydrolysis detected[5]	
laminarin	0	

[1] composition determined in our laboratory
[2] slightly contaminated with (1→4)-β-D-xylan. Published composition:fuc:gal:xyl:glc = 1:2.5:7:10 (11)
[3] supplied by Prof. D. Manners, Heriot-Watt University, Edinburgh
[4] supplied by Prof. H. Meier, University of Friburg, Switzerland
[5] suppled by Dr. R. Sturgeon, Heriot-Watt University, Edinburgh. No change in composition of mixture detected by TLC over 24h.

Table II.   SUBSTRATE SPECIFICITY OF THE β-D-GALACTOSIDASE FROM
            NASTURTIUM COTYLEDONS

Substrate	Reaction rate at 10mM substrate[1] (nKat/mg protein)	Km (mM)	Vmax (nKat/mg protein)
methyl-α-D-galactoside	0	–	–
methyl-β-D-galactoside	2.1	20	8.5
melibiose (gal α1 → 4 glc)	0	–	–
lactose (gal β1 → 4 glc)	1.7	20	4.9
p-nitrophenyl-α-D-galactoside	0	–	–
p-nitrophenyl-β-D-galactoside	657	3.7	900
guar galactomannan (gal α1 → 6 man)	0	–	–
locust bean gum (gal α1 → 6 man)	0	–	–
nasturtium xyloglucan (gal β1 → 2 xyl)	1.8	very high[2]	very high[2]
tamarind xyloglucan (gal β1 → 2 xyl)	2.6	very high[2]	very high[2]
lupin seed (1 → 4)-β-D-galactan[3]			

[1] polysaccharide concentrations adjusted to give 10mM equivalent
     terminal non-reducing D-galactose residues
[2] see text
[3] 10 mg/ml

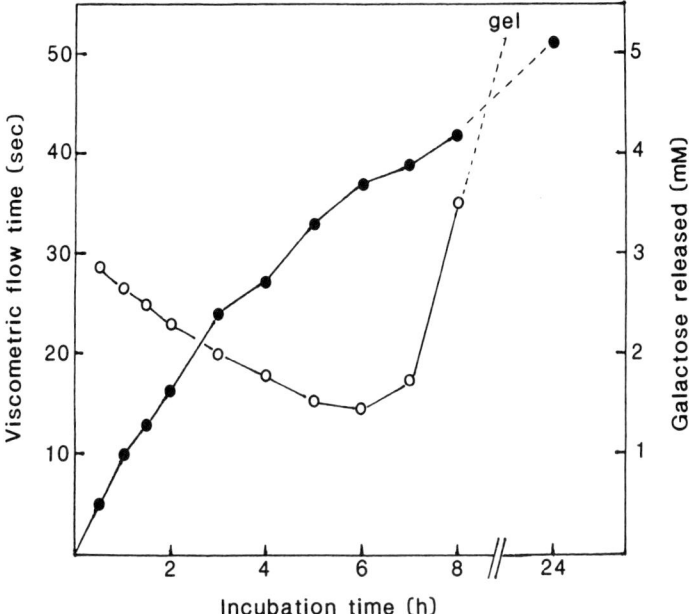

Figure 3. Digestion of tamarind seed xyloglucan (10 mg/ml) by
nasturium β-D-galactosidase. ● = free D-galactose, O = solution
viscosity, expressed as flow-time in a simple capillary viscometer.

not melt. The formation of a gel is a function of the high molecular
weight of the xyloglucans. When increasing (low) amounts of
endo-β-D-glucanase were used alongside the β-D-galactosidase in
incubations similar to that described in Fig. 3, gel formation was
suppressed and precipitation occurred.

These results indicate strongly that the terminal, non-reducing
D-galactopyranosyl residues of tamarind seed xyloglucan are the major
structural features determining the water-solubility of the molecule.
The selective removal of these residues does not, however, change the
degree of substitution of the (1→4)-β-D-glucan backbone (Fig. 1),
implying that for a xylose-substituted backbone polymer–polymer
interactions are significantly stronger than polymer–solvent
interactions. In this respect the xylose-substituted glucan backbone
has properties which are comparable with those of the unsubstituted
(1→4)-β-D-mannan backbone of the galactomannans. It has been
demonstrated clearly that the selective removal of D-galactosyl
side-chains from galactomannans increases the self-association of the
backbone leading to insolubility at degrees of substitution below
about 11% (10). Interestingly our data (Fig. 3) show that gelation of
tamarind xyloglucan occurs at a similar degree of substitution of the
xyloglucan backbone by galactose. The formation of a gel in the case
of the xyloglucan is a result of the interchain interactions. The
nature of these interactions is being investigated.

ACKNOWLEDGEMENT

We are grateful to Unilever for financial support.

REFERENCES

1. Reid, J.S.G. (1985) Adv. Bot. Res., 11, 125–155.

2. Meier, H. and Reid, J.S.G. (1982) Encycl. Plant Physiol. New Ser.
    13A, 418–471.

3. Reid, J.S.G. and Bewley, J.D. (1979) Planta 147, 145–150.

4. Campbell, J. and Reid, J.S.G. (1982) Planta 155, 105–111.

5. Kooiman, P. (1961) Recl. Trav. Chim. Pays- Bas. 80, 849–865.

6. Le Dizet, P. (1972) Carbohydr. Res. 24, 505–509.

7. Edwards, M., Dea, I.C.M., Bulpin, P.V. and Reid, J.S.G. (1985)
    Planta, 163, 133–140.

8. Edwards, M., Dea, I.C.M., Bulpin, P.V. and Reid, J.S.G. (1986) J.
    Biol. Chem. 261, 9489–9494.

9. Edwards, M., Dea, I.C.M. and Reid, J.S.G. submitted.

10. McCleary, B.V., Amado, R., Waibel, R. and Neukom, H. (1981)
    Carbohydr. Res. 92, 269–285.

11. Kato, Y and Matsudo, K. (1976) Plant Cell Physiol. 17, 1185–1198.

The effect of gamma radiation on the structure of carrageenans

W.M.Marrs

Leatherhead Food R.A., Randalls Road, Leatherhead, Surrey KT22 7RY, UK

ABSTRACT

Both kappa and iota carrageenans were irradiated as water gels and in dry powder form using dose levels up to and including 10 kiloGray. The effect of irradiation on molecular weight was investigated using intrinsic viscosity data and gel permeation elution profiles. The gel properties of irradiated kappa carrageenan gels were studied using a precision penetrometer. Depolymerisation of the carrageenans in water gel form was rapid initially and appeared to proceed by a random scission process. On continued exposure to gamma radiation, the rate of depolymerisation decreased and eventually ceased, leaving radiation-stable short-chain-length segments of carrageenan. Radiolytic depolymerisation was also observed in dry powders but the rate of molecular breakdown was slow compared with that observed in water gels. The losses in gel strength, break strength and elastic limit observed in irradiated gels are probably the result of breakdown of the macromolecular network between junction zones. Possible mechanisms for the radiolytic depolymerisation of carrageenans are discussed.

INTRODUCTION

In the food industry there is a continuing interest in the use of ionising radiation for improving the quality and extending the shelf-life of foods.[1] Worldwide, more than thirty countries have permitted the irradiation of certain specified foods within defined limits of radiation dose levels. In the United Kingdom, the irradiation of foods is not permitted except in the case of sterilised diets required by certain groups of hospital patients.[2] However, the Advisory Committee on Irradiated and Novel Foods[3] recently recommended that the irradiation of foods using dose levels up to 10 kiloGray (kGy) be permitted. This concurs with the maximum dose level recommended by the Joint FAO/IAEA/WHO Expert Committee on the Wholesomeness of Foods.[4]

Polysaccharides are susceptible to ionising radiation, the main effect of which is to initiate free radical induced scission of glycosidic links in the polysaccharide chain. The resulting breakdown of macromolecular structure has a profound effect on the functional role of polysaccharides. It is thought to be largely responsible for the softening of certain fruits and vegetables following irradiation.[5] It also reduces the ability of polysaccharides to function as thickening and gelling agents in food products.

Although much work has been done on the effects of ionising radiation on starches, very little information has been published on irradiated non-starch polysaccharides. There is clearly a requirement for more information on the properties of irradiated polysaccharides. This will enable the food industry to take account of the effects of radiation on functional properties and to compensate for these effects at the formulation stage. It will also enable the food industry to assess the potential of irradiation as a method of producing partially depolymerised polysaccharides, which may find specialised applications in foods.

The purpose of this paper is to describe the effects of gamma radiation on the molecular structure and functional properties of both iota[6] and kappa[7] carrageenans.

MATERIALS AND METHODS

The iota and kappa carrageenans used in this work were exclusive, unblended extracts of the
red seaweeds *Eucheuma spinosum* and *Eucheuma cottonii,* respectively.

Sample preparation

A solution containing iota carrageenan (1%) and calcium chloride dihydrate (0.5%) in distilled
water was prepared at 85°C. This solution was filled into 120-ml glass bottles with screw caps.
They formed a weak gel structure on cooling.

Solutions containing kappa carrageenan (0.5%, 1% and 1.5%) and potassium chloride (0.5%) in
distilled water were prepared at 85°C. These were filled into 120-ml glass bottles with screw caps
and formed firm gels on cooling.

Irradiation

Samples of iota carrageenan gel and kappa carrageenan gel were exposed to gamma radiation in
a cobalt-60 facility for varying periods of time such that sets of two iota carrageenan gels and four
kappa carrageenan gels each received doses of 0.25, 0.5, 1, 2, 5 and 10 kGy, respectively. Non-
irradiated samples of gels were set aside for testing as controls. Samples of iota and kappa
carrageenan in dry powder form were also irradiated, receiving the same dose levels as the gels.

Cation exchange

Samples of irradiated (and control) iota and kappa carrageenan gels (100 g) were heated to
around 85°C with added sodium sulphate (1.42 g) and Amberlite CG-120 ion-exchange resin (1 g)
in order to remove calcium and potassium ions from the respective carrageenan gels. After cooling
and filtration, these solutions were subjected to intrinsic viscosity determination and gel permeation
chromatography.

Viscometry

The relative viscosities of iota and kappa carrageenans in 0.1M sodium sulphate solution were
measured at carrageenan concentrations ranging from 0.025% to 0.5%, using a capillary viscometer
at 60°C ± 0.05°C. Reduced viscosities, η_{red}, were calculated from relative viscosities, η_{rel}, using
the relation:

$$\eta_{red} = (\eta_{rel}-1) / c \ \ (dl/g)$$

where c is the carrageenan concentration.

Intrinsic viscosities, $[\eta]$, were found by extrapolation of reduced viscosities to zero concentration in accordance with the definition:

$$[\eta] = \eta_{red \ c \to 0} \ \ (dl/g).$$

Gel permeation chromatography

Cation-exchanged carrageenan solutions (0.2%) were applied to a chromatographic column
(16 mm x 90 mm) containing Sepharose CL-4B using a 1-ml capacity injection loop applicator. The
samples were eluted using 0.1M sodium sulphate solution and the column was maintained at 60°C–
62°C to prevent coil-to-helix transitions in the carrageenans. The eluant was monitored using a
differential refractive index detector coupled to a chart recorder. Eluant flow rate was around
7–8 ml per hour. The void volume (V_0) was determined using a solution of Dextran Blue (0.1%)
and the total volume (V_t) was measured using sucrose (0.1%). Distribution coefficients, k_{av}, were
calculated from the elution volumes, V_e, using the relation:

$$k_{av} = \frac{V_e - V_o}{V_t - V_o} .$$

The effective molar volumes, $\overline{M}w[\eta]$, of the irradiated carrageenans were calculated using the universal calibration relation already established for this column:[8]

$$\log \overline{M}w[\eta] = 5.92 - 3.17\,k_{av}.$$

Mean molecular weights, $\overline{M}w$, were obtained by dividing effective molar volumes by intrinsic viscosities.

Penetrometry

Irradiated kappa carrageenan gels were tested using a Stevens-LFRA Texture Analyser fitted with a ½-inch diameter hemispherical probe programmed to penetrate the gel at a speed of 0.5 mm/s to a distance of 20 mm. The force on the probe was recorded as a function of penetration distance, and generally increased up to the gel break point, when a sudden drop in force was observed. Both the load and penetration distance at the gel break point were recorded and termed break load and elastic load, respectively. Gel strength was defined as the average slope of the load/penetration curve up to the break point.

RESULTS

The effect of gamma radiation on carrageenan structure

The intrinsic viscosity of iota carrageenan irradiated in 1% gel form decreased with increasing radiation dose (Figure 1). The initial fall was rapid and the intrinsic viscosity was approximately halved after exposure to 0.5 kGy. At higher dose levels, the loss was much less rapid and the intrinsic viscosity reached a nearly constant value after exposure to 10 kGy. In contrast, iota carrageenan in dry powder form lost intrinsic viscosity at a much lower rate and the initial value was roughly halved after exposure to 10 kGy.

The fall in intrinsic viscosity indicates a breakdown of the carrageenan macromolecules and this was confirmed by the corresponding gel permeation elution profiles (Figure 2). They show a progressive reduction in molecular size, and a narrowing of molecular size distribution, with increasing radiation dose. The peak at the total volume (V_t) probably included an inorganic component of the original sample as well as low molecular weight carbohydrate produced by radiolysis. The elution profiles for iota carrageenan irradiated in dry powder form (not shown) indicated that these samples contained high-molecular-weight material, which was totally excluded from the Sepharose CL-4B packing material. Only samples that had received doses of 2, 5 and 10 kGy were completely resolved on this column. In these cases, distribution coefficients were calculated from elution volume data, and are recorded in Table I together with measured intrinsic viscosities, mean molar volumes (found using the universal calibration relation) and mean molecular weights.

TABLE I
Weight average molecular weights of irradiated iota carrageenan powder

Radiation dose (kGy)	Intrinsic viscosity (dl/g)	Distribution coefficient (k_{av})	$\overline{M}w[\eta]$ (dl/mol)	$\overline{M}w$ (g/mol)
2	2.5	0.16	258,700	104,000
5	2.1	0.21	179,600	86,000
10	1.5	0.33	74,800	50,000

The mean molecular weights of the iota carrageenan irradiated as 1% gels, which were calculated using the distribution coefficients obtained from their elution profiles and their intrinsic viscosities, decreased rapidly with increasing radiation dose then levelled out and approached a lower limit of around 12,000 (Figure 3). The initial dependence of molecular weight on radiation dose took the form expected of a random depolymerisation,[9] that is the reciprocal molecular weight proportional to time. Assuming that the irradiation-induced depolymerisation process produced a random scission of glycosidic links in the polysaccharide chain, the expected dependence of molecular weight on radiation dose would have taken the form:

$$\frac{1}{\overline{M}} = \frac{1}{\overline{M}_O} + k\,t$$

where $\overline{M}_O$ is the initial molecular weight, k is a rate constant and t is the time of exposure to a constant radiation flux (hence, t in this expression is equivalent to dose). In fact, reciprocal molecular weight was proportional to radiation dose in the early stages (up to 1 kGy) but levelled off to a practically constant value at between 5 and 10 kGy (Figure 4). The rate constant (k) for the depolymerisation of iota carrageenan at low dose levels was 3.4×10^{-5} ($\overline{M}^{-1}\,kGy^{-1}$).

Kappa carrageenan in 1% gel form was also rapidly depolymerised following exposure to gamma radiation. Intrinsic viscosity values fell even faster than those of iota carrageenan with increasing radiation dose (Figure 5). In powder form, kappa carrageenan was depolymerised relatively slowly but was more susceptible to radiolysis than iota carrageenan powder.

The gel permeation elution profiles of kappa carrageenan irradiated in 1% gel form confirm the breakdown to smaller molecules with increasing radiation dose (Figure 6). Distribution coefficients obtained from these elution profiles were used together with intrinsic viscosity data to calculate mean molecular weights.

The loss in the molecular weight of kappa carrageenan was very rapid at low radiation doses but levelled out and approached a limit of around 5,000 at 10 kGy (Figure 7). The dependence of reciprocal molecular weight on radiation dose was linear up to around 5 kGy but became almost independent of dose between 5 and 10 kGy (Figure 8). The rate constant (k) for the depolymerisation of kappa carrageenan at low dose levels was 4.2×10^{-5} ($\overline{M}^{-1}\,kGy^{-1}$).

Kappa carrageenan in dry powder form was relatively resistant to radiolytic breakdown and the elution profiles (not shown) indicated that substantial amounts of material were excluded from the packing material after radiation doses of up to 2 kGy. At higher doses, the carrageenans were completely resolved and distribution coefficients were used to calculate mean molecular weights (Table II).

TABLE II
Weight average molecular weights of kappa carrageenan following irradiation
in dry powder form

Radiation dose (kGy)	Intrinsic viscosity (dl/g)	Distribution coefficient (k_{av})	$\overline{M}w\,[\eta]$ (dl/mol)	$\overline{M}w$ (g/mol)
5	3.40	0.19	207,800	61,000
10	1.75	0.29	100,700	57,000

The effect of gamma radiation on gel properties

The breakdown of the molecular structure of kappa carrageenan following irradiation resulted in a loss of gel structure in the 0.5%, 1% and 1.5% gels. Gel strength, break strength and elastic limit all decreased with increasing radiation dose (Figures 9, 10 and 11). This loss of gel structure was rapid at low doses but levelled out at higher doses. In the case of the 0.5% gels, almost no gel structure remained after a radiation dose of 0.5 kGy. Also, both gel strength and break strength were strongly influenced by carrageenan concentration, whereas elastic limit was little affected by concentration.

The depolymerisation of kappa carrageenan in powder form by gamma radiation also resulted in a loss of gelling potential; 1% gels prepared from irradiated kappa carrageenan powder had gel strengths, break strengths and elastic limits that all decreased with increasing radiation dose (Figures 12, 13 and 14). The fall in gel strength was only moderate but both break strength and elastic limit lost more than 50% of their original values after exposure to 10 kGy.

DISCUSSION

The radiation-induced depolymerisation of carrageenans is probably the result of free-radical-initiated scission at glycosidic links in the polysaccharide chain. One mechanism by which this process may proceed in aqueous solution involves hydroxyl radicals produced by the action of ionising radiation on water.[10] These radicals (together with hydrogen atoms) are able to abstract hydrogen atoms from C—H bonds in the sugar ring to give free radicals that react with water to produce scission at the glycosidic link. It is conceivable that water present as moisture in carrageenan powder may be involved in such a process but there is the additional possibility of free radical production at carbon atoms in the sugar ring by the direct action of ionising radiation at the C—H bonds.[11] It is unlikely that radiation-induced acidity is responsible for the depolymerisation of carrageenan since gels buffered at pH 7 are as susceptible to radiolytic breakdown as unbuffered gels.[7] Kappa carrageenan in gel form is less stable to radiation than iota carrageenan despite the fact that the structure of the glycosidic link is the same in the two polysaccharides. This suggests that the counter ions, potassium ions in kappa carrageenan and calcium ions in iota carrageenan, may play a role in determining the susceptibility of carrageenans to radiolytic breakdown.

The radiation-stable short-chain segments present in both iota and kappa carrageenan gels are probably those regions of the carrageenan chains that are involved in junction zones in which helical regions of polysaccharide chains form aggregates stabilised by the appropriate counter ions. Regions of the polysaccharide chain that lie outside the junction zones and that form the macromolecular network are likely to be susceptible to radiolytic breakdown. Chain scissions in these regions are probably responsible for the loss of gel structure in irradiated carrageenan gels.

When irradiated gels are melted and allowed to reset, further reduction in gel strength, break strength and elastic limit are observed.[7] This is due to a re-organisation of the depolymerised carrageenan molecules during formation of the new gel structure. The reduced gelling properties of irradiated carrageenan powder is simply the result of a reduction in mean molecular weight. The gel properties of irradiated powders are similar to those of irradiated gels of similar molecular weight after melting and resetting. Losses in gel strength and break strength of a carrageenan gel could be compensated for by increasing the concentration of carrageenan. However, no compensation could be made for losses in elastic limit, which is dependent only on carrageenan molecular weight and not on concentration.

CONCLUSIONS

Both iota and kappa carrageenans, in water gel form, are depolymerised following exposure to gamma radiation and gel structure is rapidly broken down. These carrageenans in powder form are much less susceptible to radiolytic breakdown.

Carrageenan gels contain radiation-stable short-chain segments, which are thought to be helical regions of polysaccharide present in cation-stabilised junction zones of the gel network. The loss of gel properties in irradiated gels is mainly due to the breakdown of the macromolecular network between junction zones.

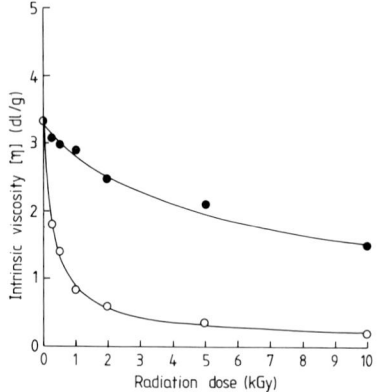

Figure 1. Effect of radiation dose on the intrinsic viscosity of iota carrageenan from irradiated 1% gels (○) and irradiated powder (●).

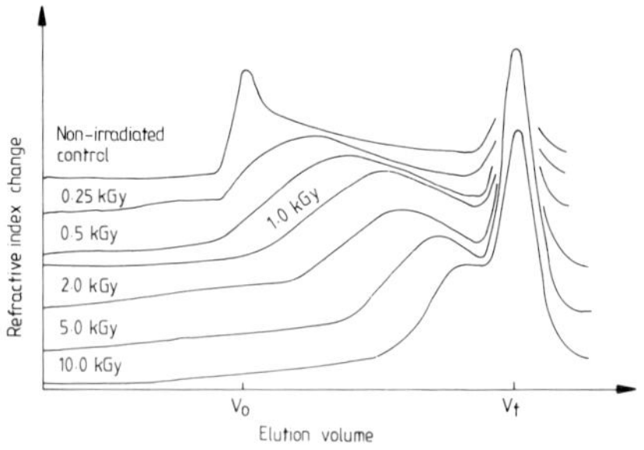

Figure 2. Gel permeation profiles of iota carrageenan from irradiated 1% gels.

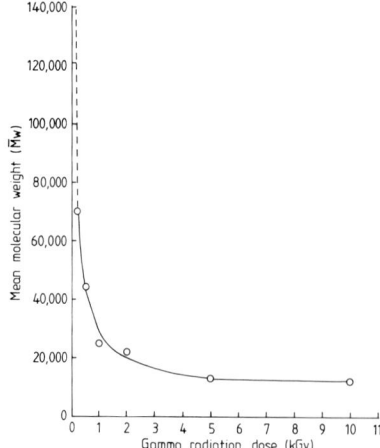

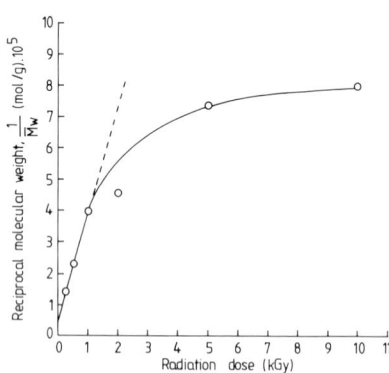

Figure 3. Effect of radiation dose on the mean molecular weight of iota carrageenan from irradiated 1% gels.

Figure 4. Dependence of reciprocal mean molecular weight upon radiation dose for iota carrageenan irradiated in the form of a 1% gel.

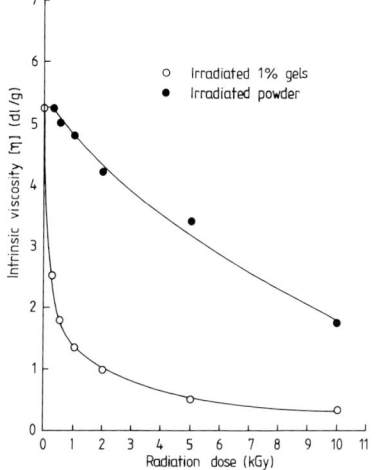

Figure 5. Effect of radiation dose on the intrinsic viscosity of kappa carrageenan from irradiated 1% gels (O) and irradiated powder (●).

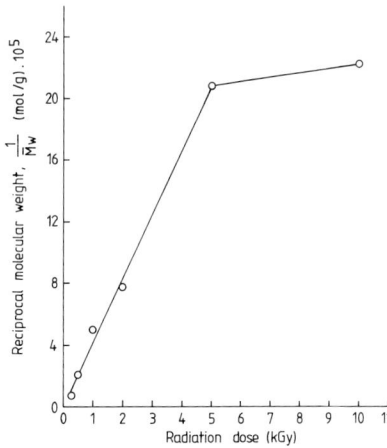

Figure 6. Gel permeation profiles of kappa carrageenan from irradiated 1% gels.

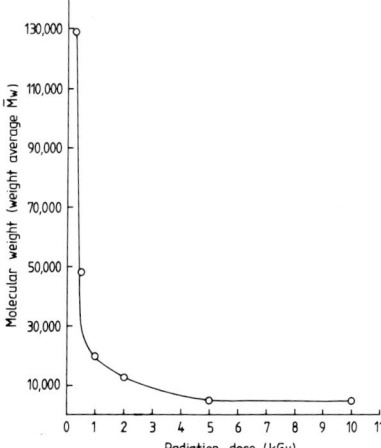

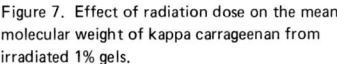

Figure 7. Effect of radiation dose on the mean molecular weight of kappa carrageenan from irradiated 1% gels.

Figure 8. Dependence of reciprocal mean molecular weight upon radiation dose for kappa carrageenan irradiated in the form of a 1% gel.

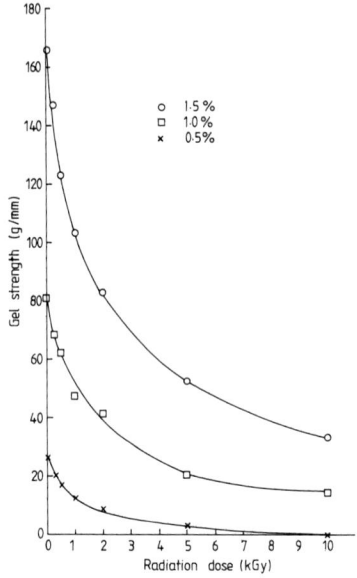

Figure 9. Effect of radiation dose on the gel
strength of kappa carrageenan gels.

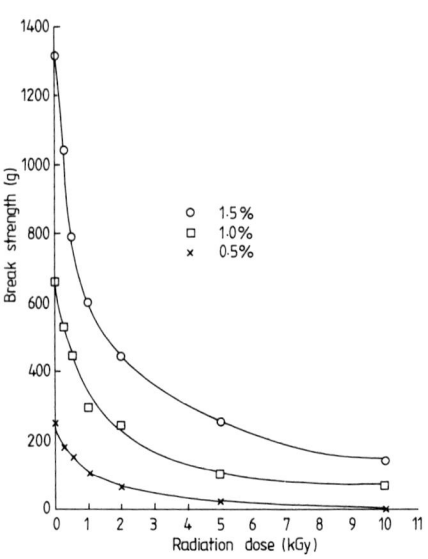

Figure 10. Effect of radiation dose on the break
strength of kappa carrageenan gels.

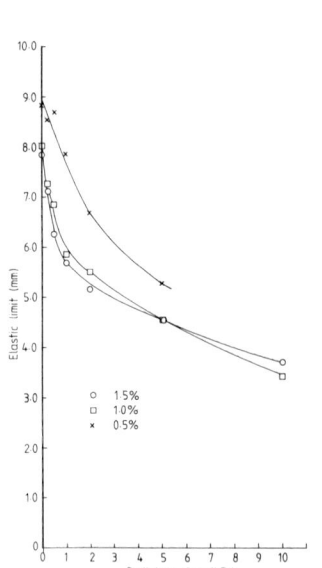

Figure 11. Effect of radiation dose on the elastic
limit of kappa carrageenan gels.

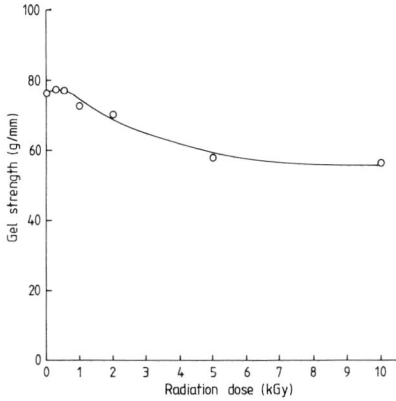

Figure 12. Effect of radiation dose on the gel strength of 1% gels prepared from irradiated kappa carrageenan powder.

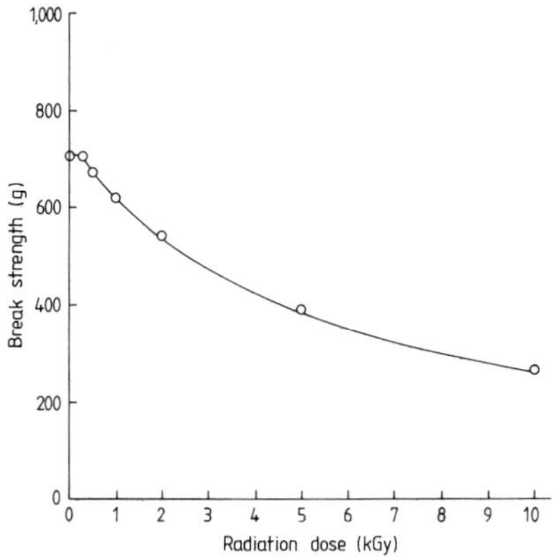

Figure 13. Effect of radiation dose on the break strength of 1% gels prepared from irradiated kappa carrageenan powder.

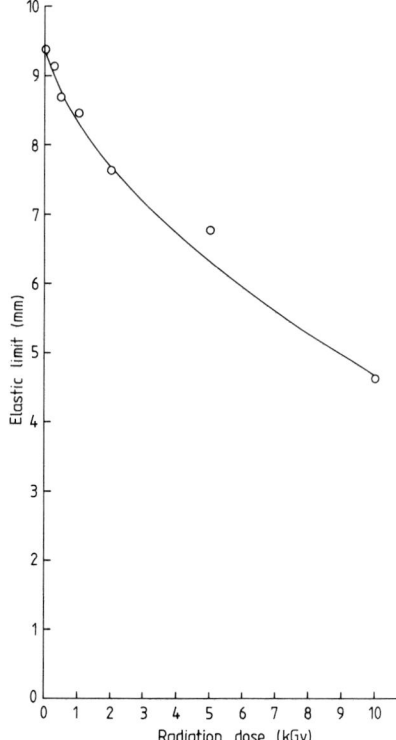

Figure 14. Effect of radiation dose on the elastic limit of 1% gels prepared from irradiated kappa carrageenan powder.

ACKNOWLEDGEMENT

This work was funded by the Gels and Thickeners Panel of the Leatherhead Food R.A. (Leatherhead, U.K.).

REFERENCES

1   Wilkinson, V.M. (1987) Fd Prod, **6** (2), 18–19.
2   Truswell, A.S. (1987) Br. med. J., **294** (6585), 1437–1438.
3   Advisory Committee on Irradiated and Novel Foods (1986) Report on the Safety and Wholesomeness of Irradiated Foods. H.M. Stationery Office, London.
4   Joint FAO/IAEA/WHO Expert Committee (1981) WHO Technical Report Series 659, WHO, Geneva.
5   Kertesz, Z.I., Glegg, R.E., Boyle, F.P., Parsons, G.F. and Massey, L.M. (1964) J. Fd Sci., **29**, 40–48.
6   Marrs, W.M. and Sworn, G. (1986) Leatherhead Fd R.A. Res. Rep. No. 573.
7   Marrs, W.M. and Sworn, G. (1987) Leatherhead Fd R.A. Res. Rep. No. 576.
8   Marrs, W.M. and Sworn, G. (1986) Leatherhead Fd R.A. Res. Rep. No. 550.
9   Tanfield, C. (1961) in Physical Chemistry of Macromolecules, pp.611–618. J. Wiley & Sons Inc.
10   Von Sonntag, C., Dizdaroglu, M. and Schulte-Frohlinde, D. (1976) Z. Naturforsch., **31b**, 857–864.
11   Schertz, H. (1974) Improvement in Food Quality by Irradiation. IAEA, Vienna.

Legislation and food starch development

D.B.Whitehouse

CPC Europe Industrial Products, Havenstraat 84, B-1800 Vilvoorde, Belgium

ABSTRACT

Legislation is one of many influences on food starch development. The ultimate consumer of a starch as a component of a processed food usually has little accurate knowledge on the subject of the components of that food but votes for the product with his or her money. The processed food manufacturer in marketing the food, attempts to meet the consumer needs profitably, whether the needs relate, for instance to convenience, or quality or price, and raw materials will be selected accordingly. Recently, a variety of pressures relating to food have arisen from consumer groups, which have exerted political pressure on governments. Governmental responses through various routes have been perceived by raw material suppliers and food manufacturers in a number of forms including fore shadowing of legislation. Legislation lags behind events and the more cumbersome the procedures, the greater the lag. From any legislative action there are winners and losers; intelligent anticipation of legislation can benefit a manufacturer. In the light of legislative and potential legislative considerations, a number of physical processes, natural products and newer techniques are reviewed, together with some consideration of well established modified food starches.

Food starch development is influenced by many of the things which affect development of all raw materials for the food industry. The producer of consumer products looks for opportunities to exploit, by marketing, his skills as a compounder, processor and packager of a series of ingredients in a form the ultimate consumer wants. If the consumer doesn't buy the product, then the product will not be made and the ingredients will not be required for that product. To achieve his objective the consumer product producer is constantly stretching his techniques to improve his products in ways which he hopes the consumer will appreciate. These improvements may be related among many other aspects, to shelf life, convenience, taste, mouthfeel, price, and to accomplish changes he often requests alternative properties from the ingredient supplier.

Legislation enters this scene in different countries in different ways, ranging from a broad legislative framework to detailed positive lists. Legislation will be written by civil servants to politicians' instructions and consumer pressure acts on the process through this route, for better or worse. In the UK in the past three years there has been a crescendo of noise on the subject of processed food, starting from a genuine desire for information and reassurance concerning E Numbers supplied with half truths from a variety of authors. A number of extremist political activists seeing some turmoil joined in, treating the whole affair as a useful anti-capitalist platform. In very diverse responses, the food industry, both manufacturers and retailers, have variously wrung their hands or spotted an 'anti-additive-natural-is-good' opportunity to exploit, often without any thought for the consequences of their actions.

The amount of noise has been sufficient to get attention from the Ministry of Agriculture, Food and Fisheries in the UK, and subsequently some excellent documents have been published by them which help put the matters in perspective. However, there is some evidence that a lot of public clamour has also caused some priorities to be shifted and for the Government to want to be seen as tough in the area of food legislation. A change in legislation always results in winners and losers. Tougher pollution laws will favour the company operating a low pollution process; a restriction on a food colour is a problem to the producer and user of that colour but not to a user of an alternative colour.

Much of the recent antagonism towards the food industry relates to "chemicals" used in processed foods, as if chemistry happens only in factories and has no place in human biology. A new "disease" has recently been described (1) called "chemophobia" - an irrational fear of chemicals which affects many consumers and politicians !

In this paper the focus is on potential legislation as far as it affects chemically modified starches, with emphasis on the word "potential". The prospect of legislation, however remote always casts a long shadow and in anticipating what might happen, in sensing the advantages of anticipating and conforming to consumer pressure, some producers will attempt to move towards alternative "natural", "physical", "non-chemical" ways of achieving what has previously been accomplished by chemical means.

Perhaps physical treatments are all acceptable. Microwave treatment is certainly respectable; very soon no home will be without one. Getting water molecules in starch granules excited under various conditions may provide some altered and beneficial properties; what about γ-irradiation ? The Chernobyl disaster has certainly (and sadly) not advanced the cause of irradiated food for preservation purposes and its use for starch treatment is not likely to have many supporters even if benefits could be proved.

When starches are heated, it is inevitable that any effects are the result of a combination of temperature and the natural

MOLECULAR WEIGHT DISTRIBUTION (G.P.C.) OF EXTRUDED STARCHES

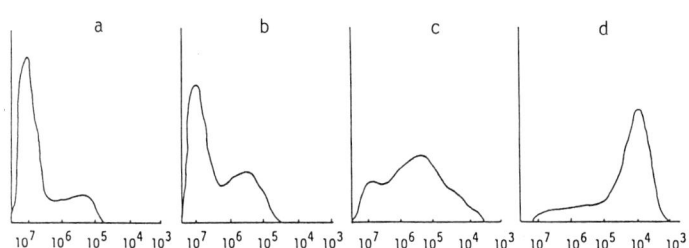

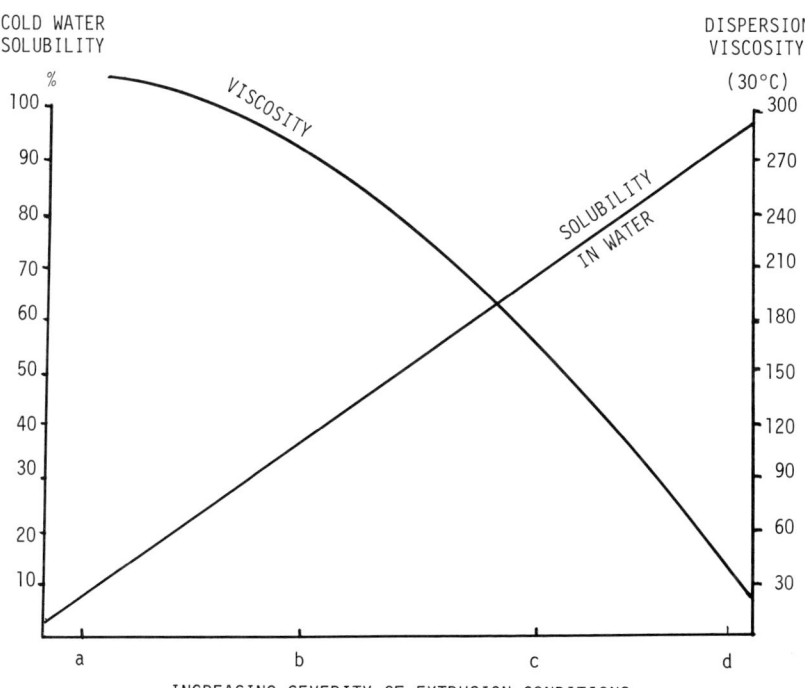

Figure 1.   Extruded starch properties.

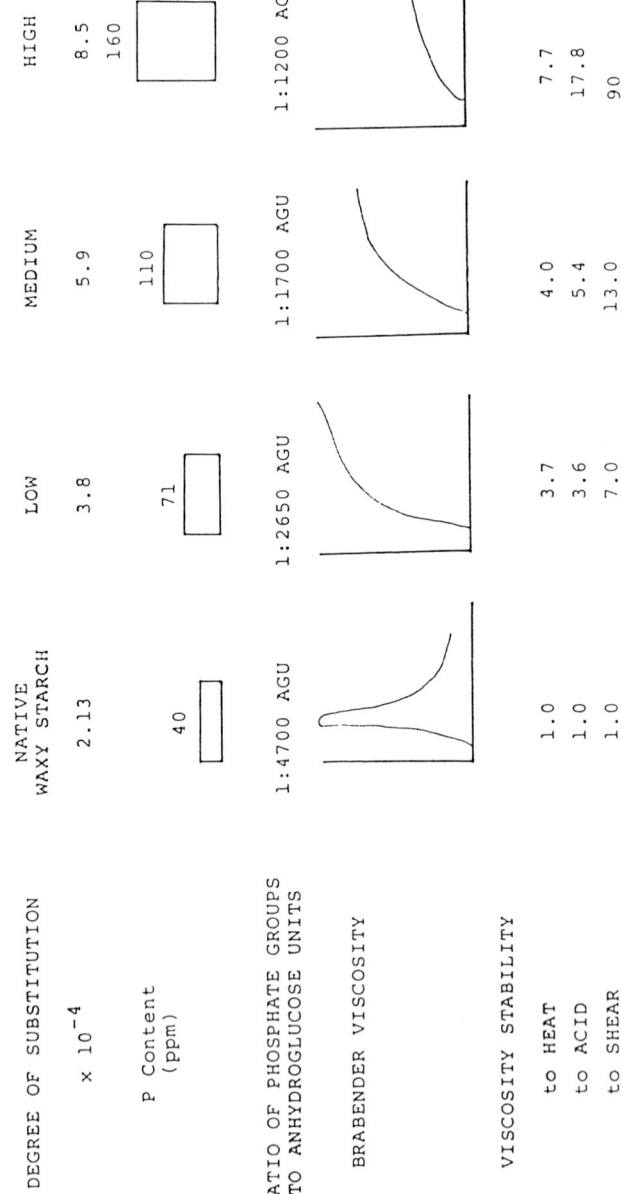

	NATIVE WAXY STARCH	LOW	MEDIUM	HIGH
DEGREE OF SUBSTITUTION $\times 10^{-4}$	2.13	3.8	5.9	8.5
P Content (ppm)	40	71	110	160
RATIO OF PHOSPHATE GROUPS TO ANHYDROGLUCOSE UNITS	1:4700 AGU	1:2650 AGU	1:1700 AGU	1:1200 AGU
VISCOSITY STABILITY				
to HEAT	1.0	3.7	4.0	7.7
to ACID	1.0	3.6	5.4	17.8
to SHEAR	1.0	7.0	13.0	90

Figure 2. Effect of phosphate crossbonding on properties of waxy corn starch.

moisture in the granule. The ratio of starch to moisture, together with the temperature, time and mechanical parameters concerned all influence the resultant "physically" modified starch.

There are a variety of techniques both ancient and more modern, through which heat and hence heat/moisture treatment can be carried out, ranging from dextrin type cooking, through roll drying to extrusion. Extrusion features shear and temperature as key variables with time being a function of the equipment in use. Viscosity, solubility, molecular weight are properties which can be affected as can be seen in figure 1.

The widespread adoption of twin screw extruders by the starch industry enables manufacturers to provide a range of physically modified starches with a wide variety of solubility and viscosity properties. Some of these starches are finding uses as replacements for enzyme converted, acid thinned and oxidised starches. The versatility of the extrusion technique results from the ability to vary the balance between the extent of gelatinisation and degradation. The higher the severity of the extrusion conditions (shear and temperature) the lower the molecular weight the lower the dispersion viscosity and the higher the solubility. In fact, "starches" with up to 100 % cold water solubility can be produced by extrusion.

Clearly there is a limit to the number of effects which can be brought about by physical processes. The chemical modifications to improve texture, reduce syneresis (by cross linking), or interrupting amylose regularity by acetylation to avoid hydrogen bond formation : these simply cannot be replicated by "physical" means.

It is worthwhile reminding ourselves of the very major changes of properties which can be brought about by very small degree of cross-bonding by phosphate groups. Figure 2 shows the changes in viscosity profile, heat, shear and pH stability which result from minor changes in phosphate content from the natural 40 ppm in waxy maize starch to 160 ppm.

In Table 1 the natural content of phosphate is shown for a series of native, unmodified starches.

Table 1. THE NATURAL PHOSPHATE CONTENT OF NATIVE STARCHES EXPRESSED AS ppm PHOSPHORUS.

STARCH ORIGIN	P (ppm)	DEGREE OF SUBSTITUTION	RATIO OF PHOSPHATE GROUPS TO ANHYDROGLUCOSE UNITS
Waxy corn	40	2.13×10^{-4}	1:4700 AGU
Corn	150	7.86×10^{-4}	1:1300 AGU
Wheat	550	2.89×10^{-3}	1:350 AGU
Potato	830	4.36×10^{-3}	1:230 AGU

The very large effects resulting from minor additions of
reagents have made analysis of finished products by traditional
techniques problematic. It is probably this situation which
has tempted legislators to regulate the production process
rather than the finished product which is an unsatisfactory
state of affairs for both industry and for regulatory bodies.

None of the physical treatments mentioned have been or are
being developed as a result of actual legislation but in
response to perceived needs with the potential of legislation
in the background. Legislation inevitably lags behind public
opinion and events, particularly positive list legislation.
There is a draft EC Directive on modified starches intended for
human consumption dated 18th December 1984 which supersedes an
original draft of 24th July 1970. The Directive is not yet
approved by the EC Council of Ministers. A three year imple-
mentation was foreseen prohibiting products not complying and
an eighteen month delay to permit products which comply. Few,
if any, chemically modified starches on the market fail to
comply with this proposed legislation. Nevertheless, it is
possible that increasing consumer awareness from labelling will
promote "chemophobia", despite legislation.

No legislator is suggesting that there is any reason why the
use of modified starches should not continue. In general the
products, which are acceptable to the Joint Experts Committee
on Food Additives of the FAO/WHO (JECFA) and the European
Community Scientific Committee for Food (ECSCF) are also
accepted by most national governments. "Chemophobia" may have
a stronger effect. If so, what alternatives are there ?

Exotic starches, natural exotic starches may have some
superficial attractions but economic factors operate to limit
their applications. The available commercial volume limits the
use and keeps the costs high. In any case the starch yield
from crops such as peas, mung beans and lentils is low compared
with the more common starch sources. No doubt genetic
engineering applied to plant breeding will produce a few
surprises. The bio-synthesis of specific polysaccharides by
microbiological routes using manipulated organisms to improve
yields, modify properties in an area of considerable interest
but which also is likely to cost a great deal of development
and safety testing time and cost. Incidentally this is a good
example of the lag of legislation behind development; most
countries are operating in this area on guidelines rather than
strict legal requirements.

The use of a variety of natural gums and in combination with
various starches has been practised for many years. Here the
producer of the finished product, guided by the raw material
suppliers is looking for a system with synergism to overcome
some of the disadvantages inherent in the native starches and
the natural gums when used alone.

The range of starch/gum combinations possible, with the many
processes, systems and products operated in the food industry
mean that model systems dealing with a minimum number of
variables are too simplistic and the possible total variations
too complex for detailed study. Consequently the development

work tends to be direct application oriented and results are very little publicized.

The effect of potential legislation through consumerist pressure has had and will have a considerable effect on the food industry, some retailers who wanted to jump on the anti-additive band wagon have found that it is not as simple as they first thought to have 'No Additives' products. If the consumer wants convenience foods, and this has been the increasing trend, there is some incompatibility with the 'freshness', 'natural', 'additive free' pressures. Food starch development must take account of consumer "preferences" seen through the eyes of the consumer product manufacturer, with an eye on potential legislation which hopefully will enshrine broad principles and not easily outdated details.

ACKNOWLEDGEMENTS

The assistance of Mr M. Fitton and Mr J. Vanhemelrijck is gratefully acknowledged.

REFERENCES

1. Chemosphere, Vol. 15, Nos 9-12 (1986)

The current toxicological status of permitted emulsifiers and stabilisers

D.M.W.Anderson

Chemistry Department, The University, Edinburgh EH9 3JJ, UK

ABSTRACT

This paper reviews the revisions of the toxicological status of the major permitted emulsifiers, stabilisers and thickeners made since this subject was last summarized at this Series of Symposia in 1983. Details will be given of some studies, in animals and in Man, providing additional evidence of safety for several permitted hydrocolloids. Initiatives taken to counteract the recent upsurge of campaigns against the use of food additives will be reviewed. There is a need for a much more vigorous and rigorous inspection system, with sanctions, to check that the consignments of additives intended for use in foodstuffs comply with the established criteria of identity and purity, for which revisions/extensions are now desirable. Only when this is done will there be reasonable grounds for confidence that the toxicological evaluations of the past 25 years, achieved at great expense, are contributing actively towards increased food safety for the consumer.

INTRODUCTION

At the Symposium held in 1983, an account was given (1) of the toxicological status of the exudate gums at that time. Since then there have been revisions of the status of several permitted hydrocolloids. Reports of studies in animals and in Man, giving increased confidence in the safety of the additives subjected to test, have also been published. The object of this paper is to present a brief review of such decisions and studies, together with some comments on the extent and justification of recent campaigns against the use of permitted food additives. It is suggested that the assurances of food safety expected by the consumer depend on a more positive system of checking that the food additives used comply with specifications of identity and purity than is given by the present system based upon the voluntary acceptance of the principles of Good Manufacturing Practice.

RECENT CHANGES IN TOXICOLOGICAL STATUS

In 1985 and 1986, upward revisions in status to "ADI not specified" were announced by the Joint FAO/WHO Committee on Food Additives (JECFA) for carrageenan (formerly 0-75 mg/kg body weight per day), and for xanthan gum (formerly 0-10 mg/kg b.w.) following assessments of new experimental evidence that had been submitted. For carrageenan, this led to the publication of a new toxicological monograph (2) and to the decision that semi-refined or unrefined native carrageenan, as exported from some producing countries, is not permitted in foodstuffs. Because JECFA's recommendations have no legal effect in any country, the mechanism by which such an important decision will be enforced presents considerable problems.

The new data submitted for xanthan gum included a dietary study in Man (3), and details (4) of the nature of the nitrogenous material present in response to requests made earlier by JECFA for further information on this point. A new Monograph for xanthan will be issued in 1988 by the W.H.O.

In 1986, JECFA also awarded the status "ADI not specified" to tara gum, a seed galactomannan from the Peruvian tree Caesalpinia spinosa; tara gum has solution properties intermediate between those of guar gum and locust bean gum. The toxicological clearance of tara gum was undertaken by a Consortium of five European companies acting as a sub-section of INEC, the international trade association established early in the seventies to protect the interests of locust bean gum and other seed galactomannans. A Monograph for tara gum will be published by the W.H.O. in 1988.

It will be of interest to learn in due course if the European Scientific Committee for Food (SCF) will adopt the "ADI not specified" status for carrageenan (E407) and for xanthan (E415) and whether the SCF will allocate an E-number to tara gum. At present, tara gum is in the anomalous position of having "ADI not specified" approval from JECFA without being included on permitted lists (for human or for animal feed-stuffs) in Europe and in the U.S.A. In turn there will have to be independent decisions regarding tara gum by the Food Additives and Contaminants Committee (FACC) and by the Committee on Toxicity (COT) in the U.K., and by the corresponding Committees in other European Member States.

Agreement was reached recently to use the number E440 to designate both pectin and amidated pectin (formerly E440(a) and E440(b) respectively).

At the end of 1985, a decision was taken to extend, for the final time, the dead-line for submissions of the final reports demanded for gums karaya and tragacanth; a final decision must be taken before the end of 1988. The reports requested were made available to the SCF and to JECFA in April/May 1987; these include an Indian 90-day dietary study of gum karaya in a non-rodent species and a 2-year combined toxicity/reproduction study of gum tragacanth in the rat. The latter study was sponsored by the American F.D.A.; the in-life phase took place in 1978-79. The results of that study will not lead to any change in the established G.R.A.S. status of gum tragacanth in the U.S.A.

RECENT STUDIES OF TOXICOLOGICAL/REGULATORY INTEREST

Electron-microscopy studies
Biological specimens, secured at autopsy from rats fed with gums arabic, karaya and tragacanth, have been processed for transmission electron microscopy. The electron micrographs have been scrutinised to confirm the absence of any abnormalities (inclusions or other pathological changes) in any of the organelles within the cells of rat heart and liver (5) and of rat jejunum, ileum, and caecum (6).

Immunogenicity studies; gums arabic, karaya and tragacanth
Allergic reactions to ingested foods or additives may lead to diseases of the gastrointestinal tract, respiratory system, skin and other tissues; a preliminary evaluation concluded (7) that gums arabic, karaya and tragacanth are capable of eliciting immune responses comparable with those elicited by hen's ovalbumin or wheat gliadin. More detailed studies have confirmed (8) these initial conclusions; purification processes reduced, but did not eliminate, the immunogenicity.
In addition, it has been established (9) that although gum arabic is a complex, proteinaceous, polysaccharide antigen, it is tolerogenic i.e. is capable of inducing oral tolerance in animals when encountered by the natural dietary route via the gut. The phenomenon of oral tolerance has not been fully established in Man but experience and circumstantial evidence (10) indicate that Man is unlikely to differ fundamentally in this respect from the other mammals examined so far (9).

The amino acids present in the major hydrocolloids
Immunogenicity is most frequently ascribed to proteinaceous components. In order to acquire additional information on various aspects of this phenomenon, the U.K. Ministry of Agriculture, Fisheries and Food sponsored a study of the amino acid components of all permitted food hydrocolloids within the period 1982-84. This has led to the publication of the amino acid compositions of gum karaya (11), gum arabic (12) gum tragacanth (13), guar gum (14), pectins (15), locust bean gum and xanthan (16). The data acquired facilitated calculations of the factors for converting nitrogen contents (by elemental analysis) to protein contents (17).

Studies of the exudate gums
At the request of a regulatory committee, the absence of rhamnose in human urine, following the ingestion of gum karaya, was confirmed (18). A dietary study of gum tragacanth in Man (19) showed faecal bulking effects but no adverse toxicological effects from the ingestion of large daily doses of gum tragacanth for 23 days. A field survey of gum production areas of Turkey led to the conclusion (20) that Turkish gum tragacanth is derived from Astragalus microcephalus, not from Astragalus gummifer as quoted in all regulatory criteria of identity. The sugar and amino acid compositions of gum tragacanth from three named Astragalus species revealed (21) large variations between them. This explains the well-known variability in commercial shipments.

The sample of gum arabic used as the Test Article in a range
of toxicological/dietary studies in animals and in Man was
completely characterized (22) and compared with samples of gum
arabic from each of the major producing countries. Gum arabic is
the natural exudate from Acacia senegal (L.) Willd. - the most
abundant Sudanese Acacia; gum arabic is not the exudate from A.
seyal or non-specified Acacia species, of which well over 900
have been identified (23). The evidence for the safety of gum
arabic derived from Acacia senegal (L.) Willd. has been reviewed
(24). The characterization of the amino acids in gum arabic (12)
has extended the long-established evidence (25) that gum exudates
are a form of proteoglycan. This has led to a reconsideration of
its structural features (26) in which the inter-relationships
between the amino acids and sugars involved in the gum macromolecules have been taken into account (27).

Gums that are not permitted as food additives

The exudate and other gums permitted as food additives are
comparatively few and have criteria of identity and purity. Many
other hydrocolloids available commercially in tropical countries
may be used for technological purposes but not as food additives,
as they have not been evaluated toxicologically and are not on
permitted lists. Analytical data for such substances are
necessary to permit their identification as adulterants for food
law compliance purposes. Gums from species of the genus
Combretum are amongst the more common adulterants present in gum
arabic, particularly from West Africa. Analytical data
characterizing the gums from eight Combretum species have been
published (28): the presence of acetyl groups, galacturonic acid,
mannose and/or xylose, and widely different amino acid compositions make Combretum gums readily identifiable.

Dietary studies in Man

In addition to the dietary studies in Man reported previously
for the exudate gums, studies were completed for carboxymethylcellulose (29) and xanthan (3); the latter played a significant
part in the recent upwards revision from ADI 0-10 mg/kg to "not
specified". Carboxymethylcellulose is a particularly effective
faecal bulking agent; xanthan gave a moderate (10%) reduction in
serum cholesterol and a significant increase in faecal bile acid
concentrations. In a review of dietary studies of hydrocolloids
in Man, it was concluded that there are no immediately discernible correlations between chemical composition and structure and
the physiological effects of several hydrocolloids that act as
soluble forms of dietary fibre (30).

Recent campaigns against food additives

Since 1984, there have been widespread activities and
campaigns, by individuals and organisations, against the use of
food additives. These appeared to have been triggered off by the
adoption of the system of E-numbers for permitted additives in
the EEC, and by their appearance on labels as required by new
legislation, which requires that only additives on permitted
lists, approved after submission of reports of detailed toxicological studies, may be used. The E-number system was intended
to present a simple method, without reference to specialised

Outlook for gum arabic production and supply

El Hag Makki Awouda

The Gum Arabic Co. Ltd, PO Box No. 857, Khartoum, Sudan

ABSTRACT

    The commodity in question is a forest product of the
genus Acacia   Senegal (L) Willd.
    It is well established in the world markets and has been
an article of commerce for some 2000 years. Its importance
lies not only on the fact that it is used in a large number
of industries but also because it is a peasant industry
providing income to the farmer at a time when it is most
needed. It is essentially an arid zone species having an
even greater importance in protection and improvement of
soils, in providing fodder, in checking desert encroachment
and indeed in facilitating further periods of agricultural
cropping in an already depleted and fragile environment. As a
result of Gum Arabic's importance to both producing and
consuming countries and because nearly all gum is produced
in Africa and consumed in the industrial world of Western
Europe and U.S.A., certain inherent problems exist. These
problems are as varied and different as are the differences
between the developing countries of Africa and those of the
already developed countries of the Western World. This
paper identifies some of these problems - and highlights
the policies adopted by the Sudan to promote production and
improve the supply situation.

INTRODUCTION

    The species has a wide distribution and remarkable
adaptability. It is essentially a semi-arid zone species
but is so adaptable that it is not only drought resistant
but is also frost hardy. It can regenerate naturally from
seeds or vegetatively from coppice. It occurs throughout
the length and breadth of Africa south of the Sahara desert
and also in some parts of India. Commercial production of
gum arabic is, however, restricted to certain zones and
almost all of it is produced in Africa north of the equator.

Further limitation of commercial production within the
African continent depends on the degree of uniformity of
the species. In the Sudan and particularly in Kordofan and
Darfur provinces the species is uniform and is found in pure
stands giving the Sudan the advantage of being the biggest
producer and exporter of the best qualities. In other
producing countries the distribution is not so uniform and
is often found mixed with other species. Another important
comparative advantage is that in the Sudan it occurs both
wild and cultivated in a wide area giving the advantages
of economies of scale. The main concentration of the species in the Sudan as given by Smith, Harrison and Jackson is
roughly the 10-15°parallel with rainfall ranging from
300 - 600 mm in sands and 600 to 800 mm in clays.

Factors Affecting Gum Production and Supply
    Categorically, factors affecting gum production may be
grouped into four main headings as in diagram No.1. The
physical factors (soils, topography, and climate) although
fairly obvious may potentially have a direct and indirect
bearing on production; directly through their effect on
growth and exudation of gum and indirectly through their
effect on agricultural production and consequently on the
social and economic conditions of the farmer. Among the
many operative physical factors, the climate may be
singled out as the most important. This is based on the
assumption that the soils are uniformly sandy and poor and
that gum grows throughout the belt only if climatic
conditions are favourable. Water, of course, is essential
to growth but the average annual rainfall within an area
seems to have little effect on total production. Its
effects may be long term but what is more important is the
local distribution of the rains. The length of the rainy
season is also a limiting factor. Longer rainy seasons
prolong the growing season for both agricultural crops and
trees. On the one hand this is beneficial because it
produces healthier trees and better agricultural harvests,
but on the other hand it is limiting to gum production by
either keeping the farmer busy on agricultural harvesting
or by prolonging tree growth up to the cold spell. In the
first case early tapping is being foregone for agricultural
production and in the latter case tapping is useless before
the trees shed their leaves.
    The biotic factors are potentially dangerous. The whole
of the gum belt could be defoliated overnight by locust
(Acridium melanorhodn), thus checking the growth on the one
hand and inducing re-growth on the other. Repeated fires
could have the same effect. Fortunately this state of affairs
does not happen frequently nor does it occur uniformly
throughout the belt. The locust attack is intermittent
and sporadic. What is more frequent, however, is the damage

Diagram No. (1)

Factors Affecting Production
of　Gum　Arabic

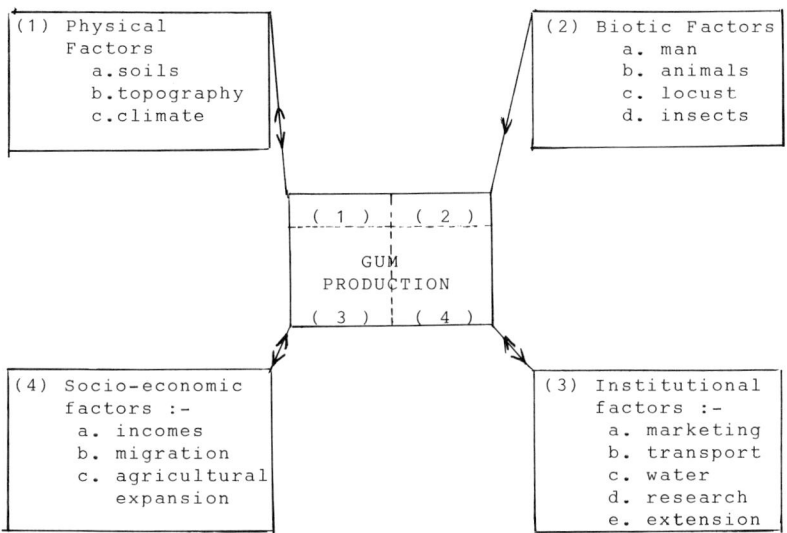

done by man and his animals. With increasing population,
greater pressures are being exerted on the land and harmful
practices, such as over-grazing, felling of immature trees
for agricultural expansion, and repeated grass fires, are
all very common.

On the significant trend in both the physical and biotic
factors is that they vary　widely from year to year and
from one locality to another. Such a variation in my opinion
suggests that these are only potentially effective and that
their actual effect on total production seems to be small,
at least in the short run, because a drop in production
attributed to these factors in one locality is counter-
balanced by an increase of production in another. It appears
that the most important factors that seriously hamper total
production are all man induced.

Most of the official reports and the work to the different
committees formed in the Sudan to that effect, attribute the

loss of production in recent years mainly to the climatic
and biotic factors. This has been an overemphasis on
factors which have been in operation for scores of years
and varied in the same manner even when production was
rising. The more important operative factors seem to be
social and economic concerning the resource and the farmer.
   The first and most important factor is that related to
the farmer's income. The general and widely held belief
is that the greater the income the farmer gets from
agriculture the less attention he gives to the production
of gum. This assumption appears to be invalid because it
is apparent from surveys that families with greater incomes
from agriculture also have greater incomes from gum. The
explanation of this is simple; the greater the income the
farmer gets from agriculture, the more likely is that he
will stick to the land and not migrate for cotton
collection in the Gezira Scheme. By not doing so he has
more time to tend his gum gardens and work them to generate
more cash. It was also evident from the survey that those
who migrate are those who generate more than 93% of their
incomes from agriculture. They are in the lower income
group from agriculture and their gum gardens, due to the
lack of supervision and maintenance, have either degenerated
or are in the process of doing so. They therefore have not
yet got enough income from agriculture to support them till
incomes from gum are realized and they have to migrate.

Distribution Channels and Marketing
   These channels are simple and governed by the social and
economic forces operating in the Sudanese economy. Like all
other Sudanese products they consist of three sub-systems,
ranging from a subsistence producing and to semi-modern
exchange sub-system and at the receiving end to a modern
export-oriented system.
   The producers' role in the system is instinctive and
traditional. The Gum Arabic is part of their wealth and
in working it they usually have to choose from two possible
courses of action. Either to work it through the family
labour or to use hired labour. The latter is done when
the areas are too big to be covered by the producer and
the practised hire system is on a half-share basis.
   The real marketing difficulties in this sub-system start
even before gum arabic is produced. If the producer is
urgently in need of cash, he will have to get his essential
supplies from the village merchant in return for the advance
sale of the expected produce.It is important to note here
that both prices for the essential supplies and the sale of
the expected produce are fixed by the village merchant, and
the producer, being a victim of the circumstances, has no
alternative but to accept. In other places where there are

water shortages, part of the gum produced is exchanged for water. Theoretically every producer should sell his gum in the auction markets but in practice there are so many hazards that they are hampered from doing so. Among these, four are evident :-

(a) Lack of cash.
(b) Lack of transport.
(c) Lack of water.
(d) Lack of labour.

The exact percentage of the producers who do not take their produce to the auction because of some of the above factors is not known. Two sources, however, give two different figures (30% and 80%). Both figures, however, seem to be off the mark, one is low and the other is high. In the opinion of the writer a figure of 50% could be closer to reality.

Marketing Flow Diagram of Gum Arabic From
Producer to Consumer

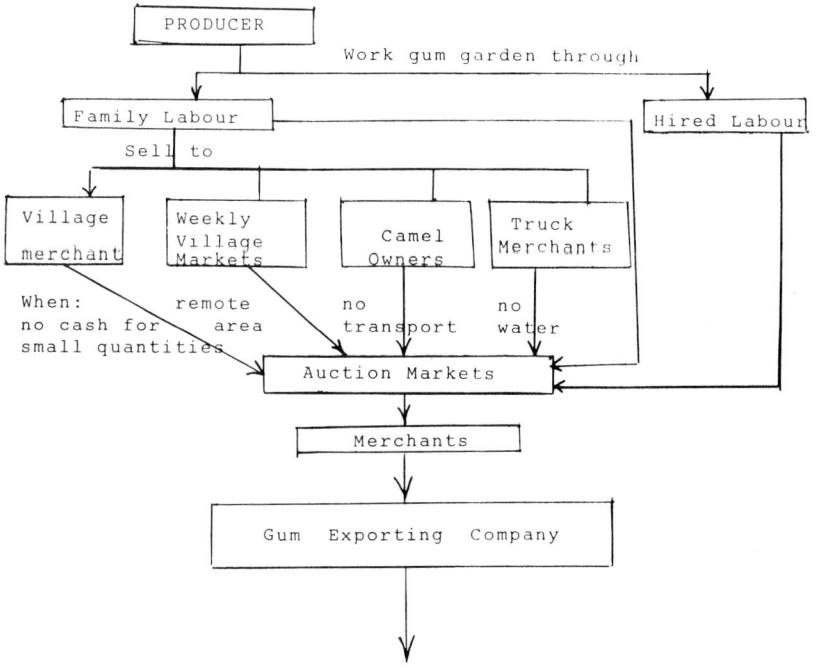

This sub-system has four main drawbacks :-

(a) The large numbers of producers, the small size of their
     individual holding and the extensiveness and remoteness
     of the production area render the concentration process
     a very difficult and expensive stage in marketing.
(b) The sub-system is rather passive and contributes very
     little to the value added to the product. It does not
     perform some of the obvious marketing functions of clean-
     ing, sorting out, or grading, . etc...
(c) The marketing functions performed are not motivated by
     the dynamic forces seen in the other parts of the system
     but they are the outcome of necessity, tradition, and
     local conditions.
(d) From the diagram it seems that there are numerous
     options open to the producer in handling and distributing
     his produce but most of them fall outside the organized
     markets for the commodity. Consequently the market prices
     are rarely reflected at the village level.

     In the second sub-system there are the auction markets
which were first introduced in 1922 and were made to bring
both sellers and buyers under one roof. These auction
markets play an important marketing role by facilitating
the concentration of the produce and achieving the benefits
of bulk handling. The essential features of these auctions
arethat the buyers and sellers come together and the government clerk auctions the gum. Buyers bid and sellers usually
settle for the highest bid. The gum is then packed, weighed,
and accounts settled.

     There are two types of market as well as two types of
buyer. The main markets in the capital of the province or
the rural council, where there are bigger and better
facilities and the buyers are usually merchants who are
resident in the area and are of reputed financial status.
The subsidiary markets are in smaller towns or bigger
villages. The buyers are usually representative for the
merchants in the main markets who perform the same job on
a commission basis.

     Both groups of markets and middlemen perform the various
marketing functions of buying, storing, cleaning, grading,
packing, transporting, and selling to the Gum Exporting
Company, ex Port Sudan.

     After the auction sales the merchants transport the gum
to large storing sheds where cleaning, grading, and
repacking is carried out. The nodules and pieces of
gum completely free from dirt and pieces of bark etc.. are
picked out. These which consist of uniformly shaped, medium
size, light-coloured nodules are kept separate and are known
as "Selected Gum". The irregular shaped or darker-coloured
nodules together with clean fragments are known as "Cleaned
Gum". The bulk of the gum is exported as either clean or
selected. The remaining gum with pieces of bark, dust, etc.
adhering to it is broken up by beating with sticks. It is
then winnowed and impurities separated and removed. The
residue is then sifted. That portion remaining in the sieve
is known as "Gum Sifting" and that passing down through the
sieve as "Gum Dust". In former days gum was exported without

cleaning and this is known as "Natural Gum" while other
cleaned gum is left out in the sun to dry till it cracks
and assumes a white colour, hence being given the name
"Bleached Gum " .
    The third and last sub-system in the distribution channel
is the exporter. All clean and graded gum is sold to the
Gum Exporting Company at Port Sudan. The Company provides
certain marketing functions especially those of time and
place and form utilities for the commodity - and carries
out all matters related to the export of the commodity to
the various parts of the world.

THE CHANGING SITUATION AND MEASURES FOR IMPROVEMENT
A) The Changing Situation
    Like all other arid zone regions of the world, the gum
belt faces more difficult problems then ever before. The
desert seems to be enlarging and moving Southwards
contributing more and more to the economic devastation
the region.
    The situation was further aggravated by the frequent
droughts that has stricken the area. Serious implications
were noted and recorded :-
    a) Because of poor pastures greater pressures were
        exerted on the natural flora and hence considerable
        damage to trees to provide fodder and other needs was
        experienced in many parts of the area.
    b) Because of successive seasons of droughts, growth was
        retarded, trees were weakened inducing failure in
        production.
    c) Because of deterioration in soils there is also a
        marked reduction in agricultural production per unit
        area in some parts of the belt.
    d) Because of failure to grow agricultural crops
        low income - people tended to leave the land and migrate
        seeking other jobs or better prospects.
B) Measures for Improvement
    (i) Improvement of Production :-
        The main problem with gum production in the Sudan is
    a planning one. Until very recently the resource was
    considered to be a peasant industry and the whole
    system was left to run itself without any intervention
    from the state.
        After feeling the effects of the successive
    seasons of drought policies had to be changed to cope
    with the changing situation.
        Before embarking on a new policy the whole of the
    gum belt was surveyed in depth. Based on the outcome
    of these surveys - specific measures were adopted as
    a standard policy, with the main aim of sustaining
    production and security of supplies.
        A yard stick was developed (See Table 1) as a
    guideline for policy decision making. The criteria
    is simple and effective classifying the gum belt of
    Sudan into zones of similar problems. Each zone would

require a special treatment to bring about the required
changes.

Table No. (1)
              Check List for policy decision making

Points to Consider	1	2	3
Availability of gum gardens	good	medium	poor
Growth conditions	good	medium	poor
Distribution in the area	good	medium	poor
Stocking per unit area	good	medium	poor
Conditions of pastures	good	medium	poor
Agricultural production	increas-ing	constant	decrea-sing
Migration of labour	neglig-able	moderate	consid-erable
Is soil production needed	no	yes	yes
Policy decision	conserve and develop product-ion	enrich and improve distrib-ution channel	introd-uce cultiv-ation cycle

As it appears from the check list the gum belt can be
divided into three zones :-
Zone (1) - Northern Part (North of Lat. 13°N)
This zone has been seriously affected by the drought -
characterized by :-
(a) Loss of agricultural production.
(b) Considerable destruction to trees.
(c) High rate of migration.

Policy decision in this respect is to restore and
rehabilitate by introduction of the gum cultivation
cycle. It is a balanced form of landuse - well
suited to the area. It offers the farmer a package deal
for the promotion of agriculture, gum arabic and pastures.

Zone (2) - Middle Part (Between Lat. 12 - 13°N)
This zone is still in tact. It has not been affected by
the drought and all production conditions are favourable
thus representing the main area for gum production policy
decisions include :-
(a) Enrichment planting to fill in gaps.
(b) Rehabilitation of mechanized farming schemes.
(c) Creation of large scale industrial plantations.
(d) Improvement of production techniques and distribution
    channels.

Zone (3) - Southern Part (South of Lat. 12°N)

The zone represents a new naturally regenerated area - it is extensive with young trees - and has a considerable potential for future production. There is a general South-ward shift of the vegetation zones - the desert is moving Southwards and so is the gum belt. What has been lost in the Northern zone has been compensated in the Southern zone. Management prescriptions in this zone include :-

(a) Demarcation, surveys and reservation.
(b) Initiation of gum production through provision of basic needs, encouragement of producers and sowing productions' problems.
(c) Promotion and improvement of gum production, through training of producers and intensification of extension service.

These three plans have been going on in all the three zones for the last seven years financed by the Sudan Government, the Gum Arabic Company and the International Aid Community.

(ii) Improvement of distribution channels

To improve the distribution channel and to facilitate the smooth running of supplies producer associations were formed in the main producing regions Kordofan, Darfur, and Kassala. They are now shareholders in the Gum Arabic Company and are represented on the Board of Directors. They take full share in policy making regarding production and marketing. A two way information flow system has been initiated from the producing end up to the exporting and visa-versa.

The suggested improvement is to shorten the distribution channel by creating a direct relationship between the producer and the Gum Arabic Company - as is shown in the supply flow chart.

The New Suggested Distribution Channel

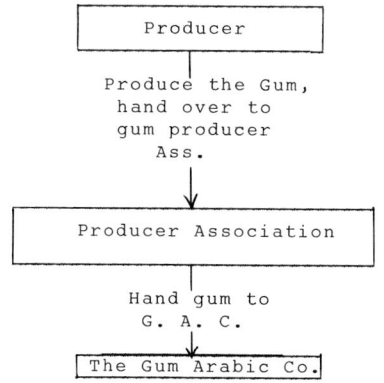

A start is being made to introduce the system this current season 1986/87. It will run concurrently with the old system and gradually will replace it.

It is apparent that the suggested system has many advantages .
   (a) It is direct, simple and quick.
   (b) No gum would be stored along the pipeline;delivery is assured and also shipment prompt.
   (c) Producers are getting the ruling market prices as a direct reflection of the International prices. This fact will motivate the whole system with regard to planting more trees, conservation of resource and promotion of production.
   (d) The producer is no longer passive in the marketing chain and can perform the marketing functions of cleaning and grading, thus getting the benefits of the value added.
   (e) The system eliminates all middle men and all the damaging practices that used to prevail.
   (f) The system gives full control of the Gum Arabic Company over the distribution channel and the smooth flow of the commodity.

CONCLUSION
   According to plan by 1990 all zones would be in full production. Each of the three zones would be capable of producing an average of 20,000 Tons. In any particular season - two zones would be productive to satisfy the world requirement while the third zone would be kept as a buffer stock of trees reserved for abnormal conditions of either increased demand or reduced supplies. By doing so a triple security system would have been initiated :
   (a) Buffer stock in the form of gum.
   (b) Buffer stock in the form of tree reserves - to be tapped when needed.
   (c) Buffer stock of trees in the form of improved varieties;high yielding - drought resistant.
   All is geared to maintain a triple long range buffer system and sustain the yield and secure supplies.

The microflora of gum arabic and their significance to gummosis

A.S.Khalid, O.H.Mohammed, S.O.Mahgoum[+] and S.A.Khalid[*]

Food Res-Centre, PO Box 213, Khartoum North, Sudan
[+]Biochemistry Department, Faculty of Agriculture, University of Khartoum,
PO Box 1996, Khartoum, Sudan
[*]Department of Pharmacognosy, Faculty of Pharmacy, University of Khartoum,
PO Box 1996, Khartoum, Sudan

ABSTRACT

Seventy-two specimens of Gum Arabic of Sudanese origin from trees of Acacia Senegal (L) Willd. of known ages have been screened for their microflora and their relevance to the gum formation process "Gummosis" is discussed. Five genera of moulds were identified. These are mainly, Asperigillus, Penicillium, Rhizopus, Gilocladium and Cladosporium. The bacteria characterized are gram positive spore formers, gram positive non-spore formers and gram negative rods. Morphological and biochemical (including immunological) evidence proved that Sudanese Gum Arabic is free from pathogenic micro-organisms specifically salmonella and coliforms.

INTRODUCTION

Gum Arabic (GA) is usually defined as a dried gummy exudate which is mainly obtained from the stems and branches of the leguminous genus Acacia Senegal (L) Willd., and occasionally from other taxonomically allied taxa within the genus Acacia. Structurally it is a complex highly branched heteropolysaccaride - in spite of its early investigations, a definitive and unambiguous structural interpretation is still lacking.
The principal source, A. Senegal, is a tree about 6 m high. This is especially true in the Sudan where 80% of the marketed gum is supplied by this species. A. Senegal, generally produces the most substantial amounts of gum between 5 and 30 years of age.
It is generally accepted that gum arabic is only produced under stress.
Many hypotheses have been put forward to elucidate the process of gum formation (i.e. gummosis). One of the most fascinating hypotheses, reported was by Blunt (1926) and later

by Malloy (1972). Both authors claimed that gummosis is
accelerated by microbial infection. They also indicated that
they had isolated micro-organisms which played certain vital
roles in gum exudation and/or formation.

Smith (1902) has earlier published an account on the
isolation of two kinds of bacteria from the twigs of A.
Penninervis. The dominant type, which he calls Bact. acaciae,
produced, when grown on artificial media, a slime from which a
gum of arabino-galactan class was obtained. Rutland (1906),
however, questioned Smith's view of the bacterial origin of gum.

A voluminous literature is available discussing the taxonomy
(Vessal, 1983) of the genus Acacia, the chemistry (e.g.
Anderson, 1966 Defaye and Wong, 1986) and functionality (Snowden
et al, 1987) of GA.

Real progress has been made by some international agencies
such as The Food and Drug Administration (FDA) since 1972
regarding the immunogenicity and, teratogencity. Most notably,
the tremendous progress we have been witnessing in the last
eight years or so with the active involvement of INGAR in
research related to the pharmacology and toxicology of GA. This
includes some safety aspects such as metabolism and
allergenicity (Anderson, 1986).

On the other hand, very little or no real attention has been
given to any systematic studies of the microbiological aspects
of GA. Moreover, most recently some questions were raised by
some GA users about the possible microbial contaminants
associated with the gum without providing any scientific
evidence.

As a response, some pharmacopocial monographs have been
introduced (e.g. USP/NF) imposing certain microbiological
specifications regarding absence of salmonella spp. and
coliforms.

Based on the above premise, it is thought desirable to embark
into a systematic long term research programme to screen
Sudanese GA for its microflora, their metabolites and their
possible role in gummosis and gum quality.

MATERIALS AND METHODS

Twelve A. Senegal (L) Willd. trees were ramdomly selected
from the El Dali area (new Acacia Senegal plantation South East
of Khartoum). Trees tapped were 3-, 5-, 10-, 15-, 20- and 25
years of age.

Every four trees of the same age group received a similar
pretapping treatment and coded by the letters a, b, and c to
designate replicates of the same age group subjected to the same
treatment.

Tapping was carried out in the traditional way using an axe
over two consecutive seasons (1986 and 1987).

The bark and stripped surface (inner cambium) of the trees designated with letter a were surface sterilized by spraying with 0.2% formaldehyde (first season) and 0.4% sodium hypochlorite solution (second season). The treated surfaces were covered with U.V. sterilized tin-foil and polythene to protect the exudate from contact with airborne organisms.

Barks from three lots of trees marked b were sterilized with 0.2% formaldehyde and 0.4% sodium hypochlorite solution in the first and second seasons respectively.

Exudates from both groups a and b were aseptically collected and packed in labelled ultra-violet sterilized polythene bags.

As a control group the c labelled trees received no sterilization. The exudates were similarly collected in sterile polyethene bags under aseptic conditions.

The exudates were consequently screened for fungi and/or bacteria on the following media:

(i)   Czapek-Dox agar (CZA) for moulds and yeast.
(ii)   Nutrient Agar (NA) for general viable bacteria.
(iii)   MacConkey's broth for coliform presumtive test.
(iv)   Selenite broth annd salmonella, shigella agar (S.SA) for salmonella.
        The isolated organisms were subcultured in brilliant green agar (BCA) for confirmation, violet red bile agar for confirmation of coliform particularly esherichia coli type I.
(v)   Cooked meat media for pathogenic anaerobic bacteria e.g. Clostridium.

All the above media were inoculated in triplicate. Bacteria were incubated at 37°C for 48 hours while moulds and yeast were incubated at 30°C for up to 10 days depending on the species growth rate. Growth is judged by production of the characteristically coloured spores. The organisms were tentatively identified at the species level, whenever appropriate (Harrigan and Moccance, 1976).

RESULTS AND DISCUSSION

It is apparant from table 1, that moulds isolated were confined only to five genera, namely Asperigillus, Penicillium, Rhizopus, Gilocladium and Cladosporium. By and large, all those organisms are soil inhabitants and can be dispersed by air.

As expected in such a tropical environment, Asperigillus Niger and Penicillium were the most dominant moulds, with a strikingly consistent pattern of distribution in the 20 years sample over the two seasons, regardless of the type of treatment. The peculiarity lies on the resistance of these organisms to the sterilizing chemicals used. However they may have been present in too high a density to be eradicated by the sterilizing agent. On the other hand, Asperigillus terrus seems to be predominantly confined to treatment a in the sample of the

TABLE (I)   PATTERN OF DISTRIBUTION OF MOULDS IN GUM ARABIC

Age in Years	Treatment	Asp-Niger		Asp-Terrus		Pencillium spp		Rhizopus spp		Gilocladium spp		Cladosporium Cladosporides	
		1st	2nd	1st	2nd	1st	2nd	1st	2nd	1st	2nd	1st	2nd
3	a	*	*		*	*	*						
	b												
	c												
5	a	*	*	*	*	*	*	*		*			*
	b												
	c												
10	a	*	*		*	*	*					*	
	b												
	c	*	*			*	*						
15	a	*	*										
	b				*	*	*						*
	c					*	*	*					
20	a	*	*			*	*						
	b	*	*		*	*	*						*
	c	*	*			*	*						
25	a	*	*				*						
	b												
	c												

younger age groups (3-, 5- and 10- years) particularily in the second season. Whereas in treatment b this species was detected only in the 15 and 20 years samples. This may indicate that Asp. terrus is very sensitive to formaldehyde, even at the low concentration (0.2%) used. Its appearance in the five years sample could possibly be as a result of contamination.

Rhizopus spp and Gilocladium appeared only rarely in two age groups. Though Cladosporium Cladospordides was isolated in the second season in samples subjected to both treatments a and b it was isolated only once in the first season. The frequent association of this species of mould with Acacia trees from our previous observation (SUNGAR report) leaves room for speculation of its possible role in initiation and/or formation of gum.

However, its absence from the first season and its entire absence from control in both seasons may rule out, herein, the suggested possible role in initiation of gum.

At this point its worth mentioning that yeasts were not detected in all samples. Previously the presence of moulds in Gum Arabic have been reported without identification (Blunt, 1926 Greig, 1908 and Vollards, 1972).

Table 2 shows the bacteria isolated from the same gum samples. It was noticed that gram +ve, short rods, spore-formers, and mesophobic bacteria were the most predominant since they were isolated from samples of all age groups over the two seasons. Surprisingly they were detected in samples from trees subjected to surface sterilization. They seem to be resistant to the chemicals applied. However it is not unusual to find these bacteria in tropical and subtropical areas.

The gram positive non-spore formers were isolated from few samples over the two seasons but not including the 3 x 25 years groups. They were also detected in samples treated with the various chemicals revealing some resistance to the surface sterilizers at the concentration used.

The gram negative (gm -ve) rods, (suspected for salmonella) were subjected to comprehensive confirmatory tests both biochemical and immuniological, such as production of hydrogen sulphide from cystine, reaction on different sugars, indole production and antibody-antigen reaction.

All these tests proved that the organism is not salmonella. Coliforms, on the other hand, were detected in samples from the first season and persistently on the first treatment a based on the Ejkman Test. These were not found to be pathogenic i.e. not to be E.Coli type onc. (Cowan and Steel 1979).

The gram +ve cocci were very scarce in appearance. When subjected to biochemical tests they were found to be micrococci.

ACKNOWLEDGEMENTS

The authors extend their sincere thanks to SUNGAR for providing material support for this work. One of us (O.H.M.) wishes to thank the (AGRENER Sudan) for the award of a scholarship.

TABLE (2)   PATTERN OF DISTRIBUTION OF BACTERIA IN GUM ARABIC

Age in Years	Treatment	Gram +ve Spore-forming Bacilli		Gram +ve non-spore forming Bacilli		Gram -ve Bacilli Samonella		Gram -ve Rods (Coliforms)		Gram +ve Cocci	
		1st	2nd	1st	2nd	1st	2nd	1st	2nd	1st	2nd
3	a	*	*					*			
	b										
	c										
5	a	*	*		*	*		*			
	b										
	c										
10	a	*	*			*		*		*	
	b										
	c			*							
15	a	*	*					*			
	b	*	*								*
	c			*							
20	a	*	*					*			
	b	*	*		*						
	c			*						*	
25	a	*	*					*			
	b										
	c										

REFERENCES

1. Anderson, D.M.W. (1986) Food Additives and Contaminants
   3, No. 3, 225-230.
2. Anderson, D.M.W., Hirst, E.L. and Stoddart, J.F. (1966),
   J. Chem. Soc. C, 1959-1966.
3. Blunt, H.S. (1926) "Gum Arabic with special reference
   to its production in the Sudan", Oxford univ. press.
4. Cowan, S.T. and Steel's (1979) "Manual for the
   Identification of Medical Bateria", Cambridge univ.
   press.
5. Defaye, J. and Wong, E, (1986) Carbohydrate Research,
   150, 221-231.
6. Harrigan, W.F. and McCance, M.E. (1976), "Laboratory
   Methods in Food and Dairy Microbiology" Academic Press.
7. Malloy, M.J. (1972) "Report on Gum Arabic Research in
   Ethiopia" cited in Awouda, E.M., 1973; Social and
   Economic problems of Gum Arabic industry" M.Sc., Thesis,
   Oxford.
8. Rutland (1906) Ber Deutsch. botan. Ges., SSIV, 393.
9. Smith, G. (1902) Proc. Linn. Soc. of N.S.W., part III,
   24th September.
10. Snowden, M.J., Phillips, G.O. and Williams, P.A., (1987),
    Food Hydrocolloids, 4, 1, 291.
11. Vassal, J. (1973) Acquisitions recentes dans les Domaines
    des hydrocolloids vegetaux naturels. Actes des purnees
    Internationales d'etude et de prospective. Presses
    universitaires d'aix, France.
12 Vollard, (1972) "Microbiological Studies on Gums and
    Hydrosoluble Natural vegetable colloids Marseilles,
    France. Third European symposium, organised by Iranex.

POSTER PRESENTATIONS

Rheological characterisation of kappa carrageenan gels

C.Rochas and S.Landry

Centre de Recherches sur les Macromolécules Végétales, CNRS-Grenoble, BP 68, 38402 Saint-Martin d'Hères Cedex, France

ABSTRACT

The stress-strain behaviour of kappa carrageenan gels was tested by compression. After the gels had been aged for one night, the elastic modulus and the yield stress were recorded. The logarithm of the modulus was linear with the logarithm of the polymer concentration but the yield stress was linear with the concentration

Key words :
Kappa carrageenan, gel, rheology, mechanical tests.

INTRODUCTION

The effect of electrolytes on the gelation of carrageenan solutions has been investigated because of the specificity of this phenomenon (1,2) and because of its importance in the food industry. The addition of alkali metal ions to a solution of kappa carrageenans induces a coil-helix transition of the macromolecule and consequently the formation of a gel. The rheological properties changed remarkably with a salt addition. Generally the rheological properties are obtained by dynamic experiments. In a first attempt we have preferred to use compression tests because these experiments are more similar to the tests used in industry. The mechanical properties for small deformations, the Young's modulus and the ultimate properties, the yield stress were recorded.

MATERIALS AND METHODS

The kappa carrageenan sample was supplied by MRS (Baupte, France). The gels were prepared as follows.The dried powder of carrageenan was heated at 100°C for 15 minutes in KCl solution. The solutions were poured into a glass tube $\emptyset$ = 17 mm. After cooling at room temperature the gel was cut into small cylinders $\emptyset$ = 17 mm and H = 17 mm, and the gels were relaxed in the KCl solution   for one night. The stress and the strain were recorded on a 4301 INSTRON machine. The stress-strain curve was linear with the stress less than 0.055. For every polymer concentration investigated ten samples were tested and the average is reported in the text. The volume of the gel after relaxation in the solvent was controlled by weighing. This volume was taken into account in the calculation of the polymer concentration inside the sample of gel. The gels prepared with a polymer concentration above 10 g/L showed no volume change after immersion for 1 night in the solvent.

RESULTS

The influence of the crosshead speed was investigated (Fig. 1).

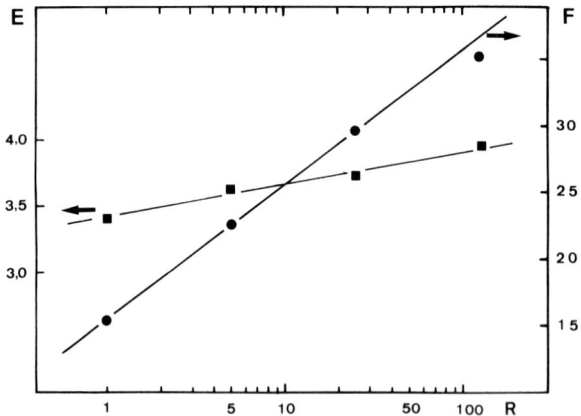

Fig. 1 : Yield stress, F, (N) and elastic modulus E (10^5 dyne/cm^2) as a
        function of the rate of compression R (mm/mn) of the gels. Polymer
        concentration 10 g/L and KCl concentration 0·1 M.

The yield stress changed drastically with the crosshead speed, but there was no significant change in the modulus. This effect is probably due to a structural change of the sample during the experiment. If the crosshead speed is very low a small amount of solvent is squeezed out during the experiment. With a crosshead speed of 1mm/mn, 25 % of solvent was drained off the gel, consequently the concentration of the gel was changed and probably the crosslinks of the network are also modified. To overcome this problem, a crosshead speed of 25 mm/mn was chosen for every experiment because a maximum of 4 % of solvent was squeezed out. The aging did not induce a very significant change of the yield modulus and of the yield stress. (Table I). The small change recorded showed a very slow structural change of the network with time. To standardize the experiments, samples were aged overnight.

Table I : Yield stress F and elastic modulus E as a function of time.

Time (h)	1	2	6	19	24	72	192
F(N)	27.7	27.5	26.7	28.8	27.7	27.8	25.3
$E(10^5$ dynes/cm^2)	3.34	3.31	3.28	3.36	3.43	3.58	3.60

Using ten samples, the reproducibility of the experiments was as follows: $\pm$ 1 % for the elastic modulus and $\pm$ 5 % for the yield stress. This difference of reproducibility is normal if we consider that the mechanical property at small deformation (elastic modulus) is a material property and that the ultimate property (yield stress) is a sample property. Consequently the value of the yield stress of one sample is dependent on the different small stress supported by the sample before the experiment. These small stresses can be due to the removal from the mould, to the cutting etc. This is easily shown by the following experiments. Ten holes, $\emptyset$ = 0.8 mm, were made with a needle perpendicular, or parallel to the axis of the cylinder of gel. The modulus (Table II) changed very slightly with or without these artificial defects, but the dependence of the yield stress was drastic.

Table II : Effects of artificial failures on the modulus E and on the yield
stress F. Polymer concentration 10 g/L and KCl concentration, 0.1 M.

Failures	$E(10^5 \ dyne/cm^2)$	$F(N)$
Without	3.40	29
Perpendicular	3.64	11
Parallel	3.68	16

A power law dependence of the modulus on concentration has been
generally reported. Very often the rubber elasticity is used to explain this
law dependence with a slope of 2. Nevertheless rubber elasticity cannot be
used to explain the mechanical properties of kappa carrageenan. Due to the
helical conformation of this polysaccharide we had to consider this gel as a
network of rigid rods. We have tested the relationship between modulus or
yield stress and polymer concentration over a large scale of concentrations
(0.25 to 5 %) with different salt concentrations to obtain basic values for
further theoretical computations.

The relationship between yield stress and the polymer concentration
(Fig. 2, Table III) was linear.

Table III : Characteristics of the linear relationship yield stress-polymer
concentration $F = kC + b$.
C_0 is the intercept of the curve with the concentration axis. The units of C
and F are   g/l and N respectively. r is the regression coefficient.

KCl concentration (M)	k	b	C_0	r
0.05	5.57	-18.7	3.35	0.993
0.10	4.76	-13.4	2.82	0.997
0.25	4.70	-13.8	2.94	0.994

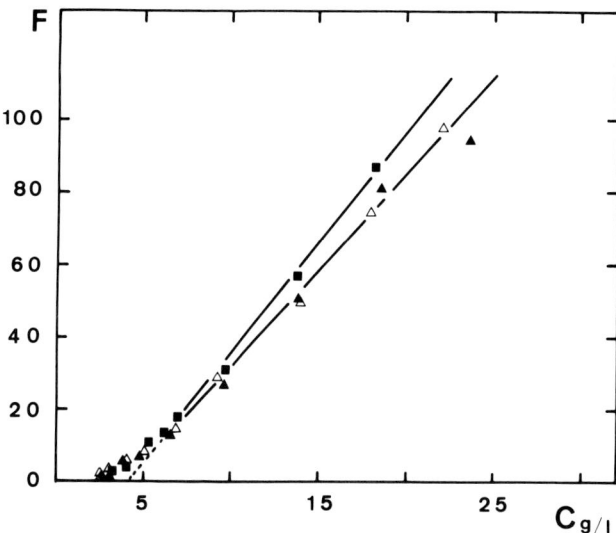

Fig. 2 : Yield stress F(N) as a function of the polymer concentration. The
KCl concentrations are respectively 0.05 ; 0.1 ; 0.25 M (■,△,▲).

Noteworthy is the very low dependence of the yield stress on the salt
concentration. Consequently the force maximum obtained before the failure
does not seem to be a good mechanical test for kappa carrageenan gels.
Nevertheless we have obtained from preliminary experiments on sharp
fractions of kappa carrageenans of different molecular weight a strong
correlation between the degree of the polymerization and the yield stress.
This result is in agreement with the first observation of Ainsworth and
Blanshard (8). Consequently, although the yield stress is not very
reproducible and not sensitive to the salt concentration, this mechanical
test can be useful in regard to the its molecular weight dependence.

Contrary to this, the elastic modulus, E is very sensitive to the
polymer concentration and the salt concentration (Fig. 3, Table IV). The
same power law was found with the three different salt concentrations
investigated. Consequently we can suppose that the nature of the crosslinks
and/or the gelation mechanism do not vary when the salt concentration is

changed. Due to the increase of the modulus with an increase of the salt
concentration, the number of crosslinks increased also. Generally the
"junction zone" of a carrageenan (9) gel is assumed to be due to linkages
between double helices. But the optical rotation of kappa carrageenan does not
change at room temperature if the KCl concentration is increased from 0.05
to 0.25M. Because the optical rotation is a probe of the conformation we
can assume that the conformation of carrageenans is the same between 0.05
and 0.25 M KCl. Then, we propose that the increases of the number of
crosslinks is due to the change of solubility of the polymer when the salt
concentration is raised.

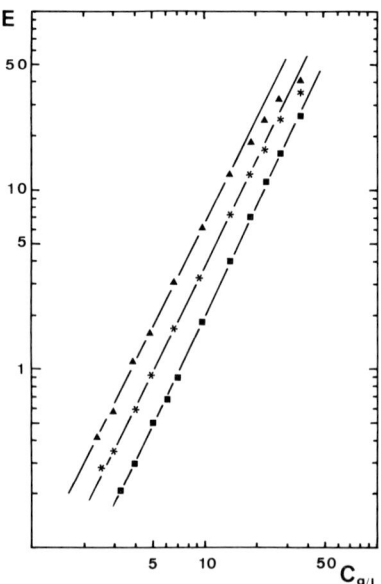

Fig. 3 : Elastic modulus E (10^5 dyne/cm^2) as a function of the polymer
concentration. The KCl concentrations of the points of the curve ▲,
∗, ■ are 0.25 ; 0.1 ; 0.05 M respectively.

Table IV : Characteristics of the relationship, modulus-polymer concentration assuming a power-law $E = k'C^n$. The units of C and E are g/L and dyne/cm^2 respectively. r is the regression coefficient.

KCl Concentration (M)	k'	n	r
0.05	1994	2.02	0.9997
0.10	3890	1.98	0.9997
0.25	7450	1.95	0.9992

ACKNOWLEDGEMENTS

ELF-AQUITAINE (France), M.R.S. (France) are gratefully acknowledged for financial support and for the gift of carrageenan.

REFERENCES

1. Rochas, C., and Rinaudo, M. (1980). Biopolymers 19, 1675-1687
2. Norton, I.T., Morris, E.R. and Rees, D.A. (1984). Carbohydr. Res. 134, 89-101.
3. Rochas, C. and Rinaudo, M. (1984) Biopolymers 23, 735-745.
4. Watase, M. and Nishinari, K. (1982a). Rheol. Acta 21, 318-324.
5. Watase, M. and Nishinari K. (1982b). Colloid and Polym. Sci. 260, 971-975.
6. Mitchell, J.R. (1976). J. Texture studies 7, 313-339.
7. Ross-Murphy, S.B. and McEvoy, H. (1986), Br. Polym. J. 17 2-7
8. Ainsworth, P.A. and Blanshard, J.M.V. (1979). J. Fd. Technol. 14, 141-147.
9. Rees, D.A. (1981), Pure and Appl. Chem. 53, 1-14.

Non-intrusive measurement of dispersion composition. Effect of xanthan on creaming profiles in emulsions

Paul A.Gunning, David J.Hibberd, Andrew M.Howe, Alan R.Mackie, Peter Richmond and Margaret M.Robins

AFRC Institute of Food Research, Norwich Laboratory, Colney Lane, Norwich NR4 7UA, UK

ABSTRACT

The effect of a range of concentrations of the anionic polysaccharide xanthan on the gravitational destabilisation (creaming and compaction) of oil-in-emulsions is presented. A non-intrusive technique to measure the velocity of ultrasound through the emulsion at a series of different heights is used to determine the oil concentration profile in these opaque dispersions. The detailed gravitational destabilisation processes can then be studied by following changes in the concentration profile with time. In the absence of polymer creaming is as expected for a dispersion of non-interacting polydisperse particles. In the presence of xanthan a number of significant differences are apparent in the evolution of the concentration profiles with time. With xanthan the lower meniscus is sharp and creams more rapidly than expected for individual droplets; while the oil concentration is lower at the top of the sample, but increases with time as the cream compacts. The compaction can be characterised by a concentration-dependent bulk modulus. The behaviour in the presence of xanthan is interpreted in terms of polymer-induced flocculation, and a rearrangement of the flocs in the cream under compression.

INTRODUCTION

Many processed foods are colloidal dispersions and their stability is of great commercial importance (1). Separation under gravity as a result of the density difference between the continuous and disperse phases is an important destabilisation mechanism. This creaming or sedimentation process leads to a non-uniform distribution of disperse phase over the height of the sample and may lead to the formation of an unsightly meniscus as the continuous phase separates. In the food and other industries polysaccharides are added to water-based dispersions to increase the continuous phase viscosity, and therefore reduce the rate of separation (2). We report studies on the effect of the commonly-used anionic polysaccharide xanthan on creaming and compaction in water-continuous emulsions.

Concentration profiles are determined using apparatus designed and constructed in our laboratory (3). The velocity of a longitudinal wave of ultrasound through a dispersion is directly related to the bulk composition of the dispersion (4). From measurements of the velocity at a series of heights a concentration profile may be determined. Collection of profiles at appropriate time intervals allows creaming/sedimentation to be followed in detail. We have investigated the creaming under gravity of alkane-in-water emulsions, stabilised by the nonionic surfactant Brij 35, containing a range of xanthan concentrations.

EXPERIMENTAL

Materials and emulsion preparation

A concentrated emulsion pre-mix and a series of xanthan solutions of varying concentration were made up separately and then mixed by stirring. We thus have the same particle size distribution in each emulsion and avoid shear degradation of the polysaccharide. The pre-mix emulsions of 60% n-alkane-in-water (Fisons, AR) were made in a constant-power Waring Commercial Blender using a set programme of shear cycles. The oils were a mixture of nine parts by volume heptane (BDH, 99.5%) to one part hexadecane (Aldrich, 99%) stabilised by Brij 35 (nominally C12E23, Sigma). By using a mixture of high and low alkanes Ostwald ripening was avoided (5) while obtaining a substantial density difference (307 kgm^{-3}) between the disperse and continuous phases. Solutions with a range of concentrations of the anionic polysaccharide xanthan (Kelco/AIL, food grade) with 0.24% w/v sodium metabisulphite (BDH, 95%) as preservative were made by stirring at room temperature. The final compositions of the emulsions were 20% v/v oil with an aqueous phase containing 0.35% w/v Brij 35, 0.20% w/v sodium metabisulphite and 0, 0.10, 0.22 or 0.33% dry w/v xanthan. The emulsions were placed in parallel-sided cells and creaming was monitored at 20.0 $^{\circ}$C.

Methods

The droplet size distribution, sample densities and continuous-phase viscosities at low shear-rate were measured as previously described (6).

The technique to measure velocity of ultrasound and hence volume fraction of a dispersion has been described in detail elsewhere (3). The propagation times of an ultrasound pulse generated from a 6.4 MHz half-sine wave are measured (at selected heights) across a cell containing the dispersion. The path-length of ultrasound across the dispersion is determined at each height from previous measurements of propagation time through the cell containing liquids of known ultrasound velocity (water, 1482.3 ms^{-1} (7) and hexadecane, 1358.9 ms^{-1} (8) or heptane, 1160 ms^{-1} (9)).

RESULTS

Calibration
The velocity of ultrasound is directly related to the relative concentrations of the disperse and continuous phases. Provided the particles are smaller than the wavelength of the ultrasound (~200 μm) a simple formula is used (4):

$$V^2 = \frac{V_c^2}{(1-\emptyset(1-\rho_d/\rho_c))\,(1-\emptyset(1-(V_c/V_d)^2\,\rho_c/\rho_d)} \qquad (1)$$

where V is the velocity through the dispersion of volume fraction $\emptyset$, and ρ_c, ρ_d, V_c, V_d are the densities of and ultrasound velocities through the continuous and dispersed phases. The calibration curve for the mixed alkane system used in this study is shown in Fig. 1. There is good agreement between measured and predicted velocities at concentrations up to ~70% oil. Above 70% there is still a simple dependence of velocity on oil content and the calibration data may be used directly.

Concentration profiles
The concentration profiles for the emulsions are shown in the absence (Fig. 2) and presence (Fig. 3) of 0.33% w/v xanthan. In both systems two menisci are present. One moves from the base of the sample to leave a layer of continuous phase, and the other from the top of the sample corresponding to a concentration build-up in the cream layer.

In the absence of polymer the lower meniscus is very diffuse. The base of the sample did not clear for approximately 12 days, although using the ultrasonic technique creaming could be detected within two hours. The droplets cream at the top of the sample to an oil concentration of ~70% which changes little with time.

In the presence of xanthan many detail differences in the evolution of the concentration profiles are apparent. The lower meniscus is sharp, while the upper meniscus is diffuse. The concentration of disperse phase increases at the top of the sample with time, approaching 70% after 87 days when the average concentration in the cream is ~65%.

DISCUSSION

Creaming - rate of rise of lower meniscus
The rate of rise U_s of a meniscus, under gravity, of a polydisperse, infinitely dilute, non-interacting droplet system is given by Stokes' law:

$$U_s = \frac{\langle d^2 \rangle g \Delta\rho}{18\,\eta} \qquad (2)$$

where g is the acceleration due to gravity, $\Delta\rho$ is the density difference between the disperse and continuous phases, η is the

medium viscosity and $\langle d^2 \rangle$ is the weight-mean value of the square
of the droplet diameter. $\langle d^2 \rangle$ is equal to the sum of the squares
of the weight-mean and standard deviations in the droplet size
distribution. In this study all the emulsions have the same
droplet size distribution, with $d = 4 \pm 2$ μm and thus $\langle d^2 \rangle$ of
20 μm^2. The dispersion viscosity is greater than the continuous
phase viscosity as a result of the hydrodynamic interactions
resulting from the presence of the droplets. The viscosity η of
a concentrated dispersion of non-interacting, hard-sphere
particles may be calculated by means of a mean-field approach
(10).

$$\eta = \eta_0 (1-\phi/\phi_M)^{2.5\phi_m} \qquad (3)$$

where ϕ_M is the close-packed volume fraction, taken here to be
0.7 (the volume fraction in the non-compacted cream for the
emulsion without polysaccharide), and η_0 the continuous phase
viscosity (at low shear rate). For emulsions of 20% oil η/η_0 is
1.8. The continuous phase viscosities and the calculated values
of U_S and the rate of rise of the meniscus U are given in Table 1.

Table I. CONTINUOUS PHASE PROPERTIES, OBSERVED AND PREDICTED
          CREAMING RATES

Xanthan Concentration % w/v	η Pas	U_S nms^{-1}	U nms^{-1}	U/U_S
0	0.002	930	155	0.17
0.10	0.20	9.3	1020	110
0.22	2.2	0.85	108	130
0.33	12	0.16	12	80

The hazy (diffuse) meniscus observed in the absence of polymer
is characteristic of the creaming of individual particles. The
creaming velocity U is lower than predicted from Stokesian
considerations. Several factors may contribute to the deviations
from the expected velocity. These include: convection currents
within the low-viscosity sample and creaming of different size
droplets at different rates (because the meniscus is so diffuse
substantially lower U values are exhibited as ϕ is decreased).
The emulsions containing xanthan show sharp lower menisci that
move typically 100 times faster than expected. Addition of
0.1% w/v xanthan to the continuous phase results in an increase
in the continuous-phase viscosity by two orders of magnitude
while creaming occurs at the predicted rate (and much faster than
the measured rate) for a polymer-free emulsion. Although there
is some uncertainty in the values of the continuous phase
viscosities owing to the non-Newtonian properties of the xanthan
solutions and the difficulty in measuring viscosity at a low
enough shear rate (< 0.01 s^{-1}), the creaming rates are clearly
much higher than expected for individual droplets. This
behaviour, coupled with the presence of the sharp lower menisci
in the xanthan-containing emulsions, is consistent with creaming
of aggregates of droplets. Therefore the hydrophilic, anionic

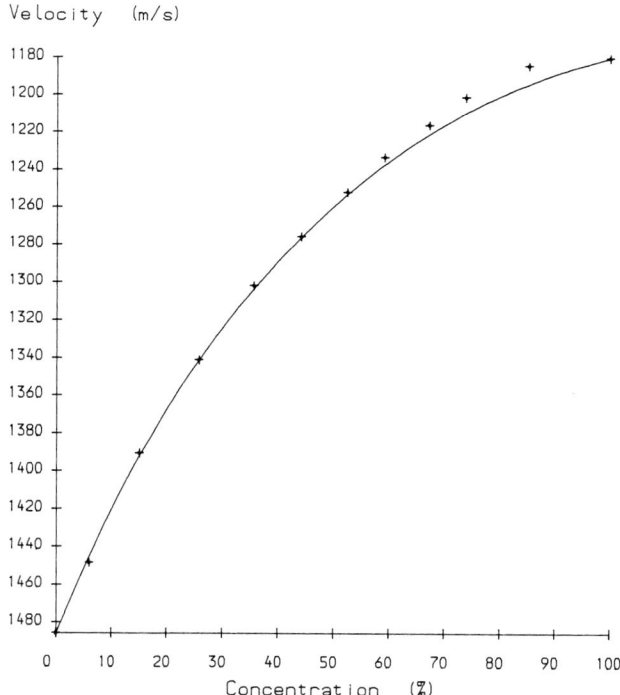

Figure 1. Ultrasonic velocity and oil concentration for alkane-in-water emulsions. Line from Equation (1).

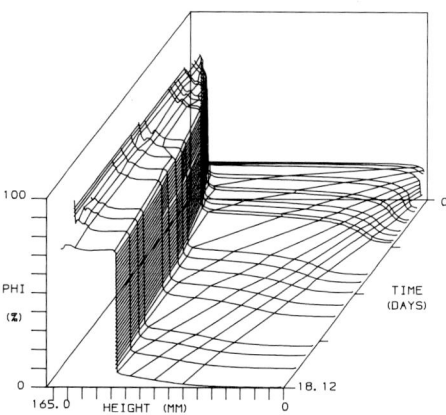

Figure 2. Concentration profiles for 20% alkane-in-water emulsion without polysaccharide.

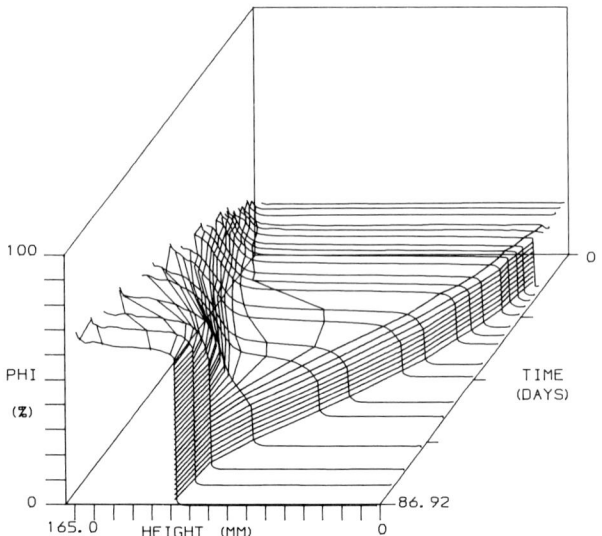

Figure 3.   Concentration profiles for 20% alkane-in-water emulsion
containing 0.33% xanthan in the continuous phase.

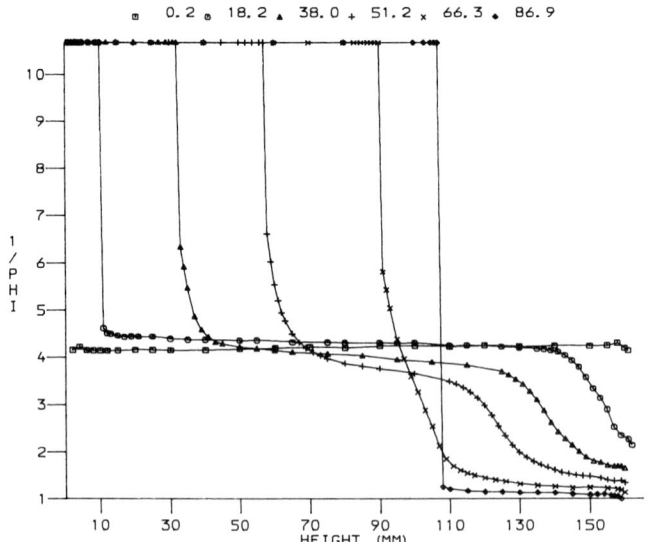

Figure 4.   Variation of 1/∅ with height for alkane-in-water
emulsion containing 0.33% xanthan in the continuous phase.

polysaccharide xanthan induces flocculation of the droplets. Flocculation does not arise via a bridging mechanism as xanthan does not lower the oil-water interfacial tension in the presence of Brij 35 (11), nor is the sub-cream layer denuded of polymer. The xanthan-induced flocculation probably arises via a depletion/ phase separation mechanism (12). Aggregates containing many droplets cream more rapidly than isolated droplets as a result of their greater buoyancy.

Compaction - packing within cream

In the emulsion without polymer (Fig. 2) the droplets cream immediately to a high concentration (~70%), near the random close-packed limit for polydisperse spheres. Such behaviour is predicted for dispersions of non-interacting spherical particles. The presence of xanthan in the emulsion has dramatic effects on the concentration profiles in the cream (Fig. 3). The oil concentration, initially much lower in the cream, increases gradually both towards the top of the sample and at the top of the sample with time. This compression behaviour is a result of the elastic properties of the cream, and may be due to the low-dimensional flocs initially packing to a low volume fraction and then rearranging under the hydrostatic pressure in the cream. Kevasamoorthy and Arora developed a theory to describe the gravity-induced concentration changes over the height of a dilute suspension of 0.1 μm polystyrene latex particles in terms of a bulk modulus (13). The hydrostatic pressure P across an element of the dispersion of thickness dh and volume fraction $\emptyset$(h) is

$$P = g \Delta \rho \emptyset dh \qquad (4)$$

but P is related to the bulk modulus B and volume v (and hence $\emptyset$).

$$P = B\frac{dv}{v} = B\emptyset d(1/\emptyset) \qquad (5)$$

Combining equations 4 and 5

$$\frac{d(1/\emptyset)}{dh} = \frac{g \Delta \rho}{B} \qquad (6)$$

The value of B in the cream of an emulsion may be determined from the reciprocal of the gradient in a plot of $1/\emptyset$ against h (depth in the cream). The $1/\emptyset$-h plot for the emulsion containing 0.33% xanthan (Fig. 4) exhibits a number of linear regions, the gradients of which decrease with height and time corresponding to a range of values of B which increase as $\emptyset$ increases during the compaction process. A similar plot for the xanthan-free emulsion (Fig. 5) exhibits a shallow gradient at high oil concentration. This indicates that there is also a bulk modulus, with a high value, in the concentrated cream of the polymer-free emulsion. This latter bulk modulus may be associated with gravitational compression of the individual deformable oil drops.

The values of the bulk moduli are plotted against the oil concentration at the base of the cream $\emptyset_B$ in Fig. 6. The bulk

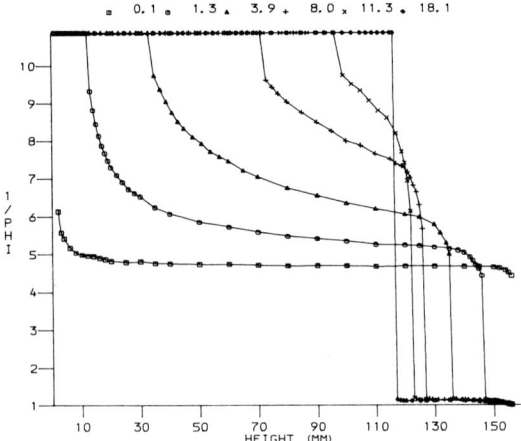

Figure 5. Variation of 1/∅ with height for alkane-in-water
emulsion without polysaccharide.

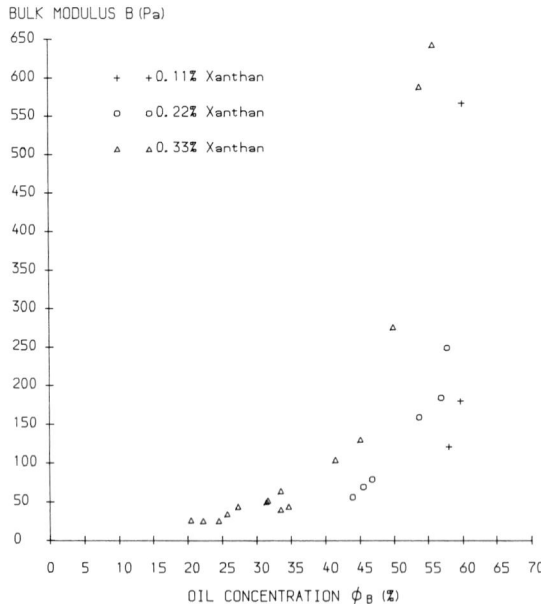

Figure 6. Bulk modulus and oil concentration in the cream layer
for alkane-in-water emulsions containing xanthan in the continuous
phase.

modulus increases dramatically from ~25 Pa to ~700 Pa as $\emptyset_B$ increases from 20% to 60%, with less dependence on xanthan concentration. The polymer-free emulsion has a bulk modulus of ~1000 Pa at ~70% oil, a value similar to that obtained at very high $\emptyset_B$ with the xanthan-containing emulsions.

SUMMARY

A non-intrusive technique to determine concentration profiles in dispersions using the velocity of ultrasound has been described. The creaming of oil-in-water emulsions in the presence and absence of the anionic polysaccharide xanthan has been followed by determination of the concentration profiles with time. Xanthan has dramatic effects on the detailed creaming behaviour: a sharp meniscus at the base of the sample which rises faster than predicted for isolated droplets and a gradual increase in the concentration at the top of the sample as the cream compacts (characterised by a $\emptyset$-dependent bulk modulus B with values in the range 25 - 1000 Pa). These effects are consistent with polymer-induced flocculation of the droplets by a depletion mechanism and the subsequent rearrangement of the flocs in the cream under hydrostatic pressure.

ACKNOWLEDGEMENTS

We are grateful to Sarah Gouldby and Chris Carter for technical assistance and to Dr Jim Mingins for helpful discussions. This work was funded by the Ministry of Agriculture, Fisheries and Food and the Department of Education and Science.

REFERENCES

1. Darling, D.F. and Birkett, R.J. (1987) in Food Emulsions and Foams, (ed Dickinson, E.) pp 1-29, RSC, London, UK.
2. Dickinson, E. and Stainsby, G. (1982) Colloids in Food, 533 pages, Applied Science, Essex, UK.
3. Howe, A.M., Mackie, A.R. and Robins, M.M. (1986) J. Dispersion Sci. Technol., 7, 231-243.
4. Urick, R.J. (1947) J. Applied Phys., 18, 983-987.
5. Buscall, R., Davis, S.S. and Potts, D.C. (1979) Colloid Polymer Sci., 257, 636-644.
6. Gunning, P.A., Hennock, M.S.R., Howe, A.M., Mackie, A.R., Richmond P. and Robins, M.M. (1986) Colloids Surf., 20, 65-80.
7. Del Grosso, V.A. and Mader, C.W. (1972) J. Acoust. Soc. Amer., 52, 1442-1446.
8. Gladwell, N.R. and Javanaud, C. (1986) unpublished results.
9. Kaye, G.W.C. and Laby, T.H. (1972) Tables of Physical Constants, 14th edn., p 65. Longman Group, London, UK.
10. Ball, R.C. and Richmond P. (1980) Chem. Phys. Liq., 9, 99-116.

11. Gladwell, N.R., Hennock, M.S.R., Howe, A.M., Mackie, A.R., Rahalkar, R.R. and Robins, M.M. (1984) Fifth International Conference Surfactants in Solution, Bordeaux, France.
12. Garvey, M.J., (1982) in The Effects of Polymers on Dispersion Properties (ed Tadros, T.F.) p 203. Academic Press, London, U.K.
13. Kevasamoorthy, R. and Arora, A.J. (1985) J. Phys. A, 18, 3389-3398.

Coalescence kinetics of protein-stabilised emulsion droplets

Eric Dickinson, Brent S.Murray and George Stainsby

Procter Department of Food Science, University of Leeds, Leeds LS2 9JT, UK

ABSTRACT

We report measurements of coalescence times of emulsion-sized oil droplets (> 2 μm) at a planar oil−water interface in an apparatus which allows direct observation and recording of both the droplet sizes and the coalescence times. Coalescence stability was found to be sensitive to ionic strength, droplet size, and protein concentration in the aqueous phase. The significant differences in coalescence times for the proteins β-casein, κ-casein and lysozyme could be correlated with the differing interfacial shear viscosities of aged films adsorbed from 10^{-4} wt% protein solutions of ionic strength 0.1 M. No coalescence at all was observed at a protein concentration of 5×10^{-4} wt% for the same ionic strength 0.1 M.

INTRODUCTION

The molecular complexity of proteins and polysaccharide gums makes it difficult to interpret the coalescence kinetics of food emulsion droplets in terms of established theories of polymeric and electrostatic stabilization (1,2). At the simplest level, such stabilizers are considered to act by forming a structural and mechanical barrier—a sort of 'protective skin' around the droplets which prevents them from coalescing. Indeed, films of gum arabic have been seen (3) to show a skin-like appearance. Various authors [e.g., Sherman and coworkers (4,5)] have claimed a correlation between emulsion stability and film viscoelasticity, although occasionally the correlation has been disputed (6). The most direct evidence has come from experiments involving the injection of large droplets beneath a planar oil−water interface. The results have generally shown (7) a correlation between droplet stability and the surface shear viscoelasticity of the planar film. Unfortunately, the droplets studied in these experiments were rarely less than 100 μm in diameter, which is much larger than those encountered in a typical food emulsion. Moreover, a strong dependence of stability on droplet size has been observed (8).

In this paper, we report experiments to test the hypothesis of a correlation between surface shear viscosity and coalescence

stability of emulsion-sized droplets (1—20 µm) at a planar oil—
water interface in the presence of three proteins which give very
different surface viscosities (9,10).

MATERIALS AND METHODS

Materials
Lysozyme (chicken egg white, grade 1, 3 X crystallized,
dialysed, lyophilized, batch no. 54F-8155) was obtained from
Sigma Chemicals. Samples of β-casein and $\varkappa$-casein had been
prepared according to standard procedures as described previously
(11). Sodium caseinate ('Scottish Pride') was obtained from the
Scottish Milk Marketing Board. The calcium contents of the
β-casein and the sodium caseinate were 0.08 g kg^{-1} and 0.1 g kg^{-1}
respectively. Proteins were dissolved in pH 7.0 phosphate buffer
made from AnalaR grade reagents and double-distilled water. The
AnalaR grade n-hexadecane (> 99 wt %) was obtained from Sigma
Chemicals.

Coalescence Measurements
The apparatus has been described previously (12). In brief, a
planar interface between an oil phase and a protein solution is
formed in a glass cell and viewed from above with a microscope.
Connected to the microscope is a video camera, an image analyser,
a video recorder, and a high-resolution monitor, as illustrated
schematically in figure 1. Floating at the oil—water interface,
and prevented from moving by four vertical stainless-steel
needles, is a thin mica strip with a small hole (diameter < 200
µm). Oil droplets, formed by vigorously mixing n-hexadecane and
buffer (no protein) in a scrupulously clean blender, are injected
just below the planar interface. The droplets rise up to that

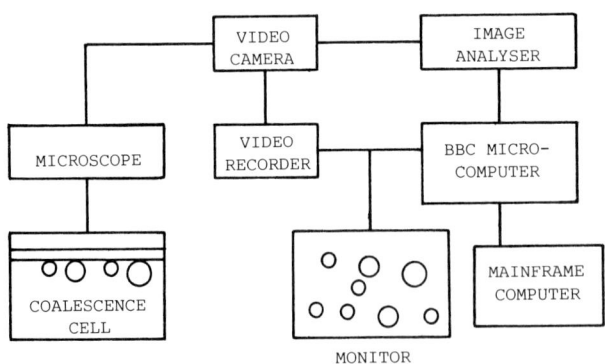

Figure 1. Arrangement of equipment for automatic monitoring
and analysis of droplet coalescence at a planar interface.

part of the oil—water interface contained within the mica hole, where they exist for some time t_C prior to coalescence with the planar interface. The boundary edge of the mica hole prevents droplets from drifting out of view before coalescence.

The image analyser system, operating with our own software, is capable of automatically recording droplet sizes and coalescence times ($t_c \geqslant 2$ min) over a period of hours or even days. For $t_c \leqslant 2$ min, however, the experiment is recorded on video tape, with coalescence times measured with a stopwatch and droplet sizes measured directly from the screen of the monitor (effective image magnification X 1100).

Interfacial Rheology and Tension Measurements

The surface rheometer is a Couette-type torsion-wire device; it is similar to that used by others [e.g., Graham and Phillips (9)] and is described in detail elsewhere (13). Surface shear viscosities are measured at the constant shear rate of 1.27×10^{-3} rad s^{-1}, which is low enough to avoid rupturing the protein film. Interfacial tensions are measured using a Wilhelmy plate method described previously (14). In making the rheology and tension measurements, care was taken to ensure that the oil—water interface was formed in exactly the same way as in the droplet coalescence experiments, with the same ratio of aqueous volume to interfacial area (~ 2.5 m^3 m^{-2}).

RESULTS

Two sets of coalescence experiments have been performed. In the first set, the planar oil—water interface was allowed to age for just 20 ± 2 minutes in contact with the protein solution (10^{-4} wt %, ionic strength 0.1 M, 25 °C), and then 'bare' oil droplets were injected into the cell. In the second set, the interface was aged for 72 ± 1 hours, all other conditions staying the same. Coalescence of droplets at 'young' (20-minute-old) and 'old' (72-hour-old) films was studied for each of the three pure proteins β-casein, $\varkappa$-casein and lysozyme. The reason for our choosing these short and long times can be seen from table I.

Table I   SURFACE VISCOSITY η_i AND TENSION γ FOR ADSORBED
PROTEIN FILMS AT THE n-HEXADECANE—WATER INTERFACE
(pH 7, 25 °C)

Protein	η_i/mN m^{-1} s		γ/mN m^{-1}	
	20 min	72 h	20 min	72 h
lysozyme[a]	< 1	220	45.0	33.4
$\varkappa$-casein[a]	< 1	60	45.0	23.0
β-casein[a]	< 0.1	< 1	41.0	22.4
caseinate[b]	< 1	15	29.0	22.4

[a] 10^{-4} wt % (0.1 M)   [b] 10^{-3} wt % (0.005 M)

After 20 minutes, the surface viscosities and tensions for the three proteins are similar. After 72 hours, however, the surface viscosities are substantially different (lysozyme ≫ χ-casein ≫ β-casein), and the tensions for β-casein and χ-casein are lower by 10 mN m^{-1} than that for lysozyme.

Figure 2 gives coalescence results for droplets injected below the young protein films. The mean coalescence time $\langle t_c \rangle$ for oil droplets of the same size is plotted against droplet diameter d. The data in figure 2 are based on over 200 coalescence events for each protein, with a standard deviation about the mean for each size interval of ca. $0.2 \langle t_c \rangle$. The scatter in the t_c values is a reflection of the stochastic nature of the coalescence process.

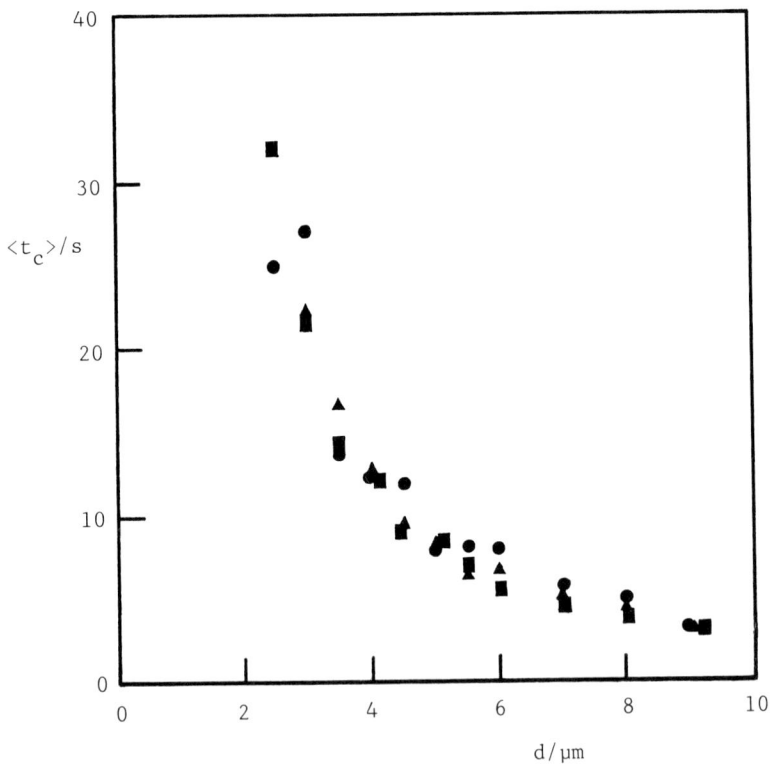

Figure 2. Droplet stability beneath 20-minute-old films of β-casein, χ-casein or lysozyme. The mean coalescence time $\langle t_c \rangle$ (± 0.3 s) is plotted against the droplet diameter d (± 0.5 μm): ▲, β-casein; ●, χ-casein; ■, lysozyme.

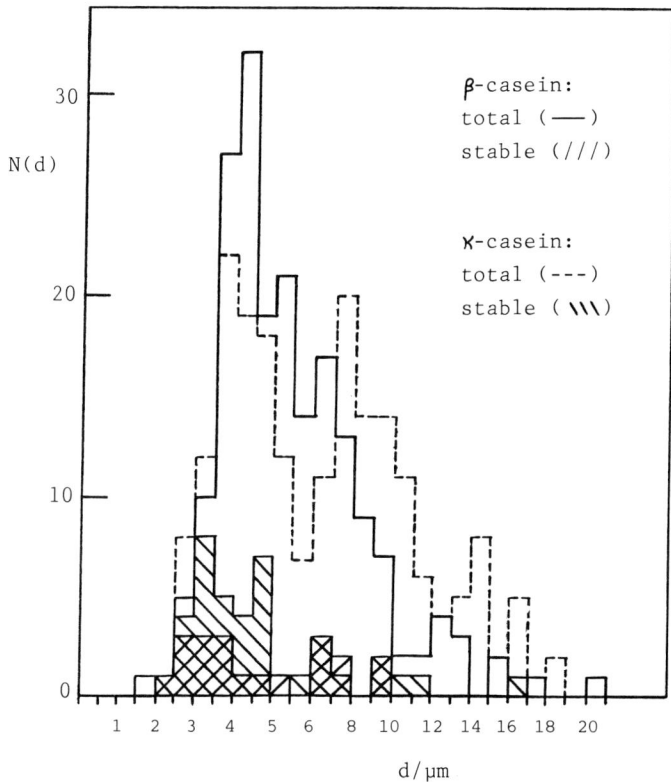

Figure 3. Droplet stability beneath 72-hour-old films of
β-casein or Χ-casein. Number of droplets, N(d), is plotted
against droplet diameter d.

We note that ⟨t$_c$⟩ is sensitive to droplet size but independent of
the nature of the protein. Values are not presented for droplets
with d < 2 µm as their Brownian motion made it difficult to
define their positions with respect to the interface. Droplets
were, however, still visible down to ca. 0.5 µm, and it was noted
that all had coalesced within 10 minutes of injection.
    Figures 3 and 4 give coalescence results for droplets injected
below the old protein films. Here, a fraction of the droplets
was found to be extremely stable, remaining at the interface for
at least 48 hours, when the experiment was terminated. The rest
of the droplets coalesced with a distribution of times not
significantly different from that shown in figure 2. The size
distributions for all droplets injected and for those which were

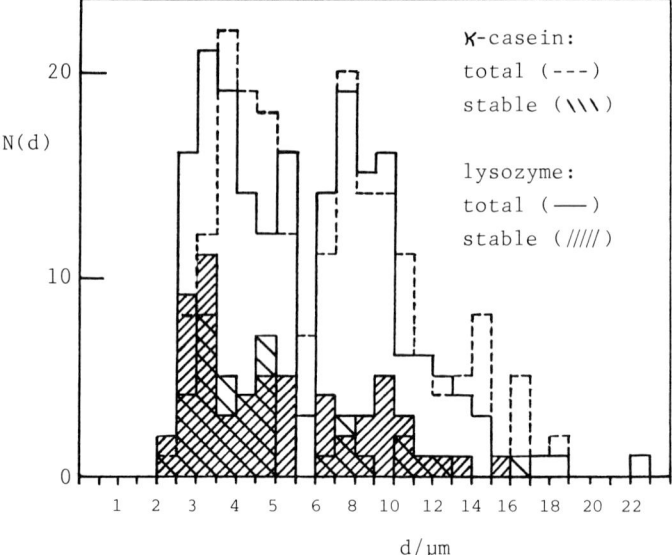

Figure 4. Droplet stability beneath 72-hour-old films of
ᴋ-casein or lysozyme. Number of droplets, N(d), is plotted
against droplet diameter d.

stable are given in figures 3 and 4. Figure 3 compares β-casein
with ᴋ-casein, and figure 4 compares ᴋ-casein with lysozyme.
The fraction of stable droplets is significantly different for
the three proteins: 11 % for β-casein, 20 % for ᴋ-casein and 32 %
for lysozyme.
    In addition to these quantitative results for the three pure
proteins, we also report some qualitative results for droplets
injected beneath a 1-hour-old film of sodium caseinate. Table II
shows the types of behaviour observed at various values of the
ionic strength I and the bulk protein concentration c_p. There
was no appreciable coalescence at $c_p \geqslant 5 \times 10^{-4}$ wt %, and at $c_p =$
10^{-3} wt % all the droplets were stable for at least 3 days,
save a very few (< 1 %) at I = 1.0 and 2.0 M. At these high ionic
strengths, two-dimensional flocs of droplets were formed at the
planar oil—water interface; and, at I = 2.0 M, the adjacent
flocculated droplets were noticeably squeezed together and were
separated by flat lamellae—but no droplet—droplet coalescence
was ever observed. At I $\leqslant$ 0.05 M, the droplets were visibly
repelled from the planar interface (and each other) as indicated
by their vigorous Brownian motion which caused them to execute
two-dimensional random walks. Under the same conditions of ionic
strength and protein concentration, droplet 'walking' was found

Table II   BEHAVIOUR OF n-HEXADECANE DROPLETS BENEATH A
           1-HOUR-OLD PLANAR INTERFACE BETWEEN n-HEXADECANE
           AND A CASEINATE SOLUTION OF CONCENTRATION c_p

c_p/wt %	Ionic strength/M				
	0.005	0.05	0.1	1.0	2.0
1×10^{-3}	NC, W	NC, W	NC	LC, F	LC, F
5×10^{-4}			NC	LC, F	
1×10^{-4}			C	C	
1×10^{-5}			C	C	

KEY: C = coalescence observed, LC = coalescence limited to a
     few early droplets, NC = no coalescence observed, W =
     droplet 'walking' observed, F = flocculation

to be slower with caseinate than with β-casein, but faster with
caseinate than with κ-casein. Droplet walking also became slower
with increasing time after injection.

DISCUSSION

   The results show that, when the planar films of β-casein, κ-
casein and lysozyme all have the same low surface viscosity, as
is the case after only 20 minutes of adsorption from solutions of
10^{-4} wt % protein, there is no significant difference in droplet
coalescence stability for the three proteins. On the other hand,
when the films have very different surface viscosities, as is the
case after 72 hours of aging, the coalescence stabilities lie in
order of increasing surface viscosity (β-casein < κ-casein <
lysozyme). Corroborative evidence for a link between protein
film surface viscosity and the dynamics of the droplet—interface
interaction comes from the observations of droplet walking: the
speed decreases in the sequence β-casein > caseinate > κ-casein,
as one would expect intuitively from the relative values of the
surface viscosities listed in table I. The decreasing walking
speed with droplet age is, moreover, consistent with a surface
viscosity that increases continuously with time (10,13,14).
   Other explanations for the coalescence results cannot be ruled
out, but the case for them, we feel, is less convincing than the
surface rheology arguments. Surface activity is certainly not a
good guide to stability: lysozyme is the least surface active but
the best stabilizer; β- and κ-casein give similar tensions after
72 hours but different coalescence stabilities. Film thickness
could be a factor, though surface concentrations for β-casein and
lysozyme would be expected to be similar (15). As for possible
electrostatic factors, our previous measurements (11) of droplet
mobilities do not suggest that charge densities on adsorbed films
of β-casein and κ-casein would be much different under these
experimental conditions.

So, for oil droplet coalescence at a planar oil—water interface, we find a clear correlation between stability and the mechanical strength of the adsorbed protein film at the planar interface (as indicated by the surface viscosity). In assessing the relevance of this result to the stability of food emulsions, we note, firstly, that the coalescence kinetics observed here is much more sensitive to protein concentration and ionic strength than it is to film age or protein type. Secondly, the protein concentrations used in food emulsions are orders of magnitude larger than those considered here (though the ratio of protein to area of available oil—water interface is similar). Thirdly, the observed dependence of coalescence time on droplet size in figure 2 is due, we believe, in large part to the strong dependence of the buoyant force on droplet size. This would not be expected to be a factor with coalescence driven by Brownian droplet—droplet encounters in the bulk of an emulsion, although such an effect may be more important when droplets push against one another in a creamed layer.

ACKNOWLEDGEMENTS

We thank Mr Richard Whyman for the samples of β- and $\varkappa$-casein, Mr Philip Nelson for technical assistance in constructing the coalescence apparatus, and the Chief Scientist's Group at the Ministry of Agriculture, Fisheries and Food for supporting the research. The results are the property of the Ministry and are Crown Copyright.

REFERENCES

1 Dickinson, E. (1986) Food Hydrocolloids, 1, 3-23.
2 Dickinson, E. and Stainsby, G. (1982) Colloids in Food, Applied Science, London.
3 Shotton, E. and Wibberly, K. (1961) Boll. Chim. Farm., 100, 802-810.
4 Rivas, H. J. and Sherman, P. (1984) Colloids Surf., 11, 155-171.
5 Doxastakis, G. and Sherman, P. (1984) Colloid Polym. Sci., 262, 902-905.
6 Phillips, M. C. (1981) Food Technol., 35, 50-57.
7 Halling, P. J. (1981) CRC Crit. Rev. Food Sci. Nutr., 15, 155-203.
8 Davis, S. S. and Smith, A. (1976) Colloid Polym. Sci., 254, 82-98.
9 Graham, D. E. and Phillips, M. C. (1980) J. Colloid Interface Sci., 76, 240-250.
10 Castle, J., Dickinson, E., Murray, B. S. and Stainsby, G. (1987) ACS Symp. Ser., in press.
11 Dickinson, E., Whyman, R. H. and Dalgleish, D. G. (1987) in Food Emulsions and Foams (ed. Dickinson, E.), pp. 40-51. Royal Society of Chemistry, London.

12 Dickinson, E., Murray, B. S. and Stainsby, G. (1987) in Food
Emulsions and Foams (ed. Dickinson, E.), pp. 286-288. Royal
Society of Chemistry, London.
13 Dickinson, E., Murray, B. S. and Stainsby, G. (1985) J.
Colloid Interface Sci., 106, 259-262.
14 Castle, J., Dickinson, E., Murray, A., Murray, B. S. and
Stainsby, G. (1986) in Gums and Stabilisers for the Food
Industry (eds Phillips, G. O., Wedlock, D. J. and Williams,
P. A.), vol. 3, pp. 409-417. Elsevier Applied Science, London.
15 Graham, D. E. and Phillips, M. C. (1979) J. Colloid Interface
Sci., 70, 415-425.

Rheological and stability properties of concentrated emulsions made with a binary mixture of proteins

Julie Castle, Eric Dickinson, Ann Murray and George Stainsby

Procter Department of Food Science, University of Leeds, Leeds LS2 9JT, UK

ABSTRACT

The properties of high-volume-fraction oil-in-water emulsions made with a binary mixture of proteins have been investigated as a function of emulsifier composition. For emulsions containing a pair of proteins chosen from the set of caseinate, gelatin and β-lactoglobulin (40 wt % vegetable oil, 0.5 wt % protein), the amount of free oil released by centrifugation and organic solvent extraction was determined. There was no obvious correlation between the amount of free oil extracted and the interfacial rheology of the corresponding protein film. For emulsions made from caseinate + gelatin (55 vol % n-hexadecane, 1.0 wt % protein), the developing time-dependent shear modulus was determined. The results suggest that the composition dependence of the emulsion rheology is closely related to the competitive adsorption of the two proteins at the oil—water interface, as reflected in the composition dependence of the surface rheology for the caseinate + gelatin system. We have found that, at low protein load under the conditions employed, it was not possible to produce stable emulsions over a certain range of emulsifier composition.

INTRODUCTION

Food emulsions typically contain a mixture of proteinaceous components with differing surface activities for the oil—water interface. This means that, if, as is usually the situation commercially, the protein load is high enough for a substantial proportion of emulsifier to remain unadsorbed in the bulk aqueous phase, the composition of the stabilizing layer around the oil droplets will generally differ from the overall composition (1). We have shown previously (2,3) that, in emulsions made with sodium caseinate + gelatin, the former protein predominates at the interface, and that, in emulsions made with gelatin, the subsequent addition of caseinate displaces gelatin from the interface to an extent which decreases with the age of the emulsion prior to the addition. Similar experiments on emulsions made with β-lactoglobulin + gelatin have shown (2) that the whey protein is much less able to inhibit gelatin adsorption than is commercial sodium caseinate. The latter is itself, of course, a

mixture of various casein monomers, the two most common of which
are α_{s1}-casein and β-casein. It has recently been shown (4) that
adding the more surface-active β-casein to an emulsion made with
α_{s1}-casein will displace α_{s1}-casein from the oil—water interface.
    In what follows, we report new results for emulsions made with
a binary mixture of proteins. Our concern here is not so much
with competitive adsorption per se, but more with the way in
which the competition affects the properties of the emulsions:
ease of formation, rheology and texture, and coalescence stability
as determined by centrifugation and subsequent solvent extraction
of the free oil. The proteins studied are sodium caseinate,
gelatin and β-lactoglobulin. We devote particular attention to
the combination of caseinate + gelatin, which we feel is uniquely
interesting because of the large difference in hydrophobicity and
surface activity of the components, the unusual gelling properties
of gelatin in bulk solution and at the interface, and the now
extensive information available (5-7) on the time-dependent film
properties of the mixed system at the oil—water interface.
    One objective of this study was to see if there is any link
between the time-dependent bulk rheology of emulsions made with
caseinate + gelatin and the time-dependent interfacial rheology
of films adsorbed from mixtures of the same two proteins. It was
anticipated intuitively that such a link, if it occurred, would
be strongest for emulsions of high volume fraction. Preliminary
experiments indicated that emulsions containing 55 vol% oil gave
rheological parameters in a range suitable for study by the two
instruments available for this work (Rank Pulse Shearometer, and
Carri-Med controlled-stress Rheometer). Difficulties were soon
encountered, however, in making such high-volume-fraction stable
emulsions in a reproducible manner using small-scale valve-type
homogenization; the main problem was air incorporation at the
pre-mix stage, but this was compounded by an inability [now well
established as a real effect (see later)] to make proper emulsions
at all at certain ratios of gelatin to caseinate. In the course
of overcoming these problems of reproducibility, we have developed
a new type of 'jet' homogenizer which will be briefly described.

MATERIALS AND METHODS

Materials
    Spray-dried sodium caseinate ('Scottish Pride') was obtained
from the Scottish Milk Marketing Board. Gelatin was food grade
(pI 5.7), and β-lactoglobulin was a gift from the AFRC Institute
of Food Research, Reading (formerly N. I. R. D.). AnalaR grade
n-hexadecane was obtained from Sigma Chemicals. The supermarket
vegetable oil contained 0.04 wt% free fatty acid. The phosphate
buffer was prepared from AnalaR grade reagents and double-
distilled water.

Emulsion Preparation
    Vegetable oil-in-water emulsions (40 wt% oil, 0.5 wt% total
protein, pH 7, 0.005 M phosphate buffer) were made at 100 bar
operating pressure using the valve mini-homogenizer described
previously (3). Droplet-size distributions were determined using
a Coulter counter model TAII with a 100 μm orifice tube and 0.18 M

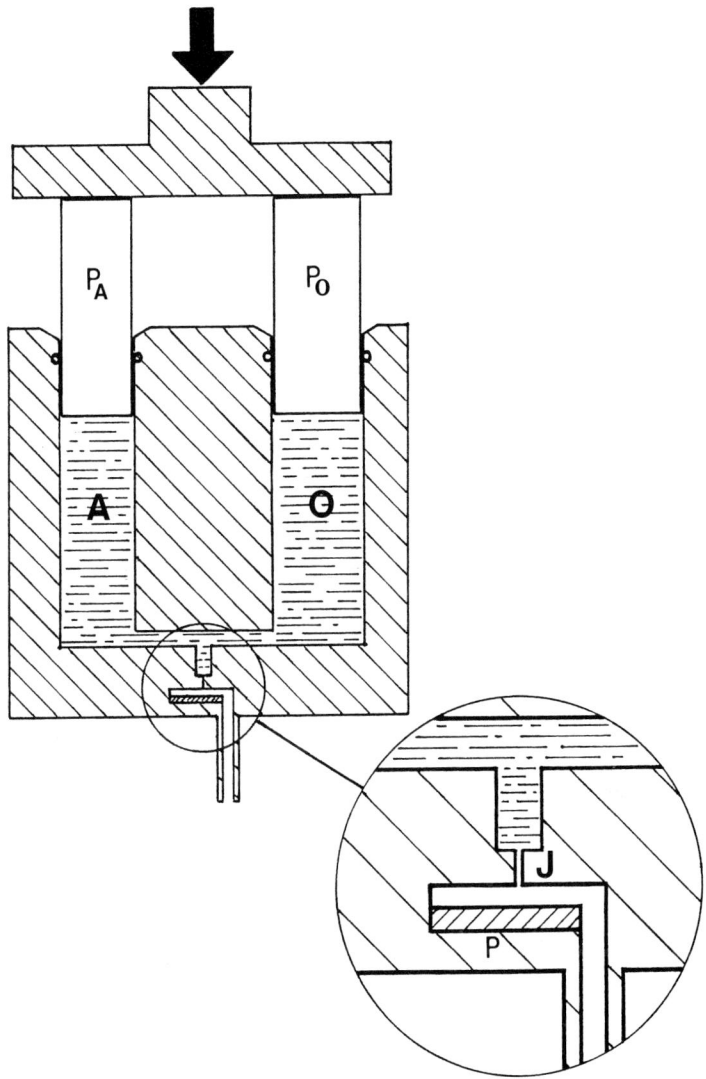

Figure 1. Schematic representation of the 'jet' homogenizer: A = aqueous protein solution; O = oil; P_A = piston A (diam. 15.875 mm); P_O = piston O (diam. 17.460 mm); P = fixed plate; J = jet hole (diam. 0.400 mm). The arrow indicates direction in which pistons are moved together by compressed-air drive.

sodium chloride as suspending electrolyte. The most-probable droplet sizes were in the range 7—10 μm for all the pure and mixed proteins investigated, except for β-lactoglobulin alone for which the value was much higher (ca. 25 μm).

The n-hexadecane-in-water emulsions (55 vol% oil, pH 7, 0.05 M phosphate buffer) were made using the 'jet' homogenizer shown schematically in figure 1. In essence, the two phases to be emulsified are forced together at uniform high speed through a small hole, and homogenization occurs as the emerging liquid jet impinges at the surface of a fixed flat plate. The controlled volume fraction of the resulting emulsion is dictated by the relative cross-sections of the pistons which are moved downwards by a common drive. The piston cylinders are lined with PTFE, and the pistons are sealed with fluorocarbon rubber 'O'-rings; the rest of the homogenizer is stainless steel. An advantage of the design is the ease of thermostatting as compared with the valve homogenizer. All the emulsions discussed below were made at 50 ± 1 °C with liquid moving through the jet at an average speed of 140 m s^{-1}. Under these conditions, the droplet-size distributions determined using the Coulter counter were found to be similar to those measured for emulsions obtained at 100 bar on the valve mini-homogenizer.

Emulsion Stability

Coalescence of vegetable oil-in-water emulsions was induced by centrifugation using a Beckman Model L2-65B Ultracentrifuge with head SW65Ti. Samples of 30-minute-old emulsions were centrifuged for 2 hours at 18 °C at a speed of 4×10^4 rpm ($8—17 \times 10^4$ g along the tube). After this treatment, the samples were observed to have separated into a clear aqueous layer and a cream layer containing some free oil. Centrifuge tubes were pierced at the bottom to remove the aqueous layer, and the cream layer was then redispersed in phosphate buffer for 20 minutes at 40 °C. The free oil was extracted with 50:50 diethyl ether/petroleum ether (40—60 °C) and dried at 80 °C to constant weight.

Emulsion Rheology

Time-dependent shear moduli of n-hexadecane-in-water emulsions were determined using a Rank Pulse Shearometer (8,9). The sample was held between a pair of stainless steel plates of diameter 25 mm and gap-width 1 mm. The test cell was surrounded by a water jacket maintained at 25.0 ± 0.2 °C. The propagation time of the shear wave (200 Hz) was measured every 30 minutes for 24 hours, and the shear modulus was calculated as described in detail elsewhere (8,9). The attenuation coefficient was taken to be zero throughout; this is a good approximation for the samples investigated here, except for the emulsions containing a low percentage of gelatin, where the approximation may lead to an error of up to 10—20%, especially at short times ($G \lesssim 500$ N m^{-2}).

RESULTS AND DISCUSSION

The effect of protein emulsifier composition on the stability with respect to accelerated coalescence by centrifugation has been studied for the binary mixtures of caseinate + gelatin,

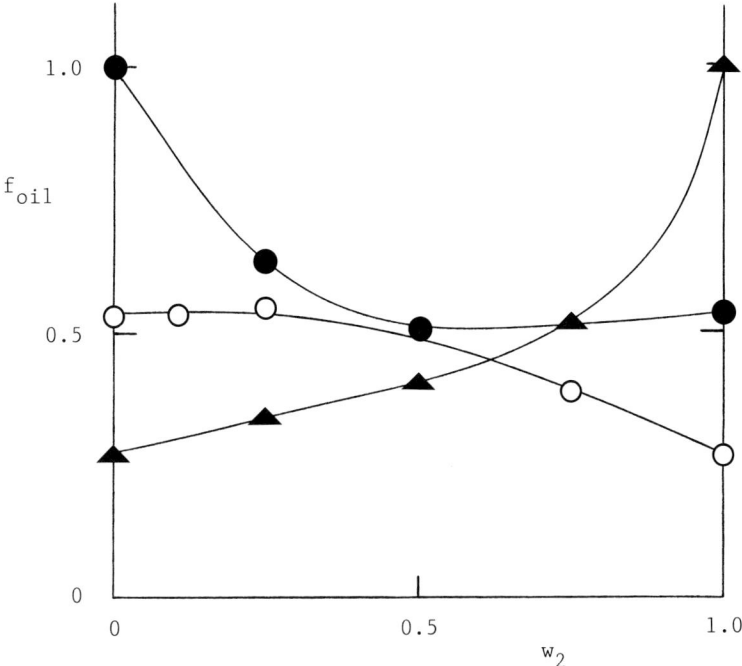

Figure 2. Accelerated coalescence stabilities of vegetable oil-in-water emulsions (40 wt % oil, 0.5 wt % total protein, pH 7). The fraction of free oil extracted (f_{oil}) is plotted against the weight fraction w_2 of protein component 2 in the emulsifier: o, gelatin (1) + caseinate (2); •, β-lactoglobulin (1) + gelatin (2); ▲, caseinate (1) + β-lactoglobulin (2).

β-lactoglobulin + gelatin, and caseinate + β-lactoglobulin. In figure 2, the fraction of free oil extracted by organic solvent is plotted against the composition of the protein emulsifier. Stabilities with the single proteins lie in the order: caseinate (27 ± 1 % free oil) > gelatin (53 ± 2 % free oil) > β-lactoglobulin (100 ± 1 % free oil). We see that the amounts of free oil released from emulsions made with the binary protein mixtures vary in a smooth monotonic manner with the composition of the emulsifier. This is particularly noteworthy for the case of caseinate + gelatin, where we know that caseinate predominates at the droplet surface (2), and for which we find very strong compositional dependences in emulsion bulk rheology and protein film surface rheology (see below). Also noteworthy is the fact that the accelerated coalescence stability with β-lactoglobulin is markedly improved by a small addition of caseinate or gelatin.

The results in figure 2 suggest no positive correlation
between the centrifugal stability of the emulsions and measured
surface rheological parameters for pure or mixed protein films at
the oil—water interface (6). Indeed, if a correlation does in
fact exist, it would seem to be negative rather than positive.
So, of the single proteins, caseinate gives the weakest and least
viscous film, but the best emulsion stability—whereas the worst
stability and the strongest film is given by β-lactoglobulin. A
similar trend was found by Graham and Phillips (10) from studies
of less stable emulsions at much lower centrifugation speeds.
In the centrifuge, far from a strong rigid film favouring droplet
stability, as it does in a quiescent emulsion, it may be quite
detrimental since it may be rather brittle and may suddenly
rupture as droplets are forced together under the influence of
the external field. By contrast, a more flexible and less viscous
film will probably more readily accommodate itself to changes in
surface shape and area caused by droplet deformation. These
considerations would tend to reinforce the view that accelerated
coalescence stability is likely to be a poor guide to the shelf-
life of protein-stabilized emulsions.

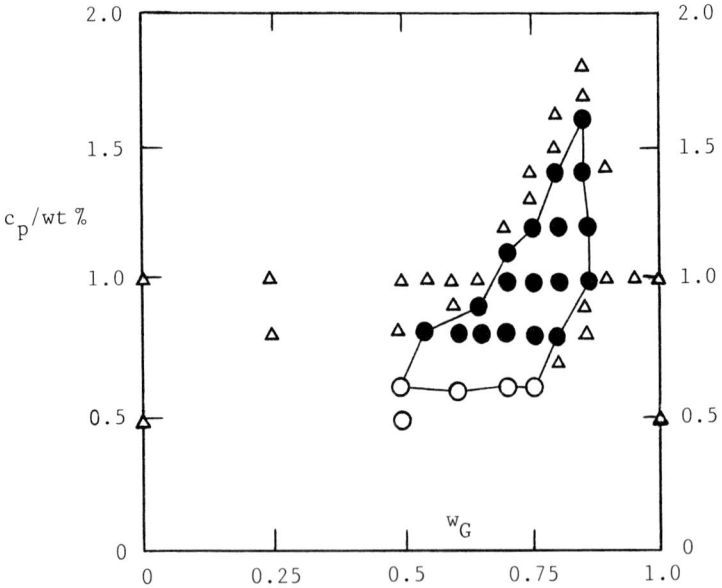

Figure 3. State diagram for n-hexadecane-in-water emulsions
(55 vol % oil, pH 7) made with caseinate + gelatin. Total
protein content c_p is plotted against gelatin weight fraction
w_G: △, normal stable emulsion ('liquid'); ●, flocculated
viscous emulsion ('cream'); o, viscoelastic emulsion ('paste').

When caseinate + gelatin is used as emulsifier, we find that there is a certain range of emulsifier composition over which it is not possible to make a stable unflocculated emulsion. Using the 'jet' homogenizer described earlier, we have been able to determine precisely the region of non-emulsification under a standard set of conditions, as illustrated in figure 3. The state diagram distinguishes three types of emulsion: normal stable liquid-like, flocculated cream-like, and highly viscoelastic paste-like. Within the closed boundary drawn in figure 3, it was not possible to make a stable emulsion. Outside the boundary, there was no problem, except for $c_p \leqslant 0.7$ wt % when the mixed systems gave paste-like emulsions, though the emulsions made with the single proteins were quite normal.

Figure 4 shows time-dependent shear moduli for a set of 55 vol % n-hexadecane-in-water emulsions made with caseinate + gelatin

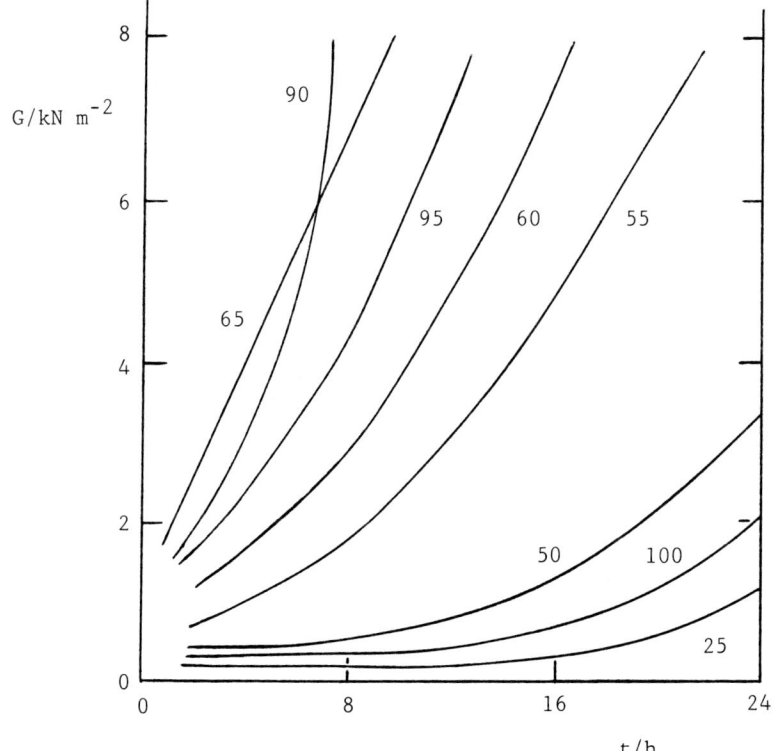

Figure 4. Shear modulus G as a function of time t for a set of emulsions (55 vol % oil, 1 wt % total protein) made with caseinate + gelatin. Numbers on curves are % gelatin in mix.

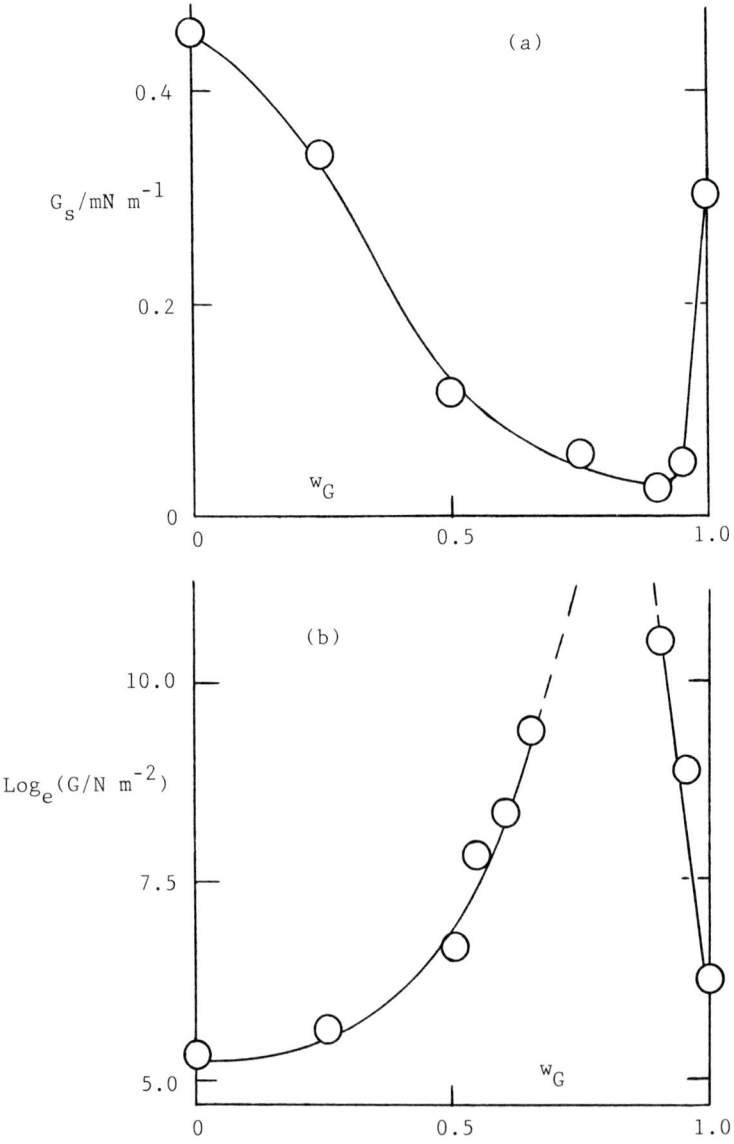

Figure 5.   Rheology of caseinate + gelatin systems after 12 h
at 25 °C. (a) Surface shear modulus G_s of adsorbed film is
plotted against gelatin weight fraction w_G. (b) Logarithm of
bulk shear modulus G of emulsion is plotted against w_G.

mixtures of variable composition but constant total protein content (c_p = 1.0 wt %). The 0 % gelatin emulsion has negligible elasticity, but for all the other compositions the modulus is finite and increases continuously with time. At this particular overall protein content, proper emulsions could not be prepared for gelatin weight fractions from 0.7 to 0.85. The results in figure 4 indicate that, as the emulsifier composition approaches the boundary between stable and unstable emulsions (see figure 3), the emulsion shear modulus becomes larger and much more strongly dependent on the age of the emulsion.

In attempting to explain these observations, we point to a striking similarity of behaviour between the bulk elasticity of the emulsion and the surface elasticity of the protein film at the planar oil—water interface. Figure 5a shows measurements by Murray (7) of the surface shear modulus of 12-hour-old films adsorbed at the n-hexadecane—water interface (pH 7, 25 °C) from mixtures of caseinate + gelatin at a constant bulk phase protein concentration of 10^{-3} wt %. Figure 5b shows measurements of the bulk shear modulus of 12-hour-old emulsions (pH 7, 25 °C) made with mixtures of caseinate + gelatin of total protein content of 1 wt %. We believe that the minimum in the surface modulus and the maximum in the bulk modulus are not directly related as such, but that both are the manifestation of competitive adsorption between caseinate and gelatin at the oil—water interface. It seems as if the presence of about 20 % caseinate in the protein mixture is enough to prevent gelatin from getting to the primary interface. This leads to a minimum in the surface rheological parameters for reasons discussed elsewhere (5-7), and a maximum in the emulsion bulk rheological parameters because nearly all the gelatin is present in the bulk phase between the droplets at a concentration well above the gelation threshold. Previously, we showed (11) that for emulsions stabilized by gelatin alone, and studied under standard conditions, the rheology is sensitively dependent on the partition of protein between surface and bulk. In the caseinate + gelatin systems, the balance of gelatin between interface and bulk is controlled by the amount of caseinate in the emulsifier mixture. The most sensitive instrument for monitoring the composition- and time-dependent changes seems to be the Pulse Shearometer, because of the negligible disturbance which it imparts to the sample. Similar general trends to those recorded in figure 4 have, however, also been obtained using a Carri-Med controlled-stress oscillatory rheometer operating at the lower frequency of 5 Hz (12).

The preceding discussion relating to the compositional aspects of the emulsion rheology also helps us to understand the poor emulsifying properties of the gelatin-rich emulsifier mixtures of caseinate + gelatin. Within the wedge-shaped closed region on the emulsion state diagram (figure 3), what came out of the 'jet' homogenizer was a highly flocculated thick cream-like product, occasionally accompanied by some free oil. Probably this is the result of caseinate going preferentially to the fresh oil—water interface immediately after droplet formation, in sufficient amounts to prevent gelatin from adsorbing, but not in sufficient amounts to properly protect the newly formed droplets against bridging flocculation or recoalescence. In general, one would expect the positions of the boundaries of the non-emulsification

region to vary a little with the conditions of homogenization and
the oil/water ratio, while the overall qualitative effect remains
the same.
    There is one important technological lesson to be drawn from
this investigation. Enriching a protein emulsifier of low
surface activity with a small proportion of protein of high
surface activity may lead to a deterioration in performance,
rather than an improvement—and increasingly so at low protein
load and high oil content.

ACKNOWLEDGEMENTS

    We thank Dr P. T. Speakman for the use of the ultracentrifuge,
Mr Philip Nelson for technical assistance in constructing the
'jet' homogenizer, and the Chief Scientist's Group at the
Ministry of Agriculture, Fisheries and Food for supporting the
research. The results are the property of the Ministry and are
Crown Copyright.

REFERENCES

1   Dickinson, E. (1986) Food Hydrocolloids, 1, 3-23.
2   Castle, J., Dickinson, E., Murray, A., Murray, B. S. and
    Stainsby, G. (1986) in Gums and Stabilisers for the Food
    Industry (eds Phillips, G. O., Wedlock, D. J. and Williams,
    P. A.), vol. 3, pp. 409-417. Elsevier Applied Science, London.
3   Dickinson, E., Murray, A., Murray, B. S. and Stainsby, G.
    (1987) in Food Emulsions and Foams (ed. Dickinson, E.), pp.
    86-99. Royal Society of Chemistry, London.
4   Dickinson, E., Whyman, R. H. and Dalgleish, D. G. (1987) in
    Food Emulsions and Foams (ed. Dickinson, E.), pp. 40-51.
    Royal Society of Chemistry, London.
5   Dickinson, E., Murray, B. S. and Stainsby, G. (1985) J.
    Colloid Interface Sci., 106, 259-262.
6   Castle, J., Dickinson, E., Murray, B. S. and Stainsby, G.
    (1987) ACS Symp. Ser., in press.
7   Murray, B. S. (1987) Ph. D. thesis, University of Leeds.
8   Buscall, R., Goodwin, J. W., Hawkins, M. W. and Cotterill,
    R. H. (1982) J. Chem. Soc., Faraday Trans. 1, 78, 2873-2887.
9   Ring, S. G. and Stainsby, G. (1985) J. Sci. Food Agric., 36,
    607-613.
10 Graham, D. E. and Phillips, M. C. (1976) in Theory and
    Practice of Emulsion Technology (ed. Smith, A. L.), pp. 75.
    Academic Press, London.
11 Dickinson, E., Stainsby, G. and Wilson, L. (1985) Colloid
    Polym. Sci., 263, 933-934.
12 Castle, J. (1987) Ph. D. thesis, University of Leeds.

Rheological properties of solutions of oat β-glucans

K. Autio

Technical Research Centre of Finland, Food Research Laboratory, Biologinkuja 1, 02150 Espoo, Finland

SYNOPSIS

Cereal β-glucans are composed of linear polysaccharide chains, whose structural components are glucose units connected with $\beta(1-3)$ and $(1-4)$ linkages. Oat β-D-glucan consists of about 70 % of $(1-4)$-linked and about 30 % of $(1-3)$-linked β-D-glucosyl residues.
In this work semi-dilute and concentrated water- and sucrose- solutions of oat β-glucans were studied by steady flow, oscillatory flow and transient flow measurements. The rheological tests cover low and high rates of deformation.
The results suggest that several relationships exist between viscoelastic data measured with different techniques.

1. INTRODUCTION

The viscoelastic properties of fluids can be characterized by classical methods, such as dynamic, relaxation and steady shear testing. Similarities have been found to exist between material functions and linear viscoelastic properties of polymeric liquids (4). For example, the dynamic viscosity of solutions of random coil polymers, is closely super-imposable on the rotational viscosity (Cox-Merz rule). This relationship is useful because where one technique runs into difficulty another can be applied.

In regard to the application of β-glucans in the food industry, the viscoelastic properties are among the most important physical characteristics of cereal β-glucans; yet they have been touched upon only in few publications (7,1).

In this paper the effects of concentration and sucrose on the viscoelastic properties of β-glucan solutions are studied and the results obtained by different techniques are compared.

2. MATERIALS AND METHODS

2.1 Preparation of β -glucan: β-glucan was extracted from oat
bran of mixed Finnish commercial varieties. The β-glucan
content of the bran was determined by the method of McCleary
and Glennie-Holmes (6) and was 8.7% (w/w).Fifty grammes of
bran was stirred in 1700 ml water for one hour. The solids
were centrifuged in a Sorvall RC-5B centrifuge at 16300 g for
20 min. and the proteins were precipitated from the clear
supernatant by lowering the pH to 4.5 with 2 M HCl. Precipitated proteins were centrifuged as above and the pH of the
clear viscous supernatant was raised to 7.0 with 2 M NaOH.β
-glucans were precipitated by rapidly mixing an equal volume
of 2-propanol with the viscous supernatant. The fibrous
precipitate was collected on a 0.105 mm sieve and washed
with 2-propanol. Before the β-glucan was dissolved in water
or sucrose solution, the 2-propanol was evaporated at 35 °C.
A 0.9 % (w/w) stock solution was prepared for the rheological
measurements and stored at -30 °C.
The β-glucan content of the freeze-dried stock solution was
determined by barley β-glucan assay procedure (6).

2.2 Determination of molecular weight:the MW of the sample was
determined at 30 °C by GP-HPLC μBondagel E-1000, E-500 and
E-125 columns having fraction ranges $2x10^6$-$5x10^4$, $5x10^5$-
$5x10^3$, and $5x10^4$-$2x10^3$, respectively were connected in series.
Dextran standards were used for the calibration of columns
and water was used as the mobile phase.

2.3 Rheological measurements: The β-glucan stock solution
was diluted with water or sucrose-solution and dispersed air
was removed in vacuum. Rheological measurements were made
with a Bohlin VOR instrument (Bohlin Rheology AB, Lund,
Sweden) operating in viscometry, oscillation and relaxation
modes. The system C25 and torsion bars ranging from 3.6 to
86 gcm were used. All measurements were carried out at 25°C.
In the oscillation and relaxation modes the experiments were
made in the linear region.

2.4 Calculations: The complex viscosity was calculated
according Cox and Merz (2):

$$\eta^* = \sqrt{\frac{G'^2 + G''^2}{\omega}}$$

where G'= storage modulus
      G"= loss modulus
      ω = 2x π xfrequency

3. RESULTS AND DISCUSSION

3.1 Viscosity measurements

Figure 1 shows the steady viscosity as measured by viscometry
as well as dynamic viscosity and complex viscosity calculated
from oscillation for β-glucan water solutions of different

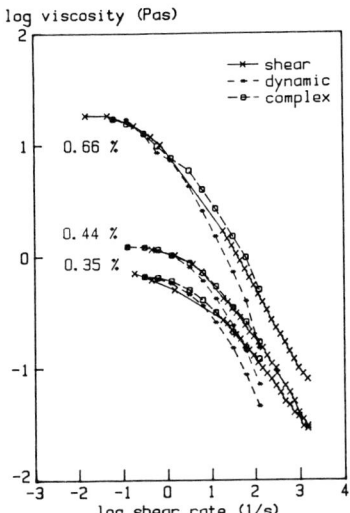

Figure 1. Viscosities of β-glucan water solutions.

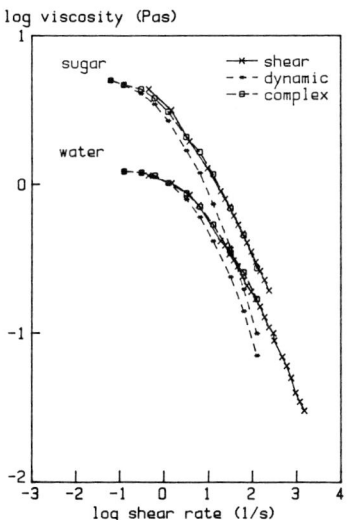

Figure 2. Viscosities of water and sucrose solutions (25%,w/w) of β-glucan (0.44 %,w/w).

concentrations. All three viscosities are similar at low
shear rates and the Cox-Merz rule is obeyed. In contrast at
high shear rates the complex and steady viscosities are higher
than the dynamic viscosity when the same numerical values of ω
and $\dot{\gamma}$ are compared. According to Leppard (5) for most polymeric materials the steady viscosity at a given high shear
rate is higher than the corresponding dynamic viscosity. The
similarity of the complex and steady viscosities at all
shear rates studied could have the practical benefit of
removing the need for steady viscosity measurements, which
are more difficult to make and less reliable than the small
amplitude dynamic viscoelastic measurements. For moderately
concentrated solutions of typical hydrocolloids the Newtonian
behaviour is observed at low shear rates and the zero shear
Newtonian value can be obtained from both oscillatory and
steady shear measurements.

Figure 2 demonstrates how sucrose (25%) increases the
viscosity of β-glucans, particularly at low shear rates.
A similar effect has been reported for guar gum and locust
bean gum solutions (3). It has been suggested that when
sugars are added at higher polymer concentrations, the
hydration and extension of polymer molecules are restricted.
In more dilute systems the increased polymer-polymer
interactions compensate for the reduction in hydration.

3.2 Oscillation test

The mechanical spectra (Figs. 3 and 4) of β-glucan solutions
are typical of concentrated solutions: at low frequencies
G">G' and at high frequencies G'>G", both moduli increase with
increasing frequency. When the polymer concentration is
increased or sucrose added, causing an increase in coil
overlap, the transition from solid-like to liquid-like
response moves to lower frequencies.

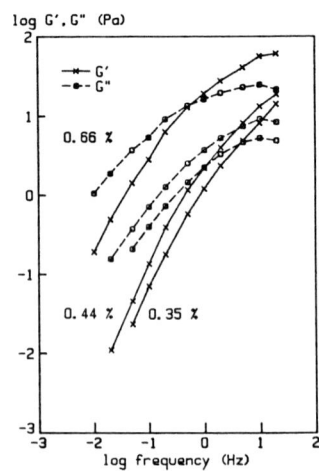

Figure 3. Mechanical spectra of β-glucan water solutions.

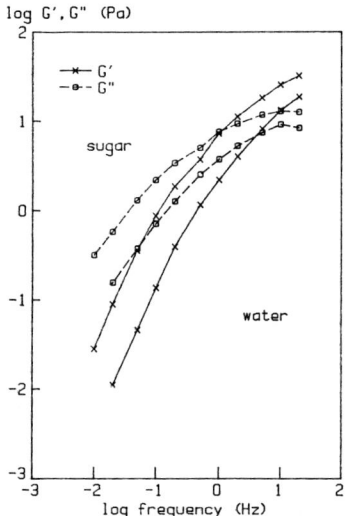

Figure 4. Mechanical spectra of water and sucrose solutions (25%,w/w) of β -glucan (0.44 %,w/w).

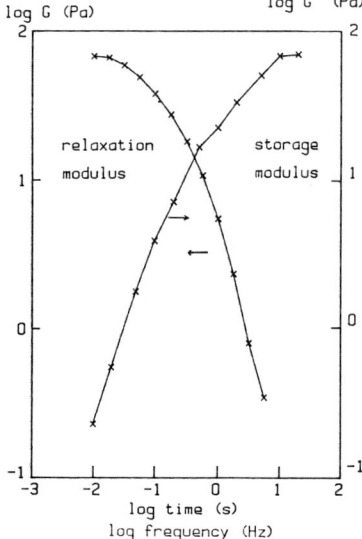

Figure 5. Storage and relaxation moduli for a β- glucan water solution (0.79%, w/w).

3.3 Relaxation test

For most of the β-glucan solutions studied the relaxation time was very short and some relaxation had taken place during the strain rise time, which was 20 ms.
For the highest β-glucan concentration (0.79 %), however, the results obtained from oscillation and relaxation tests were similar (Fig.5). Since both G(t) and G'(ω) are measures of stored elastic energy and a dynamic measurement at frequencyω is qualitatively equivalent to a transient one at t = 1/ω , these curves are approximately mirror images.

4. REFERENCES

1. Autio, K., O. Myllymäki and Y. Mälkki, Flow properties of solutions of oat β-glucans. J. Food Sci. (in press).

2. Cox, W.P. and E.H. Merz, Correlation of dynamic and steady flow viscosities. J. Polym. Sci. 28, 619-622, 1958.

3. Elfak, A.M., G. Pass, G.O. Phillips and R.G. Morley, The viscosity of dilute solutions of guar gum and locust bean gum with and without added sugars. J. Sci. Food Agric. 28, 895-899, 1977.

4. Ferry, J.D., Viscoelastic properties of polymers, 3rd edn., Wiley, New York, 1980.

5. Leppard, W.R. and E.G. Christianssen, E.G., Transient viscoelastic flow of polymer solutions. AIChEj. 21, 999-1006, 1975.

6. McClearly, B.V. and M. Glennie-Holmes, Enzymic quantification of (1-3)(1-4)-β-D-glucan in barley and malt. J. Inst. Brew. 91, 285-295, 1985.

7. Wood, P.J., Physicochemical properties and technological and nutritional significance of cereal β-glucans in "Cereal Polysaccharides in Technology and Nutrition". pp. 35-78. Am. Ass. Cereal Chem. INC. St. Paul, Minnesota, 1984.

Adsorption and stabilising properties of gum arabic samples of different origin

M.J.Snowden, G.O.Phillips and P.A.Williams

Research Division, The North East Wales Institute, Deeside, Clwyd CH5 4BR, UK

Abstract

Gel permeation chromatography (GPC) has shown that gum arabic samples from the Kordofan region of Sudan and from Nigeria have very broad molecular mass distributions. Specific analyses has indicated that the uronic acid component of the polysaccharide is contained within a relatively narrow molecular mass range illustrating the heterogeneous nature of the gum.

Studies have shown that the gums adsorb onto polystyrene latices from 0.5 mol dm^{-3} NaCl with adsorption capacities of 5.0 and 7.3 mg m^{-2} respectively.

GPC analysis for the Kordofan and Nigerian samples has also shown that fractionation occurs on adsorption. For the Kordofan sample the high molecular mass fraction is adsorbed in preference and adsorption is virtually completed within the first hour. For the Nigerian sample, although generally higher molecular mass material is adsorbed preferentially, a portion of lower molecular mass material adsorbs and this is not displaced even after four days. For both gums the fraction that does not adsorb corresponds closely to the fraction that contains the uronic acid residues. Furthermore it has been shown that all of the proteinaceous component is adsorbed.

The stabilities of dispersions of the polystyrene latex have been monitored in 0.5 mol dm^{-3} NaCl as a function of added gum arabic by turbidity measurements. In the absence of gum arabic the particles flocculate as might be expected since compression of the electrical double layer surrounding each particle in a high electrolyte environment enables van der Waals attractive forces to operate and aggregation to occur. On addition of gum arabic, aggregation is prevented at the point corresponding closely to the adsorption capacity of each of the gums as a result of steric repulsive forces between the adsorbed polymer layers.

Introduction

Gum Arabic is a naturally occurring polysaccharide obtained as the dried exudate of the acacia tree and is widely used in the

food industry in for example the encapulation of flavours, the
stabilisation of emulsions and in confectioneries[1]. It is a
heteropolymolecular substance consisting of molecules differing
not only in molecular mass but also chemically [2,3] in the
relative proportion and mode of linking of the monomer sugar
units. The gum contains galactose, arabinose, rhamnose and
glucuronic acid residues and the main structural feature is a
backbone of β-galactopyranose units linked through the 1, 3
positions with side chains of 1,6 linked galactopyranose units
terminating in glucuronic acid or 4-0-methyl glucuronic acid
residues. In addition it has been found that there is a small
amount of protein present within the material which forms an
integral part of the structure [2-4].

Despite its extensive use as an emulsifier in, for example, the
stabilisation of citrus oils and other beverage flavours its
mode of action is still not clearly understood. We have
recently reported some model experiments detailing the
adsorption of the gum onto polystyrene latex and its effect on
colloid stability [5], this work has now been extended and in this
paper we present some of our most recent findings.

Materials and Methods

Two commercial samples of Gum Arabic were used, one originating
from the Kordofan region of Sudan and the other from Nigeria.
Both were in the form of fine white crystalline particles and
their general characteristics are detailed in Table 1.

Monodisperse polystyrene latex were prepared as described
previously [5] and were found to have a mean particle diameter of
$331 \pm 6nm$ by electron microscopy. The density of the latex was
$1.056g\ cm^{-1}$ as determined by pyknometry and the electrophoretic
mobility in 10^{-4} mol dm $^{-3}$ NaCl at pH8 was $-3.33 \times 10^{-8}\ m^2\ v^{-1}$
s^{-1}.

Isotherms were dertermined for the adsorption of gum arabic
onto polystyrene latex from 0.5 mol dm $^{-3}$ NaCl as detailed
previously [5]. The molecular mass distribution was monitored
before and after adsortion from 0.5 mol dm $^{-3}$ NaCl using gel
permeation chromatography (gpc) as before [5]. Gum Arabic present
in the eluent from the g.p.c. column was detected spectrophoto-
metrically by uv absorption at 218 mm using a quartz flow cell
and also by collecting the fractions and analysing for uronic
acid residues by the carbazole reaction using an automatic
analyser [6]. The amount of protein in the supernatant was
determined as a function of time using the Bio-Rad assay[7].

Colloid stability was estimated from the wavelength dependence
of the turbidity of the dispersions as described in the previous
paper [5]. The parameter 'n', (the gradient of the log absorbance
vs log wavelength plot) was used to quantify particle size with
its magnitude decreasing as the particle (or aggregate) size
increases.

TABLE 1

Molecular characteristics of Kordofan and Nigerian gum arabic.

	Kordofan	Nigerian
Intrinsic viscosity $cm^3 g^{-1}$ at 25°c and in 1 mol dm^{-3} NaCl	23.13	23.53
Molecular weight (Mv)	1.03×10^6	1.08×10^6
Acid Equivalent Weight	1110	1380
Optical rotation	-27.5°	-30°
% Nitrogen	0.34	0.30
% Protein [N x 6.25]	2.125	1.9
% Calcium (w/w)	0.89	0.93
% Potassium (w/w)	0.98	0.59
% Sodium (w/w)	0.02	0.05
% Magnesium (w/w)	0.20	-

TABLE 2

Removal of proteinacious material within Gum Arabic as a function of adsorption time.

	Amount Adsorbed	
Time	Kordofan	Nigerian
0	0	0
15 min*	60%	55%
1 day	90%	---
4 days	100%	100%

* Note: The actual adsorption time is longer than stated due to the time required for centrifugation.

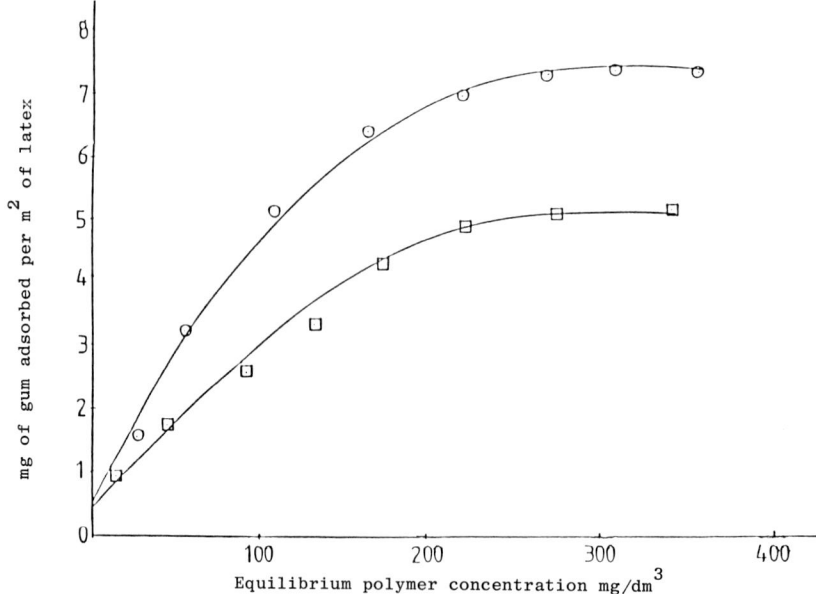

Figure 1, Isotherms for the adsorption of Gum Arabic onto polystyrene latex from 0.5 mol dm⁻³ NaCl; O Nigerian gum and ☐ Kordofan gum.

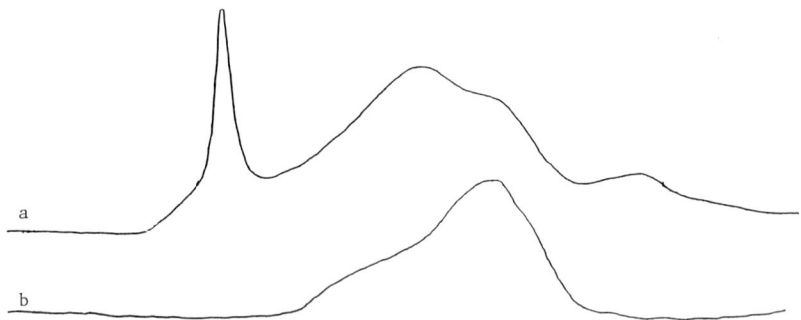

Figure 2, (a) Gel permeation chromatogram of a sample of Kordofan gum arabic as monitored by U.V. at 218 nm from a column packed with Sephacryl S500 (Pharmacia) (b) the corresponding uronic acid profile for the same sample as determined by the carbazole reaction using an automated analyser.

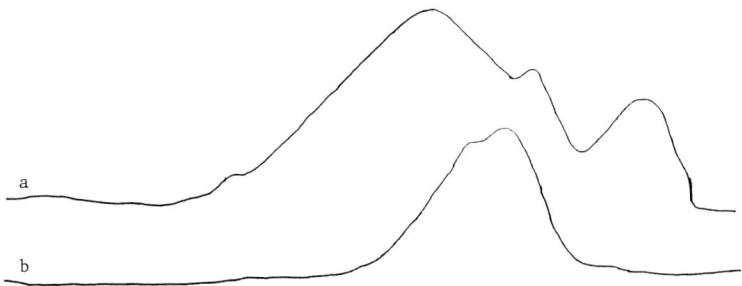

Figure 3, (a) Gel permeation chromatogram of a sample of Nigerian gum arabic as monitored by U.V. at 218 nm from a column packed with Sephacryl S500 (Pharmacia). (b) the corresponding uronic acid profile for the same sample as determined by the carbazole reaction using an automated analyser.

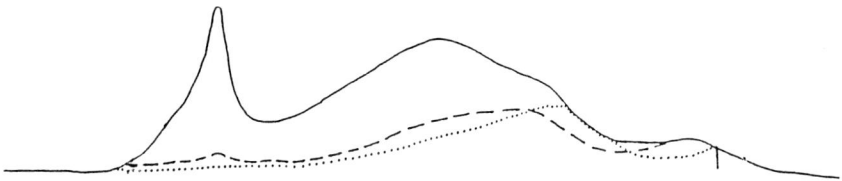

Figure 4, Gel permeation chromatograms of Kordofan gum arabic before and after adsorption onto polystyrene latex from 0.5 mol dm^{-3} NaCl _____ before adsorption, - - - after 15 minutes adsorption and ...... after 4 days adsorption.

Results and Discussion

Isotherms for the adsorption of the two gums from 0.5 mol dm^{-3}
NaCl onto polystyrene latex are given in Figure 1. The initial
slope of the two curves indicates low affinity adsorption and
the amount adsorbed at plateau coverage is 5.0 mg m^{-2} for the
Kordofan sample and 7.3 mg m^{-2} for the Nigerian sample. The
reason for the difference in the adsorption capacities is not
certain since both polymers are of similar average molecular
mass but it is possibly associated with differences in their
charge density since the Nigerian gum contains fewer uronic acid
residues than the Kordofan sample. The gel permeation
chromatograms for the two samples are given in Figures 2 and 3;
the profiles marked 'a' were obtained by monitoring the eluent
by u.v. and the profiles marked 'b' were obtained by analysis of
the fractions using the carbazole reaction which is specific for
uronic acid residues. The differences in the curves obtained by
the two techniques illustrates that the uronic acid residues are
restricted to a particular molecular mass range demonstrating
the heteropolymolecular nature of the samples. Similar results
were obtained for both gums. The chromatrograms of the gums
before and after adsorption are given in figures 4 and 5 and
indicate that fractionation occurs. For the Kordofan sample,
clearly the higher molecular mass molecules adsorb preferentially. There is very little difference in the profiles after
15 minutes and 4 days adsorption indicating that there is little
exchange taking place over this period. For the Nigerian sample
both higher and lower molecular mass molecules adsorb and again
very little exchange takes place following the initial
adsorption after 15 minutes. Generally one might expect that
lower molecular mass molecules would adsorb initially since they
would diffuse to the surface faster but these would be displaced
by higher molecular mass molecules due to their increased energy
of adsorption [8,9]. The fact that there is no or very little
exchange taking place following the initial adsorption indicates
that the molecules are strongly bound to the surface [10]. This
is in accordance with previous results which indicated that
approximately half of the adsorbed polymer segments were in
trains close to or bound to the surface [5]. One striking feature
common to both gums is that the fraction that remains unadsorbed
corresponds to the uronic acid component as established in
figures 2b and 3b. Furthermore, analysis of the supernatants
indicated that all proteinaceous material actually became
adsorbed and this is shown in Table 2.

The stability of the latex dispersions in 0.5 mol dm^{-3} NaCl is
given as a function of added polymer in Figure 6. Initially, in
the absence of polymer the particles flocculated as would be
expected since the electrical double layer surrounding each
particle would be compressed to such an extent that van der
Waal's attractive forces would predominate. At increasing
levels of added polymer, aggregation is prevented and the
particles remain dispersed. Complete stability is obtained at
polymer additions corresponding closely to the polymer
adsorption capacities as given in the adsorption isotherms in
Figure 1, and consequently more of the Nigerian gum is required

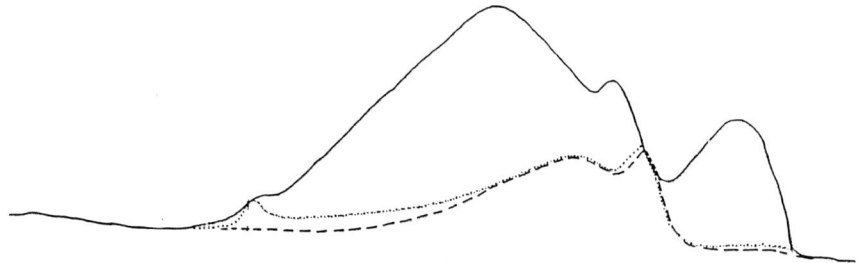

Figure 5, Gel permeation chromatograms of Nigerian gum arabic before and after adsorption onto polystyrene latex from 0.5 mol dm^{-3} NaCl _____ before adsorption, ... after 15 minutes adsorption and - - - after 4 days adsorption.

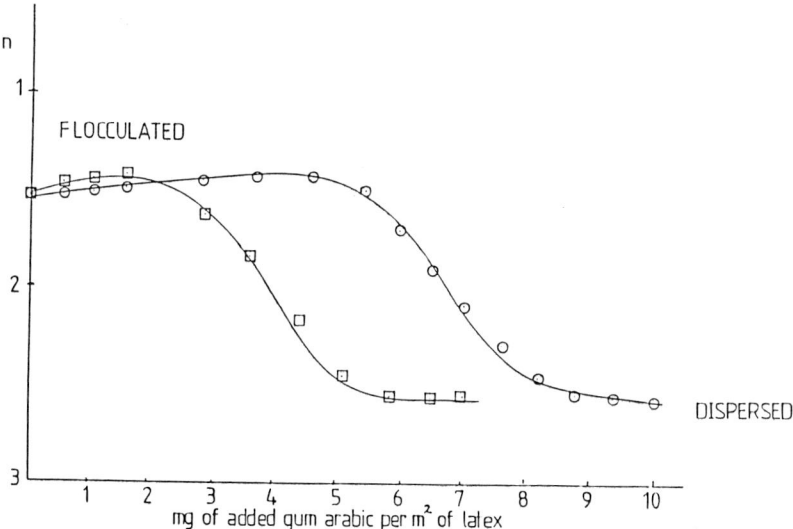

Figure 6. A plot of stability as monitored by the parameter 'n' against the amount of gum arabic added,

to achieve stability. It is apparent that stability is attained
as a result of steric repulsions between the adsorbed polymer
layers. At polymer additions where the particle surface is not
fully coated flocculation arises as a result of polymer
bridging.

The results clearly demonstrate the heterogeneous nature of gum
arabic and illustrate that particular components within the gum
are responsible for its stabilising action.

References

1. Whistler, R. L. (1973), 'Industrial Gums', Academic Press
    Inc. p197-263.
2. Anderson, D. M. W. and Street C. A., (1983), Talanta, $\underline{30}$,
    11, 887-893.
3. Vandevelde, M. C. and Fenyo, J. C. (1985), Carbohydrate
    Polymers, $\underline{5}$, 251-273.
4. Akiyama, Y., Eda, S. and Kato, K. (1984), Agric. Biol.
    Chem. $\underline{48}$ (1), 235-237.
5. Snowden, M. J., Phillips, G. O., Williams, P. A.,
    (1987) Food Hydrocolloids, $\underline{1}$, 4, 291.
6. Heinegard, D., (1973) Chemica Scripta, $\underline{4}$, 199.
7. Bradford, M., (1976) Anal Biochem, $\underline{72}$, 248.
8. Scheutjens, J. M. H. M., Fleer, G. J. (1981) ed Tadros, T.
    H. F., 'The effect of polymers on dispersion properties'
    Academic Press.
9. Dickinson, E., (1986) Food Hydrocolloids,$\underline{1}$, 1, 3.
10. Bain, D. R., Cafe, M. C., Robb, I. D., and Williams, P. A.,
    (1982), J. Colloid and Interface Sci, $\underline{88}$, 2, 467.

The use of electron spin resonance spectroscopy to study the gelation of kappa carrageenan

D.H.Day, G.O.Phillips and P.A.Williams

Research Division, North East Wales Institute, Deeside, Clwyd CH5 4BR, UK

Abstract

Electron spin resonance spectroscopy (e.s.r.) has been used to study the gelling characteristics of kappa carrageenan in the presence of potassium ions. On gelation, as monitored by photon correlation spectroscopy and also observed macroscopically, the e.s.r. spectra were found to contain both isotropic and anisotropic components and the various proportions of each were resolved by computer analysis. The results indicated that on gelation a two phase system exists consisting of precipitated aggregates dispersed within a homogeneous solution of dissolved molecules and it was found experimentally that the two phases could be separated by centrifugation. Intrinsic viscosity measurements showed that fractionation had occurred on gelation with the larger molecules constituting the gel phase. Specific analysis also indicated that fractionation had occurred as a result of differences in chemical composition.

Introduction

Carrageenans are sulphated galactans extracted from the Rhodophyceae class of red algae[1] and are widely used in the food and related industries because of their ability to form viscous solutions and gels. Their properties have been studied extensively by many workers[2-7] and a number of models have been put forward to describe their mechanism of gelation.[8-10] Although a wide variety of techniques have been employed to study their physico chemical characteristics, to date there are no reports in the literature on studies using electron spin resonance spectroscopy (e.s.r.). Using the nitroxide free radical, e.s.r. is able to monitor molecular motion in the 10^{-7} to 10^{-11} second timescale and an important advantage is that the technique is non-destructive.

Materials and methods

Kappa carrageenan was obtained from the Copenhagen Pectin Factory Ltd., and was found to contain galactose (31.2%) sulphate (26.1%) and 3,6-anhydrogalactose (25.9%). The molecular mass was found to be 250,000 as determined by intrinsic viscosity using the method of Snoren.[11] The sample was spin labelled according to the procedure of Cafe and Robb[12] using 4 amino

2,2,6,6, tetra methyl piperidine N oxyl as spin label. The degree of labelling was calculated from the intensity of the e.s.r. signal and found to be approximately 1 label per 1000 repeating units. This low degree of labelling ensured minimum perturbation of the carrageenan properties. The carrageenan was converted to the sodium salt by ion exchange and freeze dried before use.

E.s.r. spectroscopy

Spectra were recorded on a Jeol JES ME-IX X band e.s.r. spectrometer using a flat quartz cell suitable for aqueous solutions. A solution of spin labelled carrageenan (10^{-2} mol dm^{-3}) containing 4 x 10^{-2} mol dm^{-3} KCl was heated to 80°C and then introduced into the cell positioned in the cell cavity of the spectrometer which was thermostatted at 60°C. The solution was allowed to equilibrate for 10 minutes and the spectrum recorded. The temperature was lowered in 5° steps down to 5°C allowing 10 minutes equilibration time at each temperature and the spectrum recorded. The solution was held at this temperature for 1 hour and then the temperature was increased at the same rate and the spectrum recorded at each stage.

Photon correlation spectroscopy (pcs)

The diffusion coefficients (D) of the carrageenan molecules were determined as a function of temperature using p.c.s. The abrupt change in the value at a critical temperature was indicative of gelation. Measurements were carried out using the K7027 Malvern instruments p.c.s. equipped with a He-Ne laser at a measurement angle of 90°. The solutions were prepared as described above.

Results and discussion

The e.s.r. spectra for the polymer in aqueous solution at 25°C and in aqueous glycerol at 5°C are given in Fig.1. In aqueous solution the polymer molecules have a relatively high degree of mobility and this is reflected in the motionally narrowed three lined spectrum (spectrum a). In aqueous glycerol the molecules have restricted mobility due to the increased viscosity and this leads to anisotropic broadening (spectrum b). A similar type of spectrum is also obtained for the precipitated polymer. Spectra for carrageenan solutions containing 4 x 10^{-2} mol dm^{-3} KCl are given as a function of decreasing temperature in Fig.2. Above 40°C a motionally narrowed three-lined spectrum similar to 1a is obtained. Below this temperature composite spectra containing varying proportions of the isotropic (spectrum 1a) and anisotropic (spectrum 1b) components are observed indicating that some of the polymer segments have very restricted mobility as a result of aggregation and gelation. The temperature at which the composite spectra appear corresponds to the gelation temperature observed macroscopically and also found by diffusion coefficient measurements (Fig.3). Optical rotation results (not reported here) indicated also that a conformational change occurred at this temperature.

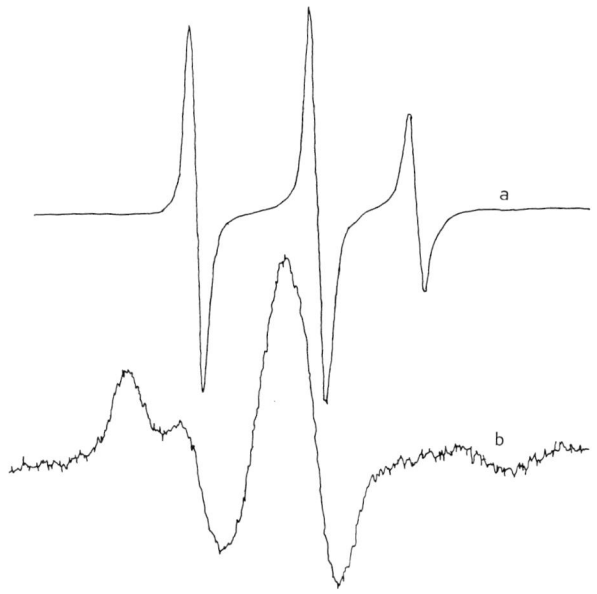

Figure 1   esr spectra for 10^{-2} mol dm^{-3} kappa carrageenan in
(a) water at 25°C, (b) 80% glycerol at 5°C.

The actual proportion of each component was resolved by computer analysis to
within ± 10% and the percentage of immobilised segments (p) as a function of
temperature is given in Fig.4. The results show that even at 5°C only 43%
of the molecular segments are in an aggregated form. The figure also gives
values of p on increasing the temperature of the system and indicates that
hysteresis occurs. The temperature at which p becomes zero i.e. 50°C
corresponds to the melting temperature noted macroscopically. Such
hysteresis effects have been reported by other workers using different
techniques.[13,14]

The results indicate that the carrageenan gel contains aggregated or
precipitated molecules (anisotropic component) dispersed throughout a
solution of dissolved molecules (isotropic component). In order to gain
further support for this interpretation a carrageenan gel was prepared at
10°C using unlabelled material and was then centrifuged at 22,000 g for 2
hours using an MSE High Speed 18 centrifuge. The supernatant liquor above
the gel was separated off and two 5 cm³ aliquots were withdrawn and dried in
an oven overnight to constant weight. It was found that approximately 29%
of the original starting material remained present in the supernatant

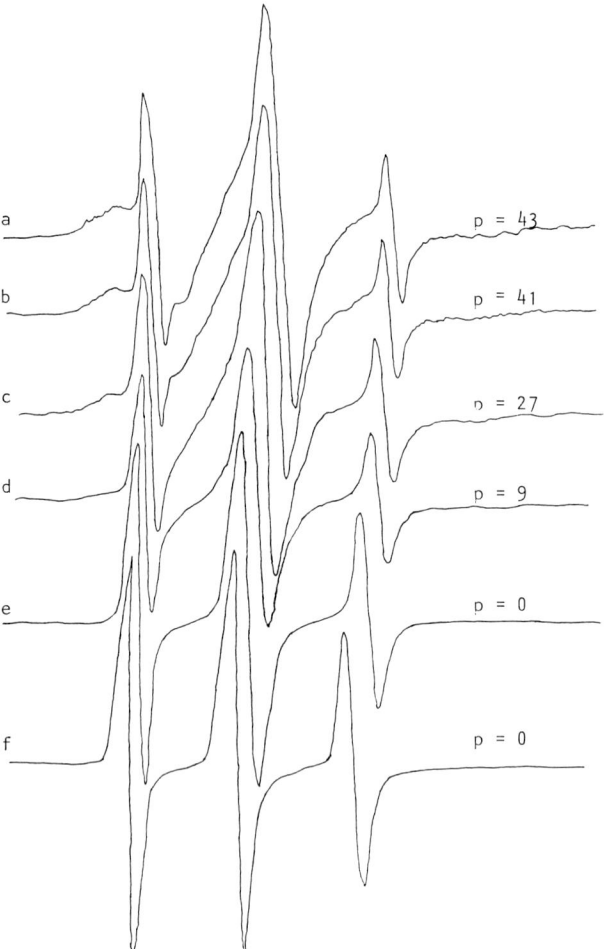

<u>Figure 2</u>   esr spectra for 10^{-2} mol dm^{-3} kappa carrageenan in
               4×10^{-2} mol dm^{-3} KCl at (a) 5°C, (b) 10°C, (c) 25°C,
               (d) 35°C,   (e) 40°C,   (f) 60°C.

solution thus providing further evidence for the existence of a two phase
system. The fact that this value does not quantitatively agree with the
e.s.r. data (58% ± 10%) may be due to the fact that dissolved molecules are
retained within the gel network during centrifugation and it is likely also
that the gelled aggregates themselves have a proportion of mobile segments
associated with them.The intrinsic viscosities of the starting material and

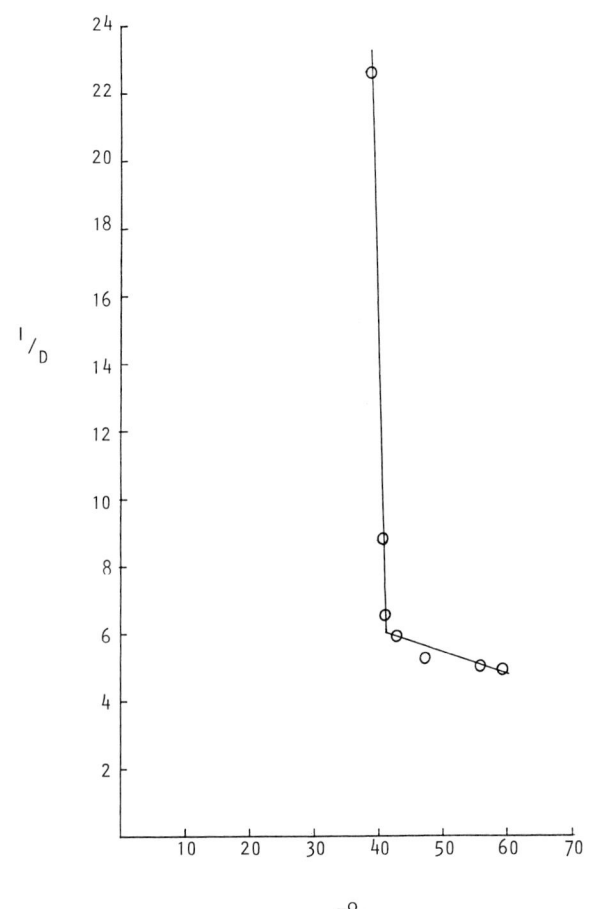

Figure 3   Plot on cooling of the reciprocal of the diffusion
coefficient as a function of temperature for 10^{-2} mol dm^{-3}
kappa carrageenan in 4 x 10^{-2} mol dm^{-3} KCl.

supernatant were determined in 0.04 mol dm^{-3} KCl at 58°C in an Ubbelohde
viscometer (Cannon Instruments Pennsylvania, USA) and were found to be 827 ±
20 and 495 ± 5 cm^{3}g^{-1} respectively indicating that the higher molecular mass
molecules are involved in gel formation. Specific analysis of the
supernatant liquor using thiobarbituric acid[15] also indicated that the
proportion of 3,6 anhydrogalactose units present was less than might be
expected from the total concentration. It is clear, therefore, that
fractionation of the carrageenan occurs on gelation and that the two phase

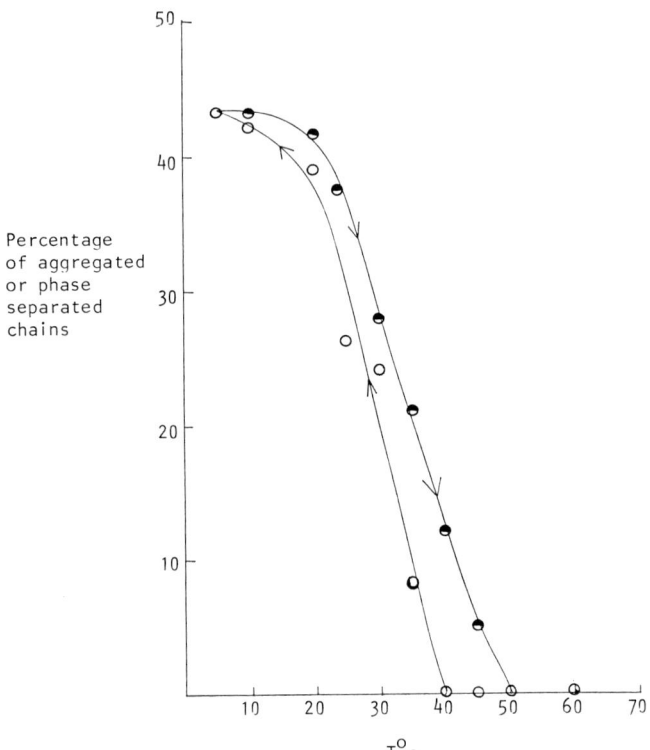

Figure 4   Percentage of aggregated or phase separated polymer
                      chains on cooling and reheating as determined by esr
                      for 10^{-2} mol dm^{-3} kappa carrageenan in 4×10^{-2} mol dm^{-3} KCl

system observed occurs as a result of variations in chemical composition and
molecular mass within the carrageenan sample.

The results are consistent with the gelation model of Rochas and
Rinaudo[10,16] who explained the process in terms of well defined thermodynamic principles. The results are also in accordance with Hayashi et
al[17-19] who showed, using fluorescence depolarisation, that agarose, a
closely related non-sulphated galactan also formed gels consisting of
isotropic and anisotropic components and who proposed a model very similar
to that of Rochas and Rinaudo.

References

1. Towle, G.A., "Industrial Gums" 1973. Ed. R.L. Whistler and J. BeMiller, Academic Press.

2. Bryce, T.A., Clark, A.H., Rees, D.A. and Reid, D.S, 1982, Eur J. Biochem 122, 63.

3. Rochas, C., Rinaudo, M. and Vincendon, M, 1980, Int.J. Biol. Macromol, 5, 111.

4. Norton, I.T., Goodall, D.M., Morris, E.R. and Rees, D.A., 1983, J. Chem. Soc. Faraday Trans. 1, 79, 2475.

5. Norton, I.T., Goodall, D.M., Morris, E.R. and Rees, D.A. 1983, J. Chem. Soc. Faraday Trans. 1, 79, 2475.

6. Watase, M., and Nishinari, K., 1981, J. Texture Studies, 12, 427.

7. Smidsrod, O., Anderson, I.L., Grasdalen, H., Larsen, B. and Pointer, T. 1980, Carbohydrate Res., 80, C11.

8. Smidsrod, O. and Grasdalen, H., 1982, Carbohydrate Polymers, 2, 270.

9. Morris, E.R., Rees, D.A. and Robinson, G.R., 1980, J. Mol. Biol. 138, 349.

10. Rinaudo, M. and Rochas, C., 1986, Polym. Prep. 27, 246.

11. Snoren, T.H., PhD. Thesis, 1976, Netherlands.

12. Cafe, M.C. and Robb, I.D., 1976, Polymer 17, 91.

13. Morris, E.R., Rees, D.A., Norton, I.T. and Goodall, D.M., 1980, Carbohydrate Research 80, 317.

14. Rochas, C. and Rinaudo, M., 1980, Biopolymers, 19, 1675.

15. Wedlock, D.J., Phillips, G.O. and Bachmann, M. 1984. "Gums and Stabilisers for the Food Industry 2". Ed. G.O. Phillips, D.J. Wedlock and P.A. Williams, Pergammon Press.

16. Rochas, C. and Rinaudo, M., 1984, Biopolymers, 23, 735.

17. Hayashi, A., Kinoshita, M. and Kuwano, M., 1977 Polym. J. 9, 219.

18. Hayashi, A., Kinoshita, M., Kuwano, M. and Nose, A., 1978, Polym. J. 10, 435.

19. Hayashi, A., Kinoshita, M. and Karueda, S., 1980, Polym. J. 12, 447.

A new method to assess the emulsifying capacity of proteins in model systems

B.Mertens and A.Huyghebaert

Laboratory of Food Technology, Chemistry and Microbiology, Faculty of Agricultural Sciences, 9000 Ghent, Belgium

1. Introduction

Despite the importance of proteins in the formation and stabilization of food emulsions, the physico-chemical properties that determine their exceptional emulsifying properties have not yet been established. One of the reasons for this gap between practical application and fundamental understanding is the lack of adequate methods to measure emulsifying properties

This abstract highlights a new method to assess the emulsifying capacity of proteins in model systems which

- (i) takes into account the fundamental difference between emulsion formation on the one side, and the different forms of emulsion instability on the other side.
- (ii) measures an essential parameter of the process .

2. The method

The model system studied is composed of 250 gram soybean oil and 750 gram protein solution of desired protein content (% w/w on aqueous phase). Both phases are previously mixed by means of a low speed mixer for 30 seconds. Emulsions are prepared using a high pressure homogenizer at 19.6 MPa (200 kg/cm^2), at 22° C.

The method is based on the turbidimetric technique as proposed by Pearce and Kinsella (1978). Emulsions of different protein concentrations are prepared under standardized conditions and the turbidity is measured at 500 nm (T$_{500}$). At low protein concentrations ($<$ 0.5 %) a linear relationship between T$_{500}$ and the protein concentration was demonstrated. The total interfacial area was found to be closely related to T$_{500}$. The slope of the regression curve can be considered as a parameter of the emulsifying capacity of the protein. This method was elaborated for caseinate.

*This work was supported by "The Institute for Scientific Research in Industry and Agriculture" - I.W.O.N.L.

WORKSHOP: EMULSION STABILISATION

CHAIRMAN: D.F.DARLING, Unilever Research, Colworth
              Laboratory, Sharnbrook, Bedford MK44 1LQ

ORGANISER: E.R.MORRIS, Cranfield Institute of Technology, Silsoe
              College, Silsoe, Bedford MK45 4DT

CONTRIBUTORS:

B. BERGENSTAHL, Institute for Surface Chemistry, Box 5607,
     S-114 86, Stockholm, Sweden.

A.H. CLARK, Unilever Research, Colworth Laboratory, Sharnbrook,
     Bedford MK44 1LQ.

A.J. CLARKE-STURMAN, Shell Research Ltd., Sittingbourne
     Research Centre, Sittingbourne, Kent ME8 8AG.

I.C.M. DEA, Leatherhead Food R.A., Randalls Road, Leatherhead,
     Surrey KT22 7RY.

E. DICKINSON, Procter Department of Food Science, University
     of Leeds, Leeds LS2 9JT.

R. HARROP, The North East Wales Institute, Deeside, Clwyd,
     CH5 4BR.

A. LIPS, Unilever Research, Colworth Laboratory, Sharnbrook,
     Bedford MK44 1LQ.

J.R. MITCHELL, University of Nottingham, School of Agriculture,
     Sutton Bonington, Loughborough LE12 5RD.

B. MURRAY, Procter Department of Food Science, University of
     Leeds, Leeds LS2 9JT.

M. RINAUDO, Centre de Recherches sur les Macromolécules
     Végétales, B.P.68-38402, Saint Martin d'Hères, France.

S.B. ROSS-MURPHY, Unilever Research, Colworth Laboratory,
     Sharnbrook, Bedford MK44 1LQ.

G. STAINSBY, Procter Department of Food Science, University of
     Leeds, Leeds LS2 9JT.

B. WALKER, Schweppes International Ltd., 105 Brook Road,
     Dollis Hill, London NW2 7DS.

P. WALSTRA, Wageningen Agricultural University, De Dieijen 12,
     6703 BC Wageningen, The Netherlands.

D.J. WEDLOCK, Shell Research Ltd., Sittingbourne Research
     Centre, Sittingbourne, Kent ME8 8AG.

P.A. WILLIAMS, The North East Wales Institute, Deeside, Clwyd
     CH5 4BR.

CHAIRMAN'S INTRODUCTION

I have tried to structure the subjects that have arisen
over the last few days and particularly this afternoon, and I
suggest we take them in this order unless there are specific
issues that you want to raise in addition.

Depletion Flocculation

One topic which seems to be of interest at the moment,
particularly in the area of hydrocolloids and macromolecules,
is so-called "Depletion Flocculation". This phenomenon is
attributed to polymers not adsorbing and, by not adsorbing,
causing flocculation. This has cropped up on several occasions
during the conference. It is an interesting subject and there
are several examples which appear to demonstrate depletion
flocculation. I would like to raise the subject of proving
whether or not depletion flocculation is present as it seems
to be relevant to so many food colloids. Furthermore, someone
raised the question of solvent quality, how to quantify it and
its effect on colloid stability. I think Eric Dickinson
raised the question of solvent quality as a single parameter
to describe macromolecules in solution. Perhaps Eric could
give some examples, or a demonstration of a calculation for
two macromolecules, say gum arabic and alginate, to illustrate
how one calculates solvent quality.

Methodology

Another subject for discussion is methodology. We have heard
about particle size analysis, the importance of particle size
analysis in colloids and emulsions, determination of protein
loading and a specific question which has been raised -
emulsifying capacity of proteins and macromolecules.

Adsorption of Macromolecules

Do polysaccharides adsorb? We have seen one or two examples
(e.g. methylcellulose) and particularly the last paper this
afternoon which suggested that many polysaccharides adsorb.
I think this is a very topical subject to debate. What is
the conformation of macromolecules at the surface, particularly
with proteins, protein aggregates, and mixed protein systems?
How do you determine the composition of interfaces and can you
predict what the composition is?

Finally, two other subjects which I haven't broken down any
further are (i) the rheology of interfaces (is it relevant and
does it relate to the bulk rheology of the same species?) and
(ii) the determination of protein polysaccharide interactions.

These are the topics which have come up this afternoon. I
suggest we start with Depletion Flocculation. Eric, you raised
the subject; would you like to start the discussion.

DEPLETION FLOCCULATION

DICKINSON: I tried to convey in my talk that I think the evidence is relatively slight in terms of concrete quantitative information in the literature. There is also something even more speculative than 'depletion flocculation', and that is the phenomenon called 'depletion stabilization', in which the theory predicts that, above a certain concentration of non-adsorbing polymer, there is stabilization rather than flocculation. You remember that I had a diagram in my talk with the two surfaces fairly close together, and the space between them smaller than the actual size of the polymer molecule in bulk solution. But, of course, the surfaces have to get to this separation in the first place and, if they start off from a long way apart, they may have to pass over a free energy barrier in order to reach this separation. The barrier is caused by the fact that polymer molecules have to be pushed out of the gap - a sort of excluded volume effect - and theory predicts a high barrier at high polymer concentration. It is possible that such concentrations of non-adsorbing polymer may be present in some of the food systems that have been studied. One problem in interpreting experiments involving hydrocolloids is how to separate the rheological contributions from the thermodynamic contributions. Resolution of the problem may require experiments on non-food systems, with polysaccharides that are not used commercially, but are designed to show up the stability effects without affecting the rheological behaviour of the continuous phase to any appreciable extent.

DARLING: Can you actually prove whether you have got depletion flocculation? Is there a method for proving it? Or is it by inference?

DICKINSON: A crucial question is whether or not the hydro-colloid is at the surface. In the talk by Dr. Bergenstahl, I was interested that there was a strong argument put forward for steric stabilization by the gums, but I do not think that there was any direct evidence presented to show that the polymers were actually at the interface in contact with the lecithin or whatever. I think that at some point we need to look at the surfaces by some appropriate techniques to ascertain whether the hydrocolloid macromolecules are there or not. If the polymer is not at the surface, and it is producing flocculation, and the effect has to have a name, then why not call it 'depletion flocculation'?

WEDLOCK: It is important to establish unequivocally whether you have adsorption on an emulsion particle surface. There is an argument that concentrated suspensions can be stabilised by the depletion mechanism. This has been proposed as the stabilis-ation mechanism for concentrated suspensions that are stabilised with hydroxyethyl cellulose and these are suspensions that have been wetted with a dispersant. However, there has been no unequivocal evidence, I think, that a depletion mechanism operates here, and in my own work at Shell, I have shown that you do in fact get adsorption of something as hydrophylic as hydroxyethyl cellulose on a hydrophobic particle in the presence of a dispersant. It is very difficult, in fact almost impossible, to remove the adsorbed polysaccharide. So, I think that there is fairly good evidence that some of the flocculation mechanisms that are proposed as being due to depletion flocculation are, in fact, bridging mechanisms.

WILLIAMS: That may be the case, but I think there are examples that I have seen myself. Dr. Dickinson referred in his paper to some silica xanthan dispersions showing phase separation. Does this refer to some work by Dr. Mike Garvey at Unilever?

DICKINSON: Yes, some work done at Unilever Research, Port Sunlight.

WILLIAMS: I have seen the effect myself. There is definitely no adsorption and the dispersion separates into two distinct phases, one rich in silica particles and the other rich in xanthan. Dr. Garvey explains the phenomenon in terms of depletion flocculation.

WEDLOCK: Andrew Howe and others at the Food Research Institute in Norwich have proposed that hydroxyethyl cellulose will flocculate paraffinic emulsions by a depletion mechanism. Considering the evidence from emulsions and suspensions together, both mechanisms may be operative for the same polymer, but depending on the whole system.

DICKINSON: Napper distinguishes between the two phenomena of phase separation and depletion flocculation in terms of the structure of the aggregates. Phase separation leads to compact aggregates with a more close-packed structure, whereas depletion flocculation is associated with lower density aggregates having a more open structure.

WILLIAMS: There are other examples in the literature, certainly. I am sure you are aware of the ones by Sperry.

DICKINSON: These were the first to receive widespread attention.

WILLIAMS: Yes, but the first evidence was by Bondy in 1939 who monitored rubber latex creaming in the presence of alginates.

WEDLOCK: I think that the two mechanisms are well established. Sperry did show definitely that there was a depletion mechanism, operating in flocculation of polystyrene latexes with hydroxy-ethyl cellulose. I think they had been sterically stabilised with polypropylene oxide - something like that - and then flocculated with hydroxyethyl cellulose. There no evidence for hydroxyethyl cellulose adsorption in that case.

ROSS-MURPHY: A lot of people are engaged in so-called surface force measurements with mica plates, etc. I was wondering whether perhaps I could ask Eric Dickinson and Pieter Walstra, or anyone else, if they feel that as a lot of effort has been invested in such experiments, can they answer any of these questions?

DICKINSON: I am not sure whether the surface force measurements can answer this particular question, though they can answer a lot of interesting questions in relation to the interactions between emulsion droplets.

WALSTRA: Depletion flocculation is a subject of which I know very little, but what I know is that calculations made by Scheutjens and Fleer show that under equilibrium conditions at a low polymer concentration there will be bridging if there is the slightest adsorption. Now usually you don't get bridging because the encounter between two particles in Brownian motion is only a brief one. But if you have concentrated systems where the particles are near to each other, the chances for bridging flocculation are very large. We should therefore be cautious in coming to the conclusion that there is depletion flocculation, unless we know for certain that there is no adsorption. That is very difficult to show.

DICKINSON: What you have raised is the question of when is 'adsorption' adsorption. If we are talking about concentrated systems (say 30-50%), the average distance between the surfaces of the droplets or particles is rather small. So, under these conditions, how close does a polymer molecule have to be to the surface to be considered to be 'adsorbed'? In a sense, you are coming down almost to semantics here as to what is the mechanism, in which case I do not think that the distinction is terribly important.

BERGENSTAHL: Does anybody believe in depletion stabilisation today? I thought that this idea was somewhat out (1,2).

LIPS: Regarding depletion flocculation and the question of adsorption vs. non-adsorption, we should in general expect that the variation in Gibbs adsorption excess with concentration of polymer can be from positive values at low to smaller and even negative values at very high concentrations. The former represents the possibility of positive adsorption and the latter the so-called depletion effect which can dominate in concentrated systems (of polymer and/or colloid). The co-existence of both effects within a unifying framework of theory has been considered for example by Silberberg and Napper.

MURRAY: May I ask a question please, really directed to Professor Walstra, because he made a comment in his talk that he really did not think the pre-wetting condition, which Dr. Dickinson mentioned in his talk, was relevant to food emulsions. If this condition is relevant, I think it is highly important because the effect that we are talking about is the growth of a polymer network at an interface on, say, an oil droplet, which could have important implications for stability. So what are his reasons for stating that?

WALSTRA: I didn't say that pre-wetting never happens; I said that it will be fairly rare, because in most food systems you will have either good solvent quality for the polymer, or a very low polymer concentration, or both.

DICKINSON: What I was referring to is the hydrocolloid stabilizer forming a thick layer on top of an existing protein layer or a low-molecular-weight surfactant layer. One of the additional complexities which we have with hydrocolloid adsorption in food emulsions is that the polysaccharide polymer molecules may be adsorbing at a surface which is itself macromolecular (i.e. consisting of adsorbed protein). Then one has to consider the interactions at the interface between the two types of polymer. That is why, in my talk, I referred to the matter of thermodynamic compatibility between different polymers. This can be investigated in the absence of particles or droplets with particular emphasis on the effect of solvent quality.

WALSTRA: I fully agree.

DARLING: I think before we move on to a different subject, specific questions were raised concerning solute and solvent effects, and solvent quality. If we accept that depletion flocculation takes place under certain circumstances, are there some rules that we can learn from our understanding of polysaccharides in solution in predicting the effect of solute and solvent on depletion flocculation?

DEA: I think I recall that the things that were mentioned this afternoon included the use of alcohol to change the solvent quality; the use of pH to change the solvent quality. Now if depletion flocculation involves a phase separation or an exclusion phenomenon, then one might expect that the changes in solvent quality of that type would counteract that type of mechanism. Now I'm not sure whether that is what the workers find; whether the presence of alcohol or lower pH does in fact promote or stop that type of depletion mechanism occurring.

DARLING: Has anyone got any observations concerning this last point?

STAINSBY: In depletion flocculation the radius of gyration of the (excluded) polymer is important, and if this changes with solvent quality then the flocculation is affected.

WEDLOCK: Sperry has shown that there is a molecular weight dependence. It is borne out.

DARLING: In food?

WEDLOCK: No, not in food.

LIPS: A problem with testing the available theories for
depletion flocculation in food systems is our lack of quantitative insight into the solution thermodynamics of food
biopolymers. It would clearly be profitable to invest in
such measurements.

ROSS-MURPHY: I think it was mentioned today that numerous and
very accurate measurements have been made by Professor
Tolstoguzov's group. I cannot remember who it was (Eric
Dickinson, I think) mentioned that that these had been largely
ignored. Maybe we should be going back and looking at them in
more detail because they have examined a large number of systems
and there is an awful lot of information there which has been
very much neglected.

WALSTRA: I would like to stress that, as Eric Dickinson said
earlier, we have to be sure about kinetics. In some cases it
has been shown that, for example, the creaming rate was faster
than one would predict and this has been ascribed to depletion
flocculation of the droplets. Björn Bergenstahl and I were
discussing today that one has to consider whether the viscosity
that a small particle perceives in a polymer solution is the
same viscosity that we measure in a macroscopic experiment.
Now, firstly, there is the matter of shear-thinning, of course,
but that can be coped with; it works the wrong way round by
the way. But the other thing might be true, that is: the
particle may feel the viscosity of the solvent rather than that
of the polymer solution. We have some vague evidence on the
deformation of particles in a flow field, that viscosity as
caused by, say, glycerol, acts different from viscosities caused
by high polymers. So there is something there that is as yet
unknown and that has to be sorted out before you can draw
conclusions from these kinds of experiments.

METHODOLOGY

DARLING: Can I suggest we change subjects and talk about
methodologies. I know that one of the issues raised in the
stabilisation of dispersions was the calculation of protein
load. How is the protein load or the thickness of the protein
layer on the surface of colloids determined? Perhaps it is
appropriate for Pieter Walstra to just mention how he measures
it. I know that there was some question that was raised by
George Stainsby about the validity and reliability of the method.

WALSTRA: If you want to determine protein load in an emulsion,
the first thing you have to do is to determine surface area, and
that is not always easy; but we have been fairly successful by
using a spectroturbidimetric technique. This has one
disadvantage, I must say: the apparatus needed to determine
turbidity spectra cannot be bought any more. Modern spectro-
photometers are not suitable, but some older ones are. You
need the right range of wavelengths including the near infrared
and you need a particular optical configuration in order to get

real turbidity measurements, excluding forward scattering. We
can determine surface area, particularly in the range of
droplet sizes that you have in emulsions, quite well. The next
thing is then to determine how much protein there is. That can
only be done by depletion studies. So what you do in fact is
take the emulsion, centrifuge it so that you get a cream and a
skim layer, and in each of these determine volume fraction,
surface area, and protein content, and then calculate the
surface load. If you are very careful and do a few parallel
determinations, you can come to an accuracy of about 20%, so
it is not easy to do - it is a lot of work, but as far as I
know it is the only way to do it in an emulsion. An additional
problem is, as I mentioned earlier, that because of differences
in molecular aggregate size present in solution, small droplets
tend to obtain thicker layers than do big droplets and that
makes the calculation awfully difficult and the outcome even
more uncertain. If you have emulsions with very small droplet
size, you just don't succeed - at least we have never succeeded.
But for droplet sizes somewhat below 1 micrometer it is just
possible. Droplet size distribution itself is very difficult to
determine exactly if the droplets are very small. Almost the
only resort you have then is to electron microscopy and that is
also a method beset with difficulties, but it can be done. It
takes an awful lot of time, effort and thus money. But the
spectroturbidimetric method gives you at least an average size
even if the droplets are very small.

DARLING: Dr. Stainsby, do you want to comment?

STAINSBY: I am interested in what Pieter Walstra has said, but I
have yet to see a complete distribution of the droplet sizes for
an emulsion made using a valve homogeniser. In food research
turbidity is often used, but at a single wavelength. Moreover,
the emulsions are generally coarse (having been made by high
speed stirring) and the turbidity is then not increasing mono-
tonically with droplet size for the larger droplets that are
present. The fact that the specific turbidity eventually
decreases with increasing size, is ignored. Moreover, they also
ignore the practical difficulties of measuring the real
turbidity, free from forward scatter. What is needed - as we
have had explained to us - is the turbidity as a function of
wavelength. To me, this is not a realistic method for most
purposes, in view of the difficulties and the time required.
The simpler and shorter procedures, however, cut off at least
one section of the droplet size distribution curve, either by
weighting the larger particles or by weighting the smallest ones.
Without the full curve you cannot get the correct protein load.

WALSTRA: I agree that just turbidity at one wavelength and then
ill-defined, does not tell you much. But I was talking about
real turbidity spectra, where you know precisely the angle of
acceptance and where you use a wide spectrum. We use a wavelength range from slightly below 400 to about 1700 nm, and then
you get good data. And by the way, I myself have determined
size distributions down to about 0.3 micrometer and I think they
were fairly reliable. But indeed, a quick test for practical
conditions does not exist, I fully agree.

STAINSBY: The so-called average particle size in a commercial food emulsion can be as low as 0.2μm, as determined by quasielastic light scattering. This is beyond the lower limit for turbidity measurements.

WALSTRA: The average particle size may be 0.2μm but that undoubtedly includes a lot of particles that are not emulsion droplets.

LIPS: Not necessarily. Quasielastic light scattering on typical food emulsions can be expected to yield moments of the size distribution that are strongly weighted to the small size tail of the distribution. This is because the wide angle scattering from emulsion droplets is relatively weak for sizes large compared to the wavelength of light; sizes of the order of the wavelength make the dominant contribution.

WALSTRA: Well, I must say that I am fairly confident that the ways we have used to determine size distribution in emulsions were pretty good. We have done it in various ways; we have got these methods matching very well but it is a difficult and tedious job. A simple method does not exist.

DARLING: Any other questions relating to methodology? I think this issue of particle size analysis has been around for a long time and I am sure it will be around for many years to come. There are techniques which are improving, but at the end of the day there isn't a single technique that a person could just walk up and use to give you the complete size distribution. It is a problem, and that is why there are very few good measurements around today.

LIPS: What would you do with the information?

DARLING: Determine the protein load from total surface area.

LIPS: Presumably the methods to which Professor Walstra has referred can provide reasonable estimates for the volume surface mean size $d_{3/2}$. For sizes comparable to or larger than the wavelength of light the determination of this mean by turbidity is not expected to vary strongly with the details of the size distribution, at least for monomodal distributions (3). This applies particularly when the particles are large compared to the wavelength of light.

WALSTRA: Well, I may have to modify that slightly. If the particle size, the volume surface average diameter comes below, let's say, 0.4 micrometers, then it breaks down in the sense that you get out another average - you get about the fourth over the third moment. In many cases we could make do with that.

LIPS: Still, if the main objective is to obtain estimates of surface area it may not be essential to have detailed knowledge of the distribution, and the single measurement of turbidity and concentration can be sufficient.

WALSTRA: That's right.

DARLING: There is another problem with spectroturbidimetry that has not really been raised; that is, you assume a distribution function. This is not always valid when working with emulsions that are undergoing flocculation or coalescence. Many emulsions are not simple normal- or log-normal distributions. The results in these circumstances are meaningless unless you know what the distribution function is.

LIPS: Not if the sizes are large, you should still obtain the volume surface mean size.

WALSTRA: I do not agree. When you take turbidity at one wavelength and apply a simple formula to that, you get something of course, but what? The way we use the spectroturbidimetric method is to match turbidity spectra with calculated spectra for certain size distributions. If you can't match them then you have no answer.

LIPS: For sizes large compared to the wavelength the scattering efficiency is a weak function of particle size approaching the constant limit of 2. In that limit clearly one should expect the determination of $d_{3/2}$ by turbidity to be insensitive to the details of the distribution.

WALSTRA: In theory the scattering efficiency should be two but in practice it rarely is because the angle of acceptance is not zero.

LIPS: I agree that is a complicating factor.

DARLING: The particles are not necessarily that large.

Any other issues related to particle size? If not, we will move on.

ADSORPTION OF MACROMOLECULES

DARLING: The next topic concerns the adsorption of macro-molecules, the conformation of macromolecules and the determination of composition at interfaces.

In the last talk there was the speculation or suggestion that multilayer structures can be formed - e.g. emulsifier layer, then a protein layer then a polysaccharide layer. I question that sort of philosophy. I am not saying it is wrong, but I would like to know if there is any evidence for the multilayer adsorption, other than just speculation.

BERGENSTAHL: The evidence for the suggested structure (Fig.2, ref.4) is mainly that the lowest oil-water interfacial tension is generated by the emulsifier. If we measure the interfacial tension with both a gum and an emulsifier present, we obtain tensions just slightly below the values obtained with just the emulsifier present. Hence, it is clear that the gum cannot squeeze the emulsifier out of the interface.

DARLING: Just because you don't change the interfacial tension doesn't mean to say it has not adsorbed. You can still have the same interfacial tension.

BERGENSTAHL: The interfacial tension measurements only tell that the emulsifier still is there. The adsorption of the gum to the oil-water interface is evident from the flocculation rate measurements (Fig. 1, ref.4). The dramatic reduction of the flocculation rate at these very low concentrations can only be explained by steric hindrance due to an adsorbed polymer layer.

DARLING: You can remove 10% of the emulsifier and replace it with 10% of the hydrocolloid and still have the same interfacial tension.

BERGENSTAHL: That would basically be the same structure that I have proposed (Fig.2, ref.4) even if we have some limited penetration into the emulsifier layer, which might be possible. But, I think, it is completely impossible to have an extended penetration of the hydrocolloid into the emulsifier layer without a significant lowering of the interfacial tension.

ROSS-MURPHY: I was going to throw a little bit of a note of caution in here because we all know the classical view of adsorption of polymers on particles: tails, loops, chains, etc. This actually does require that the polymer be fairly flexible but some of the polymers we have been talking about here may even have less than 1 statistical segment at low molecular weights. These things are worm-like chains, even rods in the short molecular weight limit. What implications does that have for such models of adsorption?

DICKINSON: I think it is just worth making the point here that the interfacial tension monitors the presence of material in a very thin layer at the interface, and that there is a role for surface rheological measurements in detecting the presence of material extending much further into the aqueous phase. Surface rheology can answer some of the questions relating to secondary layers or multilayers, because if such layers build up they will contribute to the mechanical strength of the interfacial film without changing the interfacial tension significantly.

WALSTRA: May I just point out that according to Gibbs, whose work is usually taken as gospel, if there is to be adsorption, there is a lowering of surface tension. Or to put it the other way round, you cannot have adsorption without lowering of the surface tension. The change may, however, be very small.

DICKINSON: One of the difficulties in using the Gibbs approach is to define exactly where is the surface. There has been very little work done on the statistical mechanics of multi-component surfactant systems, even simple molecules, never mind macromolecules. Most of the theories are for simple single solutes, and so we do not know what the structure will be for thick layers formed from multicomponent systems containing several components with slightly different surface activities.

LIPS: Is it reasonable to attribute the observation of apparent multiple adsorption to consequences of incompatibility in solution between protein and polysaccharide such that adsorption on to pre-adsorbed emulsifier on the droplet surface is conditioned by the raised chemical potentials of the polymers?

BERGENSTAHL: I am rather sure that this doesn't apply to these experiments. They were mostly performed at very dilute conditions. I cannot see any possible immiscibility effect in the range of 1 ppm of polymer.

CLARKE-STURMAN: I would just add a brief comment about protein adsorption at interfaces. There is some evidence (5,6) that you get conformational changes of proteins from β-sheet to α-helix and vice versa and so some of the comments about changing chemical potential of the protein would, I think, hold. You have actually got a different structure at the interface from that in the bulk. There is evidence with small molecules from spectroscopic measurements at interfaces and maybe that is one way to go and look at some of these interactions. I just put that in as a comment.

WALSTRA: Yes, there is a fair amount of evidence for conformational changes on adsorption, even if the surface excess is high. How to use this evidence is another matter.

MITCHELL: Just a brief comment. It does surprise me really that we have not done more work with radio-labelled species in these mixed surfactant systems. It seems to me we really are looking into the dark at the moment to try to get some idea about composition of these mixed interfaces and that strikes me as probably the only way you can do it.

DARLING: Does anyone want to make any comment on the use of radio-labelling techniques because a lot of work was done many years ago. Phillips and Graham have done quite a bit of work in this field.

HARROP: I don't think it is necessary to go to radio-labelling. You could use ESR probes or fluorescence probes very much more easily and just as sensitive in these systems.

DARLING: I think other techniques have been developed over the years to supercede that.

WALSTRA: But for most of these I can't see how they work in emulsions. You can do it with bulk surfaces but then you are back to the question as to what is at the bulk surface; is that the same as what has been made during emulsification? I think that there is fair evidence that it is not the same.

MITCHELL: The work on protein surface labelling using radio-labelled material at plane interfaces was very successful, and I think it would be fairly easy to extend that to mixed surfact-ant systems. Perhaps this is not directly relevant to what happens in emulsions, but at least you have somewhere to start your discussion, and can then look at ideas such as an emulsifier layer followed by a protein layer, followed by a hydrocolloid layer, because really I think we have no evidence either way at the moment.

DICKINSON: Perhaps I can put in a word in support of these planar interface studies. I accept that they have deficiencies in relation to emulsion behaviour, and one is certainly the question of timescale as Professor Walstra has pointed out. However, it has been established that if you look, for instance, at competition between two proteins, or between proteins and other things, the results obtained with emulsions are quite consistent with those obtained from studies at planar oil-water interfaces. So, it seems that, in terms of the thermodynamics of competition at the interface, there is no difference between the bulk surface and the surface of the emulsion droplet.

CLARK: In trying to answer some of these sorts of questions, perhaps you could tell me whether techniques like small angle x-ray scattering and neutron scattering have been tried, to answer this question, and do they tend to fail or is there some mileage for application in the future?

DARLING: I think you might be the expert to answer that, Allan.

CLARK: I am not a colloid man. I know of people who say that it is possible if you measure on a correct angular range with x-ray scattering to gain information about the layered structures on large particles which normally one would think were too large to study by small angle x-ray scattering, but I suspect that it is a question of making incredibly accurate measurements and picking up a tiny variation on a very large background. But of course in very much smaller microemulsions and so on, the molecular details of a concentric layered structure can be worked out by these techniques. It is just a question of whether there is some mileage in applying them to much larger particles where obviously you don't see the whole particle by the technique, but it seems one of the most direct methods of doing it without disturbing the system, as in microscopy or other related techniques. Maybe there is some future for it.

BERGENSTAHL: I get the impression that this discussion ends up with the conclusion, that the measurement of sequential flocculation rates is an efficient method to characterise these kinds of systems and structures.

DICKINSON: Agreed.

WALSTRA: Agreed.

DARLING: What about conformation of polysaccharide molecules at interfaces? We have talked about adsorption of methylcellulose. We have talked about depletion flocculation. Is there any evidence that the majority of polysaccharides will in fact adsorb even though at a small level and what is the conformation of polysaccharides on surfaces? Simon started to make the point that the shapes of polysaccharides are not necessarily simple random coils or globular structures.

WILLIAMS: We have shown using e.s.r. spectroscopy that if there is a very strong interaction between the polysaccharide and the surface then it will adopt a very flat configuration. If it is a very weak interaction then there will be a large number of loops and tails extending out into solution. Presumably the same happens with emulsions, but you can't study such systems as easily by e.s.r.

DEA: I think the point that Simon Ross-Murphy was making was that certain polysaccharides are stiffer than others. I wonder what polysaccharides you have been using in those studies.

WILLIAMS: We have used carboxymethyl cellulose for instance, but the same happens with the synthetic polymers as well.

DEA: Synthetic polymers are notorious for being fairly flexible.

WILLIAMS: Yes, I am saying the same happens with flexible as semi-flexible polymers. The same sort of information.

ROSS-MURPHY: I don't think that carboxymethyl cellulose would be regarded as particularly stiff. It would be regarded as quite a flexible polysaccharide, I would have thought, certainly compared to, for example, xanthan.

RINAUDO: I agree entirely with Dr. Williams, and would add that there is a larger dependence on molecular weight: when it increases the quantity adsorbed increases and this proves that there are larger loops. We have tested amylose and amylopectin adsorption on some solids and also chitin and chitosan and some derivatives, and the conclusion is strictly the same, and it goes in the same way as with synthetic polymers.

STAINSBY: Is collagen stiff enough?

ROSS-MURPHY: Certainly.

STAINSBY: I realise it's not a polysaccharide. Klein's group in Israel have shown that collagen adsorbs on various surfaces apparently end-on. Not along its length, but with only one end in the interface. If this really is so, then I see no problem with polysaccharides.

ROSS-MURPHY: Fair point.

MORRIS: But is it a fair point? I wonder if binding through the non-helical telopeptide end regions of collagen has more in common with binding of globular proteins than with end-on attachment of a rigid rod?

STAINSBY: As I understand it the end diameter is much smaller than that for a typical globular protein.

GENERAL DISCUSSION

DARLING: We have spent about one hour talking around 3 or 4 subjects. I would like to suggest opening up the discussion and ask for any questions on any topics, specific or general.

RINAUDO: The conformation of the polymer adsorbed on the surface and the thickness of the layer are important. People in France determine this from sedimentation and also from viscosity, and this gives good concordance.

WALKER: If I took a mixed solution, very dilute, of two polymers and then added emulsion droplets to which both polymers adsorbed, would they then behave in terms of phase separation like very much higher molecular weight polymers and de-mix?

DARLING: Two polymers, both adsorbing?

WALKER: No, I was thinking of bulk separation. I'll tell you what makes me bring the subject up. It is that in soft drinks you can sometimes see a phenomenon of sharp division between a cloudy layer and a clear layer. It looks like polymer de-mixing but it is at concentrations of ppm so I thought perhaps if you had very, very high molecular weight polymers there ....
Can anybody help me with this one?

WALSTRA: We have done some studies on such systems and it appears that if these sharply divided layers occur, there is an extremely weak gel; the gel contracts under the influence of gravity rather than by syneresis as an indigenous phenomenon. If you want to prevent it, you have to be sure that the gel is attached to the surface of the vessel, because that is usually sufficient to keep the whole system stable, unless the vessel is very large.

DARLING: Any other comments? Is it appropriate that we perhaps wind up this Workshop session? Thank you all for taking part. I hope you found it useful.

Dr. Darling was then thanked on behalf of the Organising Committee for kindly agreeing to chair the Workshop.

REFERENCES:

1. Fleer, G.J., Scheutjens, J.H.M. and Vincent, B. in Polymer Adsorption and Dispersion Stability (eds. Goddard, E.D. and Vincent, B.), ACS Symp. Ser.(1984),245-263.

2. Gast,A.P.,Hall, C.K. and Russel, W.B. (1983) Faraday Discuss. Chem. Soc., 76, 189-201.

3. Dobbins, R.A. and Jizmagian, G.S. (1966) J. Opt. Soc. Am. 56, 1345-1351.

4. Bergenstahl, B. (1988) This volume.

5. De Grado, W.F. and Lear, J.D. (1985) J. Am. Chem. Soc., 107, 7684-7689.

6. Gierasch, L.M., Cornell, D.G., Dluhy, R.A., Rafalski, M., Briggs, M.S., Hoyt, D., McKnight, C.J. and Stroup, A.N. (1987) Presented to the Division of Colloid and Surface Chemistry, 193rd. ACS meeting, Denver.

LIST OF PARTICIPANTS

A

B AHRENS	C E Roeper, Klosteralle 74, D-2000 Hamburg 13, West Germany
J C ALLEN	Research Division, The North East Wales Institute Connah's Quay, Deeside, Clwyd, CH5 4BR.
D M W ANDERSON	Department of Chemistry, University of Edinburgh, West Mains Road, Edinburgh.
K ANJOU	AB Karlshamns Oljefabriker, 292 00 Karlshamn, Sweden.
R ARMISEN	Hispanager SA, Poligono Industrial de Villa Lonquejar, Apartado 392, Burgos, Spain.
N V ASHLEY	P A Technology, Cambridge Laboratory, Melbourn Royston, Herts, SG8 6DP.
W R ASHTON	Dari-Tech Corporation, 2658 Hummingbird Drive, Duluth, GA 30316, USA.
K AUTIO	Technical Research Centre of Finland, Food Research Laboratory, Biologinkuja 1, 02150 ESPOO, Finland.

B

G BALL	Davis Germantown (Australia) Co., Private Bag 12, Botany, NSW 2019, Australia.
G BAKER	Davis Germantown (Australia) Co., Private Bag 12, Botany, NSW 2019, Australia.
G A BARBER	17 Harvey Avenue, Nantwich, Cheshire.
N M BARFORD	Grindsted Products A/s, Edwin Rahrsvej 38, DK 8220, Braband, Denmark.

R BATES

Hercules Chemical Corporation (USA)
20 Red Lion Street, London WC1R 4PB.

F BAYERLEIN

Diamalt AG, Georg-Reismuller St.34, D-8000 Munchen
50, West Germany.

J N BEMILLER

Whistler Centre for Carbohydrate Research, Purdue
University, West Lafayette, Indiana, USA.

B BERGENSTAHL

Institute for Surface Chemistry, Box 5607, S-114 86,
Stockholm, Sweden.

J A M BEVERS

DMV Campina, Hoogeindsestraat 31,
5447 PE Rijkevoort, Holland.

L BLACKWELL

Courtaulds Chemicals, P.O.Box 5, Spondon, Derby,
DE6 7BP.

W R BLAKEMORE

FMC Corporation, Marine Colloids Division, Box 308,
Rockland, Me 04841, USA.

G J BOLAND

Pedigree Petfoods, Mill Street, Melton Mowbray,
Leicestershire.

L BOTTGER

Grindsted Products AS, Edwin Rahrs Vej 38, DK-8220,
Braband, Denmark.

M M BOYAR

Quaker Oats Ltd (EPDC) PO Box 24, Bridge Road,
Southall, Middlesex UB2 4AG.

M G BRIGAND

Mero Rousselot Satia Ltd, Usine de Baupte,
50500 Carentan, France.

J BRISSON

CERMAV(CNRS), BP 68, 38402 St.Martin d'Heres,
Cedex, France.

W J BROOKS

Winthrop Laboratories, Edgefield Avenue, Fawdon,
Gosforth, Newcastle-upon-Tyne.

K BUCKLEY

Pedigree Petfoods, Mill Street, Melton Mowbray,
Leicestershire.

G BULIGA

Kraft Inc., 801 Waukegan Road, Glenview, IL 60025,
USA.

C̲

C CANDELA

P Robertet et Cie, Avenue Sidi-Brahim,
06130 Grasse, France.

C CAHILL

Kelco International Ltd, Westminster Tower,
3, Albert Embankment, London SE1 7RZ

F W CAIN

Unilever Research, Colworth House, Sharnbrook,
Bedfordshire, MK44 1LQ.

P CARTIER	Hershey Foods Ltd, 1025 Reese Avenue, PO Box 805, Hershey, PA 17033, USA.
I CHALLEN	Kelco International Ltd, Westminster Tower, 3, Albert Embankment, London SE1 7RZ.
R CHANDRASEKARAN	Whistler Centre for Carbohydrate Research, Purdue University, Smith Hall, West Lafayette, IN 47907, USA.
D D CHRISTIANSON	USDA-ARS-NRRC, 1815 N.University, Peoria, Illinois 61604, USA.
K CLARE	Kelco, Division of Merck and Co. Inc., 8225 Aero Drive, San Diego, CA 92123, USA.
A CLARK	Unilever Research, Colworth House, Sharnbrook, Bedford, MK44 1LQ.
R CLARK	Kelco, Division of Merck and Co. Inc., 8225 Aero Drive, San Diego, CA 92123, USA.
J CONGDON	Grindsted Products Ltd, Northern Way, Bury St.Edmunds, Suffolk, IP32 6NP.
A COUTANT	Rhone Poulenc, 14 Rue des Gardinoux, 93308 Aubervilliers, France.
P COWBURN	National Starch Ltd, Ashburton Road East, Trafford Park, Manchester.
B COX	Goodman Fielder Ltd, 2 Smith Street, Summer Hill, NSW 2130, Australia.
V CRESCENZI	Department of Chemistry, University of Rome, 00185 Rome, Italy.

D

L DA GAMA	British Sugar, Research Laboratories, Colney Lane, Colney, Norwich, NR4 7UB.
D DARLING	Unilever Research, Colworth House, Sharnbrook, Bedfordshire.
I DEA	Leatherhead Food Research Association, Randalls Road, Leatherhead, Surrey KT22 7RY.
J B DEEKS	Mero Rousselot Satia Ltd, Mill Reef House, 9-14 Cheap Street, Newbury, Berks.
E DICKINSON	Procter Department of Food Science, University of Leeds, Leeds LS2 9JT.

E

M EDWARDS
Department of Biological Science, University of Stirling, Stirling, FK9 4LA.

U EGGERT
Ilford AG, Industriestrasse 15, CH-1701 Fribourg Switzerland.

A EVES
Leatherhead Food Research Association, Randalls Road, Leatherhead, Surrey, KT22 7RY.

F

R FARAG
University of Cairo, Giza, Egypt.

B A FASIHUDDIN
Faculty of Science and Natural Resources, National University of Malaysia, Sabah Campus, Kota Kinabalu, Sabah, Malaysia.

T FERGUSON
Semmons, Taylor Co.(UK) Ltd, Wildmere Road, Banbury, Oxfordshire.

A FIELDING
Tunnel Avebe Starches, Avebe House, Otterham Quay, Rainham, Gillingham, Kent ME8 7UU.

T FORD
Grindsted Products Ltd, Northern Way, Bury St.Edmunds, Suffolk, IP32 6NP.

J FOX
G.C. Hahn and Co., Aegidienstrasse 22, D-2400 Lubeck 1, West Germany.

P FROMHOLT LARSEN
Grindsted Products AS, Edwin Rahrs Vej 38, DK-8220, Braband, Denmark.

D B FULLER
Department of Food Manufacture and Distribution, Manchester Polytechnic, Old Hall Lane, Manchester M14 6HR.

G

S GAISFORD
Kelco International Ltd, Pitwood Park Industrial Estate, Waterfield, Tadworth, Surrey KT20 5HQ.

W GIBSON
Kelco International Ltd, Pitwood Park Industrial Estate, Waterfield, Tadworth, Surrey KT20 5HQ.

M GIDLEY
Unilever Research, Colworth House, Sharnbrook, Bedford, MK44 1LQ.

P E GLAHN
The Copenhagen Pectin Factory, DK-4623, Lille Skensved, Denmark.

D M GOODALL   Chemistry Department, University of York, York, YO1 5DD.

D N GORE   Carri-Med "Poynings", Pilgrims Close, Westhumble, Surrey, RH5 6AR.

D GREGORY   Grindsted Products Ltd, Northern Way, Bury St Edmonds, Suffolk, IP32 6NP.

H

E S HANSEN   Orana A/S, 243 Rynkebyvej, DK-5350 Rynkeby, Denmark.

P M T HANSEN   Department of Food Science and Nutrition, Ohio State University, Columbus, OH 43210, USA.

S E HARDING   Department of Applied Biochemistry and Food Science, University of Nottingham, Sutton Bonington, LE12 5RD.

P HARRIS   Unilever Research, Colworth House, Sharnbrook, Bedfordshire.

S V HARRISON   Hercules Ltd, 20 Red Lion Street, London, WC1R 4PB.

R HARROP   The North East Wales Institute, Connah's Quay, Deeside, Clwyd, CH5 4BR.

A P M HART   Unilever Research, Colworth House, Sharnbrook, Bedford, MK44 1LQ.

A HENDERSON   Dow Chemical Europe, CH8810 Horgen, Switzerland.

A HILL   Courtaulds Chemicals, P.O.Box 5, Spondon, Derby, DE6 7BP.

M A HILL   Food Science Department, Kings College, Kensington Campus, Campden Hill Road, London W8 7AH.

I HODGSON   Kelco International Ltd, Westminster Tower, 3 Albert Embankment, London SE1 7RZ.

R HOOVER   Department of Biochemistry, University of Ottawa, 40 Somerset St.E, Ottawa, Ontario, KIN 6NS, Canada.

R M W HOPKINS   Meyhall Chemicals (UK) Ltd, 6E Church Road, Bebington, Wirral, Merseyside L63 7PG.

L HORGAN   Biocon Ltd, Kilnagleary, Carrigaline, Co.Cork, Ireland.

J C HOKES   Department of Food Science and Nutrition, Ohio State University, Columbus, OH 43210, USA.

H HUGHES   Research Division, The North East Wales Institute, Connah's Quay, Deeside, Clwyd CH5 4BR.

A HUNTER T.M.Duche and Sons(UK)Ltd, Ford Lane, Pendleton, Salford, M6 6PB.

J J HYDE-SMITH H.P. Bulmer Ltd, Plough Lane, Hereford, HR4 OLE.

I

A IMESON Kelco International Ltd, Pitwood Park Industrial Estate, Waterfield, Tadworth, Surrey, KT20 5HQ.

J

E B JACKSON CPC(UK)Ltd, Trafford Park, Manchester, M17 1PA.

F K JAISLI Obipektin AG, Mittlere Letterstr. 15, CH 9220, Bischofszell, Switzerland.

B JARVIS Express Foods Group, Victoria Road, South Ruislip, Middlesex, HA4 0HF.

F JOHNSTON-BANKS Gelatine Products Ltd, Sutton Weaver, Runcorn, Cheshire.

J JOHNSTONE Kelco International Ltd, Westminster Tower, 3 Albert Embankment, London SE1 7RZ.

B JUD Unipektin AG, CH-8264 Eschenz, Switzerland.

K

L KAABER Norwegian Food Research Institute, Box 50, 1432, AS NLH, Norway.

F KACH Mero Rousselot Satia Ltd, 15 Avenue D'Eylau, 75116 Paris, France.

J F KENNEDY Department of Chemistry, University of Birmingham, Birmingham, B15 2TT.

A E KHALID Food Research Centre, Agriculture Research Corporation, PO Box 213, Khartoum North, Sudan.

S A KHALID Faculty of Pharmacy, University of Khartoum, Khartoum, Sudan.

K KING Dept. of Agriculture (Northern Ireland), AFRCD, New Forge Lane, Belfast, Northern Ireland, BT9 5PX.

S KLEDAL United Breweries Ltd, Strandvejen 50, DK 2900 Helerup, Denmark.

R KLEPP	Jungbunzlauer Xanthan GmbH, Schwarzenbergplatz 16, 1010 Wien, Austria.
H KOCH	CPC Europe Industrial Products, Research and Development Centre, Havenstraat 84, B-1800 Vilvoorde, Belgium.
K KOHDA	Arthur Branwell & Co.Ltd, Bronte House, 58-62, High Street, Epping, Essex, CM16 4AE.
E KRAMP	Ilford AG, Industriestrasse 15, CH-1701, Fribourg, Switzerland.
D KRASOVC	ETOL-IEF, Ipavceva 18, 63000 Celje, Yugoslavia.
E KRINGELUM	Grindsted Products AS, Edwin Rahrs Vej 38, DK-8220, Braband, Denmark.

L

K LANGLEY	Institute of Food Research, Reading Laboratory, Shinfield, Reading, Berks.
P LAWSON	CPC(UK)Ltd, Trafford Park, Manchester, M17 1PA.
R LEBBAR	Setexam, BP 210, Kenitra, Morocco.
J S LEPPARD	Spillers Foods Ltd, Block C, Station Road, Cambridge, CB1 2JN.
A LIPS	Unilever Research, Colworth House, Sharnbrook, Bedford, MK44 1LQ.
T LULEY	Meyhall Chemical AG, CH-8280 Kreuzlingen, Switzerland.

M

W Y MAGAR	National Council for Research, Agricultural Research Council, PO Box 6096, Peoples Hall, Khartoum, Sudan.
F MANIEIRE	PepsiCo Inc. 100 Stevens Avenue, Valhalla, NY, 10595, USA.
M MARINO	Kitchens of Sara Lee, Carnaby Industrial Estate, Bridlington, North Humberside.
W M MARRS	Leatherhead Food Research Assoc. Randalls Road, Leatherhead, Surrey KT22 7RY.
S MARSHALL	Davis Industries (UK)Ltd, Upper Grove Street, Leamington Spa, Warwickshire.

C MATTHEWS — Kelco International Ltd, Westminster Tower, 3 Albert Embankment, London SE1 7RZ.

C D MAY — H.P. Bulmer Ltd, Plough Lane, Hereford, HR4 0LE.

S McBURNEY — Chemistry Dept, University of York, York, YO1 5DD.

B V McCLEARY — Biocon (Eire) Ltd, Kilngeary, Carrigaline, Co.Cork, Eire.

C MERKEL — PepsiCo Inc. 100 Stevens Avenue, Valhalla, NY, 10595, USA.

B MERTENS — State University of Ghent, Faculty of Agricultural Sciences, Coubure 653, 9000 Ghent, Belgium.

P D MEYER — Suiker Unie Research, PO Box 1308, 4700 BH Roosendaal, The Netherlands.

M MILAS — CERMAV-CNRS, BP 68, F38402 St.Martin d'Heres Cedex, France.

M MILES — AFRC Food Research Institute, Colney Lane, Norwich, NR4 7UA.

J R MITCHELL — University of Nottingham, School of Agriculture, Sutton Bonington, Loughborough, Leicester LE12 5RD.

J A MLOTKIEWICZ — Spillers Foods Ltd, Station Road, Cambridge, CB1 2JN.

L MOC — Meyhall Chemical AG, CH 8280 Kreuzlingen, Switzerland.

J J MODLISZEWSKI — FMC Corporation - Marine Colloids Division, 2000 Market Street, Philadelphia, Pennsylvania, 19103, USA.

K MOEBUS — Arthur Branwell Co. Ltd, 58-62 High Street, Epping, Essex CM16 4AE.

R G MORLEY — Delphi Consultant Services, 948 Cabot Court, Stone Mountain, Georgia 30083, USA.

E R MORRIS — Cranfield Institute of Technology, Silsoe College, Silsoe, Bedfordshire MK45 4DT.

V J MORRIS — AFRC Food Research Institute, Colney Lane, Norwich, NR4 7UA.

C MOULES — Rheo-Tech International Ltd, 120/122 Woodgrange Road, Forest Gate, London E7 0EW.

J C F MURRAY — Hercules Ltd, 20 Red Lion Street, London WC1R 4PB.

P MURRAY	Windsor Nutrition Ltd, 30 Lough Gall Road, Armagh, Northern Ireland.
G MUSCHIOLIK	Central Institute of Nutrition, DDR-1505 Bergholz-Rehbrucke, Arthur-Scheunert-Allee 114, German Democratic Republic.
C MYERS	IRL Press Ltd, PO Box 1, Eynsham, Oxford, OX8 1JJ.

N̲

C NEWEY	Davis Gelatine (NZ) Ltd, PO Box 19-542, Woolston, Christchurch, New Zealand.
K NISHINARI	National Food Research Institute, Tsukuba 305, Japan.

O̲

D G OAKENFULL	CSIRO Division of Food Research, P.O.Box 52, North Ryde, NSW 2113, Australia.
M O'LEARY	Arthur Branwell Co.Ltd, Bronte House, 58-62 High Street, Epping, Essex CM16 4AE.
A ONIONS	Honeywill and Stein Ltd, Greenfield House, 69-73 Manor Road, Wallington, Surrey.
E ONSOYEN	Protan A/S, Postbox 420, 3001 Drammen, Norway.
H OPPENAUER	Jungbunzlauer Xanthan GmbH, Schwarzenbergplatz, 16, 1010 Wien, Austria.

P̲

L PALLENT	AECI Ltd, Research and Development Department, PO Modderfontein 1645, South Africa.
D PANTALEONE	Nutrasweet Company, 601E Kensington Road, Mt.Prospect, Illinois 60056, USA.
V R PATEL	C Shippam, P.Box 3, East Walls, Chichester, West Sussex.
O I PAYNTER	Lyons Bakery Ltd, Fish Dam Lane, Carlton, Barnsley, South Yorkshire.
G O PHILLIPS	The North East Wales Institute, Connah's Quay, Deeside, Clwyd, CH5 4BR.

K PICKERSGILL   Investigacion Alimentaria, Larrgana 12, 0103 Vicoria, Spain

S PORRET   Sam Porret ApS, Toftekaersvej 18, DK-2820, Gentofte, Denmark.

A PROCTER   CPC (UK) Ltd, Trafford Park, Manchester, M17 1PA.

R

S RADI   Jungbunzlauer Xanthan GmbH, Schwarzenbergplatz 16, 1010 Wien, Austria.

B RAMMINGER   Bundesanstalt fur Milchforshung, Hermann Weigmann-Str.1, 2300 Kiel 1, West Germany.

J S G REID   Department of Biological Sciences, University of Stirling, Stirling, FK9 4LA.

M J RICHARDSON   IRL Press Ltd, PO Box 1, Eynsham, Oxford, OX8 1JJ.

P RICHMOND   AFRC Food Research Institute, Colney Lane, Norwich NR4 7UA.

M RINAUDO   Centre de Recherches sur les Macromolecules Vegetales, BP 68, 38402 St.Martin d'Heres, France.

G ROBINSON   Cranfield Institute of Technology, Silsoe College, Silsoe,Bedfordshire, MK45 4DT.

C ROCHAS   CERMAV-CNRS, Domaine Universitaire, BP68, 38402 St. Martin de'Heres Cedex, France.

G RODGER   Biological Products, ICI PLC, P.O.Box 1, Billingham, Cleveland, TS23 1LB.

T ROMER   Hydralco GmbH, 44 Bilker Allee, 4000 Dusseldorf, West Germany.

S ROSS-MURPHY   Unilever Research, Colworth House, Sharnbrook, Bedford, MK44 1LQ.

P L RUSSELL   Flour Milling and Baking Research Association, Chorleywood, Rickmansworth, Herts, WD3 5SH.

S

H D SABINE   Biocon(UK) Ltd, Eardiston, Nr.Tenbury Wells, Worcester.

G R SANDERSON   Kelco Division of Merck & Co. Inc., 8355 Aero Drive, San Diego, CA 92123, USA.

G SCHMIDT — Dept of Animal Science, Colorado State University, Fort Collins, CO 80523, USA.

F SHAW — Pedigree Petfoods, Melton Mowbray, Leicestershire, LE13, 1BB

D SHEPHERD — Nestec SA, Linor, Centre de Developement Alimentaire, CH 1350 Orbe, Switzerland.

B SHRIMPTON — General Foods Ltd, Banbury, Oxon.

J R SMITH — Quaker Latz GmbH, Postfach 1184, 5350 Euskirchen, West Germany.

R M SMYTHE — Rowntree Mackintosh plc, Group Products, York, YO1 1XY.

G STAINSBY — Procter Department of Food Science, The University, Leeds LS2 9JT.

R G STEAD — Protan Ltd, PO Box 8, Alton, Hants, GU34 5AX.

H STEINLIN — Polygal Ltd, CH 8560 Maerstetteu, Wienfelderstr.13, Switzerland.

O STERNBERG — Berol Kemi AB, Box 851, S-44401, Stenungsund, Sweden.

S STRAUSS — Pedigree Petfoods, Mill Street, Melton Mowbray, Leics.

T

S TAKAHASHI — The Keihin Woman's College, Chateau Bunkyo 710, 1-15-19 Nishitata, Bunkyo-ku, Tokyo 113, Japan.

N A TANNER — Red Carnation Gums Ltd, Sir John Lyon House, Upper Thames Street, London EC4V 3PA.

B THARP — Germantown Manufacturing Co., PO Box 238, Broomall, PA 19008, USA.

G TILLY — Mero Rousselot Satia, Baupte, France.

H TORGERSEN — Protan A/S, Postbox 420, 3001, Drammen, Norway.

J TORNQUIST — Berol Kemi AB, Box 851 S-444 01 Stenungsund, Sweden.

G TRUDINGER — Carri-Med Ltd, Glebelands Centre, Vincent Lane, Dorking, Surrey.

R TYE — FMC Marine Colloids Division, P.O.Box 308, Rockland, Me 04841, USA.

V

H **VAN OIJEN**	Menken Dairy Food BV, P.O.Box 6, 4750 AA Oud-Gastel, Holland.
J **VANHEMELRIJCK**	CPC Europe Industrial Products, Research Development Centre, Havenstraat 84, B-1800, Vilvoorde, Belgium.
B **VILLAUDY**	Mero Rousselot Satia, Usine de Baupte, 50500, Carentan, France.
J **VINCENT**	Mero Rousselot Satia Ltd, Mill Reef House, 9-14 Cheap Street, Newbury, Berks.
L **VINEY**	Bohlin Reologi UK Ltd, Business and Technology Centre, Bessemer Drive, Stevenage, Herts SG1 2DX.
J **VIRTANEN**	Finnish Sugar Co.Ltd. Porkkala Plant, SF-02460 Kantvik, Finland.
A G J **VORAGEN**	Dept.of Food Science, Agricultural University, De Dreijen 12, 6703 BC Wageningen, The Netherlands.
J **de VRIES**	The Copenhagen Pectin Factory, 4623 Lille Skensved, Denmark.
J G **de VRIES**	Nutricia Research, PO Box 1, 2700 AA Zoetermeer, The Netherlands.

W

B **WALKER**	Schweppes International Ltd, 105 Brook Road, Dollis Hill, London, NW2 7DS.
I **WALKER**	Reckitt and Colman Pharmaceutical Division, Dansom Lane, Hull, Humberside.
P **WALSTRA**	Dept. of Food Science, Wageningen Agricultural University, De Dieijen 12, 6703 BC Wageningen, The Netherlands.
D J **WEDLOCK**	Product Support Division, Shell Research, Sittingbourne, Kent, ME9 8AG.
D **WELSBY**	Purina Protein Europe, Excelsiorvaan 13, Zaventam, B-1930, Belgium.
D B **WHITEHOUSE**	CPC Europe Industrial Products, Research and Development Centre, 84 Haven Straat, Vilvoorde, Belgium.
W C **WIELINGA**	Meyhall Chemical AG, CH 8280 Kreuzlingen, Switzerland.

M **WILKE**   Wolff Walsrode AB, POB, D-3030 Walsrode,
                          West Germany.

C **WILLIAMS**   Leiner Gelatins Ltd, Treforest Industrial Estate,
                          Pontypridd, Mid Glamorgan CF37 5SU.

P A **WILLIAMS**   Research Division, The North East Wales
                          Institute, Connah's Quay, Deeside, Clwyd,
                          CH5 4BR.

R H **WILSON**   AFRC Institute of Food Research, Colney Lane,
                          Norwich, Norfolk, NR4 7UA.

W T **WINTER**   Polytechnic University, 333 Jay Street,
                          Brooklyn, NY 11201, USA.

J **WISNIEWSKI**   Davis Gelatine (Canada) Ltd, 230 Midwest Road,
                          Scarborough, Ontario, MIP 3A9, Canada.

P A **WOLF**   PepsiCo Inc. 100 Stevens Avenue, Valhalla, NY,
                          10595, USA.

G J M **WYNANS**   Suiker Unie Research, Oostelyke-Havendyk 15,
                          4704 BA Roosendaal, The Netherlands.

Y
─

T **YANO**   Department of Agricultural Chemistry, The
                          University of Tokyo, 1-1-1 Yayoi, Bunkyo-ku,
                          Tokyo, Japan.

SUBJECT INDEX